Theoretical Biology

Series Editor

Yoh Iwasa, Kyushu University, Fukuoka, Japan

The "Theoretical Biology" series publishes volumes on all aspects of life sciences research for which a mathematical or computational approach can offer the appropriate methods to deepen our knowledge and insight.

Topics covered include: cell and molecular biology, genetics, developmental biology, evolutionary biology, behavior sciences, species diversity, population ecology, chronobiology, bioinformatics, immunology, neuroscience, agricultural science, and medicine.

The main focus of the series is on the biological phenomena whereas mathematics or informatics contribute the adequate tools. Target audience is researchers and graduate students in biology and other related areas who are interested in using mathematical techniques or computer simulations to understand biological processes and mathematicians who want to learn what are the questions biologists like to know using diverse mathematical tools.

Hiromi Seno

A Primer on Population Dynamics Modeling

Basic Ideas for Mathematical Formulation

 Springer

Hiromi Seno
Graduate School of Information Sciences
Tohoku University
Sendai, Japan

ISSN 2522-0438   ISSN 2522-0446   (electronic)
Theoretical Biology
ISBN 978-981-19-6018-5   ISBN 978-981-19-6016-1   (eBook)
https://doi.org/10.1007/978-981-19-6016-1

This Springer imprint is published by the registered company Springer Nature Singapore Pte Ltd.
The registered company address is: 152 Beach Road, #21-01/04 Gateway East, Singapore 189721,
Singapore

Preface

This is one of the introductory books about the mathematical models of population dynamics in mathematical biology. However, the purpose of this book is not to simply give literacy about how to analyze mathematical models. This book focuses on the biological meaning/translation of mathematical structures involved in mathematical models. In some recent usages of the mathematical model simply with computer numerical calculations (especially by some researchers out of mathematical science), the model includes some inappropriate mathematical structure with respect to the reasonability of modeling for the biological problem under investigation. For students and researchers who study or use a mathematical model, it is very important and helpful to understand what mathematical structure could be regarded as reasonable for the model with respect to the relation to the biological assumptions about the problem under investigation.

Since the detailed arguments about the meaning of mathematical modeling are the principal purpose of this book, to which most of the pages are devoted, the description of mathematically detailed nature of some models is actually skipped, leaving it to some appropriate textbooks or literatures. Instead, the arguments about the mathematical modeling necessarily require some mathematical tools/techniques which correspond to those necessary for analyzing the model as well. Readers may get the knowledge of such mathematical fundamentals even just in the principal arguments of mathematical modeling described in this book and furthermore in the arguments to show the mathematical features of introduced models.

Understanding the biological meaning of mathematical modeling for the simple models treated in this book could necessarily contribute to the modification or expansion for readers to derive a specific/original mathematical model for a biological problem in an appropriate way, that is a reasonable modeling. The description of mathematical contents is not mainly for the readers with a background in mathematics but for those with the other background.

Until today, biology has extended its horizon to a wider spectrum of time and space. So has done mathematical biology, since many biological problems are emerging on the table for the theoretical and mathematical consideration in biological, medical, and social sciences. Mathematical models on population dynamics have been developed very rich in mathematical biology, and they provide the basics of mathematical modeling for a variety of new problems in biology. This book is expected to give a chance for readers to consider what is the important aspect to construct a reasonable model and how we could reasonably design mathematical modeling for a biological problem.

Mathematical biology is an interdisciplinary field emerged from a complex of mathematics, physics, chemistry, and biology. It has been trying to theoretically clarify the scientifically significant aspect of the biological problem. The research in mathematical biology mostly takes the following steps:

1. To clarify what aspect of the biological problem is considered
2. To choose the biological factors important for the theoretical consideration of the problem
3. To set up necessary hypotheses/assumptions about the nature of chosen biological factors and the relation among them
4. To give the reasonable mathematical expression of those hypotheses/assumptions
5. To construct the mathematical model with mathematically expressed hypotheses/assumptions
6. To design the mathematical analysis and choose the necessary mathematical technique according to the constructed mathematical model, taking account of the biological problem specified in the first step and the mathematical structure of the constructed model
7. To carry out the mathematical analysis and get the necessary mathematical results on the nature of model
8. To argue the relation of obtained mathematical results to the biological problem
9. To integrate the obtained mathematical results and discuss the biological meaning about the problem

The above steps from 1 to 5 are important subjects focused in this book. Usually it would not be easy to proceed with the above steps in the listed order. For example, at the step 5, it is likely that an additional factor may become necessary to be involved in the model. In such a case, we need to go back to the step 2. Further, at the step 8, it may be found that there is a logical inconsistency to the hypotheses/assumptions set up at the step 3. In such a case, it is necessary to reconsider the steps 4 and 5. Most research projects would take such a way of progress with back-and-forth steps.

In some cases, the biological conclusion derived from the obtained mathematical results may contain something inconsistent to the result obtained by the biological research. Even in such a case, it is little scientific to regard such a mathematical model as useless or nonsense. As long as the mathematical model is reasonably constructed based on the appropriate biological hypotheses/assumptions, such an inconsistent part of the obtained conclusion could be useful to imply that the applied hypotheses/assumptions may contain a biological discrepancy, or that the structure of mathematical model may not be appropriate from the viewpoint of reasonable modeling. Hence such a conclusion may be potentially able to reveal a non-trivial debating point about the biological problem or mathematical modeling.

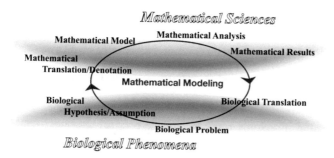

The arguments on such a subject arisen in the research in mathematical biology could provide a scientific feedback from mathematical science to biology.

There are a variety of mathematical models in mathematical biology, some of which may require a specific mathematical knowledge in order to understand them. In this book, we shall consider only classic models with simple mathematical structure, most of which can be analyzed with fundamental mathematical methods. At the same time however, all of them are the important basics for the mathematical models in modern researches, and they contain the essence of reasonable modeling common to every research in mathematical biology.

This book consists of two parts, Part I as the main contents and Part II as the important supplement to the main. Actually Part II is not the appendix to the main. It contains not only the fundamentals to understand the mathematical aspect in the main contents but also the further information to understand better the modeling in the main. However, readers may be able to read Part I without referring Part II, and most part of Part II independently of Part I. Part I and II consist of ten and five chapters, respectively.

In Chap. 1, we shall see the reasonable modeling with the geometric progression. The model is expressed as a recurrence relation as the discrete time population dynamics which generates a sequence of numbers that means the population size. This chapter is to introduce some simple ideas to bridge the biological factor to the mathematical formula. At the same time, biologically essential factors of population dynamics are mathematically introduced step by step in the model described by a recurrence relation of the seasonal change of population size, and given are some mathematical/numerical examples about the contribution of such biological factors to the dynamical nature. Readers may need to get their own idea connecting the biological factor to the mathematical formula in order to understand the modeling and models in this chapter.

In Chap. 2, we shall introduce the density effect in population dynamics, as the expansion of discrete time linear models in Chap. 1. The introduced density effect could be intraspecific, interspecific, or both. As an intraspecific density effect, we shall consider the Allee effect, and as an interspecific density effect, the interspecific competition and the prey-predator or host-parasite relation. Moreover, we shall consider the effect of harvesting/culling on population dynamics too. Then we give some arguments on the problem of maximum yield or benefit, related to the mathematical fundamental in bioeconomics. Further in the last section of Chap. 2, we shall describe the other type of modeling for discrete time population dynamics, since we do not want to make readers get only a typical idea about the modeling for discrete time population dynamics. The ideas described in the part would be expandable to the construction of another sophisticated model for a discrete time population dynamics.

Chapters 11 and 12 of Part II are closely related to Chaps. 1 and 2, and are referred section by section according to the necessity to understand the contents in Chaps. 1 and 2. Especially Chap. 11 gives the fundamentals of mathematical analysis on the discrete time nonlinear population dynamics model, that is, of the qualitative analysis on the discrete time dynamical system.

Chapter 3 is about the idea to mathematically derive continuous time models from discrete time models in Chaps. 1 and 2. Each derived model is written as an ordinary differential equation, since we focus on the model of single species population dynamics in this chapter. Logistic equation appears first in this chapter, although its modeling and nature will be described in the subsequent chapters again, Chaps. 4 and 5. Besides, as the basic idea to understand the meaning of time derivative in the population dynamics modeling, we describe the concept of momesntal velocity of population size change. It is the essential aspect for the continuous time population dynamics model and is very likely to be thought little in the biological translation of the mathematical results obtained for a continuous time model. In most cases of biological phenomena, a continuous time modeling with the momental velocity of a biological quantity like population size could be regarded as a mathematical approximation or simplification.

In Chap. 4, we shall describe the fundamental modeling with the birth-death stochastic process. The idea and concept of such a modeling are very important to understand the meaning of modeling even about the deterministic mathematical model which superficially seems not to have any relation to the stochasticity in the phenomenon. In almost all modelings for population dynamics, the deterministic model could be actually regarded as an approximation to describe an important aspect of the phenomenon under investigation. To understand the reasonability of a deterministic structure introduced in the model, it would be necessary and useful to know its relation to a stochastic process behind mathematical modeling. For this reason, this chapter contains the idea and concept essential throughout the contents of this book. As the simplest and most important stochastic process, Poisson process is introduced and used in some parts of this book. Chapter 15 of Part II serves to provide the mathematical fundamentals about it.

Chapters 5–10 of Part I are about continuous time models. Topics in these chapters may be popular in the other textbooks on mathematical population dynamics. Readers may easily refer to the corresponding topic in such a textbook in order to get some related mathematical or biological details. As mentioned above, we shall focus on the modeling itself to provide a chance for readers to consider the meaning of mathematical model.

Chapter 5 is on the continuous time modeling for single-species population dynamics. The content serves readers with the essential idea to introduce a density effect in mathematical modeling. The idea in this chapter becomes basic to understand the modelings in the subsequent chapters. As a specific topic, we describe the modeling of what is called metapopulation dynamics model too.

Chapter 6 is devoted to the description of the idea to model the interspecific reaction for the continuous time population dynamics model. The contents could be rarely found in the other textbooks, though the topic itself is very popular in mathematical biology.

Chapters 7 and 8 are about the competition and prey-predator dynamics, respectively. They are very popular topics in ecology and mathematical biology. In this book, we shall see also some related classic theoretical topics which have been rarely described in recent textbooks. They may have a potential to provide cues for readers to expand the idea for a new modeling about population dynamics.

Chapters 13 and 14 of Part II are to provide the necessary least knowledge about the fundamental mathematical theories to understand or find the dynamical nature of continuous time models. Although readers who want to know more precisely or deeply the mathematical detail about them may feel discontent with the description in these chapters, it is very easy to find some other literatures about what to be known.

Chapters 9 and 10 are about the modeling of population dynamics with a heterogeneous structure of population. After the general overview of such a heterogeneous structure of population, we shall describe the mathematical modeling and models of epidemic dynamics in Chap. 9, since it can be regarded as a typical and familiar example of population dynamics with heterogeneity within the population.

As a classic and popular heterogeneity in a population, we shall describe the modeling and models about population dynamics with age structure in Chap. 10. Both discrete and continuous time modelings are considered, and we shall give arguments on the mathematical relation between them too.

Some chapters are accompanied by some exercises shown in the related sections. They are not simply for the practice or trial about the related topic but for the further information useful to readers' understanding of the contents better. The satisfactory answers are given at the end of each chapter, which may be regarded as the supplementary description of the topic in the main text.

As mentioned at the beginning, this book focuses on the description of mathematical modelings in population dynamics, and the analysis of model is secondary for the purpose of this book. However, the dynamical nature of every mathematical model is described in a necessary manner, and we shall give the biological meaning of the mathematical results obtained by the analysis of the model. Such a translation of mathematical results to the biological meaning is regarded as an important aspect of mathematical modeling too, since it must be the inverse of modeling to construct a model. Hence we consider it important to describe the biological meaning of mathematical nature of the model for the purpose to give readers the idea of reasonable mathematical modeling. For this reason, readers must find a lot of mathematical contents even in the description only about the mathematical modeling itself.

Although this book deals with only classic models in population dynamics, I shall expect that readers could get a cue or idea about the reasonable modeling for population dynamics, following the spirit of 温故知新 (On-Ko-Chi-Shin in Japanese pronunciation; a phrase originated from the ancient Chinese book Lunyu (Analects of Confucius)), that is, "visiting old, learn new." Further I hope that readers could find something interesting about mathematical modeling.

Most of the contents in this book are based on the arguments of my books published by Kyoritsu Shuppan in Japan and also on the discussion in courses given at Hiroshima University, Nara Women's University, and Tohoku University in my career. I thank everyone for the support, encouragement, and inspiration about my accomplishment of them. Without them, I could not have completed writing this book. My specific interest in the reasonability of mathematical modeling has been nurtured through my communication with many SENPAI (elder colleagues) before and after my professional career. First, I am sincerely grateful to the SENPAI of Teramoto laboratory in Kyoto University, Nanako Shigesada, Hisao Nakajima, Norio Yamamura, Kohkichi Kawasaki, Toshiyuki Namba, Yoh Iwasa, Masahiko Higashi, Takenori Takada, Masayuki Kakehashi, Hiroyuki Matsuda, Takashi Saito, Kennosuke Wada, Tamiki Umeda, and Yasushi Harada. The discussion and communication with them during and after my student days certainly made my interest in modeling come into bud and grow. In the same period onward, Ei Teramoto and Masaya Yamaguti had been supporting and mentoring me as the great forerunners in Japanese mathematical biology. Further, during my stay at Dipartimento di Matematica ed Applicazioni, Università degli Studi di Napoli Federico II in 1987–1989, Luigi M. Ricciardi and Aniello Buonocore gave me experiences especially on some topics related to the stochastic process. On the way back to Japan from Italy in 1989, I got the chance to visit Akira Okubo, Donald L. DeAngelis, and Bernard C. Patten in USA, whose support and suggestions were instructive for my future research and direction even in my short stay at each of them. Masayasu Mimura much supported my professional career and influenced my clarifying the research interest and direction through many personal communications with him. Horst Malchow is my old friend, colleague, and collaborator with whom I had many occasions to discuss the reasonability of mathematical modeling. So are Sergei Petrovskii and Frank Hilker. Yasuhisa Saito and Kazunori Sato are Japanese colleagues and longtime friends in mathematical biology too.

Last but not least, I am especially indebted to Yoh Iwasa's invitation and continuous encouragement, and to the patience and flexibility of the editorial team at Springer for my writing of this book during the period with some hindrances by the COVID-19 pandemic and earthquakes around my living place.

Sendai, Japan   Hiromi Seno
June 2022

Contents

Part II Mathematical Equipments

Part I
Modeling Biological Factors

Chapter 1
Application of Geometric Progression

Abstract In this first chapter, we shall see the reasonable modeling with the geometric progression. The model is expressed as a sequence of numbers for the discrete time population dynamics. This chapter is to introduce some simple ideas to bridge the biological factor to the mathematical formula. At the same time, the biologically essential factors about population dynamics are mathematically introduced step by step in the model described by a recurrence relation of the seasonal change of population size, and given are some mathematical/numerical examples of the contribution of such factors to the dynamical nature. The readers may need to get an own idea connecting the biological factor to the mathematical formula in order to understand the models in this chapter.

1.1 Geometric Growth Model

Geometric progression has been used as a mathematical model for population dynamics, appeared even in WAZAN (old Japanese mathematics). It is said that it appeared first in the arithmetic book 塵劫記 (JINKOUKI) (1627) written by 吉田光由 (YOSHIDA, Mitsuyoshi) (1598–1673). For the mathematical modeling, he assumed the followings:

- In January, a couple of rats come to a house, and have twelve newborns.
- The sex ratio at birth is 1:1, that is, the half of newborns are female.
- Every newborn can mature till the next month, and have the fertility same as the parent.
- Breeding couples are necessarily formed between the newborns of same generation.
- Each couple has twelve newborns every month with even sex ratio.
- The death is ignored for one year.

The problem he gave was how many rats inhabit in this family after one year.

The assumption means a population with the inbreeding. The inbreeding increases the probability of inheritance of a lethal gene to the offspring. Although it could be regarded as disadvantageous in an evolutionary sense, it may be advantageous with respect to the probability to form the breeding couple. Actually, not a few species of animal and plant take the inbreeding strategy, which have been biologically studied. Even in human history, the inbreeding appeared for a social or cultural reason.

Let us consider the monthly variation about the number of couples under the above assumptions. In February, the initial couple and new $12/2 = 6$ couples of its offspring have newborns. In this case, the next generation consists of $12 \times 7 = 84$ newborns. They become additional $84/2 = 42$ couples. In March, $1 + 6 + 42 = 49$ couples have newborns. Although you may continue this step-by-step calculation to get the answer for the problem, let us construct a *mathematical model* by a mathematical formula to express the way of calculation.

Let us denote the number of reproductive couples in the nth month ($n = 1, 2, \ldots, 12$; the first month is January) by c_n. Then the number of newborns in the nth month is given by $12c_n$, so that the number of new couples by them becomes $12c_n/2 = 6c_n$. Therefore, the number of reproductive couples in the $(n+1)$th month satisfies the following equality: $c_{n+1} = c_n + 6c_n = 7c_n$. This recurrence relation indicates that the sequence is a geometric progression with the common ratio 7.

The general term for the recurrence relation becomes $c_n = c_1 \times 7^{n-1}$. Since $c_1 = 1$ from the assumption, we can find that $c_{12} = 7^{11}$. Hence, in the twelvth month, since each of 7^{11} couples has twelve newborns, we have $2 \times 7^{11} + 12 \times 7^{11} = 2 \times 7^{12} = 27,682,574,402$ rats in this family after one year.

As in this illustrative example, when the sequence of the temporal variation about the number of biological organisms follows a power of a certain positive constant a, it is called *geometric growth*.

Could you think the obtained number of rats more than 27 billion is unrealistic? First of all, it is a narrow view to simply compare to the reality. The modeling requires assumptions to construct the mathematical model. Even though the assumptions are set up from the knowledge about the real phenomenon, the model should not be considered as identical to the phenomenon, since the modeling is necessarily based on a certain number of assumptions chosen a priori. The mathematical model should be regarded as a reference for the scientific consideration on the reality, and as a thought experiment. The answer of a huge number for the above problem could imply the possibility of unexpectedly rapid growth in the number of rats. If the rats are pests inhabiting in ordinary houses, it is important to consider what this result would indicate, aside from the accuracy of the number as the answer.

Exercise 1.1 As a more general modeling, assume that the number of newborns per couple is $2k$ with a given positive integer k. With the same assumptions used in the above, construct the recurrence relation with respect to the number of couples c_n, and derive the formula to give the total number of rats in December. How much does it become for $k = 1$ that is the least number of newborns satisfies the above assumptions?

1.2 Immature Period

For the geometric growth model in the previous section, it was assumed that the newborn can mature till the next month. In this section, we modify this assumption, and see how the structure of mathematical model is changed.

Case of Two Months for Maturation

Let us assume that the maturation takes two months. Besides, we assume that the initial couple of rats is mature and reproductive. As in the previous section, let us consider the monthly variation about the number of reproductive couples. In February, there appear the initial mature couple and their twelve newborns which are still immature. Thus, only the initial mature couple can have the newborns. From the assumption, this mature couple can produce twelve newborns again in February. In March, the rats born in January become mature and form six reproductive couples. Therefore, seven couples can produce $12 \times 7 = 84$ newborns in this month. Although this is the similar arguments applied for the previous case, let us make the difference clearer by the mathematical modeling now.

> It is reported that the rat takes two or three months for the maturation, and the mature female can have 5–10 newborns in average. The maturation period and the number of newborns must necessarily depend on the environmental condition.

Let us denote the number of reproductive couples in the nth month by c_n as before, and additionally the number of newborns in the nth month by r_n. By these definitions, we have $r_n = 12c_n$. The number of reproductive couples in the $(n+1)$th month is given by the sum of the reproductive couples in the nth month and the new reproductive couples generated by those born in two month ago, that is, by the newborns r_{n-1} in the $(n-1)$th month. Hence, the following equation holds:

$$c_{n+1} = c_n + \frac{r_{n-1}}{2}.$$

From the relation $r_n = 12c_n$, we can obtain the following recurrence relation in terms of the number of reproductive couples:

$$c_{n+1} = c_n + 6c_{n-1}. \tag{1.1}$$

Note that this equation can be applied only for $n \geq 2$.

Since this is a *homogeneous second order linear difference equation*, we can solve and obtain the general term, making use of the characteristic equation (Sect. 11.1). From the characteristic equation $\lambda^2 - \lambda - 6 = (\lambda + 2)(\lambda - 3) = 0$, we can transform the Eq. (1.1) to $c_{n+1} + 2c_n = 3(c_n + 2c_{n-1})$ and equivalently to $c_{n+1} - 3c_n = -2(c_n - 3c_{n-1})$ which are in the form of geometric recurrence with respect to $c_n + 2c_{n-1}$ and $c_n - 3c_{n-1}$. Since $c_1 = 1$ and $c_2 = 1$, we can easily obtain the following expression of general term, making use of these transformed recurrence equations:

$$c_n = \frac{3^n - (-2)^n}{5}.$$

Now, from this general term, we can immediately obtain the number of reproductive couples in November as $c_{11} = (3^{11} + 2^{11})/5 = 35,839$, and that in December as $c_{12} = (3^{12} - 2^{12})/5 = 105,469$. Hence, by summing up the mature rats of reproductive couples, the immature rats born in November, and the newborns in December, we have the total number of rats in December as $2 \times c_{12} + 12 \times c_{11} + 12 \times c_{12} = 1,906,634$. This is about 0.007% of the case in the previous section. We note how the maturation speed would be significant about the reproduction.

Case of Three Months for Maturation

Next, let us consider the case where it takes three months for the maturation after the birth. It is easy to derive the following recurrence relation by the same way as that for (1.1):

$$c_{n+1} = c_n + 6c_{n-2}, \tag{1.2}$$

where $n \geq 3$. From this recurrence relation with the early sequence of $c_1 = c_2 = c_3 = 1$, we can obtain the sequence of the temporal variation $\{c_n\}$: $\{1, 1, 1, 7, 13, 19, 61, 139, 253, 619, 1453, 2971\}$. Therefore, in December, the family of rats becomes $2 \times c_{12} + 12 \times c_{10} + 12 \times c_{11} + 12 \times c_{12} = 66,458$ which is obtained by summing up the mature rats of reproductive couples, the immature rats born in October and November, and the newborns in December. This is about 3.5% of the number in the previous case of two months for the maturation. We see again how the maturation speed would be significant about the reproduction.

There are some different derivation ways and mathematical expression of the general term for (1.2). Since the characteristic equation for (1.2) becomes $\lambda^3 - \lambda^2 - 6 = 0$, which has one real and two imaginary roots, the expression of the general term must become less simple than that for the previous case. For instance, we can give the following expression:

$$c_n = b_1 \lambda_r^n + \left(\frac{6}{\lambda_r}\right)^{n/2} (b_2 \cos n\theta + b_3 \sin n\theta) \qquad (n \geq 3),$$

where $\lambda_r = (\alpha^2 + \alpha + 1)/(3\alpha) \approx 2.21878$, $\alpha = (82 - 9\sqrt{83})^{1/3} \approx 0.182694$, and $\tan\theta = -\sqrt{3}(1 + \alpha)/(1 - \alpha)$ where $\pi/2 < \theta \approx \pi - 1.19117 < \pi$. Coefficients b_1, b_2 and b_3 are determined by $c_4 = 7$, $c_5 = 13$ and $c_6 = 19$ with α. Since $\sqrt{6/\lambda_r} \approx 1.64444 < \lambda_r$, the first term of right side becomes principal for sufficiently large n. That is, for sufficiently large n, the sequence $\{c_n\}$ can be approximated well by a geometric growth with the common ratio λ_r.

1.3 Fibonacci Sequence

Here let us consider again the case of two months for the maturation, and change the assumption about the number of newborns per couple from twelve to two, which is the least number that does not violate the other assumptions (Fig. 1.1).

In this case, we have the following recurrence relation instead of (1.1):

$$c_{n+1} = c_n + c_{n-1} \qquad (1.3)$$

This indicates that the sequence $\{c_n\}$ is what is called *Fibonacci sequence*. From $c_1 = 1$ and $c_2 = 1$ of the assumption for modeling, it becomes $\{1, 1, 2, 3, 5, 8, 13, 21, 34, 55, 89, 144\}$.

This kind of sequence was considered by Leonardo Fibonacci[1] (c. 1170–c. 1240), and he used the sequence for an example about the temporal variation about the number of rabbits with the assumptions corresponding to those set up by us here. The general term for the recurrence relation (1.3) was given by a famous

---

[1] The actual name was Leonardo da Pisa, while Fibonacci was his nickname. It is said that a historian of mathematics in the nineteenth century made the mistake about his name and it has become popular.

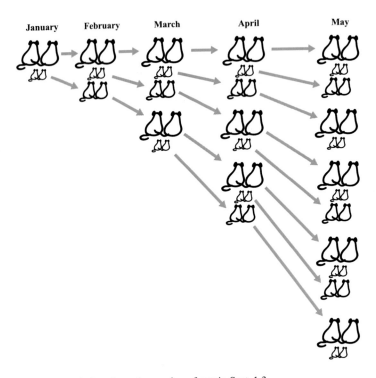

January   February   March   April   May

Fig. 1.1 Illustrative variation about the number of rats in Sect. 1.3

mathematician and physicist Daniel Bernoulli (1700–1782) [1]:

$$c_n = \frac{1}{\sqrt{5}}\left(\frac{1+\sqrt{5}}{2}\right)^n - \frac{1}{\sqrt{5}}\left(\frac{1-\sqrt{5}}{2}\right)^n. \tag{1.4}$$

You can easily derive this formula with roots of the characteristic equation $\lambda^2 - \lambda - 1 = 0$ (Sect. 11.1). In this case, the number of rats (or rabbits) in the family becomes $2 \times 144 + 2 \times 89 + 2 \times 144 = 754$ in December.

Fibonacci sequence has been attracting many scientists, because there are a variety of examples observed in nature. For well-known example, we can find it for the spatial configuration of seeds of pineapple, pinecone, and sunflower, or the number of leaves or branches in their arrangement per around stem. You can find the other examples about the Fibonacci sequence and the golden ratio in [2]. We can see from (1.4) that $c_{n+1}/c_n \to (1+\sqrt{5})/2 \approx 1.61803$ as $n \to \infty$. The number $(1+\sqrt{5})/2$ is what is called "golden ratio" observable

(continued)

in many parts of architecture and art as it is consciously used or eventually appears. Related to this fact, we can find many artificial structures containing a Fibonacci sequence.

The general term (1.4) for the recurrence relation (1.3) can be written as

$$c_n = \frac{1}{\sqrt{5}} \left(\frac{1+\sqrt{5}}{2} \right)^n \left[1 - \left(\frac{1-\sqrt{5}}{1+\sqrt{5}} \right)^n \right].$$

Since $\left| (1-\sqrt{5})/(1+\sqrt{5}) \right| < 1$, we note that

$$c_n \approx \frac{1}{\sqrt{5}} \left(\frac{1+\sqrt{5}}{2} \right)^n$$

for sufficiently large n, and the sequence $\{c_n\}$ can be approximated well by a geometric growth with the golden ratio as the common ratio.

1.4 Life Span

In this section, let us consider the influence of life span on the population dynamics by the geometric progression. Up to now, we assumed to ignore the death for one year, though it is not negligible when we consider the geometric growth model for the longer period.

Assume again that it takes only one month for the maturation as we did about the original model. The newborn in a month becomes mature and reproductive in the next month. First, let us consider the simplest case of two months for the life span. This means that a mature rat one month after its birth produces twelve newborns per couple, and does so again one month later. Then it dies. Now let us assume in addition that the initial couple consists of mature rats one month after their birth.

The biologically estimated life span of rat is 1–3 years. It significantly depends on the environmental condition. If there are some operations for the extermination, like the introduction of an enemy (e.g., cat), the 'averaged' life span could become shorter. We reconsider such an influence of environmental condition in the next section.

With these assumptions, we have reproductive couples of rats one month after their birth and of those two months after it in every month. Let denote the number of the former couples by x_n, that of the latter by y_n, and the number of newborns by r_n for the nth month. Repeated calculations with these assumptions give the following result:

n	1	2	3	4	5	6	7	8	9	10	11	12
x_n	1	6	42	288	1980	13,608	93,528	642,816	4,418,064	30,365,280	208,700,064	1,434,392,064
y_n	0	1	6	42	288	1980	13,608	93,528	642,816	4,418,064	30,365,280	208,700,064
r_n	12	84	576	3960	27,216	187,056	1,285,632	8,836,128	60,730,560	417,400,128	2,868,784,128	19,717,105,536
a_n	14	96	660	4536	31,176	214,272	1,472,688	10,121,760	69,566,688	478,130,688	3,286,184,256	22,585,889,664

In this table, a_n denotes the number of rats in the family at the end of nth month. Since any rat dies after making reproduction twice, we have $a_n = 2x_n + r_n$. The above result indicates that the number of rats becomes near 22.6 billions in December. This is smaller than the number in the case without death, while it is still huge, in contrast to the case with a longer period for the maturation. So it may seem that the influence of life span on the population dynamics with the geometric progression would be weaker than that of maturation period, though we will see that this result significantly depends on the number of newborns per couple. The number 12 is too big as argued later.

From the assumptions, we can obtain the following relations among x_n, y_n and a_n: $x_{n+1} = r_n/2$; $y_{n+1} = x_n$; $r_n = 12(x_n + y_n)$. Thus, since the number of reproductive couples in the nth month is $c_n = x_n + y_n$, we can derive the following recurrence relation about the number of couples from these relations:

$$c_{n+1} = 6(c_n + c_{n-1}). \tag{1.5}$$

With $c_1 = 1$ and $c_2 = 7$, the general term becomes

$$c_n = \frac{5 + \sqrt{15}}{60}(3 + \sqrt{15})^n + \frac{5 - \sqrt{15}}{60}(3 - \sqrt{15})^n$$

(Sect. 11.1). Besides, from the relation $a_n = r_{n-1} + r_n = 12(c_{n-1} + c_n)$ which can be derived by the relations shown in the above, we can obtain the following general term about the total number of rats in each month:

$$a_n = \frac{15 + 4\sqrt{15}}{15}(3 + \sqrt{15})^n + \frac{15 - 4\sqrt{15}}{15}(3 - \sqrt{15})^n \quad (n \geq 2).$$

As easily seen, the recurrence relation (1.5) is mathematically similar to (1.3) for Fibonacci sequence. Actually, if we change the assumption on the number of newborns per couple from 12 to 2 (Fig. 1.2), that is the same as about Fibonacci sequence model, the population dynamics model considered here becomes (1.3) again, so that the monthly variation about the number of couples becomes a Fibonacci sequence. Remark however that we have now $c_1 = 1$ and $c_2 = 2$, differently from the previous case of Fibonacci sequence with $c_1 = c_2 = 1$, and the

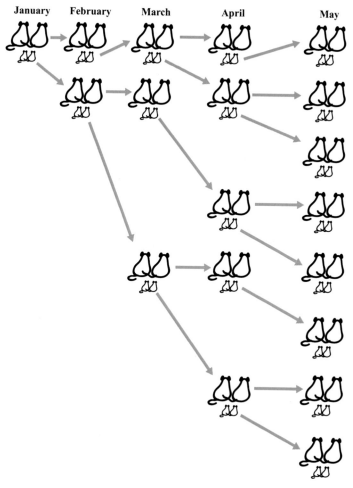

Fig. 1.2 Illustrative variation about the number of rats with two month life span and two newborns per couple in Sect. 1.4

sequence $\{c_n\}$ becomes $\{1, 2, 3, 5, 8, 13, 21, 34, 55, 89, 144, 233\}$ and consequently $a_{12} = 2 \times (233 + 144) = 754$. The general term for this sequence is necessarily different from (1.4).

In this section, we needed to take account of age structure in the family. In such a case, we must consider the monthly variation of age distribution, that is, the numbers of newborns (age 0), mature rats one month after their birth (age 1) and those two months after it (age 2) respectively. This is because we

(continued)

must take account of how many rats disappear from the family by the death. Therefore, the temporal variation of age distribution in the family determines the nature of the temporal variation of family size. As shown below, we can consider the temporal variation of the age distribution itself with vector and matrix formulas, while we used some relations of the age structure to derive the recurrence relation about $\{c_n\}$ in the above description.

As before, let denote the number of rats of age 0 by r_n, that of age 1 by x_n, and that of age 2 by y_n for the nth month. We can express the age distribution by the following column vector $\mathbf{f}_n$

$$\mathbf{f}_n := \begin{pmatrix} x_n \\ y_n \end{pmatrix}$$

Once the numbers x_n and y_n are determined, we can find the number r_n from the relation $r_n = 12(x_n + y_n)$. Hence, it is satisfactory to consider the age distribution defined by the two dimensional vector $\mathbf{f}_n$. From those relations obtained before, we can derive the following recurrence relation with a 2×2 matrix to give the monthly variation of the age distribution:

$$\mathbf{f}_{n+1} = \begin{pmatrix} 6 & 6 \\ 1 & 0 \end{pmatrix} \mathbf{f}_n. \tag{1.6}$$

Hence, denoting the above matrix by A, we have $\mathbf{f}_n = A^{n-1} \mathbf{f}_1$.

We have the characteristic equation $\lambda^2 - 6\lambda - 6 = 0$ with respect to the eigenvalue λ for the matrix A. This characteristic equation is the same with that for the recurrence relation (1.5). It is mathematically reasonable consequence, since the arguments is for the same mathematical model. The eigenvalues are $\lambda_\pm = 3 \pm \sqrt{15}$. Therefore, using the right eigenvectors $^\mathsf{T}(6, -\lambda_-)$ and $^\mathsf{T}(6, -\lambda_+)$ for eigenvalue λ_+ and λ_- respectively, we can define the 2×2 matrix

$$P := \begin{pmatrix} 6 & 6 \\ -\lambda_- & -\lambda_+ \end{pmatrix},$$

and have

$$P^{-1}AP = \begin{pmatrix} \lambda_+ & 0 \\ 0 & \lambda_- \end{pmatrix}.$$

(continued)

From this equation, we can derive

$$A^n = P \begin{pmatrix} \lambda_+^n & 0 \\ 0 & \lambda_-^n \end{pmatrix} P^{-1}.$$

Since $\mathbf{f}_1 = {}^T(x_1, y_1) = {}^T(2, 0)$ from the assumption for the modeling, we can derive the general term of $\mathbf{f}_n$, that is, those of x_n and y_n. In this way, the use of vector and matrix may simplify the construction and analysis for the more general model. We shall revisit such a modeling with vector and matrix in Sect. 10.1.

1.5 Survival Probability

Differently from the previous section to introduce the 'life span' in the modeling, we introduce the 'survival probability' now. It means at the same time to introduce the death probability. Some readers might have felt strange or unrealistic about the previous assumption that the rat must necessarily die just two months after its birth. Such feeling is acceptable because we are considering the ecology of a rat. However, in reality, it is known that there are many animals and plants which have such a life history with a definite life span. There are a variety of plant species categorized as the annual plant, for example. They have definitely one year life span after germination, and every individual dies after the reproduction with seedling etc.

In contrast, it may be possible to think that the biological population would generally consist of individuals, some of which live longer than others. In such a population, the 'life span' could be defined as the 'mean life span' or 'expected life span' which is determined consequently from the accumulation of life histories about all members of the population. Such a 'life span' should be clearly distinguished from the 'life span' in the previous section.

In biology, there are concepts of *physiological life span* and *ecological life span*. The physiological life span means the maximal period in which the organism could physiologically maintain the homeostasis as living state. The ecological life span means the period in which the organism could keep the life, determined by all environmental factors to affect the life/death. Although the actual physiological life span may be defined under the environment specified for each organism, it may be defined in the most rigorous sense as the maximal period for the life in the ideal environmental condition for the survival of the organism.

To keep the living state, any organism needs the homeostasis. The homeostasis requires to maintain the activity of cells in vivo. Some cells such as blood cells (hematocyte) and skin cells (chrotoplast) are consumed for the homeostasis. They are thus to be renewed by repeating cell division. It is known for some cells that there is a physiological system to limit how many times it could produce the daughter cell by the division. It is clear that, if the stem cell to renew the above-mentioned consumed cells would be a cell that has such a limit for the number of division times, the limit would determine the physiological life span. The human marrow cell is a stem cell for the blood cells with such a nature. Some cancer cells, like famous HeLa cell, are known as those in which the physiological system to limit the number of division times has broken down.

Modeling in the previous sections introduced only the physiological life span. In this section, we are going to consider the ecological life span with the introduction of survival probability. As mentioned in the above, this is to introduce a death probability as well, while we shall not care about the cause of death. The decrease of population size due to the death is essential for the biological population dynamics.

Ecological Assumptions
To introduce the survival probability in the modeling, we need to generalize the modeling of typical geometric growth model and add an ecological assumption first.

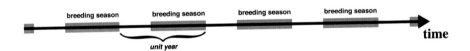

As shown in the above figure, let us assume that the unit year consists of the alternate breeding and non-breeding seasons, like that for most of biological populations, while the time unit was a month for the geometric growth model in the previous sections.

Mathematical modeling in this section will be based on the following assumptions:

- Each reproductive couple produces m newborns at the breeding season.
- The *sex ratio at birth* is 1:1.
- Newborn becomes mature and reproductive until the next breeding season.
- Formation of the reproductive couple occurs just before the breeding season.
- All reproductive couples at each breeding season disappear (die) in the subsequent non-breeding season and before the next breeding season.
- The death in the breeding season is ignored.

- The death probability for the newborn during the non-breeding season is constant, independently of year and sex.
- The survival and reproduction are independent of the population size, that is, are not affected by the population density (i.e., no *density effect*).

The assumptions except for those related to the death probability are similar to those for the geometric growth model considered in the previous sections. It is important to set up the breeding and non-breeding seasons for the following modeling. The assumptions about the period for the maturation of newborn and the disappearance of reproductive couples mean that the physiological life span is set up as a year. Since we ignore the death in the breeding season, it may be regarded as an assumption that only the death of immature individual is taken into account.

The last assumption in the above is to ignore the dependence of death probability and fertility on the shortage of food or on the degradation of environment under the high population density. We will consider the further modeling taking account of such an effect by modifying the assumption in the next chapter.

Mathematical Modeling

Let us denote the survival probability in the non-breeding season by σ ($0 < \sigma \le 1$). The death probability is given by $1 - \sigma$ in the non-breeding season. When $\sigma = 1$, the model becomes the previous geometric growth model. With c_n defined as the number of reproductive couples at the nth season, we have the number of newborns at the season mc_n from the above assumptions. The *expected* number of these newborns which can survive and mature until the next breeding season is given by σmc_n, so that they forms $\sigma mc_n/2$ reproductive couples. Since their parents die out before the subsequent breeding season from the assumption, the reproductive couples at the $(n+1)$th breeding season are formed only by those mature newborns.

As a result, we have the following recurrence relation of geometric growth model in this section:

$$c_{n+1} = \frac{\sigma mc_n}{2} \qquad (n = 1, 2, \cdots). \tag{1.7}$$

The general term is given by

$$c_n = \left(\frac{\sigma m}{2}\right)^{n-1} c_1 \qquad (n = 1, 2, \cdots).$$

Therefore, we find that, if $\sigma m/2 < 1$, the population goes extinct since c_n approaches zero as n gets larger.

The *net reproduction rate* or *net replacement rate* is defined in ecology as the expected number of mature females produced by a mature female. The value $\sigma m/2$ in the above mathematical model means the net reproduction rate. The condition that $\sigma m/2 < 1$ means that the expected number of reproductive couples produced by a reproductive couple or that of mature females produced by a mature female is smaller than one. Therefore, if the condition holds every year, the population eventually goes extinct.

There is the other ecological rate called *gross reproduction rate*, which is defined as the expected number of mature females produced by a mature female without taking account of the death. The gross reproduction rate can be regarded as the supremum of the net reproduction rate in terms of the death. The value $m/2$ corresponds to it in the above mathematical model.

Mean Life Span

How can we find the mean life span for the population governed by the mathematical model (1.7)? Let us take year as the time unit, and consider that the individual who successfully becomes mature has the unity life span, since the mature individual dies after the reproduction. For an individual who dies before the maturation, let us simply consider its life span zero. With the survival probability σ and the death probability $1 - \sigma$ in the period for the maturation, we have the expected number of successfully mature individuals σm and that of individuals which die before the maturation $(1 - \sigma)m$ for m newborns. Hence, we can define the mean life span for the population governed by the mathematical model (1.7) by $\{1 \times \sigma m + 0 \times (1 - \sigma)m\}/m = \sigma$ in the unit of year. Therefore, the ecological life span for the population is now given by σ.

Generationally Overlapping Reproduction

Let us consider the modeling without the above assumption of the physiological life span. This means to omit the assumption "Reproductive couples at each breeding season disappear (die) before the next breeding season." Thus, every individual of the population is exposed only to the death in the non-breeding season. Some of reproductive individuals can survive through the non-breeding season subsequent after their reproduction. Differently from the previous model in which every individual can pass the breeding season only once, some individuals can now pass several breeding seasons in their life. In biology, the organism which makes reproduction only once in the life is called *semelparity*, while one which can make reproduction more than once is called *iteroparity*.

Now we have to take account of the other matters related to this modification in the modeling for the iteroparity. At the breeding season, the reproductive individuals have different generations, that is, different ages, while they had the same generation/age in the previous model. Such a reproduction by individuals of different generations/ages is called *generationally overlapping reproduction* in biology. In comparison, the reproduction only by individuals of the same generation/age is sometimes called *generationally non-overlapping reproduction*. We need some new assumptions for the generationally overlapping reproduction at each breeding season of our model.

Additionally to previous assumptions except for that omitted in the above, we take the following new assumptions:

- The death probability and fertility are independent of age.
- Every reproductive couple is dissolved just after the breeding season.
- The formation of reproductive couple is independent of the age of mature individual, so that the couple formation between different generations is possible.

With these assumptions, the number of reproductive couples at the nth breeding season is given by the half of the number of mature individuals alive just before the nth breeding season. Note that every alive individual just before the breeding season is now assumed to be mature.

From the assumptions that the sex ratio at birth is 1:1 and the death rate is independent of the sex, we can estimate the sex ratio of mature individuals, that is, the *operational sex ratio* as 1:1 in the following way. Let us denote the number of female newborns at the nth breeding season by F_n. When they mature at the beginning of the $(n + 1)$ th breeding season after one year, the *expected* number becomes σF_n. The same arguments can be applied for the male newborn, and it is concluded that the expected number of mature males is the same with that of mature females, since the number of male newborns at the nth breeding season is equal to F_n from the assumption about the sex ratio at birth 1:1. Further, these arguments can be applied also for the older mature females/males at each breeding season. Consequently, with the mathematical induction, we can find that the operational sex ratio in the present model is necessarily given as 1:1 for any breeding season.

Individuals alive just before the $(n + 1)$ th reproduction season consist of those which become mature after their birth at the nth reproduction season and those which are mature individuals surviving after their reproductive activity at the nth reproduction season. With the number of reproductive couples at the nth reproduction season c_n, the number of mature individuals at the nth reproduction season is given by $2c_n$. Thus we have the expected number of mature individuals which survive until the $(n + 1)$ th reproduction season as $\sigma_A \cdot 2c_n$, where σ_A is the survival probability for the mature individual in the non-breeding season. The survival probability of the immature individual in the non-breeding season is now given by σ_J, different from σ_A in general. Since the number of newborns at the nth reproduction season is mc_n, the number of those which can survive till the next reproduction season and become mature is given by $\sigma_J mc_n$. Therefore, the sum $2\sigma_A c_n + \sigma_J mc_n$ gives the number of individuals alive just before the $(n + 1)$ th reproduction season, and finally we obtain the following recurrence relation:

$$c_{n+1} = \frac{1}{2}\left(2\sigma_A c_n + \sigma_J mc_n\right) = \left(\sigma_A + \frac{\sigma_J m}{2}\right)c_n. \tag{1.8}$$

Consequently, from the arguments same as for the previous model, we find that the population goes extinct if and only if $\sigma_A + \sigma_J m/2 < 1$. Note that the present model (1.8) becomes equivalent to the previous one with $\sigma_A = 0$. Thus, we can regard the present model (1.8) as a model including and generalizing the previous (1.7).

At the end of this section, let us consider the mean life span for the population governed by the model (1.8). As before, we regard the life span for the individual died before the maturation as zero. The probability that an individual dies after passing k reproduction seasons is given by

$$\left(1 - \sigma_A\right)\sigma_A^{k-1}\sigma_J \qquad (k \geq 1). \tag{1.9}$$

We regard the life span of such an individual as k. Hence, by the standard mathematical definition of the expected value, we can calculate the mean life span as follows:

$$\sum_{k=1}^{\infty} k\left(1 - \sigma_A\right)\sigma_A^{k-1}\sigma_J = \frac{\sigma_J}{1 - \sigma_A}. \tag{1.10}$$

Exercise 1.2 Show that the net reproduction rate for the population governed by the model (1.8) becomes

$$\frac{\sigma_J m}{2} \frac{1}{1 - \sigma_A}. \tag{1.11}$$

1.6 Sexual Difference

Up to now, we have been considering the mathematical modeling with the assumption that there is no difference between female and male with respect to the survival and death. For most organisms, this is not applicable even in any sense of approximation. Moreover, there are many animals and plants which has the sex ratio at birth or the operational sex ratio significantly different from 1:1. Biologists try to understand such a biased sex ratio as an evolutionary strategy established in the evolutionary process to have the larger number of offsprings. In this section, without stepping into such an argument from the viewpoint of evolutionary biology, we are going to consider the mathematical modeling of a population dynamics introduced such a biased sex ratio.

Additional Ecological Assumptions
To consider the case of a biased sex ratio, we need an additional assumption about the formation of reproductive couple, since the female and male numbers are not

the same. We add the following assumption to the previous assumptions about the formation and dissolution of reproductive couple:

- Reproductive couples are generated among mature individuals as many as possible.

From the previous assumptions, the formation of reproductive couples is at the beginning of the reproduction season. By the above new assumption, if the mature males are more than the mature females, the number of reproductive couples becomes equal to the number of mature females.

To avoid much complicatedness, let us assume here the followings:

- The survival probability in each non-breeding season is independent of whether the individual could form a couple or not at the previous breeding season.
- The survival probability of immature individual is independent of the sex.

Let us denote the survival probabilities of mature female and male respectively by σ_F and σ_M. The sex ratio at birth is now given by $\omega : 1 - \omega$, where ω is the female ratio at birth. When $\omega > 1/2$, the number of female newborns is greater than that of male.

Mathematical Modeling

For a reasonable step of mathematical modeling, let us denote the numbers of mature females and males at the nth breeding season by F_n and M_n respectively. Since the number of newborns at the nth breeding season is given by mc_n as before, the number of female newborns is ωmc_n, and that of male is $(1-\omega)mc_n$ with the female ratio ω at birth.

Note that the number of mature individuals at the nth breeding season must be greater than or equal to $2c_n$, that is, $F_n + M_n \geq 2c_n$. There may be some mature (only female or male) individuals which could not form couples from the assumption given in the above. Some such uncoupled mature individuals may join the reproduction at the subsequent breeding season if they can survive until it. From the assumption added in this section, the number of reproductive couples is equal to the smaller in the numbers of mature females and males: $c_n = \min[F_n, M_n]$. Thus, the number of uncoupled mature (only female or male) individuals is given by $\max[F_n, M_n] - c_n$.

The mature individuals at the nth breeding season consist of those which were born at the $(n-1)$th breeding season and became mature and those which were mature at the $(n-1)$th breeding season and could survive to the nth breeding season. Thus, with the survival probabilities σ_J, σ_F and σ_M for the immature individual, the mature female and male respectively, we can obtain the following mathematical model by the recurrence relations:

$$F_{n+1} = \sigma_F F_n + \sigma_J \omega m \min[F_n, M_n];$$
$$M_{n+1} = \sigma_M M_n + \sigma_J (1 - \omega) m \min[F_n, M_n], \tag{1.12}$$

with the initial mature individuals $2c_1$ as in the previous geometric growth model. This initial number $2c_1$ indicates that the population dynamics begins with the initial number of reproductive couples c_1, and $F_1 = M_1 = c_1$ since we assume that the numbers of mature females and males are equal to c_1 at the initial.

Population Dynamics

As shown by numerical examples in Fig. 1.3, the population dynamics governed by (1.12) becomes approximated well by a geometric growth for sufficiently large n. To consider this nature, let us put $F_n \approx F^*\lambda^n$ and $M_n \approx M^*\lambda^n$ with undetermined positive constants F^* and M^*. From (1.12), we can easily find that

$$\lambda = \begin{cases} \lambda_F := \sigma_F + \sigma_J\omega m & \text{if } F^* < M^*; \\[2mm] \lambda_M := \sigma_M + \sigma_J(1 - \omega)m & \text{if } F^* > M^*, \end{cases}$$

and that $\lambda_F < \lambda_M$ if and only if $F^* < M^*$, while $\lambda_F > \lambda_M$ if and only if $F^* > M^*$. Besides, $F^* = M^*$ if and only if $\lambda_F = \lambda_M$. From these results, the population dynamics for sufficiently large n can be regarded as following

$$(F_n, M_n) \approx \left(F^*\{\min[\lambda_F, \lambda_M]\}^n, M^*\{\min[\lambda_F, \lambda_M]\}^n\right). \tag{1.13}$$

Therefore, the population geometrically grows if and only if both of λ_F and λ_M are greater than 1 (Fig. 1.3b). If one of λ_F and λ_M is less than 1, the population geometrically goes extinct (Fig. 1.3a).

The parameter λ_F (resp. λ_M) means the expected number of mature females (resp. males) at the next breeding season, which are produced by a mature female (resp. male) at each breeding season. It is defined by adding the expectation σ_F (resp. σ_M) of a mature individual surviving until the next breeding season to the number of newborns produced by it. From this definition, we can see that λ_F does not correspond to the net reproduction rate (see p. 16), but does to the sum of the net reproduction rate and the expectation σ_F. Hence, let us call hereafter λ_F and λ_M respectively *female replacement rate* and *male replacement rate*.

When the female replacement rate $\lambda_F < 1$, each mature female at every breeding season can produce mature females less than one in mean at the next breeding season, so that the number of mature females is expected to decrease season by season. From the assumption about the couple formation, the number of reproductive couples decreases at the same time, independently of the value of the male replacement rate λ_M. This situation causes the seasonal decrease in the total number of newborns to make the population go extinct. These arguments can be applied for the case where $\lambda_M < 1$ (see Fig. 1.3a).

Let us see the sex ratio in the mature individuals including uncoupled at the breeding season. As Fig. 1.3 indicates, it gradually approaches a certain constant, independently of which the population can persist or goes extinct. From the above feature of the population dynamics, we can obtain the following mathematical

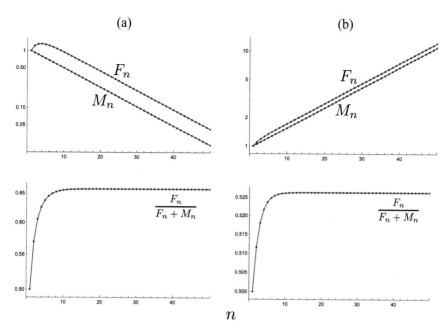

Fig. 1.3 Numerical example of the temporal variation about the numbers of mature females and males, F_n and M_n, and that about the female ratio in the mature individuals $F_n/(F_n + M_n)$ for the population dynamics governed by (1.12). $(\lambda_F, \lambda_M) =$ (a) $(1.225, 0.925)$; (b) $(1.1, 1.05)$, with (a) $\omega = 0.5$; (b) $\omega = 0.4$. Commonly, $(F_1, M_1) = (1, 1)$; $m = 5$; $\sigma_J = 0.25$; $\sigma_F = 0.6$; $\sigma_M = 0.3$. The ordinate of the upper graphs for F_n and M_n has a common logarithmic scale

result:

$$\frac{F_n}{F_n + M_n} \to \frac{\sigma_J m}{\sigma_J m + \sigma_M - \sigma_F}\, \omega \qquad \text{if and only if } \lambda_F > \lambda_M;$$

$$\frac{F_n}{F_n + M_n} \to 1 - \frac{\sigma_J m}{\sigma_J m + \sigma_F - \sigma_M}\, (1 - \omega) \quad \text{if and only if } \lambda_F < \lambda_M,$$

as $n \to \infty$. This result shows how the operational sex ratio is determined as the consequence of the population dynamics with given sex ratio at birth and survival probabilities. Note that, when λ_F or λ_M is less than one, the population goes extinct over time. In such a case, the above convergence of the operational sex ratio is just a mathematical result, because the reproduction becomes impossible so that the population dynamics breaks down.

As for the special case where $\lambda_F = \lambda_M = \lambda$, the model (1.12) becomes mathematically simpler. Especially when $F_1 = M_1 = c_1$, we can easily find with the mathematical induction that $F_n = M_n = c_1 \lambda^{n-1}$ for any $n \geq 1$. Thus, the operational sex ratio is kept as 1:1 at every season. For more general case where $F_1 \neq M_1$, we can prove that, if $F_n > M_n$ for some n, then $F_k > M_k$ for any $k > n$. Further, in such a case, since we have $F_{k+1} - M_{k+1} = \sigma_F(F_k - M_k)$ for $k > n$, it

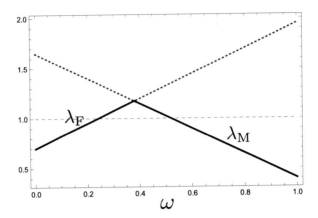

Fig. 1.4 Dependence of the female replacement rate λ_F and the male replacement rate λ_M on the sex ratio at birth ω for the model (1.12). The solid line indicates $\min[\lambda_F, \lambda_M]$. Numerically drawn with $m = 5$; $\sigma_J = 0.25$; $\sigma_F = 0.7$; $\sigma_M = 0.4$

is concluded that $F_n - M_n \to 0$ as $n \to \infty$. With these arguments, we can find that the operational sex ratio necessarily approaches 1:1 when $\lambda_F = \lambda_M = \lambda$. This result is consistent with the meaning of the female/male replacement rates λ_F and λ_M.

In these arguments, we have seen that the population dynamics is determined by the female/male replacement rates λ_F and λ_M. These replacement rates linearly depend on the female ratio at birth ω. Indeed, λ_F is linearly increasing while λ_M is linearly decreasing in terms of ω (Fig. 1.4). When the population is persistent, that is, if $\lambda_F > 1$ and $\lambda_M > 1$, its geometric growth has a velocity determined by the smaller of λ_F and λ_M.

This result can be regarded as consistent with the meaning of the female/male replacement rates λ_F and λ_M, and the assumptions for the model (1.12). When the population grows, the total number of newborns is determined by the number of reproductive couples, while the number of reproductive couples is determined by the smaller number of mature females and males. Therefore, the velocity of the population growth is determined by the sex of which the number of mature individuals is smaller.

Dependence on Sex Ratio at Birth
With those results obtained in the previous sections, we are briefly going into the discussion from the viewpoint of evolutionary biology on the nature of population dynamics governed by the model (1.12).

From the viewpoint of *selfish gene* in biology, the appearance of a characteristic (phenotype) dominant in a population is regarded as the establishment of a gene which governs the phenotype, or as the spread of the gene over the population. The phenotype different from the dominant one is going to disappear in generational time scale without its establishment in the population. The process of such an establishment and disappearance leads to the evolution, which is called *natural selection.* The kinetics of the natural selection relies on the population dynamics. The disappearance of a phenotype can be translated as the extinction of individuals with the corresponding gene from the population, while the establishment can be done as the persistent frequency of individuals with the corresponding gene in the population. If the individual with a phenotype can produce offsprings more than those produced by the others with any phenotype different from it, the phenotype can be established in the population.

Now let us briefly go into such an evolutionary viewpoint for the dependence of the population dynamics governed by (1.12) on the sex ratio at birth. The sex ratio at birth is certainly one of phenotypes characterizing a biological population. As we have seen in the previous section, the persistence/extinction of a population significantly depends on it. Only when the sex ratio at birth satisfies the condition that $\lambda_F > 1$ and $\lambda_M > 1$, the population can persist and grow.

Further, supposing two different strains with respect to the replacement rates λ_F and λ_M, we can say that the strain with the larger $\min[\lambda_F, \lambda_M]$ could have the larger subpopulation than the other, because the velocity of the growth is determined by it. From the evolutionary viewpoint, this means that the strain with the sex ratio at birth to make the value $\min[\lambda_F, \lambda_M]$ larger could be established in the population, provided that any feature other than the sex ratio at birth is common.

The strain is similar to the family lineage in human society. We are now considering the family line inheriting a gene which determines the sex ratio at birth. A different strain may appear in the same population, for example, by a mutation stochastically occurred in the genetic process, or by a mixture (for example, caused by a change of the geographic configuration) of different strains which were independent before.

Based on these arguments, we could assume that the evolutionary process favors to select the larger $\min[\lambda_F, \lambda_M]$. By the result obtained in the previous section, it becomes the largest when $\lambda_F = \lambda_M$. A strain with the sex ratio at birth to satisfy this condition can be regarded as one having the growth velocity larger than any other strain. Thus such a strain can have the frequency larger than the others, and evolutionarily become established in the population.

For the present model given by (1.12), when every member has the sex ratio at birth to satisfy $\lambda_F = \lambda_M$, the population is said to be at the *evolutionarily stable state* (ESS), because any other strain with different sex ratio at birth cannot increase its frequency, as long as any other phenotype is common.

Let us consider the feature of such a population at the ESS. As already seen, the operational sex ratio becomes 1:1 in the population. From the assumptions for the model (1.12), this is the situation such that there does not exist any uncoupled mature individual at the breeding season. This can be regarded as ideal from the standing point of the reproductive individual and the viewpoint of selfish gene, since any offspring can make the reproduction when it survives and matures. This kind of evolutionary consideration on the sex ratio has been dealt by what is sometimes called "Fisher's theory" in biology, which essentials were discussed and established by Sir Ronald A. Fisher (1890–1962). It explains that the operational sex ratio under the simplest situation becomes 1:1 at the ESS.

As for the model (1.12), we can find that the condition $\lambda_F = \lambda_M > 1$ is fulfilled by the sex ratio at birth satisfying

$$\frac{1 - \sigma_F}{\sigma_J m} < \omega = \frac{1}{2}\left(1 - \frac{\sigma_F - \sigma_M}{\sigma_J m}\right) < 1 - \frac{1 - \sigma_M}{\sigma_J m}.$$

It is intuitively obvious that the sex with the lower survival probability needs to have the larger sex ratio at birth in order to lead to the operational sex ratio 1:1. From the above condition, we can see that the bias in the sex ratio at birth becomes smaller as the expected number of survival and mature offsprings per couple $\sigma_J m$ gets larger. In contrast, as the expected number of survival and mature offsprings per couple is smaller, the bias becomes more striking. The former case could correspond to a case where the number of newborns is huge even with their small survival probability, or a case where the survival probability of newborn is sufficiently large, for example, due to the parental care, even with a small number of newborns.

We must note that the above conclusion significantly depends on the assumption about the reproduction for the model (1.12): The reproduction is possible only by the formation of couple with a female and a male. About many species, the energetic requirement for the reproduction is bigger for the female than that for the male, since the energetic cost to produce the gamete is generally smaller for the male (e.g., sperm or pollen) than for the female (e.g., egg or ovule). For this reason, from the viewpoint of selfish gene aimed to produce offsprings as many as possible, it would be likely for a male to have children with more than one females. Even when the reproduction is done only with the couple formation, there are ecological examples to allow the existence of uncoupled individuals, like what is called "helper" in some bird species. Therefore, the operational sex ratio 1:1 is not always established in the evolution, even though the reproduction is possible only by the formation of reproductive couple.

In this section, we briefly went into the evolutionary problem on the sex ratio at birth. The modern theory on the sex ratio has been systemized and sophisticated with the accumulated researches on a variety of species in history. In general, we need to take account of a *trade-off* relation of one phenotype to the other phenotype. The benefit by one phenotype may cause a loss for the other. It is likely that the difference in the sex ratio at birth would be reflected to a difference in the survival probability for the immature individual (σ_J). For example, the sex ratio at birth could have a close

(continued)

relation to the energetic investment by the parent to the offspring (for example, embodied with the egg size). In the modern biology, the evolutionarily stable state is considered in general as attainable under such a trade-off relation between phenotypes. For the readers interested in the mathematical modeling for this kind of evolutionary aspect, for example, see [3–5].

Answer to Exercise

Exercise 1.1 (p. 5)

Along the same arguments in the main text, we can get the recurrence relation $c_{n+1} = (k + 1)c_n$, that leads to $c_n = (k + 1)^{n-1}$. Thus, in December, the family consists of $2 \times (k + 1)^{11} + 2k \times (k + 1)^{11} = 2 \times (k + 1)^{12}$ individuals. For $k = 1$ that gives the least number of newborns satisfies the above assumptions, it becomes $2^{13} = 8,192$. How do you think about this result? More realistic?

Exercise 1.2 (p. 18)

As defined at p. 16, the net reproduction rate is given by the expected number of mature females produced by a mature female. Let us consider a mature female at a breeding season, which is born at the previous breeding season and successfully survives to become mature in the subsequent breeding season.

From the same arguments as for the probability (1.9), we have the probability that the mature female dies after passing k reproduction seasons is given by

$$\left(1 - \sigma_A\right)\sigma_A^{k-1} \qquad (k \geq 1). \qquad (1.14)$$

This is the probability that a female makes the reproduction k times in the life span.

The female which makes the reproduction k times produce km newborns, that is, $km/2$ female newborns. Thus, the expected number of produced mature females by such a female is given by $\sigma_J km/2$ because of the death probability $1 - \sigma_J$ before the maturation. Hence, with the probability (1.14), we can derive the net reproduction rate as the expected number of mature females produced by a mature female, making use of the mathematical definition of expected value given by

$$\sum_{k=1}^{\infty} \sigma_J \frac{km}{2} \cdot \left(1 - \sigma_A\right)\sigma_A^{k-1},$$

and get the result (1.11).

The net reproduction rate (1.11) can be translated as the product of the expected number of mature females produced by a mature female per breeding season and the expected survival duration of a female after the maturation. Actually, from (1.14), the latter expected value can be given by

$$\sum_{k=1}^{\infty} k \cdot (1 - \sigma_A)\sigma_A^{k-1} = \frac{1}{1 - \sigma_A},$$

and the former is given by $\sigma_J m/2$.

Same as the argument in p. 16 for the previous model (1.7), the condition that the net reproduction rate (1.11) is less than one is equivalent to that $\sigma_A + \sigma_J m/2 < 1$ derived from the recurrence relation (1.8).

References

1. N. Bacaër, *A Short History of Mathematical Population Dynamics* (Springer, London, 2011)
2. P. Ball, *Patterns in Nature: Why the Natural World Looks the Way It Does* (University of Chicago Press, Chicago, 2016)
3. M. Bulmer, *Theoretical Evolutionary Ecology* (Sinauer Associates Publishers, Sunderland, 1994)
4. J.M. Smith, *Evolutionary Genetics*, 2nd edn. (Oxford University Press, Oxford, 1998)
5. T.L. Vincent, J.S. Brown, *Evolutionary Game Theory, Natural Selection, and Darwinian Dynamics* (Cambridge University Press, New York, 2005)

Chapter 2
Influence From Surrounding

Abstract In this chapter, we shall introduce the density effect in the population dynamics, as the expansion of discrete time linear models in Chap. 1. The introduced density effect could be intraspecific, interspecific, or both. As the intraspecific density effect, we shall consider the Allee effect, and as the interspecific one, the interspecific competition and the prey-predator or host-parasite relation. Moreover, we shall consider the effect of harvesting/culling on the population dynamics too. Then we give some arguments on the problem of maximum yield or benefit, related to the mathematical fundamental in bioeconomics. Further, in the last section of this chapter, we shall describe the other type of modeling for the discrete time population dynamics, since we do not want to make the readers get a fixed idea about the modeling for the discrete time population dynamics. The ideas shown in the part would be expandable to the construction of another sophisticated model for a discrete time population dynamics.

Every mathematical model in the previous chapter characteristically had a mathematical nature of geometric progression, even though they had different formulas depending on different assumptions for the modeling. Such a mathematical nature arose fundamentally from the feature that the number of newborns per couple m and the survival probability σ are constant independently of the situation in the population.

In reality, the life of biological population necessarily affects the environment surrounding it. While the population dynamics clearly depends on the environmental condition, the environmental change by the life of the population has a feedback effect on the population dynamics. It is very likely that such a feedback effect appears on the number of newborns or the survival probability. Investigation and analysis on such an effect would be one of important research subjects in ecology.

In this chapter, we shall consider some basic mathematical modelings about the influence on the reproduction and survival from the surrounding condition within

---

The original version of this chapter was revised: Footnote 1 have been updated. The correction to this chapter can be found at https://doi.org/10.1007/978-981-19-6016-1_16

© The Author(s), under exclusive license to Springer Nature Singapore Pte Ltd.2022, corrected publication 2022
H. Seno, *A Primer on Population Dynamics Modeling*, Theoretical Biology,
https://doi.org/10.1007/978-981-19-6016-1_2

the same population, the interaction with another population, and the external force
to affect the population dynamics.

2.1 Negative Density Effect

Let us consider again the mathematical modeling for the geometric growth
model (1.7) with the generationally non-overlapping semelparity, introduced at
p. 15 in the previous chapter. To simply introduce an influence of the surrounding
in the modeling, we modify the last assumption "The survival and reproduction are
independent of the population size" introduced at p. 14 in Sect. 1.5 as follows:

- The reproduction is influenced by the population size, while the survival is
  independent of it.

A cause of such an influence of the population size on the fertility is the
dependence of the energy gain for the reproduction on the population size, as
mentioned in many textbooks on ecology (for example, see [3, 17, 24, 29, 48,
56, 66]). For example, under a situation that the amount of food is limited in the
habitat, the larger population size leads to the smaller expected amount of food
per individual, which necessarily makes the portion of gained energy to allocate for
the reproduction smaller, subsequently declining the fertility. The similar arguments
would be applicable for the death (or survival) probability. However, the correlation
of the environment to it must be different from that to the fertility. For the sake of
simplicity, we shall introduce here only the influence on the fertility, ignoring that
on the death probability in the modeling of this chapter.

It is more accurate in an ecological sense to consider that the influence is not
from the number of individuals in the population but from the population density.
This is because the influence of the surrounding is mainly determined by the number
of neighboring individuals, so that it gets stronger as the distance from neighbor
individuals is smaller. Briefly, the per capita fertility significantly depends on the
population density. Such an influence of the population density on the population
dynamics (not only about the reproduction) is generally called *density effect* in
ecology.

Now, with the new assumption given in the above, we modify the geometric
growth model (1.7) as follows:

$$c_{n+1} = \frac{\sigma m(a_n) c_n}{2} \qquad (n = 1, 2, \cdots), \qquad (2.1)$$

where a_n denotes the total number of mature individuals at the nth breeding season.
From the assumption for the geometric growth model (1.7), we have $a_n = 2c_n$,
so that we introduce the number of newborns m per couple in this modeling by a
function of the total number of mature individuals a_n.

In this book, we have been using the word "number" to explain the variable in the population dynamics from the first chapter. More precisely, it should be "number density" from the theory of population dynamics. This means that we had better mostly translate "number" into "number density", for example, about the definitions of c_n and a_n. The number density is defined as the expected or averaged "number" per area or per volume in general. Its value naturally depends on the unit of area or volume. For the mathematical modeling, we remark that the value of number density becomes fractional in general, and it is generally treated as a real number if we would need to consider the value itself.

The reason why the number of newborns m is given by a function not of the number of couples c_n but of the number of mature individuals a_n is as follows: The amount of energy available for the reproduction is determined by the surrounding condition before the breeding season. As mentioned in the above, the higher population density makes the energy gain available for the reproduction per individual smaller, so that it is reasonable to consider that the number of newborns per couple is given by a decreasing function of the population density. This negative correlation gives the *negative density effect* on the number of newborns per couple. In the following part, we shall see some mathematical models with a given function for it.

In this type of modeling to describe a sequence of temporally changing value, the value at the next time step would depend on the whole past sequence as the most general influence from the population size. When a model is given by a recurrence relation to determine the next value from the past sequence, the recurrence relation can be said that it defines a *discrete dynamical system* if the recurrence relation determines a unique sequence with a given appropriate initial value. The recurrence relation itself may be called discrete dynamical system. For the introduction of discrete dynamical system, the reader can refer to [16, 19, 20, 36], and for the more advanced mathematical description, can see [11, 53].

2.1.1 Beverton-Holt Type Model

First let us introduce the following function defining the influence of the number of mature individuals a_n on the number of newborns per couple m:

$$m(a_n) = \frac{m_0}{1 + (a_n/\alpha)^\theta} = \frac{m_0}{1 + (2c_n/\alpha)^\theta}, \tag{2.2}$$

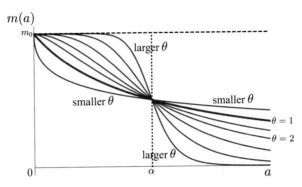

Fig. 2.1 Parameter dependence of the density effect (2.2) from the number of mature individuals *a* on the number of newborns per couple *m*

where $m(0) = m_0$ means the physiological upper bound for the number m, that is, the number of newborns produced when no density effect works. Parameters α and θ characterize the relation between the numbers of newborns and mature individuals, as explained more in the below.

> Some readers getting accustomed with the modeling may feel strange on the above explanation "the number of newborns produced when no density effect works". It is very natural. Because the situation without density effect corresponds to the case of $a_n = 0$, which means that there is no mature individual, that is, no couple. In such a case, the reproduction is impossible. As a reasonable way to understand it, we must consider the case where $a_n = 2c_n$ is sufficiently small. In a mathematical sense, we need to consider the limit of $c_n \to 0$ instead of $c_n = 0$. In a biological sense, it takes natural to suppose a certain upper bound for the number of newborns per couple under some density effect to suppress the number. In the mathematical modeling with (2.2), it is introduced by m_0.

As shown by Fig. 2.1, the parameter θ characterizes the sensitivity of the number of newborns per couple m to the number of mature individuals a. For sufficiently small θ, the value of m becomes much small even for a range of small a, while it has relatively little relation to a for a range of large a. In such a case, the number of newborns per couple m has strong sensitivity to the number of mature individuals a when a is small, and has weak sensitivity to it when a is large.

In contrast, for sufficiently large θ, the value of m has the weak density effect from the mature individuals when a is less than α, and is relatively near to the physiological upper bound. It becomes much small by the strong density effect when a is more than α. In such a case, the parameter α can be regarded as the critical value for the number of mature individuals a with respect to the density effect from the mature individuals. If we consider the mathematical limit of $\alpha \to \infty$ for (2.2), we can find that m converges to the value m_0 in case of no density effect from the mature individuals. This is consistent with the meaning of α as the critical value for a with respect to the density effect.

For the intermediate value of θ (for example, $\theta = 1$), the number of newborns per couple m has an intermediate sensitivity to the number of mature individuals a, and it changes necessarily little for any little change of a.

With the density effect function (2.2), the recurrence relation (2.1) becomes

$$c_{n+1} = \frac{\mathscr{R}_0 c_n}{1 + (2c_n/\alpha)^\theta}, \tag{2.3}$$

where $\mathscr{R}_0 := \sigma m_0/2$ means the net reproduction rate expected when the density effect is absent (refer to the definition of the net reproduction rate at p. 16 of Sect. 1.5). More precisely it means the physiological upper bound for the net reproduction rate under the density effect given by (2.2), since the actual net reproduction rate is determined under the density effect.

Especially the mathematical model (2.3) with $\theta = 1$ is frequently called *Beverton-Holt model* after the work applied for the fishery problem by Raymond (Ray) J.H. Beverton (1922–1995) and Sidney J. Holt (1926–2019) in 1957 [5]. It is investigated more mathematically by M.P. Hassell in 1975 [23].

Substituting $x_n = 2c_n/\alpha$ for (2.3), we can obtain

$$x_{n+1} = \frac{\mathscr{R}_0 x_n}{1 + x_n^\theta}. \tag{2.4}$$

This is the discrete time model investigated by Maynard et al. [59] and Bellows [4] as a modification of Beverton-Holt model. On the other hand, the substitution of $\zeta_n = (2c_n/\alpha)^\theta$ for (2.3) leads to

$$\zeta_{n+1} = \frac{\mathscr{R}_0^\theta \zeta_n}{(1 + \zeta_n)^\theta}. \tag{2.5}$$

This is a special case of the model investigated by M.P. Hassell [23] as the other modification of Beverton-Holt model. Since α is a positive constant, the sequences of x_n, ζ_n, and c_n qualitatively have the equivalent nature in a mathematical sense. The nature of the sequences $\{x_n\}$ and $\{\zeta_n\}$ is determined by parameters $\mathscr{R}_0$ and θ, and is independent of α. Thus the qualitative nature of the sequence $\{c_n\}$ is independent of α too. In the above arguments, the parameter α is an important constant to characterize the density dependence of the population dynamics given by (2.3), while it does not play any role to determine the qualitative nature of the sequence $\{c_n\}$.

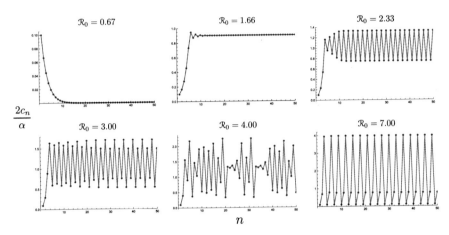

Fig. 2.2 Temporal sequences by the population dynamics model (2.3). Numerical calculations for different value of $\mathcal{R}_0$ with $\theta = 4.0$ and $2c_1/\alpha = 0.1$

Mathematical Features

As shown by numerical calculations in Fig. 2.2, the population dynamics by the recurrence relation (2.3) does not cause any unbounded increase of the population size. With the *local stability analysis* and the *cobwebbing method* introduced in Sect. 12.1, it is easy (whereas required a careful investigation of different cases) to mathematically show the following features of the model (2.3):

- The population size monotonically decreases to go extinct if and only if $\mathcal{R}_0 \leq 1$.
- When $\theta \leq 1$, c_n monotonically approaches $c^* := (\mathcal{R}_0 - 1)^{1/\theta}\alpha/2$ if and only if $\mathcal{R}_0 > 1$.
- When $\theta > 1$, c_n monotonically approaches c^* if and only if $1 < \mathcal{R}_0 \leq \theta/(\theta-1)$.
- When $1 < \theta \leq 2$, c_n approaches c^* with a damped oscillation if and only if $\mathcal{R}_0 > \theta/(\theta - 1)$.
- When $\theta > 2$, c_n approaches c^* with a damped oscillation if and only if $\theta/(\theta - 1) < \mathcal{R}_0 \leq \theta/(\theta - 2)$.
- When $\theta > 2$, c_n does not approach any specific value but keep changing if and only if $\mathcal{R}_0 > \theta/(\theta - 2)$.

The *damped oscillation* means the oscillation which oscillatory range gradually subsides toward zero as time goes. The above features about the temporal variation of the number of reproductive couples can be correspondingly applied for the temporal variation of the population size itself.

The above description about the features of the model (2.3) with $\mathcal{R}_0 > 1$ is not exact in a mathematically rigorous sense. Because there are countable set of initial values c_1 from which the sequence $\{c_n\}$ remains a constant (i.e.,

(continued)

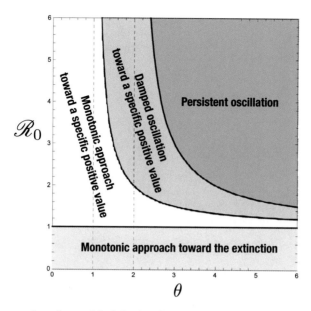

Fig. 2.3 Parameter dependence of the behavior of the population dynamics by (2.3)

the initial value) or becomes a repeated number of constants (i.e., a periodic solution). However, any such sequence is *unstable*, and any perturbation makes the sequence have a corresponding feature given in the above. Since we are considering the recurrence relation (2.3) as a mathematical model of population dynamics, any such unstable state is generally negligible for the theoretical discussion with the biological meaning of the mathematical feature about the model, even though some mathematician might criticize for it.

Figure 2.3 shows the above features in the parameter space of $(\theta, \mathscr{R}_0)$. When $\theta \leq 1$, the variation is necessarily monotonic, while, when $\theta > 1$, an oscillatory variation appears for a large $\mathscr{R}_0$ (see Fig. 2.2). Further, we can find that, when the population does not go extinct but persists with $\mathscr{R}_0 > 1$, an oscillatory variation appears for a large θ.

These results imply that the appearance of an oscillatory variation requires large θ and $\mathscr{R}_0$. Large θ corresponds to the case where the number of newborns is much sensitive to the density of mature individuals, as described before. Therefore, we may consider the kinetics of such an oscillatory variation as a process accompanying the decrease of population size due to the severe density effect, the subsequent acute moderation in the density effect with the high sensitivity to it, and the fast recovery of population size with the large reproductive faculty. Especially, since the

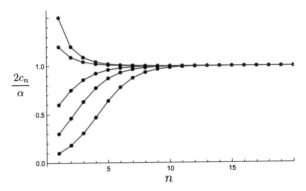

Fig. 2.4 Sequences $\{c_n\}$ by Beverton-Holt model (2.3) with $\theta = 1$. Numerically drawn for five different initial values of $2c_1/\alpha$ when $\mathscr{R}_0 = 2.0$. The carrying capacity is given as $2c^*/\alpha - \mathscr{R}_0 - 1 = 1.0$

oscillatory variation cannot occur with small θ, we find that the high sensitivity of the reproduction to the density effect is essential for the oscillatory variation.

For Beverton-Holt model (2.3) with $\theta = 1$, the sequence $\{c_n\}$ monotonically approaches $c^* := \max[0, (\mathscr{R}_0 - 1)\alpha/2]$ from any positive initial value c_1 (Fig. 2.4). That is, the number of mature individuals a_n in Beverton-Holt model monotonically approaches $a^* := \max[0, (\mathscr{R}_0 - 1)\alpha]$ for any positive initial value a_1.

Exercise 2.1 For Beverton-Holt model (2.3) with $\theta = 1$, show by the cobwebbing method that the sequence $\{c_n\}$ monotonically approaches zero when $\mathscr{R}_0 \leq 1$, and $(\mathscr{R}_0 - 1)\alpha/2$ when $\mathscr{R}_0 > 1$, independently of the initial value $c_1 > 0$.

Carrying Capacity
In an ecological sense, the equilibrium value a^* for the population governed by Beverton-Holt model (2.3) with $\theta = 1$ can be regarded as *carrying capacity* for the population dynamics. The carrying capacity is defined as the upper limit of the population size supported by the environment in which the population inhabits, following its own population dynamics. As illustrated by Fig. 2.4, the population with the size beyond the carrying capacity cannot be supported by the environment, so that it gradually declines toward the carrying capacity by the reduction of the surplus.

In most cases, the definition of carrying capacity does not take account of the influence by the other populations (as described in the later sections) in the same environment. This means that the carrying capacity is defined under the condition without any influence by the other populations. However, any

(continued)

biological population dynamics in nature is necessarily under such influences and affected also by some other factors related to the population dynamics. For example, the influence of some enemies may make the population size smaller than the carrying capacity, and the availability of resource (e.g., light, water, food, etc.) for the reproduction may not be stationary. The actual population size must be determined under such influences, while it would be theoretically worth while to consider the upper bound even for such a population size. Thus the carrying capacity is used to be theoretically defined with the ideal condition without any influence from the other populations. It should be remarked that the population dynamics may not keep the population size at a certain value even under such an ideal condition, and may cause a persistent oscillatory variation of the population size, when the carrying capacity cannot be defined well as a certain value.

Actually, the carrying capacity cannot be defined in some theoretical population dynamics. The geometric growth model discussed in the previous section does not have any carrying capacity, since the growth of population size is unbounded, which might be mathematically said that the carrying capacity is infinite. As mentioned above, the natural population size necessarily has the upper bound. However, it is not appropriate to regard such a model without it as nonsense. It is certainly thoughtless for the theoretical consideration on the population dynamics. Each theoretical consideration focuses on an aspect of the population dynamics, for example, illuminating a biological problem. From this standpoint, the theoretical model like the geometric growth model could not be regarded as null only for the reason that it causes the unbounded growth of population size, or that the carrying capacity cannot be defined. Such a way of thought itself is nonsense from the viewpoint of science.

Bifurcation Diagram

For the population dynamics model (2.3), we have seen the appearance of a persistently oscillatory variation as shown by Fig. 2.2. The nature of such an oscillatory variation significantly depends on parameters θ and $\mathscr{R}_0$. When $\mathscr{R}_0 = 2.33$, it approaches what is called *period-2 solution* which repeats two different values A and B as $ABABAB \cdots$ (see Sect. 12.1.4). When $\mathscr{R}_0 = 3.00$, it approaches a *period-4 solution*, and when $\mathscr{R}_0 = 7.00$, it does a *period-3 solution*. Further, when $\mathscr{R}_0 = 4.00$, it approaches an aperiodic oscillatory variation.

We can obtain a mathematical diagram which shows such a parameter dependence of the characteristics of oscillatory solutions. It is called *bifurcation diagram*. In this book, we do not go into the mathematically rigorous treatment of the bifurcation diagram, but use the numerically (i.e., approximately) drawn bifurcation

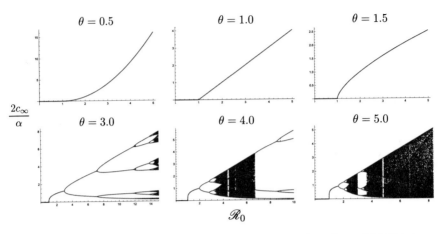

Fig. 2.5 Numerically drawn bifurcation diagram for the population dynamics model (2.3) in terms of the bifurcation parameter $\mathscr{R}_0$

diagram as a useful information to discuss the mathematical nature of population dynamics model.

The lower three bifurcation diagrams of Fig. 2.5 for the population dynamics model (2.3) include the above-mentioned persistently oscillatory behavior of the sequence $\{c_n\}$ in the range of $\mathscr{R}_0 > \theta/(\theta - 2)$. The outline of the numerical calculation to draw the bifurcation diagram for (2.3) in Fig. 2.5 is as follows:

1. Choose a value of $\mathscr{R}_0$.
2. Give the initial value $2c_1/\alpha = 0.1$.
3. Determine the value $2c_{200}/\alpha$ by recurrently using (2.3).
4. Make a set of values $\{2c_n/\alpha\}$ from $n = 201$ to $n = 400$ again by recurrently using (2.3).
5. Plot all values of the set against the value of $\mathscr{R}_0$.
6. Choose a different value of $\mathscr{R}_0$, and repeat this process from step 2.

The essence of this process is to plot 200 points against each chosen value of $\mathscr{R}_0$. When the population size approaches a unique positive value, they apparently appears a single point. When it approaches a period-2 solution, they apparently appears two points, and when it does a period-4 solution, they does four points, and so on.

As already seen, there does not appear any persistently oscillatory variation when $\theta \leq 2$. Hence in such a case, the plot in the bifurcation diagram appears a single point against each value of $\mathscr{R}_0$, as shown in the upper three diagrams of Fig. 2.5. The digram is composed only with a curve or line. In contrast, when $\theta > 2$, the sequence $\{c_n\}$ may approach a persistently oscillatory variation (Fig. 2.3). Actually the bifurcation diagrams in Fig. 2.5 indicate that the population dynamics model (2.3) brings periodic solutions with doubling the period like period-2, period-4, period-8, period-16, and so on, as $\mathscr{R}_0$ gets larger. This type of parameter dependence of the periodic solutions is called *period-doubling bifuration* (see Sect. 12.1.5).

Further we can see a part of dense plots for further large $\mathcal{R}_0$ in Fig. 2.5, which may include a *chaotic variation* which is an aperiodic and persistent oscillatory variation. In the chaotic variation, the sequence $\{c_n\}$ consists of all different values. That is, mathematically it does not approach any periodic solution but continues to change without taking the value same as any of past values. This is the case of $\mathcal{R}_0 = 4.00$ in Fig. 2.2. The population dynamics with the parameter values to cause such a chaotic variation can be characterized by a specific feature called *initial value sensitivity*. It means that arbitrary slight difference in the initial value eventually generates a significant difference in the variation.

The above description about the chaotic variation follows the mathematical theory of the chaos, which has been developed in the dynamical system theory of mathematics and related mathematical sciences. In any numerical calculation, we cannot avoid the round error which may reveal the limitation of numerical approach to the mathematical problem. While the chaotic dynamical system has the feature of initial value sensitivity, a numerical calculation of number sequence like the above may hit a value same as one in the past due to such a numerical error. We know that the modern computer provides a very high accuracy out of our ordinary sense. Hence such a case caused by the round error would be considered as very rare case. Naturally from the scientific viewpoint, we could not deny such a possibility, and hence it is an important mathematical problem to show the existence of chaotic variation by mathematical analysis.

Moreover, the numerically drawn bifurcation diagram in Fig. 2.5 cannot show periodic solutions with any much long period. Especially for $\mathcal{R}_0 > \theta/(\theta - 2)$ when $\theta > 2$, it is likely that the numerical calculation could not plot every value appearing in the sequence. For this reason, such a numerical bifurcation may be regarded as just an approximation with experimental numerical calculations, and thus it is sometimes called *orbit diagram* in contrast with the bifurcation diagram.

On the other hand, Fig. 2.5 may be regarded as an approximated expression of the *ω-limit set* drawn by numerical experiments too. The ω-limit set for the initial value c_1 means the set of values which the sequence $\{c_n\}$ approaches as $n \to \infty$. When it approaches a unique value, the ω-limit set consists of only the value. When it approaches a period-k solution, it consists of k different values composing the periodic solution. When the sequence shows a chaotic variation, the ω-limit set consists of the infinite number of values and is called *strange attractor*. The reader can get the further mathematical detail about these kinds of feature in the dynamical system, for example, in [11, 16, 25, 62].

As numerically exemplified by the sequence when $\mathscr{R}_0 = 7.00$ in Fig. 2.2 and by the bifurcation diagram when $\theta = 4.0$ and 5.0 in Fig. 2.5, the existence of a period-3 solution is implied. The period-3 solution has an important meaning in mathematical science, since a theorem called *Sharkovskii theorem* proves that any periodic solution exists if a period-3 solution exists (see Sect. 12.1.5). It must be remarked that the existence of a periodic solution does not necessarily mean the approach of sequence to it. It is possible that the sequence does not approach an existing periodic solution. Such a periodic solution is mathematically identified as *unstable*. On the other hand, if there is a periodic solution which the sequence approaches, it is mathematically identified as *asymptotically stable*.

The description of "unstable" and "asymptotically stable" here is much rough in a mathematical sense, while it would be very important for the reader to get a closer image about such concepts of stability. As for the more mathematical concept of stability, refer to Sect. 12.1.1.

2.1.2 Ricker Model

Next let us consider the following function of negative density effect on the number of newborns per couple m:

$$m(a_n) = m_0\, e^{-\gamma a_n}, \tag{2.6}$$

by which m is monotonically decreasing in terms of the density of mature individuals a_n. Parameter m_0 means the physiological upper bound for the number of newborns per couple, same as for Beverton-Holt type model. Parameter γ characterizes the sensitivity of m to the density of mature individuals according to the density effect. As shown in Fig. 2.6, the larger γ means the stronger sensitivity of m such that the higher density of mature individuals causes the steeper decrease in the number of newborns per couple.

The population dynamics model (2.1) with the density effect function (2.6) is today called *Ricker model* after its application for the fishery problem by William E. Ricker (1908–2001) in 1954 [52]:

$$c_{n+1} = \mathscr{R}_0 c_n e^{-2\gamma c_n}, \tag{2.7}$$

where $\mathscr{R}_0 := \sigma m_0/2$ means the net reproduction rate expected when the density effect is absent, as before.

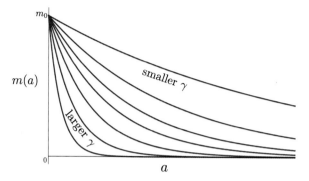

Fig. 2.6 Parameter dependence of the density effect function (2.6) for the number of newborns per couple m

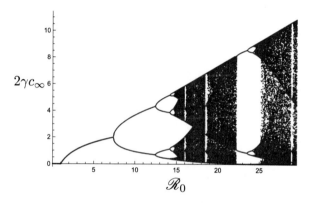

Fig. 2.7 Numerically drawn bifurcation diagram for Ricker model (2.7) in terms of the bifurcation parameter $\mathscr{R}_0$

In 1950, Patrick A.P. Moran (1917–1988) considered the same model [41]. Hence, this model is sometimes called *Ricker-Moran model* too. Its mathematical nature was studied earlier by MacFadyen [35], and later in a well-known work by May and Oster [39].

As clearly seen from the bifurcation diagram of Fig. 2.7, Ricker model (2.7) shows the parameter dependence of a period-doubling bifurcation toward the chaotic variation, same as for Beverton-Holt type model (2.3) with $\theta > 2$. The essential features of population dynamics by Ricker model (2.7) are as follows:

- The population size monotonically decreases to go extinct if and only if $\mathscr{R}_0 \leq 1$.
- c_n monotonically approaches $c^* := (1/2\gamma) \ln \mathscr{R}_0$ if and only if $1 < \mathscr{R}_0 \leq e \approx 2.71828$.

- c_n approaches c^* with a damped oscillation if and only if $e < \mathscr{R}_0 \le e^2 \approx$ 7.38906.
- c_n does not approach any specific value but keep changing if and only if $\mathscr{R}_0 > e^2$.

These features can be mathematically shown with the *local stability analysis* and the *cobwebbing method* (refer to Chap. 12.1).

It is remarked that the parameter γ has no relation to the above features. That is, for each value of $\mathscr{R}_0$, the sequence $\{c_n\}$ has the same qualitative nature for any different value of γ. The reason is the same as that mentioned in p. 31 for the parameter α of Beverton-Holt type model (2.3).

Ricker model is sometimes referred as a mathematical model for the fish population dynamics with the filial cannibalism, whereas it may not be necessarily related to such a cannibalism. We can derive Ricker model from a population dynamics modeling with the filial cannibalism as follows.

Let A_k denote the density of mature (reproductive) individuals just before the kth breeding season, and J_k do the density of immature individuals just after the kth breeding season, which now corresponds to the offsprings born at the kth breeding season. We assume that the immature individual can become mature just before the next breeding season. Let T denote the length of non-breeding season, and t do time in the non-breeding season after the kth breeding season. $J(t)$ and $A(t)$ respectively denote the densities of immature and mature individuals at time t. Now let us consider the following ordinary equation as the population dynamics model within the non-breeding season:

$$\frac{dJ(t)}{dt} = -\delta_J(t)J(t) - \kappa(t)A(t)J(t), \qquad (2.8)$$

where $\delta_J(t)$ and $\kappa(t)$ are positive functions of time t which respectively denote the natural death rate and the coefficient of filial cannibalism by mature individuals at time t. Although we give them as time-dependent functions, they may be time-independent constants as the simpler modeling. In this continuous time model by the ordinary differential equation (2.8), the interaction between immature and mature individuals is introduced by the mathematical modeling with the *mass action assumption* described in Sect. 6.1. With a different assumption about the interaction, the result of the following arguments could be changed more or less.

The temporal change of the density of mature individuals is assumed to be given by a certain function of time t, $A(t) = \widetilde{A}_k \psi(t)$ with a monotonically decreasing non-negative function $\psi(t)$ such that $\psi(0) = 1$ and $\psi(T) \ge 0$, where $\widetilde{A}_k = A(0)$ denotes the density of mature individuals just at the end of the kth breeding season. Due to the natural death, the density of mature individuals must monotonically decrease in the non-breeding season.

(continued)

By substituting $A(t)$ for (2.8), we can easily (and mathematically) solve the ordinary differential equation (2.8), and get the following solution (refer to Sect. 13.1):

$$J(t) = J_k \exp\left[-\int_0^t \delta_J(\tau)d\tau - \int_0^t \kappa(\tau)A(\tau)d\tau\right]$$

$$= J_k \exp\left[-\int_0^t \delta_J(\tau)d\tau - \widetilde{A}_k \int_0^t \kappa(\tau)\psi(\tau)d\tau\right].$$

As a result, we can obtain the density of immature individuals $J(T)$ just before the $k+1$ th breeding season:

$$J(T) = J_k e^{-\widehat{\delta}_J T - \widehat{\kappa}\widetilde{A}_k T}, \tag{2.9}$$

where

$$\widehat{\delta}_J = \frac{1}{T}\int_0^T \delta_J(\tau)d\tau; \quad \widehat{\kappa} = \frac{1}{T}\int_0^T \kappa(\tau)\psi(\tau)d\tau.$$

Now let σ_A the survival rate of mature individual within the breeding season, and then we have $\widetilde{A}_k = \sigma_A A_k$. Besides, with the averaged reproduction rate per mature individual $\bar{r}_k$ at the kth breeding season, we have $J_k = \bar{r}_k A_k$. Since $A_{k+1} = A(T) + J(T)$ from the assumption about the maturation, we find the following equation from $A(t)$ and (2.9):

$$N_{k+1} = \sigma_A A_k \psi(T) + \bar{r}_k A_k e^{-\widehat{\delta}_J T - \widehat{\kappa}\sigma_A A_k T}. \tag{2.10}$$

If we assume the generationally non-overlapping reproduction, then it must be satisfied that $A(T) = 0$, that is, $\psi(T) = 0$. In this case, the recurrence relation (2.10) becomes mathematically equivalent to Ricker model (2.7).

2.1.3 Logistic Map Model

As a mathematical modeling to introduce the negative density effect on the number of newborns, let us consider here the following piece-wise linear function:

$$m(a_n) = \begin{cases} m_0\left(1 - \dfrac{a_n}{a_c}\right) & (a_n < a_c); \\ 0 & (a_n \geq a_c). \end{cases} \tag{2.11}$$

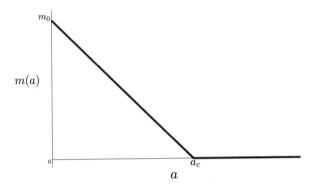

Fig. 2.8 The density effect function (2.11) for the number of newborns per couple m

As clearly indicated by Fig. 2.8 of the density effect function (2.11), this modeling introduces an assumption that the reproduction becomes impossible with a strong negative density effect when the density of mature individuals a is beyond a threshold a_c.

Some may consider the smooth function for the density effect as for Beverton-Holt type model and Ricker model better than the piece-wise and non-smooth function like (2.11). Such a way of thought could not be regarded as reasonable. It is most important what function would be reasonable from the biological viewpoint to introduce the essential nature of density effect. The smoothness of the function is not necessarily important for the reasonability of mathematical modeling.

For some plant population, it is a well-known example that the population becomes extinct without any successful flowering or seedling when the planted density is too high. For Beverton-Holt type model and Ricker model, a positive reproductive success, that is, a positive number of newborns is assumed for any density of mature individuals. This may be regarded as a strong mathematical simplification/idealization/approximation from the reasonability of modeling, in comparison with the modeling with a certain threshold density like (2.11). It depends on the biological nature of considered/assumed population which modeling for the density effect would be more reasonable.

The population dynamics model (2.1) with the density effect function (2.11) is given by the following recurrence relation:

$$
c_{n+1} =
\begin{cases}
\mathscr{R}_0 \left(1 - \dfrac{c_n}{c_c}\right) c_n & (c_n < c_c); \\
0 & (c_n \geq c_c),
\end{cases}
\tag{2.12}
$$

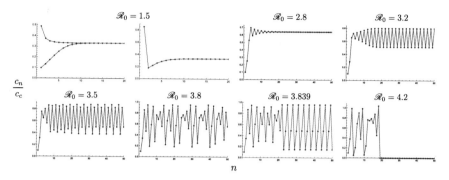

Fig. 2.9 Temporal sequences by the population dynamics model (2.12). Numerical calculations for different values of $\mathcal{R}_0$. Only in case of $\mathcal{R}_0 = 1.5$, the sequences for three different initial value $c_1/c_c = 0.1, 0.49$, and 0.867 are shown, while the other cases are commonly for $c_1/c_c = 0.1$

where $\mathcal{R}_0 := \sigma m_0/2$ means the net reproduction rate when the density effect is absent as before, and $c_c := a_c/2$. It is biologically nonsense unless $0 < c_1 < c_c$ for the initial value c_1. When $c_1 \geq c_c$, that is, when $a_1 \geq a_c$, the population immediately goes extinct since it cannot make reproduction at all.

As the fundamental features of the sequence $\{c_n\}$ by the recurrence relation (2.12), we can easily find the followings:

- When $\mathcal{R}_0 < 4$, it holds that $0 < c_n < c_c$ for any finite n if $0 < c_1 < c_c$.
- When $\mathcal{R}_0 > 4$, we have $c_n = 0$ for some $n > 1$ for any positive $c_1 \neq (1 - 1/\mathcal{R}_0)c_c$.

Therefore, when $\mathcal{R}_0 > 4$, the population necessarily goes extinct (Fig. 2.9).[1] From the above mathematical features, as long as considering the model (2.12) for the biologically reasonable initial value c_1 such that $0 < c_1 < c_c$ when $\mathcal{R}_0 \leq 4$, the model can be described in the following simpler form:

$$c_{n+1} = \mathcal{R}_0\left(1 - \frac{c_n}{c_c}\right)c_n. \tag{2.13}$$

It is necessary to consider the model of this recurrence relation under the constraint that $\mathcal{R}_0 \leq 4$, because it is mathematically proved that c_n cannot be beyond c_c from the initial value c_1 such that $0 < c_1 < c_c$ if and only if $\mathcal{R}_0 \leq 4$ (Exercise 2.2 below). The recurrence relation (2.13) is equivalent to what is today called *logistic map*.

Exercise 2.2 About the logistic map (2.13), show that c_n is non-negative for any $n > 1$ from any positive initial value such that $0 < c_1 < c_c$ if and only if $\mathcal{R}_0 \leq 4$.

---

[1] Rigorously saying in a mathematical sense, the extinction occurs except when the initial value belongs to a measure–zero set. However, in numerical calculations to get the sequence $\{c_n\}$, it eventually occurs (refer to the arguments in p. 37).

Logistic map has become well-known by some pioneer works by a mathematical biologist Robert M. May (1936–2020) on the mathematical nature of the recurrence relation (2.13) in the early 1970s [37, 38]. His works inspired many researchers in mathematical sciences, and a variety of his and the others' subsequent works contributed to the great development in dynamical system theory [2].

> The logistic map appears in most of text books on mathematical biology and on dynamical system theory. In such literatures, only the recurrence relation mathematically equivalent to (2.13) is described, and there does not make any mention of the mathematical model (2.12) in general. Instead they may remark that the case of $c_n > c_c$ is nonsense as a mathematical model for population dynamics since the right side of (2.13) then becomes negative. Only such a description would be unsatisfactory about the reasonability as a mathematical modeling. When $\mathscr{R}_0 > 4$, the recurrence relation (2.13) with the initial value satisfying $0 < c_1 < c_c$ generates a sequence $\{c_n\}$ such that a negative number c_n appears for a certain finite $n = n^\star$, and it becomes negative for any $n > n^\star$, diverging toward negative infinity. However, as long as considering the logistic map (2.13) as a mathematical model with the confinement for the initial value c_1 and parameter $\mathscr{R}_0$ as given in the above Exercise 2.2, it could be a reasonable mathematical model for population dynamics.

The logistic map model (2.12) has the following characteristics about the population dynamics (for the detail, refer to Sects. 12.1.3, 12.1.4, and 12.1.5):

- When $\mathscr{R}_0 \leq 1$, the population size monotonically decreases to go extinct.
- When $1 < \mathscr{R}_0 \leq 2$, the number of reproductive couples c_n monotonically approaches $(1 - 1/\mathscr{R}_0)c_c$.
- When $2 < \mathscr{R}_0 \leq 3$, c_n approaches $(1 - 1/\mathscr{R}_0)c_c$ with a damped oscillation.
- When $3 < \mathscr{R}_0 \leq 4$, c_n does not approach any specific value and keep changing.
- When $\mathscr{R}_0 > 4$, c_n goes extinct after a finite number of irregular changes.

As a specific case, when $3 < \mathscr{R}_0 \leq 1 + \sqrt{6} \approx 3.44949$, c_n approaches a period-2 solution which repeats the following two values (refer to Sect. 12.1.4 and Exercise 12.2 therein):

$$\frac{c_c}{2}\left[1 + \frac{1}{\mathscr{R}_0} \pm \sqrt{\left(1 + \frac{1}{\mathscr{R}_0}\right)\left(1 - \frac{3}{\mathscr{R}_0}\right)}\right]. \qquad (2.14)$$

As expected from numerically obtained temporal sequences in Fig. 2.9, and as indicated by the numerically obtained bifurcation diagram in Fig. 2.10, the logistic map model (2.12) has the parameter dependence with the period-doubling bifurcation as well as Ricker model, in which the chaotic variation can appear. The

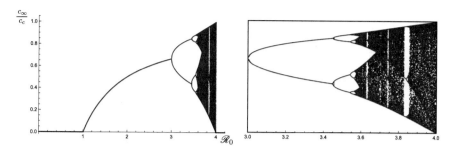

Fig. 2.10 Numerically drawn bifurcation diagram for the logistic map model (2.12) in terms of the bifurcation parameter $\mathscr{R}_0$. The right figure is the magnification about the range [3, 4] of $\mathscr{R}_0$

previous mathematical works have shown that the chaotic variation can appear only for $\mathscr{R}_0 > 3.569945 \cdots$ (for example, see [11, 16, 25, 53, 62]). As a specific case, a period-3 solution can appear for $\mathscr{R}_0 = 3.839$ (see also Fig. 12.5 in p. 393 of Sect. 12.1.5).

The popular naming *logistic map* for the recurrence relation (2.13) is associated with the works by the famous mathematical biologist Robert M. May in which the following population dynamics model of ordinary differential equation (a continuous time model) was referred as the origin for it [37, 38]:

$$\frac{dN(t)}{dt} = r\left\{1 - \frac{N(t)}{K}\right\}N(t), \tag{2.15}$$

where $N(t)$ is the population density at time t. Parameters r and K are positive constants which mean the *intrinsic (natural) growth rate* and the carrying capacity respectively (the detail explanation will be given in Sect. 5.3). This model is well-known as *logistic equation* (refer also to Sect. 3.3.2). The recurrence relation (2.13) can be derived by a simple time-discretizing approximation of the ordinary differential equation (2.15), as described in Sect. 5.5. The time-discretization is an approximation of the time derivative in the left side of (2.15) by replacing it with the simple difference $\{N(t + h) - N(t)\}/h$ which appears in the mathematical definition of the derivative. This replacement leads to a recurrence relation which determines the value $N(t + h)$ at the future by period h from the value $N(t)$ at time t. It is mathematically equivalent to (2.13). It was known in the numerical theory that the simple time discretization with a much large value of time step h results in the sequence of value N not only quantitatively but also qualitatively much different from the solution $N(t)$ of the ordinary differential equation (2.15). Such a qualitatively different behavior by the sequence of value N was

(continued)

regarded just as inappropriate as the approximation for the solution $N(t)$ of the ordinary differential equation (2.15). Robert M. May was interested in such a variety of different behaviors, and developed the research on the recurrence relation from the viewpoint of mathematical biology.

Moreover, the ordinary differential equation (2.15) can be rewritten as

$$\frac{d \ln N(t)}{dt} = r\left\{1 - \frac{N(t)}{K}\right\}.$$

The same time-discretizing approximation of this equation leads to the recurrence relation mathematically equivalent to Ricker model, as argued in Sect. 5.5. For this reason, we may understand that both of the logistic map and Ricker model correspond to the recurrence relations derived by the simple time-discretization of the logistic equation 2.15. This fact implies the qualitative coincidence about the characteristics of the logistic map model and Ricker model.

Today many literatures refer to the logistic map as the discrete time model corresponding to the logistic equation (continuous time model) (2.15). This would be because of the naming itself, and we must pay attention to such a reference in order to avoid a misunderstanding.

The solution $N(t)$ of logistic equation (2.15) monotonically approaches the value K for any positive initial value $N(0)$ (see Fig. 2.11). In contrast, the recurrence relation derived by the simple time-discretization may show a periodic sequence or chaotic variation with sufficiently large value of time step h, as seen for (2.13). Such a behavior of the sequence cannot be regarded as any approximation of the solution $N(t)$, as mentioned above. It can be mathematically shown that the sequence monotonically approaches K only when $h \leq 1/r$.

As shown in Sect. 5.4, Beverton-Holt model can be regarded as the discrete time model corresponding to the logistic equation (2.15) in a mathematically exact sense. That is, the sequence generated by Beverton-Holt model has the mathematical characteristics same as the solution of logistic equation, as seen from Figs. 2.4 and 2.11. It is actually possible to make the mathematical correspondence between parameters of their equations (Sect. 5.4). Therefore, it is the most reasonable to say that the discrete time model corresponding to the logistic equation is Beverton-Holt model, while the logistic map model (2.13) and Ricker model (2.7) are discrete time models *derived* from the logistic equation, which are not exactly what correspond to it.

In this Sect. 2.1, we have seen the mathematical model (2.1) with the density effect on the number of newborns m by three classic types of the function. Mathematically, for any density effect function m that is monotonically decreasing in terms of a_n, it can be shown that the model (2.1) has the following features:

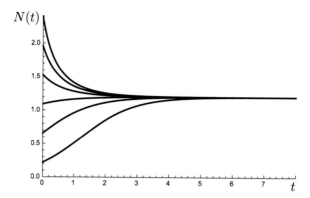

Fig. 2.11 Numerically drawn temporal variations of $N(t)$ for the logistic equation (2.15) with six different initial values of $N(0)$. $r = 1.2$; $K = 1.2$. Note the similarity to Fig. 2.4

- When $\mathscr{R}_0 := \sigma m(0)/2 \le 1$, the population size monotonically decreases to go extinct.
- The number of reproductive couples approaches a certain positive value only when $\mathscr{R}_0 > 1$.

Therefore, in fact these features appear commonly for three mathematical models considered in this section.

2.2 Positive Density Effect

In some cases, the reproduction rate per individual or per reproductive couple is increasing in terms of the population density. For example, the low density of mature individuals makes harder the formation of reproductive couple or the possibility to find a preferable feeding place. These factors possibly cause the positive correlation between the population density and reproduction rate. It is likely also that the higher population density may decrease the risk of fatal event like the predation. The alarm call becomes more effective for escaping from the predator as the population density gets larger. In general, the behavior of nearby individuals may be helpful to get an information about the preferability of surrounding environment. In contrast, as argued in the previous section, the higher population density necessarily causes the degradation of environmental quality and the stronger interaction between individuals which could result in the decrease of the reproduction rate per individual/couple.

In this Chap. 2, we are considering the population dynamics with the reproduction only by the formation of reproductive couple. In our modeling, a positive density effect may be reflected to the number of newborns per reproductive couple as well as the negative density effect. The positive density effect is then given by the positive correlation such that the higher (resp. lower) density of mature individuals results in the larger (resp. smaller) number of newborns per couple. The reproduction success could be regarded as the consequence of an integration of such positive and negative density effects.

2.2.1   Allee Effect

From the arguments in the above, we can assume the density effect function for the number of newborns per couple such that it takes the maximum at an intermediate density below which it is increasing (the case of *undercrowding*), and beyond which it is decreasing (the case of *overcrowding*) in terms of the density.

Following this assumption, we shall consider here the density effect function given by

$$m(a_n) = m_{\mathrm{op}} \left(\frac{a_n}{a_{\mathrm{op}}} \right) e^{1 - a_n/a_{\mathrm{op}}}. \tag{2.16}$$

As shown in Fig. 2.12, the number of newborns per couple takes the maximum m_{op} when the number of mature individuals is a_{op}.

For this type of density dependence, the increase of the reproduction rate per individual in terms of the population density is classically called *Allee effect*. In other words, if the reproduction rate per individual decreases as the population density gets smaller, it can be called Allee effect. This naming is after the works by a famous animal ecologist Warder C. Allee (1885–1955). For the density effect function (2.16), the Allee effect appears for the density of mature individuals less than a_{op}. It may be sometimes called *Allee's principle* that there is an *optimum density* about the density effect, which is now given by a_{op} for the modeling with (2.16).

The Allee effect indicates a positive correlation of the reproduction rate per individual to relatively lower population density. The positive correlation of the velocity of the density change to the population density is not the Allee effect. Besides the Allee effect does not necessarily mean that the density effect could be expressed by a unimodal function of the population density. Moreover, the Allee effect does not necessarily induce the extinction for sufficiently lower population density, although the Allee effect is frequently mentioned as a factor to cause such an extinction as we will see also about our model in this section.

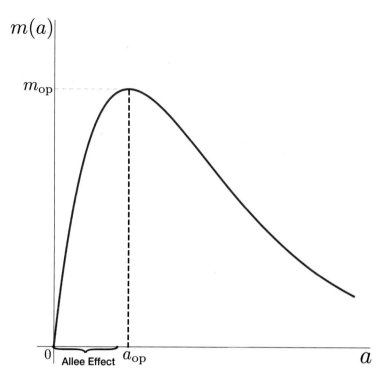

Fig. 2.12 The density effect function (2.16)

The recurrence relation (2.1) with the density effect function (2.16) leads to the following population dynamics model:

$$\frac{c_{n+1}}{c_{op}} = \mathscr{R}_{op} \left(\frac{c_n}{c_{op}} \right)^2 e^{1-c_n/c_{op}}, \tag{2.17}$$

where $c_{op} = a_{op}/2$. The parameter $\mathscr{R}_{op} := \sigma m_{op}/2$ means the maximal net reproduction rate attained when the number of reproductive couples is c_{op}, that is, when the number of mature individuals is a_{op}.

For the population dynamics model (2.17), the population extinction depends on the initial value of c_1. As shown by numerical example in Fig. 2.13, too small or too large initial value of c_1 causes the population extinction. The population can persist only when the initial value of c_1 is in an intermediate range. This nature can be regarded as what is called *bistability*, with which the dynamics has alternative characteristics depending on the initial value. [2]

---

[2] More generally, if there are more than two different characteristics depending on the initial value, such a nature is called *multi-stability*.

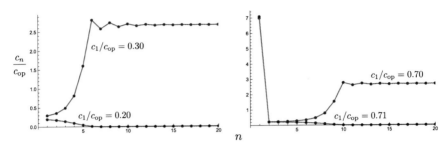

Fig. 2.13 Temporal sequences by the population dynamics model (2.17). Numerical calculations for different initial values of c_1/c_{op} when $\mathcal{R}_{op} = 2$

The population dynamics model (2.17) has the following features:

- When $\mathcal{R}_{op} < 1$, the population size monotonically decreases to go extinct.
- When $\mathcal{R}_{op} = 1$, the population size monotonically decreases to go extinct if $c_1 < c_{op}$ or $c_1 > \kappa_1 c_{op} \approx 3.51286\, c_{op}$, while the number of reproductive couples c_n monotonically approaches c_{op} if $c_1 \geq \kappa_1 c_{op}$. Here κ_1 is given by the root more than 2 about the equation $\kappa_1^2 e^{1-\kappa_1} = 1$.
- When $1 < \mathcal{R}_{op} \leq e/2 \approx 1.35914$, the population size monotonically decreases to go extinct if $c_1 < \kappa_u c_{op}$ or $c_1 > \kappa_c c_{op}$, while c_n monotonically approaches $\kappa_s c_{op}$ if $\kappa_u c_{op} < c_1 \leq \kappa_c c_{op}$. Here κ_u and κ_s are roots for the equation $\mathcal{R}_{op}\kappa\, e^{1-\kappa} = 1$ such that $0 < \kappa_u < 1 < \kappa_s$. κ_c is the root more than 2 for the equation

$$2\ln(\mathcal{R}_{op}\kappa_c) + 2 - \kappa_c - \mathcal{R}_{op}\kappa_c^2 e^{1-\kappa_c} = 0.$$

- When $e/2 < \mathcal{R}_{op} \leq e^2/3 \approx 2.46302$, the population size monotonically decreases to go extinct if $c_1 < \kappa_u c_{op}$ or $c_1 > \kappa_c c_{op}$, while c_n approaches $\kappa_s c_{op}$ with a damped oscillation (see Fig. 2.13) if $\kappa_u c_{op} < c_1 \leq \kappa_c c_{op}$.
- When $e^2/3 < \mathcal{R}_{op} \leq \overline{\mathcal{R}}_{op} \approx 7.22207$, the population size monotonically decreases to go extinct if $c_1 < \kappa_u c_{op}$ or $c_1 > \kappa_c c_{op}$, while c_n does not approach any specific value but keeps changing (see Fig. 2.14) if $\kappa_u c_{op} < c_1 \leq \kappa_c c_{op}$. $\overline{\mathcal{R}}_{op}$ is the root of the following equation:

$$\ln\left[\frac{e\ln(2\overline{\mathcal{R}}_{op}) - \overline{\mathcal{R}}_{op}}{4(\overline{\mathcal{R}}_{op})^3}\right] + \frac{4\overline{\mathcal{R}}_{op}}{e} = 0.$$

- When $\mathcal{R}_{op} > \overline{\mathcal{R}}_{op}$, c_n becomes less than $\kappa_u c_{op}$ after a finite number of changes, and approaches zero, that is, the population goes extinct (see Fig. 2.14).

These features can be mathematically shown with the local stability analysis and the cobwebbing method (refer to Chap. 12.1). Figure 2.15 shows an integration of the above features. As seen from the figure, the population can persist only for an

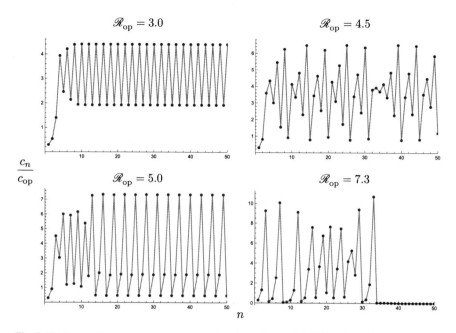

Fig. 2.14 Temporal sequences by the population dynamics model (2.17). Numerical calculations for different values of $\mathscr{R}_{op}$ with the initial value $c_1/c_{op} = 0.3$

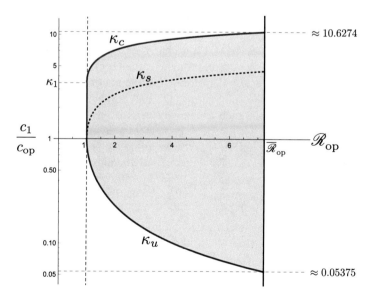

Fig. 2.15 Dependence of the population dynamics by (2.17) on the parameter $\mathscr{R}_{op}$ and initial value of c_1/c_{op}. The vertical axis is in the logarithmic scale. The population can persist only for the filled region

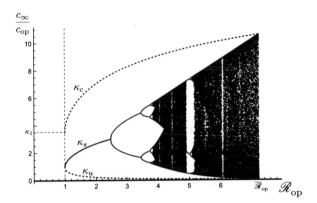

Fig. 2.16 Numerically drawn bifurcation diagram according to the persistent case for the population dynamics model (2.17) in terms of the bifurcation parameter $\mathscr{R}_{op}$

intermediate range of the parameter $\mathscr{R}_{op}$, $1 < \mathscr{R}_{op} \leq \overline{\mathscr{R}}_{op}$, and the initial value of c_1, $\kappa_u c_{op} < c_1 \leq \kappa_c c_{op}$. Moreover, from the numerically obtained bifurcation diagram shown in Fig. 2.16, we can find the period-doubling nature again about the parameter dependence.

2.2.2 Scenarios of Extinction with Allee Effect

The population dynamics model (2.17) is characterized by the nature such that the population size monotonically decreases to go extinct for the initial value of c_1 less than $\kappa_u c_{op}$ or larger than $\kappa_c c_{op}$. For example, if the hunting results in the number of reproductive couples less than $\kappa_u c_{op}$, the Allee effect causes the population extinction (Fig. 2.17a-3). In such a case, the population extinction is driven by the population dynamics itself once a human interference generates the causal condition. For the other example, suppose an introduction of excessive number of grazing animals into an island. The population dynamics by (2.17) results in the earlier death of most introduced animals and the subsequent extinction (Fig. 2.13) unless there is no appropriate management of the number of animals.

On the other hand, such a scenario implies the possible artificial management for the conservation of an endangered biological population. If the likelihood of extinction follows the Allee effect, the artificial introduction of individuals to increase a population size could lead to its autonomous persistence (Fig. 2.17b-3). From the other viewpoint, for a pest population, a human operation to reduce the population size to less than a certain level once could induce its autonomous population extinction (Fig. 2.17a-3), if the population dynamics follows the Allee effect. In such a case, when the human operation cannot make the population size

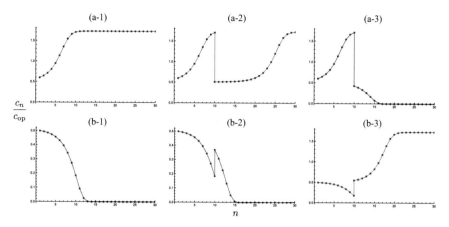

Fig. 2.17 Temporal sequences by the population dynamics model (2.17). Numerical calculations with $\mathcal{R}_{op} = 1.2$. (a) $c_1/c_{op} = 0.6$; (b) $c_1/c_{op} = 0.5$. (a-1) with no human interference, the population approaches an equilibrium value; (a-2) the population size is artificially reduced by 70 % at $n = 10$; (a-3) the population size is artificially reduced by 75 % at $n = 10$. (b-1) with no human interference, the population goes extinct; (b-2) the population size is artificially doubled at $n = 10$; (b-3) the population size is artificially tripled at $n = 10$

sufficiently small, the population recovers to the previous size after the operation (Fig. 2.17a-2).

In contrast, the scenario of population extinction when $\mathcal{R}_{op} > \overline{\mathcal{R}}_{op}$ must take account of the other factor. For such a large value of $\mathcal{R}_{op}$, we can suppose a sufficiently large physiological upper bound for the number of newborns per couple m_{op}. It would be the population with a sufficiently high fertility. Although such a high fertility could cause the effective growth of population, the net growth rate is determined with the density effect. Too high fertility could cause a rapid increase of the population size, which leads a strong density effect to induce the subsequent drastic decrease of the population size (the case of $\mathcal{R}_{op} = 7.3$ in Fig. 2.14). Such a drastic decrease of the population size could not be compensated by the high fertility but cause the way to the extinction by the Allee effect.

To make the conservation of an endangered population with such a high fertility, it would not be the fundamental resolution to artificially increase the population even though it could become a recovery of the population size only for a while. For the population dynamics model (2.17), if it would be possible to make $\mathcal{R}_{op}$ sufficiently small, the population can persist. Thus one choice is to reduce the survival probability of immature individual σ. For example, the artificial culling of individuals before the breeding season by an appropriate proportion would be effective to induce the population

(continued)

persistence. Such a persistence could be regarded as a sort of coexistence of human and wild organism, because the coexistence could become possible only with an artificial operation for the wild population. Today many wild animals live in or around the environment of human community. It is likely that the persistence of such an animal population would rely on the human activity.

2.2.3 Weak Allee Effect

The population dynamics by (2.17) with the Allee effect has a bistable nature such that the population goes extinct when the initial value of c_1 is sufficiently small. However, generally the Allee effect itself does not necessarily bring such a bistable nature in the population dynamics. The Allee effect introduced by the density effect function (2.16) can be regarded as strong enough to cause such a bistable nature which may cause the extinction depending on the initial value.

In this section, let us consider a mathematical modeling for the density effect function about the number of newborns per couple, which can introduce a weak Allee effect, given by

$$
m(a_n) = \begin{cases} m_0 + (m_{op} - m_0)\left(\dfrac{a_n}{a_{op}}\right) & (0 < a_n \le a_{op}); \\[2mm] -\dfrac{m_{op}}{a_c - a_{op}}(a_n - a_c) & (a_{op} < a_n < a_c); \\[2mm] 0 & (a_n \ge a_c), \end{cases} \tag{2.18}
$$

where $m_{op} > m_0 > 0$ (Fig. 2.18). This modeling for the density effect function is similar with that of the logistic map model in the previous section. It is assumed that the fertility is lost due to the overcrowding and the population extinction occurs when the density of mature individuals is beyond a threshold value a_c.

The recurrence relation (2.1) with the density effect function (2.18) makes the following population dynamics model:

$$
\frac{c_{n+1}}{c_{op}} = \begin{cases} \mathcal{R}_0 \left\{ 1 + \left(\dfrac{\mathcal{R}_{op}}{\mathcal{R}_0} - 1\right)\left(\dfrac{c_n}{c_{op}}\right) \right\}\left(\dfrac{c_n}{c_{op}}\right) & (0 < c_n \le c_{op}); \\[3mm] \dfrac{\mathcal{R}_{op}}{c_c/c_{op} - 1}\left(\dfrac{c_c}{c_{op}} - \dfrac{c_n}{c_{op}}\right)\left(\dfrac{c_n}{c_{op}}\right) & (c_{op} < c_n < c_c); \\[3mm] 0 & (c_n \ge c_c), \end{cases} \tag{2.19}
$$

where $c_{op} = a_{op}/2$ and $c_c = a_c/2$. The parameter $\mathcal{R}_0 := \sigma m_0/2$ means the net reproduction rate when the density effect is absent, and $\mathcal{R}_{op} := \sigma m_{op}/2$ does the

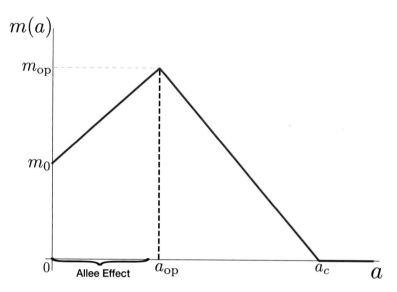

Fig. 2.18 The density effect function with a weak Allee effect (2.18)

maximal net reproduction rate when the number of reproductive couples is c_{op}. Note that $c_c > c_{op}$ and $\mathscr{R}_{op} > \mathscr{R}_0$.

The population dynamics model (2.19) has the following features:

- When $\mathscr{R}_{op} < 1$, the population size monotonically decreases to go extinct.
- When $\mathscr{R}_0 \geq 1$, the population goes extinct after a finite number of changes for $\mathscr{R}_{op} \geq c_c/c_{op}$ if $c_c/c_{op} \leq 2$, or for $\mathscr{R}_{op} \geq 4(1-c_{op}/c_c)$ if $c_c/c_{op} > 2$. Otherwise when $\mathscr{R}_0 \geq 1$, the population persists.
- When $\mathscr{R}_0 < 1 < \mathscr{R}_{op}$, the population goes extinct after a finite number of changes for $\mathscr{R}_{op} \geq c_c/c_{op}$ if $c_c/c_{op} \leq 2$, or for $\mathscr{R}_{op} \geq 4(1 - c_{op}/c_c)$ if $c_c/c_{op} > 2$. Besides, there is a critical value $\mathscr{R}_c$ such that the number of reproductive couples monotonically decreases to go extinct after a finite number of changes for $\mathscr{R}_c < \mathscr{R}_{op} < c_c/c_{op}$ if $c_c/c_{op} \leq 2$, or for $\mathscr{R}_c < \mathscr{R}_{op} < 4(1 - c_{op}/c_c)$ if $c_c/c_{op} > 2$. Otherwise when $\mathscr{R}_0 < 1 < \mathscr{R}_{op}$, the population dynamics is in a bistable situation, so that the population persists or monotonically goes extinct after a finite number of changes, depending on the initial value of c_1.

As seen in Fig. 2.19, there are two scenarios of the population extinction. One is by the Allee effect leading to a much low fertility due to a much small population size. This is similar to that for the previous model (2.17). The other is by the overcrowding to cause the loss of fertility, like that of the logistic map model (2.12) with $\mathscr{R}_0 > 4$.

For the population dynamics model (2.19) with $\mathscr{R}_0 \geq 1$, there is no possibility of the former case about the population extinction, that is, of the extinction caused by the Allee effect (upper numerical examples in Fig. 2.19). Since the Allee effect

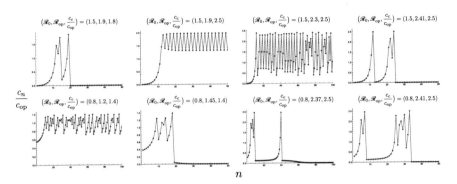

Fig. 2.19 Temporal sequences by the population dynamics model (2.19)

works in the population dynamics, this is regarded as the situation with a weak Alee effect in comparison with the previous model (2.17).

As for the latter case about the population extinction, it arises from the modeling of the density effect function (2.18). Therefore, the reason why the latter case of the population extinction does not occur but only the former does in the previous model (2.17) is for the modeling of the density effect function given by (2.16). It is important to distinguish one from the other of these two scenarios of the population extinction. Without such a distinction, those two population dynamics models would appear only to have similar characteristics.

In contrast to the previous model (2.17), the model (2.19) is composed with the density effect function (2.18) that has the higher degree of freedom according to the number of involved parameters. Actually the characteristics of the density effect function (2.18) are determined by four independent parameters m_0, m_{op}, a_{op} and a_c, while that of (2.16) are by two parameters m_{op} and a_{op}. Such a higher degree of freedom leads for the model (2.19) to have a diversity of nature wider than that for the model (2.17). This is clearly expressed also by the bifurcation diagram given in Fig. 2.20.

When the population goes extinct with $\mathscr{R}_{op} > 1$ for $\mathscr{R}_0 < 1$ and $c_c/c_{op} < 3/2$, the temporal variation is necessarily chaotic, and there does not occur the variation toward neither periodic oscillation nor equilibrium, as seen from the bifurcation diagram given in Fig. 2.20. Further, we note that there does not appear the period-doubling bifurcation for $\mathscr{R}_0 < 1$ and $c_c/c_{op} < 2$. This nature is different from that for the previous model (2.17), and can be regarded as one due to the difference in the mathematical modeling about the density effect function with a different degree of freedom.

Such a bifurcation without period-doubling similarly appears for what is today called *tent map* described in Sect. 12.1.6 (see also Fig. 12.10 therein).

(continued)

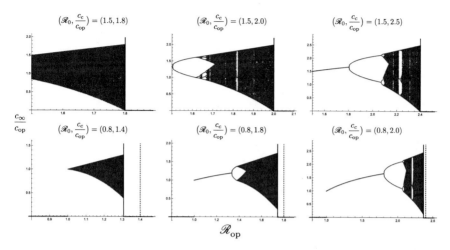

Fig. 2.20 Numerically drawn bifurcation diagram for the population dynamics model (2.19) in terms of the bifurcation parameter $\mathscr{R}_{op}$

Actually, the tent map mathematically corresponds to the model (2.19) with $\mathscr{R}_{op} = \mathscr{R}_0$ and $c_{op} = c_c/2$. Moreover, we will see another similar bifurcation for a prey-predator population dynamics model in Sect. 2.4.3 too.

2.2.4   Reproduction Curve

The curve of the relation between population sizes of the kth (parent) and $k + 1$ th (offspring) generations is called *reproduction curve*. When the reproduction curve is monotonically increasing like Fig. 2.21a, the density dependence for such a reproduction curve is called *contest type*. In contrast, when it is a single humped curve that shows the maximal population size of the $k + 1$ th generation for a specific intermediate size of the kth, it is called *scramble type* [46].

The contest type of density dependence may be observed in the case when the population growth can be regarded as a type of "playing musical chairs". If the survival and reproduction require a certain least condition, and if every individual uses the environment to satisfy the condition, the environment limits the number of individuals which can satisfy it. This means that there exists an upper bound N_{max} for the number of survival and reproductive individuals. If the population size is beyond N_{max}, the surplus number of individuals beyond N_{max} cannot satisfy the condition for the survival and reproduction. This means that the population can grow up over the upper bound N_{max} while the surplus cannot survive or make reproduction in the same environment. The environment allows at most N_{max}

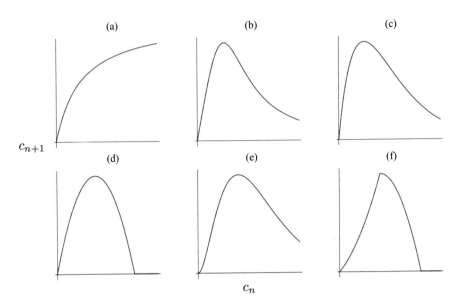

Fig. 2.21 Shapes of the reproduction curve (return map) for different models, (**a**) Beverton-Holt type model (2.3) with $\theta \leq 1$; (**b**) Beverton-Holt type model (2.3) with $\theta > 1$; (**c**) Ricker model (2.7); (**d**) logistic map model (2.12); (**e**) model with Allee effect (2.17); (**f**) model with weak Allee effect (2.19). Only the reproduction curve of (**a**) is of the contest type, while the others are of the scramble type

individuals the survival and reproduction. Therefore, in other words, the survival and reproduction follow the rule of all-or-none for each individual's survival and reproduction in the population under the contest type of density dependence. In this case, only when the parent population size is less than the upper bound $N_{\max}$, the offspring population size becomes larger as the parent population size gets larger. When the parent population size is beyond $N_{\max}$, the offspring population size cannot beyond a certain supremum.

The scramble type of density dependence may appear when the above assumption that every individual tends to use the environment for satisfying a certain least condition about the survival and reproduction could not be applied for the population dynamics. For instance, we can consider the case when the fertility is significantly affected by the density effect to reduce the number of offsprings for every individual, even if every individual could get the same amount of resource from the environment. In case of the contest type of density dependence, this cannot necessarily hold, since some individuals can get the resource for the survival and reproduction while the other cannot. In case of the scramble type of density dependence, the population may grow even when the per capita fertility is reduced by the density effect, because the population growth is based on the total number of offsprings produced by all parents. However, too high density induces a severe effect on the per capita fertility and results in the decline of population size at the next generation, since the mean number of offsprings per parent becomes much small even with a large number of parents in the population. This is the case of scramble type of density dependence.

2.3 Competition

Until here we have considered the density effect from the population density within the same population. There may be the other density effect from a different population, which is caused by a certain direct or indirect interaction between individuals belonging to different populations. In such a case, the individual of each population undergoes the density effect not only from the density of its own population but also from the density of the other.

Now let us consider two populations which uses a *common* resource relevant to the reproduction. As such a resource, the reader may imagine a food or some other environmental factors. In case of plant, the water or light is the limiting factor for the growth and reproduction, and can be regarded as such resources common among different plant species in the same habitat. For some nesting animals, the place favorable for the nesting would be such a common resource.

In such a situation, *the use of the resource by one population could lower the efficiency of its use by the other.* Thus, these two populations are under the *competition* about the resource. This kind of competitive relation is called *exploitative competition* in ecology. Such a competitive relation is induced by the indirect interaction between individuals of two populations through the use of a common resource. For this reason, the exploitative competition is regarded as a sort of *indirect effect* in a wide sense, and is an *indirect competition*.

In contrast, the *direct competition* is defined by a direct interaction between individuals of different populations, and frequently called *interference competition* in ecology. For example, the fight between individuals about a common resource may affect the survival and reproduction rates. In the environment with rich resource, the effect of exploitative competition could be weak, whereas that of interference competition would be still considerable between populations which have strong direct interaction such as fight.

For example, when a waste or chemical substance (e.g., allelopathic, pheromonic or hormonic matter) produced by one population can affect the growth rate of the other population, such an interspecific relation could not be regarded as direct interference in general. Because the former population does not have any direct interference to the latter, but only the substance produced by the former affects the latter. It is clear that this relation is not exploitative. In a wider sense, this interspecific relation may be regarded as an interference

(continued)

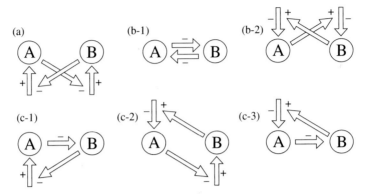

Fig. 2.22 Schematic figure of competitive relation between two populations. (**a**) (indirect) exploitative competition; (**b-1**) direct interference competition; (**b-2**) indirect interference competition; (**c-1**) population A poses a direct interference effect to B, while B does an indirect (exploitative) effect to reduce the gain for A; (**c-2**) population A poses an indirect (exploitative) effect to reduce the gain for B, while B does an indirect interference effect to increase the loss for A; (**c-3**) population A poses a direct interference effect to B, while B does an indirect interference effect to increase the loss for A. In (**c-1**) and (**c-2**), the competitive relation itself is neither exploitative nor inference, but their combination. The relation of (**c-3**) gives an example of interference competition consisting of direct and indirect effects

competition (see Fig. 2.22). Although the interspecific competition is one of important topics in the textbook of ecology, its mathematical modeling would not have been discussed with respect to this kind of qualitative difference about the competition.

In more general sense, the competitive relation between populations can be regarded as the relation such that *the interaction causes the reduction of the reproduction rate for each individual in both populations.* The effect from the other population can be treated as a negative density effect to cause the reduction of the reproduction rate. In this section from this viewpoint, we shall consider the mathematical modeling for the population dynamics of two competing populations along the framework in the previous sections.

Let us apply the assumptions same as in Sect. 2.1 for the population dynamics without the other population. Same as before, we assume that the number of newborns per couple is determined by the density effect also from the other population, by the following mathematical modeling:

$$
\begin{cases}
u_{n+1} = \dfrac{\sigma_1 m_1(a_n, b_n)u_n}{2} \\[2mm]
v_{n+1} = \dfrac{\sigma_2 m_2(a_n, b_n)v_n}{2}
\end{cases}
\quad (n = 1, 2, \cdots), \qquad (2.20)
$$

where u_n and v_n are the expected numbers of reproductive couples belonging to the population 1 and population 2 respectively in the nth breeding season. m_1 and m_2 are the expected numbers of newborns per couple for each population. Variables a_n and b_n are the expected numbers of mature individuals belonging to each population in the nth breeding season. From the assumption for the geometric growth model (1.7), we have $a_n = 2u_n$ and $b_n = 2v_n$. The survival probability per individual in the non-breeding season is given by σ_1 and σ_2 respectively for each population. Since we are going to consider the competition between two populations, the number of newborns per couple has a negative correlation to the density of the other population.

2.3.1 Leslie-Gower Model

In this section, we shall consider the following mathematical modeling which can be regarded as an expansion of Beverton-Holt model (2.2) with $\theta = 1$ in Sect. 2.1.1 to the competitive relation between two populations:

$$\begin{cases} m_1(a_n, b_n) = \dfrac{m_{10}}{1 + \beta_{11}a_n + \beta_{12}b_n}; \\ m_2(a_n, b_n) = \dfrac{m_{20}}{1 + \beta_{21}a_n + \beta_{22}b_n}, \end{cases} \tag{2.21}$$

where m_{01} and m_{02} are respectively the physiological upper bound for the number of newborns per couple of each population, which means the number of newborns per couple when the density effect is absent. The positive parameter β_{ij} $(i, j = 1, 2)$ reflects the severity of density effect on the number of newborns in population i from the density of mature individuals in population j, or the sensitivity of the number of newborns in population i to the density effect from population j.

The population dynamics model is given by (2.20) with (2.21) and expressed as the following paralleled recurrence relations:

$$\begin{cases} a_{n+1} = \dfrac{\mathscr{R}_{01}a_n}{1 + \beta_{11}a_n + \beta_{12}b_n}; \\ b_{n+1} = \dfrac{\mathscr{R}_{02}b_n}{1 + \beta_{21}a_n + \beta_{22}b_n}, \end{cases} \tag{2.22}$$

where $\mathscr{R}_{01} := \sigma_1 m_{01}/2$ and $\mathscr{R}_{02} := \sigma_2 m_{02}/2$ are the net reproduction rate for each population when any density effect is absent. Note that these recurrence relations are about the number of mature individuals instead of the number of reproductive couples in the previous sections.

This population dynamics model is called *Leslie-Gower model*. The parameters β_{12} and β_{21} are sometimes called *interspecific competition coefficient*. In contrast, the parameters β_{11} and β_{22} are called *intraspecific competition coefficient*. Following to this meaning of the parameter β_{ii} $(i = 1, 2)$, the parameter α in Beverton-Holt model (2.2) with $\theta = 1$, the parameter γ in Ricker model (2.7), and the parameter a_c in the logistic map model (2.12) of Sect. 2.1 may be called intraspecific competition coefficient.

The model (2.22) gives the population dynamics governed by Beverton-Holt model when two populations are independent of each other or when one of them is absent. As described in Sect. 2.1, when $\mathscr{R}_{01} \leq 1$, population 1 monotonically goes extinct when population 2 is absent. When population 2 presents, the density effect from it further reduces the number of newborns per couple in population 1, so that population 1 must monotonically go extinct. This arguments imply that, when $\mathscr{R}_{0i} \leq 1 \ (i = 1, 2)$, the population i must monotonically go extinct. On the other hand, it is easily understood that, when one of two populations goes extinct, the other population approaches the population dynamics with the other population's absence, which is governed by Beverton-Holt model. Therefore, when one of two populations goes extinct, the size of the other population must approach a non-negative value determined by the parameters, as shown about Beverton-Holt model in Sect. 2.1.1.

In contrast, when $\mathscr{R}_{01} > 1$ and $\mathscr{R}_{02} > 1$, population 1 approaches $a^* := (\mathscr{R}_{01} - 1)/\beta_{11}$, while population 2 does $b^* := (\mathscr{R}_{02}-1)/\beta_{22}$ if they are independent of each other. However, it is not trivial what occurs when they are in the competitive relation through the interspecific density effect. When $\mathscr{R}_{01} > 1$ and $\mathscr{R}_{02} > 1$, the two species population dynamics model (2.22) has the following features, depending on the strength of interaction between two populations:

- When $\beta_{11}a^* > \beta_{12}b^*$ and $\beta_{22}b^* > \beta_{21}a^*$, where it holds that $\beta_{11}\beta_{22} - \beta_{12}\beta_{21} > 0$, the numbers of mature individuals a_n and b_n respectively approach the following positive values a^{**} and b^{**}:

$$(a^{**}, b^{**}) = \left(\frac{\beta_{22}(\beta_{11}a^* - \beta_{12}b^*)}{\beta_{11}\beta_{22} - \beta_{12}\beta_{21}}, \frac{\beta_{11}(\beta_{22}b^* - \beta_{21}a^*)}{\beta_{11}\beta_{22} - \beta_{12}\beta_{21}} \right)$$

  In this case, two population coexist.
- When $\beta_{11}a^* > \beta_{12}b^*$ and $\beta_{22}b^* < \beta_{21}a^*$, population 2 goes extinct, while population 1 approaches a^*. In this case, population 2 is induced to go extinct by the density effect from population 1. Population 1 beats population 2 at the competition, and persists.
- When $\beta_{11}a^* < \beta_{12}b^*$ and $\beta_{22}b^* > \beta_{21}a^*$, population 1 goes extinct, while population 2 approaches b^*.
- When $\beta_{11}a^* < \beta_{12}b^*$ and $\beta_{22}b^* < \beta_{21}a^*$, where it holds that $\beta_{11}\beta_{22} - \beta_{12}\beta_{21} < 0$, one of two populations goes extinct while the other approaches a positive equilibrium (a^* or b^*), depending on the initial values of a_1 and b_1.

In the first case, two populations approach a coexistent state (Fig. 2.23c), and in the second and third cases, one of them is forced to go extinct by the competition while the other wins the competition and persists (Fig. 2.23a, b). In the fourth case, the population dynamics has a bistable nature, and the fate of each population depends on the initial condition given by (a_1, b_1) (Fig. 2.23d).

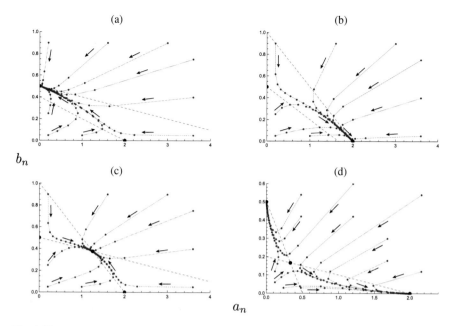

Fig. 2.23 Orbits with the sequence of (a_n, b_n) for two populations by the Leslie-Gower model (2.22). Numerical calculations with some different initial conditions. $(\beta_{12}, \beta_{21}) =$ (**a**) $(5.0, 0.2)$; (**b**) $(2.0, 0.6)$; (**c**) $(2.0, 0.2)$; (**d**) $(10.0, 2.0)$; $(\beta_{11}, \beta_{22}) = (1.0, 2.0)$; $(\mathcal{R}_{01}, \mathcal{R}_{02}) = (3.0, 2.0)$

Focusing on the interspecfic competition coefficients β_{12} and β_{21} which reflect the strength of density effect on the other population, the above nature implies that the coexistence can be established when they are sufficiently small, that is, when the interspecific competition is sufficiently weak. In contrast, when they are sufficiently large, that is, when the interspecific competition is severe, the coexistence is impossible and the bistable situation appears. In this way, the density effect of the competitive relation between two populations may cause the extinction of one of them. [3]

Ecological researchers have studied the difficulty of the coexistence of competing species in many cases. Joseph Grinnell (1877–1939) was one of them to clearly mention it [18]: Two species with the completely same niche cannot coexist at the same habitat, where *niche* is defined as an index

(continued)

---

[3] The nature of Leslie-Gower model (2.22) qualitatively corresponds to that of Lotka-Volterra two species competition model in Sect. 7.1.

to indicate the population's characteristics with respect to the use of the non-organic and biological environment, and is sometimes explained as the functional role played by the population (species) within the ecosystem. Georgy F. Gause (1910–1986) published the experimental result supporting the competitive species' exclusion implied by the theoretical discussion on the Lotka-Volterra competition system (see Sect. 7.1), making use of well-designed experiments with paramecium and yeast species [14]. Gause's claim is well-known today as *Gause's competitive exclusion principle* or *Gause's law*.

2.3.2   Ricker Type of Competition Model

Leslie-Gower model (2.22) is regarded as an expansion of Beverton-Holt model (2.2) with $\theta = 1$ to the two species competition system. In this section, let us see the other model of two species competition system with a different type of density effect function as a comparison to it. We shall consider here the following density effect functions as the expansion of that for Ricker model described in Sect. 2.1.2:

$$\begin{cases} m_1(a_n, b_n) = m_{10}\, e^{-\gamma_{11}a_n - \gamma_{12}b_n}; \\ \\ m_2(a_n, b_n) = m_{20}\, e^{-\gamma_{21}a_n - \gamma_{22}b_n}, \end{cases} \tag{2.23}$$

where parameters γ_{12} and γ_{21} are the coefficients of interspecific competition, while γ_{11} and γ_{22} are those of intraspecific competition. With these density effect functions, the population dynamics model (2.20) leads to the following two species competition system:

$$\begin{cases} a_{n+1} = \mathscr{R}_{01}\, a_n\, e^{-\gamma_{11}a_n - \gamma_{12}b_n}; \\ \\ b_{n+1} = \mathscr{R}_{02}\, b_n\, e^{-\gamma_{21}a_n - \gamma_{22}b_n}, \end{cases} \tag{2.24}$$

where $\mathscr{R}_{01} := \sigma_1 m_{01}/2$ and $\mathscr{R}_{02} := \sigma_2 m_{02}/2$ are respectively the net reproduction rates for each population when any density effect is absent, as before. For this model, when the other population is absent or two populations are independent of each other, the population dynamics has characteristics by Ricker model.

Since a chaotic variation may occur with the period-doubling structure in Ricker model as described in Sect. 2.1.2, it is expected that the population dynamics model (2.24) could have similar nature. Actually the bifurcation diagram shown in Fig. 2.24 indicates that two competing populations could coexist with a periodic or chaotic variation in some cases.

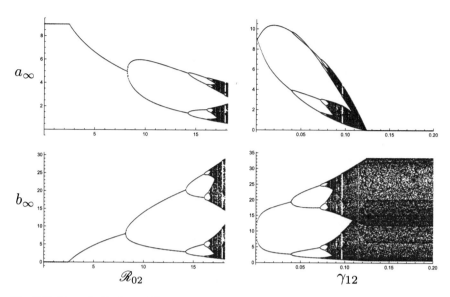

Fig. 2.24 Numerically drawn bifurcation diagram for the two species competition model (2.24) in terms of the bifurcation parameter $\mathscr{R}_{02}$ ($\gamma_{12} = 0.1$) and γ_{12} ($\mathscr{R}_{02} = 18.17$). $\mathscr{R}_{01} = 6.05$; $(\gamma_{11}, \gamma_{21}, \gamma_{22}) = (0.2, 0.1, 0.2)$

Mathematically cumbersome details aside, the population dynamics model (2.24) has characteristics similar to Leslie-Gower model (2.22) according to the coexistence of two competing populations. The coexistence can be established only when the competitive relation is sufficiently weak. When it is much strong, a bistability appears and the coexistence becomes impossible. In contrast to the coexistence for Leslie-Gower model, the model (2.24) shows a variety of coexistent patterns.

Figure 2.25 shows a numerical example of the temporal sequence for the number of mature individuals (a_n, b_n) when their coexistence is possible. At the coexistent state, they approach a period-4 solution (Fig. 2.25c). In this numerical example, if two populations are independent of each other, population 1 approaches a period-2 (Fig. 2.25a), while population 2 leads to a chaotic variation (Fig. 2.25b). Like this case, it is not trivial at all what pattern appears at the coexistent state under the competitive relation. It is an interesting mathematical problem.

Figure 2.26 shows a numerical example of the temporal sequence for the number of mature individuals (a_n, b_n) when the population dynamics is bistable. Depending on the initial condition, one of two population goes extinct. When population 2 goes extinct, population 1 approaches the positive value $(1/\gamma_{11}) \ln \mathscr{R}_{01}$, while, when population 1 goes extinct, population 2 approaches a chaotic variation. Like this bistability, the population dynamics model (2.24) could have a variety of combination of locally asymptotically stable states (in a wide sense). It would be an interesting mathematical problem what combination of states can appear at the bistable situation too.

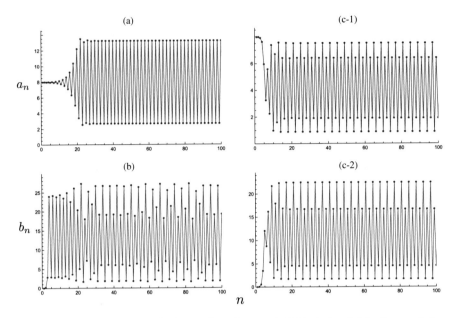

Fig. 2.25 Temporal sequences for the number of mature individuals (a_n, b_n) by the population dynamics model (2.24). $(\mathscr{R}_{01}, \mathscr{R}_{02}) = (11.02, 14.88)$; $(\gamma_{11}, \gamma_{12}, \gamma_{21}, \gamma_{22}) = (0.3, 0.1, 0.1, 0.2)$; (**a**) $(a_1, b_1) = (7.992, 0.0)$; (**b**) $(a_1, b_1) = (0.0, 0.0135)$; (**c**) $(a_1, b_1) = (7.992, 0.0135)$

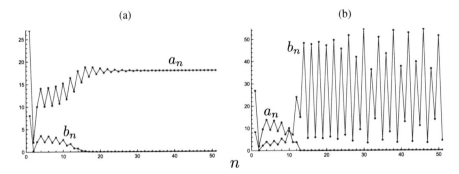

Fig. 2.26 Temporal sequences for the number of mature individuals (a_n, b_n) by the population dynamics model (2.24) when it has a bistable structure. It depends on the initial condition which population goes extinct. $(\mathscr{R}_{01}, \mathscr{R}_{02}) = (6.05, 14.88)$; $(\gamma_{11}, \gamma_{12}, \gamma_{21}, \gamma_{22}) = (0.1, 0.2, 0.2, 0.1)$; (**a**) $(a_1, b_1) = (27.0, 8.10)$; (**b**) $(a_1, b_1) = (27.0, 8.37)$

2.4   Enemy

The relation between *prey* and *predator* is one of typical interaction in ecosystem. The enemy for one species is defined as the other species which causes the death of individual of the former species. The enemy is not necessarily a predator. A *parasite* or *parasitoid* may be the enemy. For a parasite or parasitoid, the prey is in general called *host*.

Prey-predator or host-parasite relation can be modeled by a density effect with a biased nature. Predator/parasite can get an energetic benefit from the prey/host, while the prey/host is exploited by the predator/parasite with the death or a loss of energy. In contrast, two competing species are damaging to each other.

In this section, we shall consider a mathematical modeling of a prey-predator or host-parasite relation by introducing the effect of enemy in the survival rate for the prey, making use of the geometric growth model (1.7). Let us consider the modeling such that the survival probability for the population of prey or host (say prey hereafter) σ is negatively correlated to the number of mature enemy individuals p_n as a result of the interspecific reaction:

$$c_{n+1} = \frac{\sigma(p_n) m c_n}{2},$$
(2.25)

where the survival probability is given by $\sigma(p_n)$, a monotonically decreasing function of p_n. With the same assumptions in Sect. 1.5, we are going to consider a population dynamics with generationally non-overlapping reproduction. Since σ is defined as the survival probability in non-breeding season, it is now assumed that the prey becomes the victim of enemy in the non-breeding season. The value $\sigma(0) = \sigma_0$ means the survival probability when the enemy is absent.

Although the survival probability for the prey population σ is given above as a function of the number of mature enemy individuals only, the survival probability σ could have a relation to the population size of prey c_n. In such a modeling, σ must be given as a function of p_n and c_n: $\sigma = \sigma(p_n, c_n)$. The higher prey density could attract the enemies and increase the risk of enemy's attack. On the other hand, the higher prey density would cause the faster spread of an alarm signal over the prey population, for example, by an alarm call against the enemy. Such the faster spread of an alarm signal could reduce the probability of the success of enemy's attack.

Furthermore, such a much high prey density would lead to the saturation of the frequency of enemy's attacks per unit time. This is because each attack must take time so that the number of attacks per enemy per unit time has an upper bound. Hence, when the prey density is much high, the frequency of enemy's attacks would be large, while it is saturated with an upper bound. Such a relation of the frequency of enemy's attacks to the prey density is what is called *functional response* of the enemy in ecology [61]. In this section, for the simplicity, we do not introduce any factor related to the functional response of enemy in the modeling. We shall consider its modeling later in Sect. 8.1.

2.4.1   Dynamics of Enemy Population

From (2.25), the death probability in the non-breeding season between the nth and $n + 1$th breeding seasons is now given by $1 - \sigma(p_n)$. Thus, the expected number of deaths of the immature preys becomes $\{1 - \sigma(p_n)\}mc_n$ in the non-breeding season. Let us consider the death due to some reasons other than the enemy, and here call it the natural death. The natural death probability can be defined by $1 - \sigma_0$, since σ_0 means the survival probability when the enemy is absent. Hence the expected number of deaths with some reasons other than the enemy is given by $(1 - \sigma_0)mc_n$. Therefore, the expected number of deaths by the enemy is given by

$$\mathcal{Z}_n := \{1 - \sigma(p_n)\} mc_n - (1 - \sigma_0) mc_n = \{\sigma_0 - \sigma(p_n)\} mc_n. \tag{2.26}$$

Now we shall consider the modeling for the dynamics of enemy population, taking the simplest assumptions for it:

- The enemy reproduction is possible only by victimizing the prey.
- The growth rate of enemy population is determined by the number of victimized preys.

With the first assumption, the enemy population is assumed to go extinct if the prey population is absent. For this reason, the enemy may be regarded as a kind of predator, called *specialist*,[4] or monophagous predator especially in ecology. In contrast, if a predator uses some different species of prey, it is called *generalist* or polyphagous predator.

With the second assumption, let $g(\mathcal{Z}_n, p_n)$ denote the reproduction rate per enemy by predation/parasitism in the non-breeding season between the prey's nth and $n + 1$th breeding seasons. We take here a general assumption that the reproduction rate per enemy depends on the density p_n with the intraspecific density effect. Thus we have $g(\mathcal{Z}_n, p_n)p_n$ as the expected total number of enemy newborns produced by the predation/parasitism in the non-breeding season between the prey's nth and $n+1$th breeding seasons. Now the population dynamics of enemy is given by

$$p_{n+1} = (1 - \mu)p_n + g(\mathcal{Z}_n, p_n)p_n, \tag{2.27}$$

where μ is the death probability for the mature enemy until the next season ($0 < \mu \le 1$). As described in Sect. 1.5, this assumption indicates that the mean life span of enemy after the maturation is given by $(1 - \mu)/\mu$. For a simplicity of formula, let us consider that the survival probability of enemy newborn is incorporated in the function g, while we shall still call g the reproduction rate per enemy hereafter.

---

[4] In the more precise definition, a predator which eats a specific part of prey is called specialist too.

> The function g may be consider to provide a modeling for what is called *numerical response* of the enemy population in ecology [61], in contrast to *functional response* mentioned in the first part of this section. The numerical response means here the change in the population size of enemy as a result of the predation/parasitism for the prey (refer to Sect. 8.1).

The recurrence relation (2.27) implies that the enemy population dynamics obeys the following assumptions:

- Enemy newborn can become mature until the next season for the predation/parasitism;
- Mature enemy can repeatedly make the reproduction if it survives until the next season;
- The fertility is the same even at the repeated reproduction.

Although the reproduction rate per enemy g may depend on the digestive efficiency and the prey's physiological resistance, we consider here the simplest modeling. Since the averaged amount of feed per enemy is given by $\mathcal{Z}_n/p_n$, we now define the function g as

$$g(\mathcal{Z}_n, p_n) := \rho \frac{\mathcal{Z}_n}{p_n}. \tag{2.28}$$

This definition means the modeling assumption that the expected reproduction rate per enemy is proportional to the averaged amount of feed per enemy. The parameter ρ is the coefficient to convert the amount of feed to the fertility, and is sometimes called (energy) *conversion coefficient* in the arguments of population dynamics modeling for the prey-predator relation.

By these mathematical modelings with (2.25)–(2.28), we can derive the following population dynamics model governing the temporal variation of enemy and prey populations:

$$\begin{cases} a_{n+1} = \dfrac{\sigma(p_n)ma_n}{2}; \\ p_{n+1} = (1-\mu)p_n + \dfrac{\rho}{2}\{\sigma_0 - \sigma(p_n)\}ma_n, \end{cases} \tag{2.29}$$

where $a_n = 2c_n$ means the number of mature individuals in the prey population as before. This population dynamics model is about the temporal sequence of mature individuals for the enemy and prey. In the subsequent sections, we are going to consider the mathematical modeling about the interaction between the enemy and prey populations, and give the survival probability $\sigma(p_n)$.

2.4.2 Nicholson-Bailey Model

Let q denote the probability that an enemy succeeds in the predation/parasitism when it finds a prey individual. Then $(1 - q)^j$ gives the probability that a prey individual can escape from the enemy after it is found by the enemy only j times. Now for simplicity, we assume that the prey individual is not damaged by the attack of enemy when it can escape from it. Then $\sigma_0 (1 - q)^j$ gives the probability that a prey individual can make the reproduction in the subsequent breeding season after being found by the enemy only j times in the non-breeding season before it, since the natural survival probability that the prey individual survives except for the death by the predation/parasitism is given by σ_0. The case of $j = 0$ means when a prey individual does not encounter the enemy at all in the non-breeding season, so that the survival probability of such a prey individual is given by σ_0.

Let $\Pi_n(j)$ denote the probability that a prey individual encounters the enemy only j times in the non-breeding season between the nth and $n + 1$th breeding seasons. From the above arguments, we can define the survival probability $\sigma(p_n)$ for the prey in the non-breeding season as

$$\sigma(p_n) := \sigma_0 \sum_{j=0}^{\infty} (1 - q)^j \Pi_n(j). \tag{2.30}$$

We shall introduce here a *Poisson distribution* for the probability $\Pi_n(j)$:

$$\Pi_n(j) := \frac{(\lambda p_n)^j}{j!} e^{-\lambda p_n} \quad (j = 0, 1, 2, \dots), \tag{2.31}$$

where $\Pi_n(0) = e^{-\lambda p_n}$. This is the Poisson distribution with the intensity λp_n (see Fig. 15.1 of Sect. 15.2) [33]. As described in Chap. 15, the introduction of the above Poisson distribution represents the assumption that the prey's encounter with the enemy in the non-breeding season occurs with the likelihood same at any moment. That is, the prey's encounter with the enemy follows a *Poisson process* with the intensity λp_n. Poisson process is regarded as one of the simplest assumptions for randomly occurred events [1, 12].

Substituting (2.31) for (2.30), we can get

$$\sigma(p_n) = \sigma_0 e^{-\lambda p_n} \sum_{j=0}^{\infty} \frac{\{(1 - q)\lambda p_n\}^j}{j!} = \sigma_0 e^{-q\lambda p_n}. \tag{2.32}$$

As a result, the population dynamics model (2.29) becomes

$$
\begin{cases}
a_{n+1} = \mathscr{R}_0 e^{-q\lambda p_n} a_n; \\
p_{n+1} = (1 - \mu)p_n + \rho\mathscr{R}_0\big(1 - e^{-q\lambda p_n}\big)a_n,
\end{cases}
\tag{2.33}
$$

where $\mathscr{R}_0 := \sigma_0 m/2$ is the net reproduction rate of prey when the enemy is absent.

This population dynamics model (2.33) is today called *Nicholson-Bailey model*, which is usually referred in case of $\mu = 1$ when the enemy follows the generationally non-overlapping reproduction. This naming is after the work on the prey-predator population dynamics model by Australian entomologist Alexander J. Nicholson (1895–1969) and physicist Victor A. Bailey (1895–1964) in 1930s [45, 47].

Nicholson-Bailey model has the following nature:

- When $\mathscr{R}_0 \leq 1$, both of prey and enemy populations go exinct.
- When $\mathscr{R}_0 > 1$, both of prey and enemy populations show an excited oscillation.

Since the prey population goes extinct when $\mathscr{R}_0 \leq 1$ even though the enemy is absent, the enemy population must go extinct at the same time. In contrast, when $\mathscr{R}_0 > 1$, the temporal variation of population size has an oscillatory behavior in which the amplitude gradually becomes larger, that is called *excited oscillation*. Especially for Nicholson-Bailey model (2.33), the excitation of oscillation is unbounded so that the population diverges. Such a divergence of the population size originates from the nature of prey population dynamics. When the enemy is absent, the prey population dynamics follows the geometric progression and unboundedly increases.

When $\mathscr{R}_0 > 1$ for Nicholson-Bailey model (2.33), the population sizes of prey and enemy do not become zero, that is, do not go extinct in a mathematical sense. However, as seen in Fig. 2.27, their population sizes become much small in some seasons. In an ecological sense, such a much small population size implies a high risk of population extinction due to some causes unexpectedly or stochastically arisen (e.g., climate or artificial event) in the environment, which especially lead a significant influence on the reproduction. Such an extinction is sometimes mentioned as the extinction due to the *demographic stochasticity* or the *ecological disturbance*. From this nature of the temporal variation by Nicholson-Bailey model (2.33), the population dynamics is characterized by a high risk of extinction. According to this arguments on the likeliness of extinction, there are mathematical concepts called *persistence*, *uniform persistence*, and *permanence*, which are sometimes significant for the theoretical discussion in ecology. For their

(continued)

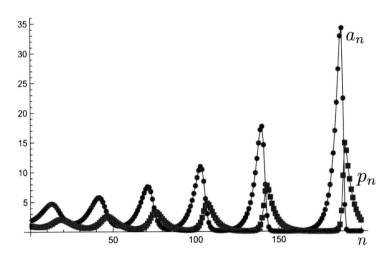

Fig. 2.27 Temporal sequence by Nicholson-Bailey model (2.33). Numerically drawn with $(a_1, p_1) = (2.0, 1.0)$; $\mathcal{R}_0 = 1.3$; $q\lambda = 0.2$; $\mu = 0.9$; $\rho = 0.3$

mathematically rigorous definitions in the dynamical system theory, the reader can refer to [27, 60, 65] for instance.

When their definitions are applied for the population dynamics model, the persistence means that the temporal variation of population size from the positive initial value keeps positive for any finite time. Thus, the population dynamics with Nicholson-Bailey model (2.33) has the persistence. In contrast, the uniform persistence requires that the population size keeps beyond a certain positive value for any finite time after sufficiently long transition phase. The permanence requires that the population size keeps staying in a certain finite range for any finite time after sufficiently long transition phase. The population dynamics with the uniform persistence or permanence has the nature of persistence. But the inverse does not necessarily hold. The population dynamics with Nicholson-Bailey model (2.33) does not have the nature of the uniform persistence or permanence.

It should be remarked that the mathematically defined permanence does not necessarily deny the risk of extinction mentioned in the above. The population dynamics with the uniform persistence must be regarded as having relatively smaller risk of the above-mentioned extinction in comparison to one only with the persistence, whereas it depends on the detail of the permanence itself characterizing the population dynamics.

2.4.3 Synergy with the Density Effect for Prey

From the standing point that the divergence of the temporal variation of population size for Nicholson-Bailey model (2.33) is caused by the nature of geometric growth for the prey population, a theoretical question reasonably arises: What nature could appear when the prey population dynamics is self-regulated by a density effect within the population (for example, by the intraspecific competition)?

In this section, we consider the following mathematical model with the density effect on the number of newborns for the population dynamics model (2.29):

$$
\begin{cases}
a_{n+1} = \dfrac{\sigma(p_n)m(a_n)a_n}{2} \\[2ex]
p_{n+1} = (1-\mu)p_n + \dfrac{\rho}{2}\{\sigma_0 - \sigma(p_n)\}m(a_n)a_n
\end{cases}
\tag{2.34}
$$

Further, we shall now introduce (2.32) for the survival probability $\sigma(p_n)$ of prey individual in the non-breeding season, and (2.2) with $\theta = 1$ for the density effect function $m(a_n)$ about the number of prey newborns. Then we have the following population dynamics model:

$$
\begin{cases}
a_{n+1} = \mathscr{R}_0 e^{-q\lambda p_n}\dfrac{a_n}{1+a_n/\alpha}; \\[2ex]
p_{n+1} = (1-\mu)p_n + \rho\mathscr{R}_0\left(1 - e^{-q\lambda p_n}\right)\dfrac{a_n}{1+a_n/\alpha},
\end{cases}
\tag{2.35}
$$

where $\mathscr{R}_0 := \sigma_0 m_0/2$ which means the net reproduction rate of prey when the enemy and the density effect are absent. This model can be regarded as a combination of Nicholson-Bailey model (2.33) and Beverton-Holt model (2.3) with $\theta = 1$. The number of mature prey individuals monotonically approaches $a^* := \max[0, (\mathscr{R}_0 - 1)\alpha]$ when the enemy is absent, as described in Sect. 2.1.1. Now in the population dynamics governed by (2.35), the prey population undergoes a random predation/parasitism (introduced as Poisson process). It must be remarked that the smaller α means the stronger density effect for the prey population.

The population dynamics model (2.35) has the following nature:

- When $\mathscr{R}_0 \leq 1$, both of prey and enemy populations go extinct.
- When $1 < \mathscr{R}_0 \leq 1 + \mu/(q\lambda\rho\alpha)$, the enemy population goes extinct, while the prey population persists and the number of mature prey individuals a_n approaches $(\mathscr{R}_0 - 1)\alpha$.
- When $\mathscr{R}_0 > 1 + \mu/(q\lambda\rho\alpha)$, prey and enemy populations coexist.

As shown in Fig. 2.28 about the parameter dependence of population dynamics, too strong density effect in the prey population makes the enemy population go extinct.

Further from the bifurcation diagrams in Fig. 2.29, it can be seen that the numbers of prey and enemy can show oscillatory sequences at the coexistent state, which amplitude is suppressed more by the stronger density effect. Since Nicholson-

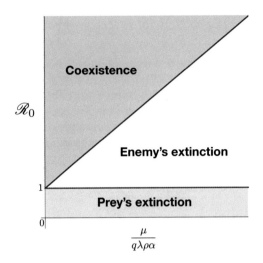

Fig. 2.28 Parameter dependence of the population dynamics by (2.35)

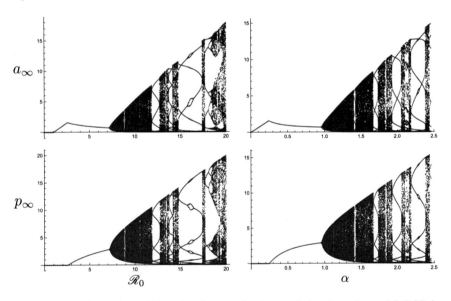

Fig. 2.29 Numerically drawn bifurcation diagrams for the population dynamics model (2.35) in terms of the bifurcation parameter $\mathscr{R}_0$ with $\alpha = 1.0$, and α with $\mathscr{R}_0 = 7.5$. Commonly, $q\lambda = 0.5$; $\mu = 0.8$; $\rho = 1.0$

Bailey model (2.33) shows a divergent amplitude of oscillatory sequence which corresponds to the model (2.35) without the density effect in the prey population (i.e., mathematically the case of $\alpha \to \infty$), this result implies that the density effect in the prey population gives a secondary effect counteracting the oscillatory feature of population dynamics to suppress such an excitation of oscillation. Moreover, the

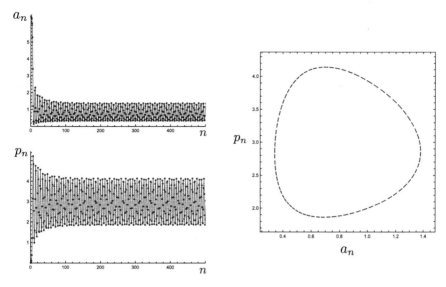

Fig. 2.30 Temporal sequences by the population dynamics model (2.35). The right figure shows the plots of (a_n, p_n) from $n = 500$ to $n = 1000$ in the phase plane. The plots appears on a certain closed curve. Numerically drawn with $\mathcal{R}_0 = 7.5$; $q\lambda = 0.5$; $\alpha = 1.0$; $\mu = 0.8$; $\rho = 1.0$

bifurcation diagram indicates that only an intermediately strong density effect can stabilize the population dynamics to make the prey and enemy populations approach a stationary state with positive values.

The bifurcation diagram in Fig. 2.29 implies a specific bifurcation structure from the equilibrium to a chaotic variation, which is called *Neimark-Sacker bifurcation*, *Saker-Neimark bifurcation*, *Naimark-Sacker bifurcation* (spelled differently), *secondary Hoph bifurcation*, or *torus bifurcation* [55]. It is different from the bifurcation structure with period-doubling appeared in Sect. 2.1. As shown by Fig. 2.30, the chaotic variation appeared for the population dynamics model (2.35) is characterized by a bounded irregular oscillation with group-wave-like repeated swellings. The repeated swellings are not periodical in the rigorous mathematical sense, but may be regarded as quasi-periodic. We can see such a group-wave-like (quasi-periodic) repeated swellings in plots of (a_n, p_n) in Fig. 2.30.

2.5   Harvesting/Culling

Human activities have been affecting the population dynamics to reduce a number of individuals in a biological population. Hunting, fishery, and harvesting are such examples. The other example is the application of a pesticide. Differently from the influence by the enemy population, such an artificial reduction of population size does not have any feedback effect on the strength of the effect itself in principle. Aside from some industrial or ethical reasons, there is no autonomous feedback relation between the strength of human activity and the size change of an affected population. We know in history not a few examples such that a unilateral reduction of biological population caused the extinction or endangered situation.

On the other hand, the operation of harvesting/culling may have an aspect to manage the biological resource. Some may be for the purpose of conserving an endangered population, and the other may be for the purpose of controlling the size of another population interacting the targeted one.

As seen in the previous section, the feedback relation between the effect to reduce the prey population size and the enemy population dynamics would be an important factor to cause an oscillatory variation in the temporal sequence. In this section, we shall consider how the artificial harvesting/culling could affect the characteristics of the population dynamics.

Let us consider again the framework of modeling for the geometric growth model (1.7). We shall consider the population dynamics with the generationally non-overlapping reproduction. For the harvesting/culling, we add the following assumption:

- The harvesting/culling occurs only once just before the breeding season.

With this assumption for the harvesting/culling, let us proceed our modeling with the density effect on the number of newborns per couple m, as applied for the geometric growth model (2.1) in Sect. 2.1.

As known well for the cultivation of vegetable and fruit in agriculture, the appropriate culling can make the growth or the yield (amount of net reproduction) larger. In such a case, the culling operation could be regarded as an effective control of the negative density effect considered in Sect. 2.1. The operation of harvesting/culling serves to reduce the population density and suppress the negative density effect, so that the net reproduction rate per couple or per individual increases.

As in Sect. 2.1, the density effect on the number of newborn per couple m is regarded as an effect on the gained energy per individual for the reproduction. The gained energy per individual could depend on the influence from surroundings, for example, competition about some resources (e.g., food, light, and water), fight between individuals about the territory or the mate, or stress to keep a caution around, etc.

In this section, we shall consider the following model with the above assumption:

$$a_{n+1} = \frac{\sigma m\big((1-h)a_n\big)(1-h)a_n}{2},\tag{2.36}$$

where the parameter h $(0 < h < 1)$ is the reduction ratio of mature individuals by the harvesting/culling. Now a_n denotes the number of mature individuals without the harvesting/culling just before the nth breeding season. The expected number of individuals reduced by the harvesting/culling just before the nth breeding season is given by ha_n, and the expected number of mature individuals in the nth breeding season becomes $(1-h)a_n$. Hence the expected number of couples is given by $(1-h)a_n/2$. When the population is a biological resource, a_n is its amount subject to the harvesting. The recurrence relation (2.36) expresses the effect of the harvesting/culling at each year on the amount of the biological resource at the next year.

Mathematically, with the replacement by $b_n = (1-h)a_n$, the relation (2.36) can be expressed as that about the sequence $\{b_n\}$, which becomes equivalent to the recurrence relation (2.1) in Sect. 2.1. However, as a mathematical model for the consideration on the effect of harvesting/culling, it is still necessary to clarify the dependence of the nature of population dynamics on the parameter h, and discuss the effect of harvesting/culling on the population dynamics.

2.5.1 Beverton-Holt Type Model with Harvesting/Culling

Let us consider here the following population dynamics model of the recurrence relation (2.36) with the density effect function (2.2) of Beverton-Holt model (2.3) with $\theta = 1$ in Sect. 2.1.1:

$$a_{n+1} = \frac{\mathscr{R}_0(1-h)a_n}{1+(1-h)a_n/\alpha},\tag{2.37}$$

where $\mathscr{R}_0 := \sigma m_0/2$ as before. From the nature of Beverton-Holt model, we know that this population dynamics model (2.37) has the following nature:

- When $(1-h)\mathscr{R}_0 \leq 1$, the population size monotonically decreases to go extinct.
- When $(1-h)\mathscr{R}_0 > 1$, the number of mature individuals a_n monotonically approaches $a^* := \big\{\mathscr{R}_0 - 1/(1-h)\big\}\alpha$.

As indicated by the bifurcation diagram in Fig. 2.31, the larger reduction ratio h by the harvesting/culling makes the equilibrium value for a_n smaller. If the reduction rate h is beyond the critical value $h_c := 1 - 1/\mathscr{R}_0$, the population extinction is induced by the harvesting/culling.

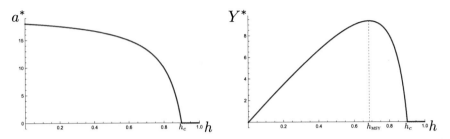

Fig. 2.31 Numerically drawn bifurcation diagrams for the population dynamics model (2.37) in terms of the bifurcation parameter h. $Y^* := ha^*$; $\mathscr{R}_0 = 10.0$; $\alpha = 0.5$. Here $h_c = 0.9$ and $h_{\mathrm{MSY}} = 1 - 1/\sqrt{10} \approx 0.6838$

2.5.2 Maximum Sustainable Yield

The yield by harvesting/culling may be important in some arguments on its efficiency. For our model (2.37), the yield at the nth season is defined by $Y_n := ha_n$, where we regard the amount of reduced number of individuals as the yield itself. From the above mentioned nature of the population dynamics, the yield at equilibrium $Y^*(h)$ is given by $Y^*(h) := ha^* = h\{\mathscr{R}_0 - 1/(1 - h)\}\alpha$. The bifurcation diagram in Fig. 2.31 clearly indicates the existence of the harvesting/culling operation to maximize the yield at equilibrium. We can easily find that the yield at equilibrium Y^* takes the maximal value $Y^*(h_{\mathrm{MSY}}) = (\sqrt{\mathscr{R}_0} - 1)^2\alpha = \alpha/(1 - h_{\mathrm{MSY}})^2$ when the reduction rate h is

$$h_{\mathrm{MSY}} := 1 - \frac{1}{\sqrt{\mathscr{R}_0}}. \qquad (2.38)$$

In the theory of bioeconomics, the maximum yield $Y^*(h_{\mathrm{MSY}})$ is especially called *maximum sustainable yield* (MSY).

2.5.3 Cost for the Harvesting/Culling

With the results about the model (2.37), let us go into the further modeling, taking account of the cost for the harvesting/culling operation. For example, the efficiency of the pesticide use significantly depends on the used amount and the quality which requires a cost. Such a cost counteracts the agricultural purpose of the pesticide use to reduce the damage on the yield gained by the cultivation. In fishery, the cost depends on the fishing period and manpower. The larger yield necessarily increases the cost for the harvesting. Hence, in order to maximize the profit, the maximization of yield would not be necessarily suitable for the agricultural purpose, and it is necessary to take account of the cost about the yield.

As we now consider the maximization of the profit, let us call h *harvesting rate* hereafter. We shall consider the modeling with the following simple assumptions:

- The expected earnings per unit yield are given by a constant p.
- The cost for the harvesting operation is proportional to the harvesting rate h.

From these assumptions, the total earnings from yield Y are given by pY, and the cost for the harvesting operation with h is now given by ηh. The positive constant η may depend on the characteristics of harvesting operation.

Now, with these modelings for Beverton-Holt type model with harvesting (2.37), the profit $\mathcal{P}^*(h)$ by the harvesting with harvesting rate h at the equilibrium is formulated as

$$\mathcal{P}^*(h) := pY^*(h) - \eta h = ph\left(\mathcal{R}_0 - \frac{1}{1-h}\right)\alpha - \eta h. \tag{2.39}$$

Since it is reasonable to consider this problem when the population is persistent by itself, we will consider hereafter only the situation with $\mathcal{R}_0 > 1$.

Mathematical investigation of the profit $\mathcal{P}^*(h)$ brings us the following results:

- When $\mathcal{R}_0 - \eta/(p\alpha) \leq 1$, we have $\mathcal{P}^*(h) \leq 0$ for any $h > 0$.
- When $\mathcal{R}_0 - \eta/(p\alpha) > 1$, $\mathcal{P}^*(h)$ takes the maximum $\mathcal{P}^*(h_{\text{MEY}}) = p\{\sqrt{\mathcal{R}_0 - \eta/(p\alpha)} - 1\}^2\alpha = p\{h_{\text{MEY}}/(1 - h_{\text{MEY}})\}^2\alpha$ for

$$h = h_{\text{MEY}} := 1 - \frac{1}{\sqrt{\mathcal{R}_0 - \eta/(p\alpha)}}.$$

- When $\mathcal{R}_0 - \eta/(p\alpha) > 1$, we have $\mathcal{P}^*(h) > 0$ for

$$h < h_{\text{s}} := 1 - \frac{1}{\mathcal{R}_0 - \eta/(p\alpha)},$$

while we have $\mathcal{P}^*(h) < 0$ for $h > h_{\text{s}}$.

The first result indicates that the harvesting results in a deficit when the cost for the harvesting is relatively large (with large η), or when the earnings is expected poor (with small p). Further, if the population has unsatisfactory fertility (with small $\mathcal{R}_0$), the profit cannot be obtained. Such a biological population cannot be regarded as a profitable natural/agricultural resource.

Figure 2.32 shows a numerical example of the h-dependence of the profit $\mathcal{P}^*(h)$ at the equilibrium for Beverton-Holt type model with harvesting (2.37). It is a case where $\mathcal{R}_0 - \eta/(p\alpha) > 1$ when the earnings is expected. As shown in Fig. 2.32, we find a specific harvesting rate h_{MEY} with which the profit takes the maximum. As the harvesting rate h gets larger than h_{MEY}, the profit $\mathcal{P}^*(h)$ becomes smaller. In the theory of bioeconomics, the yield with the harvesting rate h_{MEY} to maximize the profit is called *maximum economic yield* (MEY). The harvesting with $h > h_{\text{MEY}}$ is

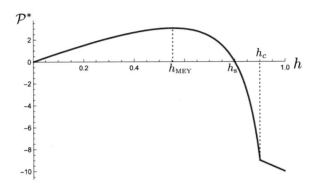

Fig. 2.32 The h-dependence of the profit $\mathcal{P}^*(h)$ at the equilibrium for Beverton-Holt type model with harvesting (2.37). Numerical example with $\mathcal{R}_0 = 10.0$; $\alpha = 2.0$; $p = 1.0$; $\eta = 10.0$. In this figure, $h_s = 0.8$; $h_c = 0.9$; $h_{MEY}^* \approx 0.5528$

further called *economic overexploitation*. For the population dynamics model (2.37), the maximum economic yield $Y^*(h_{MEY})$ at the equilibrium is given by

$$Y^*(h_{MEY}) = h_{MEY}\left(\mathcal{R}_0 - \frac{1}{1 - h_{MEY}}\right)\alpha$$

$$= \left(1 - \frac{1}{\sqrt{\mathcal{R}_0 - \eta/(p\alpha)}}\right)\left(\mathcal{R}_0 - \sqrt{\mathcal{R}_0 - \frac{\eta}{p\alpha}}\right)\alpha. \tag{2.40}$$

With the harvesting rate large enough to satisfy that $h > h_s$, the profit becomes negative even though the population does not go extinct. With such a too large harvesting rate, the population size a^* at the equilibrium becomes much small, so that the earnings become smaller than the harvesting cost.

From the analysis in the above, we can find the nature of the profit $\mathcal{P}^*(h)$ at the equilibrium for Beverton-Holt type model with harvesting (2.37), and summarize it as Fig. 2.33. It can be mathematically proved that $h_{MEY} < h_{MSY} < h_c$, as shown by Fig. 2.33. This means that the harvesting with h_{MSY} to maximize the maximum sustainable yield $Y^*(h_{MSY})$ is necessarily an economic overexploitation. The harvesting with h_{MSY} can maximize the yield Y^* at the equilibrium, while it cannot maximize the profit $\mathcal{P}^*$. Although the larger yield would be usually regarded as the better result of the harvesting, it is not appropriate from the viewpoint of the profit maximization.

As long as the harvesting is controlled to maximize the yield ($h \to h_{MSY}$), the population would be less likely to go extinct with the harvesting for the model (2.37), and we may keep using it as the biological resource. However, it may cause the situation of deficit ($\mathcal{P}^*(h_{MSY}) < 0$). Such a case occurs for the

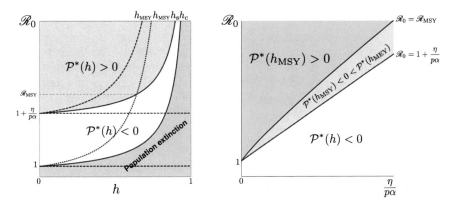

Fig. 2.33 Parameter dependence of the profit $\mathcal{P}^*(h)$ at the equilibrium for Beverton-Holt type model with harvesting (2.37)

model (2.37) under the condition that

$$\frac{\eta}{p\alpha} < \sqrt{\mathcal{R}_0}(\sqrt{\mathcal{R}_0} - 1),$$

that is,

$$\mathcal{R}_0 < \mathcal{R}_{\text{MSY}} := \frac{1}{4}\left(1 + \sqrt{1 + \frac{4\eta}{p\alpha}}\right)^2. \tag{2.41}$$

In contrast, if $\mathcal{R}_0 > \mathcal{R}_{\text{MSY}}$, then the harvesting for the purpose to maximize the yield becomes a sustainable resource use to get a profit $\mathcal{P}^*(h_{\text{MSY}}) > 0$.

Consequently, from the condition (2.41), we find the following three conditions necessary for the sustainable resource use with the harvesting purposed to maximize the yield:

1. The population has a sufficiently large net reproduction rate $\mathcal{R}_0$.
2. The cost for the harvesting is sufficiently small (sufficiently small η).
3. The expected earnings per unit yield p are sufficiently large.

The first biological condition may be controlled, for example, by the environmental conservation or the artificial incubation. The improvement or innovation for the harvesting can reduce the cost. The expected earnings per unit yield may be increased by improving the process of commercialization of the yield or adding some value to the final good. It is clear that the sustainable use of a biological resource requires a well-designed management integrating biological, agricultural, engineering, and economical researches.

2.5.4 Ricker Type Model with Harvesting/Culling

Clearly from the arguments in Sect. 2.1, the characteristics of the population dynamics significantly depend on the nature of density effect involved in it. As a simple example, let us see here the following population dynamics with a harvesting effect (2.36), accompanied by the density effect function (2.6) of Ricker model (2.7):

$$a_{n+1} = \mathscr{R}_0(1-h)a_n e^{-\gamma(1-h)a_n}. \tag{2.42}$$

The population dynamics governed by (2.42) has some interesting characteristics different from those for (2.37):

- When $h \geq h_c := 1 - 1/\mathscr{R}_0$, the population goes extinct.
- When $1 - e/\mathscr{R}_0 \leq h < h_c$, the number of mature individuals a_n monotonically approaches $a^* := \ln\{(1-h)\mathscr{R}_0\}/\{\gamma(1-h)\}$.
- When $1 - e^2/\mathscr{R}_0 \leq h < 1 - e/\mathscr{R}_0$, a_n approaches a^* with a damped oscillation.
- When $h < 1 - e^2/\mathscr{R}_0$, a_n keeps changing without approaching any specific value.

These follows the characteristics of Ricker model (2.7) in Sect. 2.1.2.

As indicated by Fig. 2.34, we can further find the following results on the population size at $n \to \infty$:

- When $1 < \mathscr{R}_0 \leq e \approx 2.71828$, the number of mature individual a_n monotonically approaches a^*, which is decreasing in terms of h.
- When $\mathscr{R}_0 > e$, the supremum of a_n as $n \to \infty$ becomes maxumum for

$$h = h_P^* := 1 - \frac{e}{\mathscr{R}_0}. \tag{2.43}$$

The value a_∞ for $h = h_P^*$ corresponds to the case where a_n monotonically approaches a^*, and coincides to the maximum of a^* in terms of h.

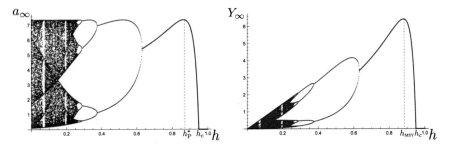

Fig. 2.34 Numerically drawn bifurcation diagrams for the population dynamics model (2.42) in terms of the bifurcation parameter h. $Y_\infty := ha_\infty$; $\mathscr{R}_0 = 20.0$; $\gamma = 1.0$. Here $h_P^* \approx 0.8641$; $h_c = 0.95$; $h_{\mathrm{MSY}} \approx 0.8795$

It is now shown that the harvesting operation could maximize the equilibrium population size at a specific harvesting rate $h = h_{\mathrm{P}}^*$ for the model (2.42), differently from the model (2.37) for which the equilibrium population size is monotonically decreasing in terms of h as shown by Fig. 2.31.

The existence of such a specific harvesting to maximize the population size could be related to the resurgence problem in the pest control with a pesticide. As an important problem for the agriculture, the pest control has been studied experimentally and theoretically [21, 26, 28, 32, 40, 44, 49, 51, 64]. In the application of a pesticide, there are some cases such that the pesticide is effective only in the early stage of its application, and the pest density revives later to become higher than before. Such a phenomenon is called *resurgence*, and has been studied in different aspects [8–10, 15, 22, 64]. DeBach [9] defined it in a narrow sense as the pest population's unexpected and rapid revival by the decrease of the enemy population due to the pesticide applied for the pest. Today it frequently means a paradoxical increase of the pest density under an operation of the pest control. In this meaning, there are different causes of the resurgence such as the pesticide-resistance appeared in the pest, the increase in the preferability of the crop by the pesticide (called trophobiosis) [7], and the pest's fertility risen by the stimulus of sublethal pesticide dose (called hormesis or homoligosis) [30, 34, 43]. In contrast, for the population dynamics by (2.42), a possible cause of a resurgence is implied not by such a specific cause but by a balance of the density effect.

When the harvesting is for the yield of a biological resource, it is important to consider the yield $Y_n(h) := ha_n$ for the population dynamics (2.42). We can find the following nature of the yield $Y_\infty(h)$ as $n \to \infty$ (see Fig. 2.34):

- The supremum of $Y_\infty(h)$ becomes maximum for $h = h_{\mathrm{MSY}}$ which is determined by the unique positive root of the following equation with $\mathscr{R}_0 > 1$:

$$\ln\left\{(1 - h_{\mathrm{MSY}})\mathscr{R}_0\right\} - h_{\mathrm{MSY}} = 0.$$

- It is necessarily satisfied that $h_{\mathrm{MSY}} > h_{\mathrm{P}}^*$.
- The number of mature individuals a_n monotonically approaches equilibrium a^* for the harvesting rate $h = h_{\mathrm{MSY}}$.

The profit $\mathcal{P}_\infty(h)$ as $n \to \infty$ with the harvesting rate h has characteristics as shown in Fig. 2.35. The earnings are possible only for an intermediate range of harvesting rate. With smaller or larger harvesting rate out of the range, the profit results in a deficit. This is different from the nature of the model (2.37) for which the earnings is necessarily obtained for the harvesting rate less than a critical value, that is, for $h < h_s$ (see Fig. 2.32).

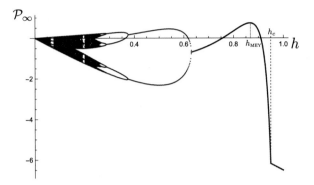

Fig. 2.35 The h-dependence of the profit $\mathcal{P}_\infty(h)$ at the equilibrium for the model (2.42). Numerical example with $\mathcal{R}_0 = 20.0$; $\gamma = 1.0$; $p = 1.0$; $\eta = 6.5$. In this figure, $h_c = 0.95$; $h_{\text{MEY}}^* \approx 0.8665$

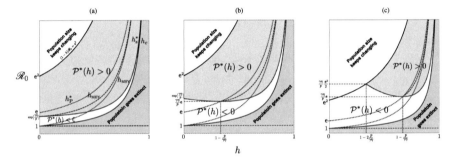

Fig. 2.36 Parameter dependence of the profit $\mathcal{P}^*(h)$ at the equilibrium for the model (2.42). **(a)** $\gamma\eta/p \leq 1$; **(b)** $1 < \gamma\eta/p < 2$; **(c)** $\gamma\eta/p \geq 2$ (drawn for the case of $\gamma\eta/p > $ e)

The profit when the population dynamics (2.42) approaches the equilibrium, that is, when $a_n \to a^*$ as $n \to \infty$, becomes

$$\mathcal{P}^*(h) := pY^*(h) - \eta h = \frac{ph}{\gamma}\left[\frac{1}{1-h}\ln\left\{(1-h)\mathcal{R}_0\right\} - \frac{\gamma\eta}{p}\right]. \qquad (2.44)$$

The analysis on this profit $\mathcal{P}^*(h)$ brings the following specific results on the parameter dependence of the profit $\mathcal{P}^*(h)$ (see Fig. 2.36):

- When $\gamma\eta/p \leq 1$, if $\mathcal{R}_0 \leq e^{\gamma\eta/p}$, then the profit $\mathcal{P}^*(h)$ becomes negative for any harvesting rate $h > 0$.
- When $\gamma\eta/p \geq 1$, if $\mathcal{R}_0 < (\gamma\eta/p)e$, then the profit $\mathcal{P}^*(h)$ becomes negative for any harvesting rate $h > 0$.
- When $1 < \gamma\eta/p < 2$, if $(\gamma\eta/p)e < \mathcal{R}_0 < e^{\gamma\eta/p}$, then $\mathcal{P}^*(h)$ is positive only for an intermediate range of h.
- When $\gamma\eta/p \geq 2$, if $(\gamma\eta/p)e < \mathcal{R}_0 < (\gamma\eta/p)e^2/2$, then $\mathcal{P}^*(h)$ is positive only for an intermediate range of h.

The profit $\mathcal{P}^*(h) > 0$ becomes maximum for $h = h_{\mathrm{MEY}}$ which is given by the root greater than $1 - \gamma\eta/p$ for the following equation:

$$\ln\left\{(1 - h_{\mathrm{MEY}})\mathcal{R}_0\right\} - h_{\mathrm{MEY}} - \frac{\gamma\eta}{p}(1 - h_{\mathrm{MEY}})^2 = 0.$$

When the profit $\mathcal{P}^*(h)$ becomes positive for some $h > 0$, there exists a harvesting rate $h = h_{\mathrm{MEY}}$ to maximize $\mathcal{P}^*(h)$. In contrast, when the population size keeps changing, so does the profit $\mathcal{P}_\infty(h)$ as $n \to \infty$. In such a case, the situation would be regarded as undesirable for the stable commerce, so that the harvesting operation with such a harvesting rate h is not appropriate for the commercial reason. Besides, when the profit $\mathcal{P}_\infty(h)$ keeps changing its supremum cannot be beyond $\mathcal{P}^*(h_{\mathrm{MEY}})$ (see Fig. 2.35). Thus we can regard the harvesting rate $h = h_{\mathrm{MEY}}$ as one which maximizes the profit $\mathcal{P}_\infty(h)$.

The critical values for the harvesting rate h for the model (2.42) satisfy that $h_{\mathrm{P}}^* < h_{\mathrm{MEY}} < h_{\mathrm{MSY}} < h_c$ (see Fig. 2.36), where h_{MEY} can be defined only when $\mathcal{P}^*(h_{\mathrm{MEY}}) > 0$. Since the critical value h_{P}^* is smaller than the others, a sufficiently weak harvesting rate would be appropriate when the harvesting/culling purposes the conservation of the population to increase its size, differently from the harvesting rate to maximize the yield or the profit. In contrast, the harvesting/culling to maximize the yield or the profit necessarily makes the population size smaller than its possible maximum.

For both models (2.37) and (2.42), when $\mathcal{R}_0$ is sufficiently large, the difference between h_{MEY} and h_{MSY} is much small. Besides, for the model (2.42), the difference of h_{P}^* from them is much small too. This means that, for the population with sufficiently high fertility, the harvesting purposed to maximize the yield would be almost appropriate also with respect to the profit and the conservation of the population. In contrast, for the population with poor fertility, the difference among those critical values could become significant, and hence the use of such a biological resource must require a careful assessment and discussion.

2.6 Semi-Spatial Modeling

In this section, we introduce the idea of mathematical modeling for a single-species population dynamics, taking account of a spatial heterogeneity of population. Differently from the model with an explicit population distribution in space, for example, making use of the reaction-diffusion equation, the lattice space, the cell-automaton, or the network, we shall discuss here a mathematical modeling in which a relation to the spatial heterogeneity of population could be embedded in the model as a mathematical structure. Such a modeling could be regarded as following a statistical sense. Briefly, it is an idea to introduce the relation of such a spatial heterogeneity to the interaction between individuals in a way of the mean over the population. Although the mathematical modeling introduces such a spatial heterogeneity, the constructed model includes only the relation to it as a structure of the density effect. In this sense, such a model may be called *semi-spatial* model [13].

2.6.1 Royama's Idea of Modeling

This section describes an idea of mathematical modeling for the population dynamics with the intraspecific reaction within it. Its essence is based on the idea by Tomoo Royama [54].

Let us consider a population which uses a limited resource. Individuals must compete for the resource with the others in the same population. The reproduction is assumed generationally non-overlapping. Now we define the per capita reproduction rate $r_k(i)$ under the competition for a common resource among i individuals inhabiting at a same local habitat. In this idea of mathematical modeling, the whole habitat is assumed to consist of a number of local regions in which some individuals settle down and compete for the resource there. It must be remarked that the per capita reproduction rate $r_k(i)$ corresponds to the net reproduction rate $\sigma m(a_k)/2$ in the modeling for (2.1) of Sect. 2.1. Especially, $r_k(1)$ means the per capita reproduction rate when there is no competitor for the resource. Hence we may give it as $r_k(1) = \mathcal{R}_0$ which means the net reproduction rate when the resource competition is absent, as before (see the definition given at p. 31 in Sect. 2.1.1).

Now we introduce the probability that an individual settles in a local habitat where there are the other $i - 1$ individuals, and denote it by $P_k(i)$ at the kth generation. The probability $P_k(i)$ can be defined at the same time as the probability that a local population size is i with respect to the competition for the resource. The difference in the resource competition due to the spatial heterogeneity of local habitats is now introduced by the probability distribution $\{P_k(i)\}$. The probability $P_k(1)$ can be regarded as the frequency of individuals each of which uses the resource to oneself with no competitor.

We can define the following mean (expected) per capita reproduction rate $\langle r \rangle_k$ about the whole population at the kth generation:

$$\langle r \rangle_k = \sum_{i=1}^{\infty} r_k(i) P_k(i). \tag{2.45}$$

Under the generationally non-overlapping reproduction, the mean (expected) per capita reproduction rate $\langle r \rangle_k$ at the kth generation is defined on the other hand by a_{k+1}/a_k with the expected number of mature individuals a_k and that of a_{k+1} at the next generation. Therefore, we have

$$a_{k+1} = a_k \sum_{i=1}^{\infty} r_k(i) P_k(i). \tag{2.46}$$

In general, the strength of competition depends on the population density, that is, the mean distance between individuals. The competition becomes severer as the density gets larger. Since we are now assuming a spatially restricted region as the habitat of considered population, the population density becomes larger for the larger population size. This means that the probability $P_k(i)$ now depends on the population size, that is, the expected number of mature individuals a_k at the kth generation.

As a simple modeling, let us assume here a Poisson distribution for $\{P_k(i) \mid i = 1, 2, \ldots\}$ with the intensity depending on the expected number of mature individuals (refer to Chap. 15):

$$P_k(i) = \frac{\gamma_k^{i-1} e^{-\gamma_k}}{(i-1)!} \qquad (i = 1, 2, \ldots), \tag{2.47}$$

where $\gamma_k = \gamma_k(a_k)$ is a positive function of the expected number of mature individuals a_k at the kth generation. With Poisson distribution (2.47), the expected number of individuals under the resource competition at the kth generation, $\langle a \rangle_k$, is given by

$$\langle a \rangle_k = \sum_{j=1}^{\infty} j \, P_k(j) = \gamma_k + 1.$$

If the function $\gamma_k(a_k)$ is increasing in terms of the population size a_k, the resource competition becomes severer as the population size gets larger.

Royama [54] assumed that the population is distributed at random in space, and the resource competition occurs among the individuals within the neighborhood of a specific finite range around each individual. With a Poisson distribution of spatial heterogeneity about a population, the number of individuals within such a fixed range follows a Poisson distribution. The similar idea was applied for the mathematical modeling of the interaction between individuals by some other researchers, for example, Skellam [57, 58], Morisita [42], and Pielou [50].

Firstly let us consider the per capita reproduction rate $r_k(i)$ given by

$$r_k(i) = \mathscr{R}_0 \rho^{i-1}, \tag{2.48}$$

where $r_k(1) = \mathscr{R}_0 > 0$. Parameter ρ reflects the strength of the resource competition to reduce the reproduction rate ($0 < \rho \leq 1$). It is smaller as the influence of the resource competition gets stronger on the per capita reproduction rate. Substituting Poisson distribution (2.47) and the per capita reproduction rate (2.48) for (2.45), we derive the expected per capita reproduction rate $\langle r \rangle_k$ over the whole population:

$$\langle r \rangle_k = \mathcal{R}_0 \, e^{-(1-\rho)\gamma_k} = e^{\ln \mathcal{R}_0 - (1-\rho)\gamma_k}. \tag{2.49}$$

As a result, we get the following population dynamics model from (2.46):

$$a_{k+1} = a_k \, e^{\ln \mathcal{R}_0 - (1-\rho)\gamma_k}. \tag{2.50}$$

If the function $\gamma_k(a_k)$ is proportional to the expected number of mature individuals a_k as $\gamma_k(a_k) = \kappa a_k$ with a positive constant κ, the recurrence relation (2.50) becomes Ricker model in Sect. 2.1.2. In the present modeling, the parameter κ characterizes the severity of the resource competition in the population. As an extremal case with $\kappa = 0$, this modeling includes the case where there is no resource competition in the population. In such an extermal case, the distribution given by (2.47) is a singular one with $P_k(1) = 1$ and $P_k(i) = 0$ ($i = 2, 3, \dots$).

2.6.2 Skellam Model

As the other assumption for the per capita reproduction rate, we may consider a simple case where it is inversely proportional to the number of individuals in the local habitat:

$$r_k(i) = \frac{\mathcal{R}_0}{i}. \tag{2.51}$$

This may be regarded as the case such that the resource is evenly shared among individuals by the competition in the local habitat, and the per capita reproduction rate is assumed proportional to the resource gain. Substituting Poisson distribution (2.47) and the assumption (2.51) for (2.45), we derive the expected per capita reproduction rate $\langle r \rangle_k$:

$$\langle r \rangle_k = \frac{\mathcal{R}_0}{\gamma_k} \left(1 - e^{-\gamma_k} \right), \tag{2.52}$$

and subsequently the population dynamics model:

$$a_{k+1} = \frac{\mathcal{R}_0}{\gamma_k} a_k \left(1 - e^{-\gamma_k} \right). \tag{2.53}$$

As before, with the additional assumption that $\gamma_k(a_k) = \kappa a_k$, the population dynamics model (2.53) becomes

$$a_{k+1} = \frac{\mathcal{R}_0}{\kappa} \left(1 - e^{-\kappa a_k} \right). \tag{2.54}$$

This is the model proposed by John G. Skellam (1914–1979) [57], making use of the idea same as Royama's [54], and sometimes called *Skellam model*. The population dynamics by Skellam model has characteristics similar with those of Beverton-Holt model (2.3) with $\theta = 1$. For $\mathcal{R}_0 \leq 1$, the population size monotonically decreases to become extinct, while, for $\mathcal{R}_0 > 1$, it monotonically approaches a positive equilibrium, which may be regarded as the carrying capacity for the population.

Skellam [57] discussed the population dynamics by (2.54) in comparison with the continuous time population dynamics by the logistic equation (2.15). As long as κa_k is sufficiently small, the model (2.54) can be approximated by

$$a_{k+1} = \frac{\mathcal{R}_0 a_k}{1 + \frac{1}{2}\kappa a_k}.$$

This recurrence relation clearly corresponds to Beverton-Holt model (2.3) with $\theta = 1$. Making use of this approximated recurrence relation, Skellam discussed the similarity of the model (2.54) with the logistic equation (2.15).

2.6.3 Site-Based Model

Sumpter and Broomhead [63], Johansson and Sumpter [31], and Br'annstr'om and Sumpter [6] modified and expanded the idea by Royama [54] to what is called *site-based model*. Their modeling may be regarded as fundamentally same as Royama's, while theirs focuses on the frequency of local habitats in the whole habitat with respect to the number of individuals in the local habitat. They called the local habitat "site".

The frequency of sites inhabited by i individuals is now denoted by $p_k(i)$ at the kth generation. The frequency $p_k(i)$ generally depends on the number of sites in the whole habitat and the whole population size of mature individuals a_k. The per capita reproduction rate at the site with i individuals is denoted by $r_k(i)$, which depends on the number of individuals inhabiting at the same site. It is clear that these assumptions correspond to the idea by Royama [54], described in the previous section.

Now we define the mean (expected) growth rate for the subpopulation at the site containing i individuals by $\phi_k(i) = r_k(i) \cdot i$. In [6, 31] on the site-based model, $\phi_k(i)$ is given a priori in the modeling, and called *interaction function*.

For a descriptive convenience, let denote the total number of sites in the habitat by K. Then the expected number of sites containing i individuals at the k generation $\langle i \rangle_k$ is given by $\langle i \rangle_k = K p_k(i)$. From the above-mentioned assumption on the reproduction, the population size at the $k + 1$ th generation is now assumed to be

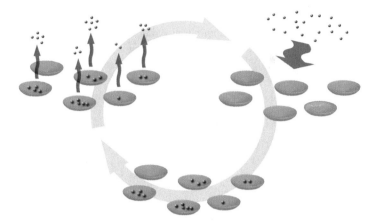

Fig. 2.37 Schematic figure of the site-based model

determined by the following recurrence relation:

$$a_{k+1} = \sum_{i=0}^{\infty} \phi_k(i)\langle i\rangle_k = K \sum_{i=1}^{\infty} i \cdot r_k(i) p_k(i). \tag{2.55}$$

This is the essential equation of the site-based model. Similarly with the recurrence relation (2.46) in the previous section by Royama's idea [54], the variation of the population size in the site-based model is determined also by the number of individuals settling in the same site, where the expected per capita reproduction rate is determined by the situation of each site. Differently from (2.46) to take account of the mean per capita reproduction rate over the whole population, the site-based model takes account of that over each local habitat (i.e., site) with $\sum_{i=0}^{\infty} i \cdot r_k(i) p_k(i)$ in (2.55). As schematically shown in Fig. 2.37, every newborn is equivalent with respect to its settlement of a site though the per capita reproduction rate depends on which local habitat it will settle to, that is, how many individuals will settle at the same site at the breeding season. This assumption is equivalent to the idea of Royama [54] about the local competition about the resource.

Now we shall consider the simplest case where K sites have the equivalent environmental condition, and the settlement of each individual randomly occurs for every site. In this case, the probability that a site is settled by i individuals $P(i)$ is given by

$$P(i) = \binom{N-i+K-2}{K-2} \Big/ \binom{N+K-1}{K-1}$$

$$= \left(1 - \frac{1}{K}\right) \prod_{j=0}^{i-1} \left(\frac{N}{K} - \frac{j}{K}\right) \Big/ \prod_{j=1}^{i+1} \left(\frac{N}{K} + 1 - \frac{j}{K}\right) \qquad (i = 1, 2, \ldots, N),$$

where N is the total number of individuals in the population. The above formula of this probability is derived from the number of ways to distribute N individuals over K sites, allowing some sites to which no individual settles. When N and K are sufficiently large, this probability can be approximated by the following geometric distribution:

$$P(i) \approx \frac{1}{\langle n \rangle + 1} \left(\frac{\langle n \rangle}{\langle n \rangle + 1} \right)^i \qquad (i = 0, 1, 2, \ldots),$$

where $\langle n \rangle = N/K$ is the mean (expected) number of individuals settling to the same site.

From this modeling, let us assume that the frequency of sites containing i individuals $p_k(i)$ follows the following geometric distribution:

$$p_k(i) = \frac{1}{\langle a \rangle_k + 1} \left(\frac{\langle a \rangle_k}{\langle a \rangle_k + 1} \right)^i \qquad (i = 0, 1, 2, \ldots), \tag{2.56}$$

where $\langle a \rangle_k = a_k/K$ is the mean number of individuals settling to the same site at the kth generation.

First, let us consider the case of the per capita reproduction rate $r_k(i)$ given by (2.48) in the previous section. In this case, the population dynamics model becomes

$$a_{k+1} = K \sum_{i=1}^{\infty} i \cdot r_k(i) p_k(i) = \frac{\mathcal{R}_0 a_k}{\left\{ 1 + (1 - \rho) a_k / K \right\}^2}. \tag{2.57}$$

This corresponds to the model (2.5) in Sect. 2.1.1, that is, Beverton-Holt type model (2.3) with $\theta = 2$. The reproduction curve for (2.57) is of the scramble type (refer to Sect. 2.2.4).

In contrast, with $r_k(i)$ given by (2.51) related to the Skellam model (2.54), the population dynamics model becomes

$$a_{k+1} = K \sum_{i=1}^{\infty} \mathcal{R}_0 \frac{1}{\langle a \rangle_k + 1} \left(\frac{\langle a \rangle_k}{\langle a \rangle_k + 1} \right)^i$$

$$= K \mathcal{R}_0 \frac{\langle a \rangle_k}{\langle a \rangle_k + 1} = \frac{\mathcal{R}_0 a_k}{1 + a_k / K}. \tag{2.58}$$

This is Beverton-Holt model (2.3) with $\theta = 1$, which has the reproduction curve of the contest type as described in Sect. 2.2.4.

For the same $r_k(i)$, when the frequency $p_k(i)$ follows a Poisson distribution instead of the geometric distribution (2.56), the same arguments leads to the Skellam model (2.54) again [6, 31].

As a last example in this section, let us consider a special assumption that the reproduction is possible only when the site contains only an individual, while it is impossible at the site with more than one individuals. This is a specific contest type of density dependence. In [6, 31], they discussed the case where $p_k(i)$ follows a Poisson distribution, and the mean (expected) growth rate for the subpopulation at the site containing i individuals $\phi_k(i) = r_k(i) \cdot i$ is given by

$$r_k(i) = \begin{cases} \mathcal{R}_0 & \text{if } i = 1; \\ 0 & \text{otherwise.} \end{cases} \tag{2.59}$$

In this case, from (2.55), we immediately get

$$a_{k+1} = K \mathcal{R}_0 p_k(1).$$

Thus, when $p_k(i)$ is given by (2.56), we have

$$a_{k+1} = \frac{\mathcal{R}_0 a_k}{(1 + a_k/K)^2}. \tag{2.60}$$

This corresponds again to Beverton-Holt type model (2.3) with $\theta = 2$ in Sect. 2.1.1. On the other hand, when $p_k(i)$ follows a Poisson distribution, we can get Ricker model (2.7) [6, 31].

> As seen from the schematic figure of the site-based modeling in Fig. 2.37, this modeling may be easily applied for the population dynamics of animals like barnacles or corals which have a floating planktonic juvenile stage and subsequent reproductive stage with the settlement to a patchy habitat in the life history, or like plants with widely dispersing seeds.

Answer to Exercise

Exercise 2.1 (p. 34)

The right side of the recurrence relation for Beverton-Holt model (2.3) with $\theta = 1$ is a convex and monotonically increasing function of c_n as shown in Fig. 2.38. Its curve is upperbounded by $\alpha \mathcal{R}_0/2$. When $\mathcal{R}_0 \leq 1$, the curve does not have any intersection with the line $c_{n+1} = c_n > 0$ other than the origin (Fig. 2.38a). When and only when $\mathcal{R}_0 > 1$, it has only one other intersection at $c_{n+1} = c_n = (\mathcal{R}_0 - 1)\alpha/2$ (Fig. 2.38b). Therefore, the cobwebbing method in Sect. 12.1.2 can clearly show that the sequence $\{c_n\}$ monotonically approaches zero when $\mathcal{R}_0 \leq 1$, and

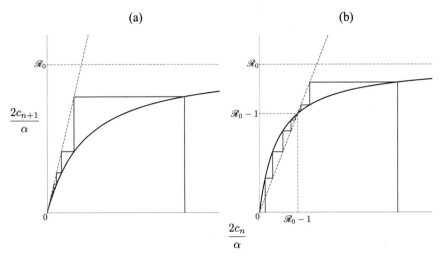

Fig. 2.38 Application of the cobwebbing method for Beverton-Holt model (2.3) with $\theta = 1$. (a) $\mathscr{R}_0 \leq 1$; (b) $\mathscr{R}_0 > 1$. In (b), the trajectories from two different initial values are shown

$(\mathscr{R}_0 - 1)\alpha/2$ when $\mathscr{R}_0 > 1$, independently of the initial value $c_1 > 0$, as illustrated by Fig. 2.38 (also see Fig. 2.4).

Exercise 2.2 (p. 43)

First, let $x_n = c_n/c_c$ in (2.13), then we can get the following recurrence relation mathematically equivalent to (2.13):

$$x_{n+1} = \begin{cases} \mathscr{R}_0(1 - x_n)x_n & (0 \leq x_n < 1) \\ 0 & (x_n \geq 1), \end{cases} \tag{2.61}$$

For $0 \leq x_n < 1$, the right side of the above recurrence relation is the function which has the maximal value $\mathscr{R}_0/4$. That is, as long as $0 \leq x_n < 1$, x_{n+1} cannot become beyond $\mathscr{R}_0/4$. If a positive value $x_{n+1} < \mathscr{R}_0/4$ makes x_{n+2} negative, then x_{n+2} must become negative when $x_{n+1} = \mathscr{R}_0/4$. Inversely, if x_{n+2} becomes non-negative for $x_{n+1} = \mathscr{R}_0/4$, so it does for any non-negative $x_{n+1} < \mathscr{R}_0/4$. This condition can be expressed as

$$\mathscr{R}_0\left(1 - \frac{\mathscr{R}_0}{4}\right)\frac{\mathscr{R}_0}{4} \geq 0.$$

Therefore, in order to make x_k non-negative for any $k \geq 2$ with any initial value x_1 that makes x_2 positive, it is necessary and sufficient that $\mathcal{R}_0 \leq 4$. The initial value x_1 that makes x_2 positive must satisfy that $0 < x_1 < 1$, that is, $0 < c_1 < c_c$.

References

1. L.J.S. Allen, *An Introduction to Stochastic Processes with Applications to Biology*, 2nd edn. (Chapman & Hall/CRC, Boca Raton, 2010)
2. N. Bacaër, *A Short History of Mathematical Population Dynamics*. (Springer, London, 2011)
3. M. Begon, M. Mortimer, D.J. Thompson, *Population Ecology: A Unified Study of Animals and Plants*, 3rd edn. (Blackwell Science, Oxford, 1996)
4. T.S. Bellows, Jr., The descriptive properties of some models for density-dependence. J. Anim. Ecol. **50**, 139–156 (1981)
5. R.J.H. Beverton, S.J. Holt, On the dynamics of exploited fish population. Fish. Invest. Lond. Ser. **19**, 533 (1957)
6. Å. Brännström, D.J.T. Sumpter, The role of competition and clustering in population dynamics. Proc. R. Soc. B **272**, 2065–2072 (2005)
7. F. Chaboussou, Nouveaux aspects de la phytiatrie et de la phytopharmacie. Le phênom'ene de la trophobiosse, in *Proceedings of FAO Symposium of Integrated Pest Control* (1966), pp. 33–61
8. E. Cohen, Pesticide-mediated homeostatic modulation in arthropods. Pestic. Biochem. Physiol. **85**, 21–27 (2006)
9. P. DeBach, *Biological Control of Insect Pests and Weeds* (Chapman and Hall, London, 1964)
10. P. DeBach, D. Rosen, C.E. Keffett, Biological control of coccids by introduced natural enemies, in *Biological Control*, ed. by C.B. Huttaker (Plenum Press, New York, 1971), pp. 165–194
11. R.L. Devaney, *An Introduction to Chaotic Dynamical Systems*, 3rd edn. (Chapman and Hall/CRC, Boca Raton, 2021)
12. W. Feller, *An Introduction to Probability Theory and Its Applications*, vol. 1, 3rd edn. (Wiley, New York, 1968)
13. J.A.N. Filipe, G.J. Gibson, C.A. Gilligan, Inferring the dynamics of a spatial epidemic from time-series data. Bull. Math. Biol. **66**, 373–391 (2004)
14. G.F. Gause, *The Struggle for Existence* (Williams and Wilkins, Baltimore, 1934)
15. U. Gerson, E. Cohen, Resurgence of spider mites (Acari: Tetranychidae) induced by synthetic pyrethroids. Exp. Appl. Acarol. **6**, 29–46 (1989)
16. J. Gleick, *Chaos — Making A New Science* (Penguin Books, New York, 1987)
17. N.J. Gotelli, *A Primer of Ecology*, 3rd edn. (Sinauer, Sunderland, 2001)
18. J. Grinnell, The niche relationship of the California thrasher. Ark **34**, 427–433 (1917)
19. R. Haberman, *Mathematical Models: Mechanical Vibrations, Population Dynamics, and Traffic Flow* (Prentice-Hall, New Jersey, 1977)
20. R. Haberman, *Mathematical Models: Mechanical Vibrations, Population Dynamics, and Traffic Flow*, Classics in Applied Mathematics, vol. 21 (Society for Industrial and Applied Mathematics (SIAM), Philadelphia, 1998)
21. A.E. Hajek, M.L. McManus, I. Delalibera, Jr., A review of introductions of pathogens and nematodes for classical biological control of insects and mites. Biol. Control **41**, 1–13 (2007)
22. M.R. Hardin et al., Arthropod pest resurgence: an overview of potential mechanisms. Crop Prot. **14**, 3–18 (1995)
23. M.P. Hassell, Density-dependence in single-species populations. J. Anim. Ecol. **44**, 283–295 (1975)
24. A. Hastings, *Population Biology: Concepts and Models* (Springer, New York, 1997)

25. M.W. Hirsch, S. Smale, Devaney, R.L. *Differential Equations, Dynamical Systems, and an Introduction to Chaos*, 3rd edn. (Elsevier/Academic Press, Waltham, 2012)
26. M.E. Hochberg, R.I. Anthony, *Parasitoid Population Biology* (Princeton University Press, Princeton, 2000)
27. J. Hofbauer, K. Sigmund, *Evolutionary Games and Population Dynamics*. (Cambridge University Press, 1998, Cambridge)
28. C.B. Huffaker, *New Technology of Pest Control* (Wiley, New York, 1980)
29. G.E. Hutchinson, *An Introduction to Population Ecology* (Yale University Press, New Haven, 1978)
30. D.G. James, T.S. Price, Fecundity in twospotted spider mite (Acari: Tetranychidae) is increased by direct and systemic exposure to imidacloprid. Ecotoxicology **95**, 729–732 (2002)
31. A. Johansson, D.J.T. Sumpter, From local interactions to population dynamics in site-based models of ecology. Theor. Pop. Biol. **64**, 497–517 (2003)
32. S.D. Lane, N.J. Mills, W.M. Getz, The effects of parasitoid fecundity and host taxon on the biological control of insect pests: the relationship between theory and data. Ecol. Entomol. **24**, 181–190 (1999)
33. G. Ledder, *Mathematics for the Life Sciences: Calculus, Modeling, Probability, and Dynamical Systems*. Springer Undergraduate Texts in Mathematics and Technology. (Springer, New York, 2013)
34. T.D. Luckey, Insecticide hormoligosis. J. Econ. Entomol. **61**, 7–12 (1968)
35. A. MacFadyen, *Animal Ecology* (Pitman, London, 1963)
36. M. Martelli, *Discrete Dynamical Systems and Chaos*, 1st edn. (CRC Press LLC, Boca Raton, 1992)
37. R.M. May, *Stability and Complexity in Model Ecosystems*, 2nd edn. (Princeton University Press, Princeton, 1973)
38. R.M. May, Biological populations with non-overlapping generations: stable points, stable cycles, and chaos. Science **186**, 645–647 (1974)
39. R.M. May, G.F. Oster, Bifurcations and dynamics complexity in simple ecological models. Am. Nat. **110**, 573–599 (1976)
40. R.L. Metcalf, W.H. Luckmann, *Introduction to Insect Pest Management*. (Wiley, New York, 1975)
41. P.A.P. Moran, Some remarks on animal population dynamics. Biometrics **6**, 250–258 (1950)
42. M. Morisita, Estimation of population density by spacing method. Memoir. Faculty Sci. Kyushu Univ. Ser. E **1**, 187–197 (1954)
43. J.G. Morse, Agricultural implications of pesticide-induced hormesis of insects and mites. Hum. Exp. Toxicol. **17**, 266–269 (1998)
44. W.W. Murdoch, C.J. Briggs, Theory for biological control: recent developments. Ecology **77**, 2001–2013 (1996)
45. A.J. Nicholson, The balance of animal populations. J. Anim. Ecol. Suppl. **2**, 132–178 (1933)
46. A.J. Nicholson, An outline of the dynamics of animal populations. Austrl. J. Zool. **2**, 9–65 (1954)
47. A.J. Nicholson, V.A. Bailey, The balance of animal populations. Part I. Proc. Zool. Soc. Lond. **1935**(3), 551–598 (1935)
48. R. Pearl, *The Biology of Population Growth* (Alfred A. Knopf, New York, 1925)
49. J.H. Perkins, *Insects, Experts, and the Insecticide Crisis* (Plenum Press, New York, 1982)
50. E.C. Pielou, *Mathematical Ecology*, 2nd edn. (Wiley, London, 1977)
51. R.E. Plant, M. Mangel, Modeling and simulation in agricultural pest management. SIAM Rev. **29**, 235–261 (1987)
52. W.E. Ricker, Stock and recruitment. J. Fish. Res. Board Can. **11**, 559–623 (1954)
53. R.C. Robinson, *An Introduction to Dynamical Systems: Continuous and Discrete*. Pure and Applied Undergraduate Texts, vol. 19, 2nd edn. (American Mathematical Society, Providence, 2012)
54. T. Royama. *Analytical Population Dynamics* (Chapman & Hall, London, 1992)

55. R. Seydel, *Practical Bifurcation and Stability Analysis: From Equilibrium to Chaos*. Interdisciplinary Applied Mathematics, vol. 5, 2nd edn. (Springer, New York, 1994)
56. J.W. Silvertown, D. Charlesworth, *Introduction to Plant Population Ecology*, 4th edn. (Blackwell Scientific Publications, Oxford, 2001)
57. J.G. Skellam, Random dispersal in theoretical populations. Biometrika **38**, 196–218 (1951)
58. J.G. Skellam, Studies in statistical ecology: I. Spatial pattern. Biometrika **39**, 346–362 (1952)
59. J.M. Smith, M. Slatkin, The stability of predator-prey systems. Ecology **54**, 384–391 (1973)
60. H.L. Smith, P. Waltman, *The Theory of the Chemostat — Dynamics of Microbial Competition* (Cambridge University Press, Cambridge, 1995)
61. M.E. Solomon, The natural control of animal populations. J. Anim. Ecol. **2**, 235–248 (1949)
62. S.H. Strogatz, *Nonlinear Dynamics and Chaos: With Applications to Physics, Biology, Chemistry, and Engineering*, 2nd edn. (Westview Press, Boulder, 2015)
63. D.J.T. Sumpter, D.S. Broomhead, Relating individual behaviour to population dynamics. Proc. R. Soc. B **268**, 925–932 (2001)
64. M. Takagi, Perspective of practical biological control and population theories. Res. Popul. Ecol. **41**, 121–126 (1999)
65. H.R. Thieme, *Mathematics in Population Biology* (Princeton University Press, Princeton, 2003)
66. J.H. Vandermeer, D.E. Goldberg, *Population Ecology: First Principles*, 2nd edn. (Princeton Univerisity Press, Princeton, 2013)

Chapter 3
From Discrete Time Model to Continuous Time Model

Abstract This chapter is about the idea to mathematically derive the continuous time model from the discrete time model in Chaps. 1 and 2. The derived model is written as an ordinary differential equation, since we focus on the model of single species population dynamics in this chapter. Logistic equation appears first in this chapter, although its modeling and nature will be described in the subsequent chapters again, Chaps. 4 and 5. Besides, as the basic idea to understand the meaning of time derivative, we describe the concept of momental velocity of population size change. It is the essential aspect for the continuous time population dynamics model, and is very likely to be forgotten in the biological translation of the mathematical results obtained for the continuous time model. In most case of biological phenomena, the continuous time modeling with the momental velocity of a biological quantity like population size could be regarded as a mathematical approximation or simplification.

In this chapter, we are going to see some mathematical arguments to derive the mathematical model described by a *differential* equation, corresponding to that described by the *difference* equation. The former model is a kind of what is called *continuous time model* in contrast to the latter, what is called *discrete time model*. Differently from most of textbooks containing a description about the relation between continuous and discrete time models, this section is not going to describe the derivation of the latter from the former, making use of a mathematical discretization for time as mentioned in Sects. 2.1.3 and 5.5. We are going to consider here a reasonable way to make a relation of the discrete time model to a continuous time one, especially about the model described by a first-order difference equation.

3.1 Geometric Growth to Exponential Growth

In the previous chapters, the time step size between subsequent numbers in the sequence generated by the recurrence relation was not explicitly given, since it was given and fixed as a part of modeling assumptions. Now let us explicitly give it by a constant positive parameter h. That is, in the following mathematical arguments, we regard the sequence $\{x_n\}$ as the numbers which appear or are observed with a time step size h.

In this section, let us consider the geometric growth model given by the recurrence relation

$$x_{n+1} = \mathscr{R}_0 \, x_n \tag{3.1}$$

with the common ratio $\mathscr{R}_0$ which can be regarded as the net reproduction rate for the population dynamics from the meaning of the modeling (refer to Sect. 1.5). The general term of the sequence $\{x_n\}$ can be given by $x_n = x_0 \mathscr{R}_0^n$.

Introducing the time step size h, we can regard the value x_n as the population size of x at time $t = nh$. Subsequently, with $x(t) = x_n$ and $x(0) = x_0$, we have

$$x(t) = x_0 \mathscr{R}_0^{t/h} = x_0 \exp\left[\frac{\ln \mathscr{R}_0}{h} t\right]. \tag{3.2}$$

Therefore, as the continuous curve fitting to the sequence of geometric progression $\{x_n\}$ determined by (3.1), we have derived now the exponential curve given by (3.2). This means that the exponential function as the change of a variable continuous in terms of time. can approximate the temporal change of the population size by a geometric growth.

With such an approximation with the exponential function, we can obtain the population size at any time t of real value. However, from the viewpoint of a mathematical approximation with an interpolation during the time step, the continuity of the variable does not necessarily indicate that the population size has the value of continuous function at any moment. In fact, in our modeling described in Chap. 1, we assumed the non-breeding season in which the population size could be unchanged or decrease due to the death even when the growth of population size in the seasonal sequence is the geometric one. The sequence of population sizes by the discrete time model does not provide any information on the actual temporal change of population size during the time step.

In contrast, when the population size may be regarded as changing at any time with the same dynamics like the growth of a bacteria population with the division, the exponential growth model would be reasonable in the modeling sense. Even in such a case, the temporal change of population size can be usually observed as a sequence of its values with a certain time step, and the geometric growth could be regarded as a reasonable mathematical model, which is really an approximation for the population dynamics.

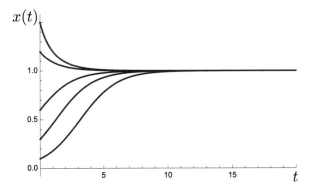

Fig. 3.1 The logistic growth curves by (3.5) for different five initial value $x(0) = x_0$. Numerically drawn with $\mathcal{R}_0 = 2.0$, $\beta = 1.0$, $x^* = 1.0$ and $h = 1.0$. Compare to Fig. 2.4 in Sect. 2.1.1

3.2 Beverton-Holt Model to Logistic Growth

Beverton-Holt model (2.3) with $\theta = 1$ in Sect. 2.1.1, now rewritten for the sequence $\{x_n\}$,

$$x_{n+1} = \frac{\mathcal{R}_0 x_n}{1 + \beta x_n} \tag{3.3}$$

is solvable[1] to get the following general term of x_n:

$$x_n = x^* \left\{ 1 + \left(\frac{x^*}{x_0} - 1 \right) \mathcal{R}_0^{-n} \right\}^{-1} = x^* \left\{ 1 + \left(\frac{x^*}{x_0} - 1 \right) e^{-n \ln \mathcal{R}_0} \right\}^{-1}, \tag{3.4}$$

where $x(0) = x_0$ and $x^* = (\mathcal{R}_0 - 1)/\beta$. $\mathcal{R}_0$ is the upper bound for the net reproduction rate under the density effect, and β a positive constant to mean the coefficient of the strength of density effect. Therefore, introducing the time step size h as before, we can get the following function of time t:

$$x(t) = x^* \left\{ 1 + \left(\frac{x^*}{x_0} - 1 \right) \exp\left[-\frac{\ln \mathcal{R}_0}{h} t \right] \right\}^{-1} \tag{3.5}$$

with $t = nh$ and $x(t) = x_n$. As a temporally continuous change of the population size, the function (3.5) gives so-called *logistic growth*. The growth curve in terms of t is shown in Fig. 3.1. As described in Sect. 2.1.1, the value $x(t)$ eventually converges to $x^* = (\mathcal{R}_0 - 1)/\beta$ for any positive initial value $x(0)$ when $\mathcal{R}_0 > 1$, while it monotonically decreases toward zero when $\mathcal{R}_0 < 1$.

---

[1] For instance, consider the recurrence relation with respect to $y_n = \mathcal{R}_0^n/(\beta x_n)$, and find the general term of y_n.

3.3 Time-Step-Zero Limit

Let us consider the general time discrete model described by a first-order difference equation:

$$x_{n+1} = x_n + \varphi(x_n). \tag{3.6}$$

Next, introducing the time step size h, the recurrence relation (3.6) can be rewritten as

$$x(t + h) = x(t) + \varphi(x(t); h), \tag{3.7}$$

where $x(t) = x_n$, $x(t + h) = x_{n+1}$ and $x(0) = x_0$ with $t = nh$.

In (3.7), we introduced h into the function φ too. This is because the function φ must depend on the time step size h for a mathematical consistency in the generalized equation (3.7) with h. From the meaning of the time step size h, the value $x(t + h)$ approaches $x(t)$ as $h \to 0$, that is, $x_{n+1} \to x_n$ as $h \to 0$. Thus, φ in (3.6) must converge to zero as $h \to 0$. This mathematical arguments show that the function φ must depend on the time step size h, and satisfy that $\varphi \to 0$ as $h \to 0$: $\lim_{h \to 0+} \varphi(x(t); h) = 0$. This is reasonable from the meaning of the modeling. Since the smaller time step size could be regarded as the shorter duration for the change in the population size, the zero time step size means no possibility to cause any change in the population size.

Now, let us consider the limit of $h \to 0$, that is, the limit of zero time step size, say, the *time-step-zero limit* for any fixed t. Provided that, for sufficiently small time step size h, we can apply Taylor expansion for φ in terms of h, then we can get

$$\varphi(x(t); h) = \varphi(x(t); 0) + \varphi_h(x(t); 0)h + o(h), \tag{3.8}$$

where $o(h)$ denotes the higher order terms and $\varphi_h = \partial\varphi/\partial h$. Since $\lim_{h \to 0+} \varphi(x(t); h) = 0$, we have $\varphi(x(t); 0) = 0$ and the following equation from (3.7) and (3.8):

$$\frac{x(t + h) - x(t)}{h} = \varphi_h(x(t); 0) + \frac{o(h)}{h}. \tag{3.9}$$

Therefore, taking the limit as $h \to 0$, we can derive the following ordinary differential equation for the function $x(t)$ of time t:

$$\frac{dx(t)}{dt} = \varphi_h(x(t); 0). \tag{3.10}$$

Thus the solution of (3.10) could be regarded as a continuous time model corresponding to the discrete time model (3.7).

3.3.1 Geometric Growth Model to Malthus Model

In the case of the geometric growth model (3.1), we can mathematically define the function φ as $\varphi(x(t); h) = (\mathscr{R}_0 - 1)x(t)$. From the above-mentioned mathematical constraint for the function φ about its dependence on h, we must suppose that the value $\mathscr{R}_0$ depends on the time step size h, that is, $\mathscr{R}_0$ is a function of h: $\mathscr{R}_0 = \mathscr{R}_0(h)$.

This is not surprising at all. To consider the temporally continuous change in the population size $x(t)$, we must suppose the reproduction with a net reproduction rate $\mathscr{R}_0$ for any given time step size h when we observe the sequence of population sizes with the time step size h. That is, the time step size for the observation must be reflected to the value of the net reproduction rate $\mathscr{R}_0$ because the temporally shorter interval of observations leads to the smaller difference of the observed subsequent population sizes. Since $\varphi(x(t); h) \to 0$ as $h \to 0$, it is now supposed that $\mathscr{R}_0(h) \to 1$ as $h \to 0$. From the meaning of the modeling, this is reasonable too.

The net reproduction rate $\mathscr{R}_0$ may include the survival of reproductive individuals (i.e., parents). The smaller time step size leads to the smaller number of died parents during the time interval, and the population size must become the same as the number of parents as $h \to 0$ since no recruitment is possible with no time passing at $h = 0$. This means that $\mathscr{R}_0 \to 1$ as $h \to 0$.

Supposing the differentiability of $\mathscr{R}_0(h)$ in terms of h, we obtain $\varphi_h(x(t); 0) = \mathscr{R}_0'(0)x(t)$, where $\mathscr{R}_0'(0)$ is the derivative of $\mathscr{R}(h)$ at $h = 0$. It is clear that the detail of the function $\mathscr{R}_0$ is not essential now. Finally, the ordinary differential equation (3.10) now becomes

$$\frac{dx(t)}{dt} = \mathscr{R}_0'(0)x(t), \tag{3.11}$$

which solution is given by

$$x(t) = x_0 e^{\mathscr{R}_0'(0)t} \tag{3.12}$$

(as for the fundamentals on the linear differential equation, see Sect. 13). This result shows again that a continuous time model corresponding to the geometric growth model is the exponential growth one, as already shown in Sect. 3.1.

The continuous time population dynamics governed by the type of ordinary differential equation (3.11) is sometimes called *Malthus growth*, and the model given by (3.11) is frequently called *Malthus model*. These names are after an English economist Thomas R. Malthus (1766–1834) who is famous from his theoretical arguments on the geometric population growth in his book "An Essay on the Principle of Population". The more detail description about Malthus model from the viewpoint of modeling will be given in Sect. 5.1.

The function $x(t)$ given by (3.12) can be derived directly from the general term (3.2) with the time step size h. Since $\mathcal{R}_0$ must be regarded as a function of h when we apply the time-step-zero limit for it, we need to mathematically consider the limit of $\{\ln \mathcal{R}_0(h)\}/h$ in (3.2) as $h \to 0$. Making use of de l'Hôpital's law since $\ln \mathcal{R}_0(h) \to 0$ as $h \to 0$, we have

$$\lim_{h \to 0+} \frac{\ln \mathcal{R}_0(h)}{h} = \lim_{h \to 0+} \frac{\mathcal{R}_0'(h)}{\mathcal{R}_0(h)} = \mathcal{R}_0'(0).$$

Therefore we can find that the function $x(t)$ given by (3.2) converges to (3.12) as $h \to 0$.

3.3.2   Beverton-Holt Model to Logistic Equation

For Beverton-Holt model (3.3), we can define the function φ as

$$\varphi(x(t); h) = \frac{\mathcal{R}_0(h) - 1 - \beta(h)x(t)}{1 + \beta(h)x(t)} x(t), \tag{3.13}$$

where we need to suppose that both of $\mathcal{R}_0$ and β are functions of h such that $\mathcal{R}_0 \to 1$ and $\beta \to 0$ as $h \to 0$. This is because the smaller time step size would lead to the weaker influence from the density effect during the time interval. Since $\mathcal{R}_0$ is the upper bound for the net reproduction rate, we could assume that $\mathcal{R}_0 \to 1$ as $h \to 0$ for the reason same as for the geometric growth model in the previous section. From the constraint for the function φ, $\varphi(x(t); h) \to 0$ as $h \to 0$, we have $\beta(h) \to 0$ as $h \to 0$ in order to satisfy that $x(t + h) = x_{n+1}$ converges to $x(t) = x_n$ in (3.3) as $h \to 0$.

Supposing the differentiability of $\mathcal{R}(h)$ and $\beta(h)$, we can get

$$\frac{\partial \varphi(x(t); h)}{\partial h} = \frac{\mathcal{R}_0'(h)\{1 + \beta(h)x(t)\} - \mathcal{R}(h)\beta'(h)x(t)}{\{1 + \beta(h)x(t)\}^2} x(t). \tag{3.14}$$

Therefore we can obtain the following ordinary equation from (3.10):

$$\frac{dx(t)}{dt} = \{\mathcal{R}_0'(0) - \beta'(0)x(t)\} x(t). \tag{3.15}$$

This ordinary differential equation is frequently called *logistic equation* in the theory of population dynamics, as already mentioned in the last part of Sect. 2.1.3 (the detail explanation will be given in Sect. 5.3).

In the early nineteenth century, a Belgian mathematician Pierre F. Verhulst (1804–1849) presented the following mathematical model for the population growth [4]:

$$\frac{dN(t)}{dt} = aN(t) - b\{N(t)\}^2, \tag{3.16}$$

where $N(t)$ is the population size at time t with positive parameters a and b. It is said that he introduced the second term in the right side of (3.16) as a suppressive effect from the environment on the population growth, intuitively making use of the analogy to the physical law such that a subject moving in a fluid medium has a resistance force proportional to the square of the velocity. Moreover, it is said that the name "logistic equation" conventionally used today would have the origin that Verhulst used the word *logistique* for his description about the Eq. (3.16) in his papers of 1845 and 1847 [5, 6], whereas he did not call it "logistic equation", and the definite origin has not been known.

Verhulst's works became well-known later with famous researches [1–3] by an American zoologist Raymond Pearl (1879–1940) and a biometrician Lowell J. Reed (1886–1966), in which they applied the same equation for the population dynamics. For this reason, the ordinary differential equation (3.16) is sometimes called *Verhulst-Pearl logistic equation*, and the parameter b is occasionally called *Verhulst coefficient* or *Verhulst-Pearl coefficient*.

The ordinary differential equation (3.15) is nonlinear, but is one of exactly solvable equations, for example, by the method of separation of variables (refer to Sect. 13). The solution becomes

$$x(t) = x^* \left\{ 1 + \left(\frac{x^*}{x_0} - 1 \right) e^{-\mathscr{R}_0'(0)t} \right\}^{-1}, \tag{3.17}$$

where $x^* = \mathscr{R}_0'(0)/\beta'(0)$ (refer to (5.11) in Sect. 5.3). We can see a clear correspondence to (3.5). Thus, as shown for (3.5), we can see that the solution of the logistic equation necessarily shows a monotonic approach to x^* from any positive initial value $x(0)$ as $t \to \infty$. Further, it is easy to derive (3.17) from (3.5), taking the time-step-zero limit with $\mathscr{R}_0 \to 1$ and $\beta \to 0$ as $h \to 0$.

3.3.3 Logistic Map Model to Logistic Equation

For the logistic map model in Sect. 2.1.3:

$$x_{n+1} = \mathscr{R}_0(1 - \beta x_n) x_n, \qquad (3.18)$$

we can mathematically define the function φ as

$$\varphi(x(t); h) = \{\mathscr{R}_0(h) - 1 - \mathscr{R}_0(h)\beta(h)x(t)\} x(t), \qquad (3.19)$$

where we suppose that $\mathscr{R}_0$ and β are functions of h, which satisfy that $\mathscr{R}_0 \to 1$ and $\beta \to 0$ as $h \to 0$ because of the constraint for the function φ. Lastly, we can obtain the following logistic equation again from (3.10):

$$\frac{dx(t)}{dt} = \{\mathscr{R}_0'(0) - \beta'(0)x(t)\} x(t).$$

3.3.4 Skellam Model to Logistic Equation

For the Skellam model in Sect. 2.6.2:

$$x_{n+1} = \frac{\mathscr{R}_0}{\kappa}\left(1 - e^{-\kappa x_n}\right), \qquad (3.20)$$

we can mathematically define the function φ as

$$\varphi(x(t); h) = \mathscr{R}_0(h) \frac{1 - e^{-\kappa(h)x_n}}{\kappa(h)} - x(t), \qquad (3.21)$$

where we suppose that $\mathscr{R}_0$ and κ are functions of h as before. Further, same as before, we can assume from the constraint for the function φ that $\mathscr{R}_0 \to 1$ and $\kappa \to 0$ as $h \to 0$.

Now we have

$$\frac{\partial \varphi(x(t); h)}{\partial h} = \mathscr{R}_0'(h) \frac{1 - e^{-\kappa(h)x_n}}{\kappa(h)}$$

$$+ \mathscr{R}_0(h) \frac{x_n e^{-\kappa(h)x_n}\kappa(h) - (1 - e^{-\kappa(h)x_n})}{\{\kappa(h)\}^2} \cdot \kappa'(h), \qquad (3.22)$$

and can easily find that

$$\lim_{\kappa \to 0} \frac{1 - e^{-\kappa x_n}}{\kappa} = x_n; \qquad \lim_{\kappa \to 0} \frac{x_n e^{-\kappa x_n}\kappa - (1 - e^{-\kappa x_n})}{\kappa^2} = -\frac{x_n^2}{2},$$

making use of de l'Hôpital's law. Thus, from (3.10), we can obtain the following logistic equation again:

$$\frac{dx(t)}{dt} = \left\{ \mathscr{R}_0'(0) - \frac{\kappa'(0)}{2} x(t) \right\} x(t). \tag{3.23}$$

Exercise 3.1 With the time-step-zero limit, show that the logistic equation is derived again for Ricker model in Sect. 2.1.2.

As we have seen in these sections, there is no one-to-one correspondence from a discrete time model to a continuous time model. This is true also about the correspondence from a continuous time model to a discrete time one, because the correspondence must depend on the derivation way of a discrete time model from a continuous time model.

Note that the nature of the population dynamics for the discrete time model is not necessarily conserved in the continuous time model derived from it by the time-step-zero limit. Exceptionally in the case of Beverton-Holt model and Skellam model, it appears conserved at least qualitatively, while it cannot in the case of Ricker model and logistic map model in which a bifurcation to a chaotic variation is involved. Especially on the logistic equation, we will revisit its correspondence to those discrete time models in Sects. 5.4 and 5.5.

3.4 Momental Velocity of Population Size Change

For most of organisms which have the distinctive reproductive season, the temporal change of population size could be considered with a discrete time model in the reasonable modeling sense, like a geometric growth model in Chap. 1. In contrast, the population growth of microorganism (bacteria, cell, etc.) or human could be regarded as having no distinct reproductive season, so that it may be approximated as a continuous temporal variation as most of theoretical works did. Moreover such a continuous temporal variation has been applied even for many cases of the organism with a distinct reproductive season. Such an application may be regarded as a kind of what is called *continuum approximation* in the time line.

As treated in the previous chapters, the population size varies by the natural reproduction process, the natural death process within the population, and the migration (immigration and emigration) process between the population and its surrounding environment. Therefore, we have

(Change of the population size in Δt) =

$\quad\quad$ + (Increase by the natural reproduction $\mathcal{B}_{\Delta t}$ in Δt)

$\quad\quad$ − (Decrease by the natural death $\mathcal{D}_{\Delta t}$ in Δt)

$\quad\quad$ + (Immigration from the outside of population $\mathcal{I}_{\Delta t}$ in Δt)

$\quad\quad$ − (Emigration to the outside of population $\mathcal{E}_{\Delta t}$ in Δt).

These four elements $\mathcal{B}_{\Delta t}$, $\mathcal{D}_{\Delta t}$, $\mathcal{I}_{\Delta t}$, and $\mathcal{E}_{\Delta t}$ may have some interrelation, but we could usually decompose the change of population size in Δt into these four.

$\quad$ When there is neither immigration nor emigration, that is, when it is satisfied that $\mathcal{I}_{\Delta t} = \mathcal{E}_{\Delta t} = 0$ for any time t and Δt, the population is called *closed population*. In the wider sense about the population dynamics, the population may be regarded as *closed* when the migration has no contribution to the temporal change of population size, that is, when $\mathcal{I}_{\Delta t} - \mathcal{E}_{\Delta t} = 0$ for any time t and Δt, even in an approximated sense.

$\quad$ The word "growth" is generally used in the population dynamics to mean the temporal change of population size by the reproduction process. In the wider sense, it can be used to mean the temporal change of population size by the above four elements. Note that it does not mean only the increase but also the decrease, which may be mentioned as the negative growth.

$\quad$ The net variation rate of the population size by the above four elements can be called *net growth rate*. Dividing it by the population size, *the growth rate per unit population size* is defined. Typically when the population size is given by the density of the number of individuals in the population, it is called *growth rate per individual\seeper capita growth rate* or *per capita growth rate*. For the reasonable modeling, it is very important to distinguish the growth rate per individual from the net growth rate. The latter indicates the velocity of the population size itself, which can be affected by the *ecological disturbance* (for example, natural disaster like fire, flood, etc. or human interference). The former usually indicates the velocity averaged over the population, which must reflect the contribution of every individual to the change of the population size.

$\quad$ Now $N(t)$ denotes the population size (density) at time t. The change of population size $\Delta N = N(t + \Delta t) - N(t)$ in a period Δt after time t could be decomposed into the above-mentioned four elements, the increase by the natural reproduction $\mathcal{B}_{\Delta t}$, the decrease by the natural death $\mathcal{D}_{\Delta t}$, the immigration from the outside of population $\mathcal{I}_{\Delta t}$, and the emigration to the outside of population $\mathcal{E}_{\Delta t}$ as follows:

$$\Delta N = \mathcal{B}_{\Delta t} - \mathcal{D}_{\Delta t} + \mathcal{I}_{\Delta t} - \mathcal{E}_{\Delta t}. \tag{3.24}$$

The mean change of population size *per unit time* in the period Δt after time t is given by

$$\frac{\Delta N}{\Delta t} = \frac{N(t + \Delta t) - N(t)}{\Delta t}.$$

From the definition of the differential, we know that

$$\frac{\Delta N}{\Delta t} \longrightarrow \frac{dN(t)}{dt} \qquad (\Delta t \to 0),$$

which means the *momental velocity* for the change of population size $N(t)$ at time t. On the other hand, the mean change of population size *per individual* in the period Δt after time t can be defined by

$$\frac{\Delta N}{N(t)} = \frac{N(t + \Delta t) - N(t)}{N(t)},$$

which indicates the averaged individual contribution to the change of population size in the period Δt after time t. For a bacteria or another microorganism population cultivated in vitro under a suitable condition, ΔN becomes positive, since it is a closed population with negligible effect of the natural death. In such a case, the mean value $\Delta N / N(t)$ can be regarded as the mean number of newborns produced per individual in the period Δt after time t.

Dividing the mean change per individual $\Delta N / N(t)$ by Δt, we get the mean change of population size per individual per unit time in the period Δt after time t, and can define the *per capita momental velocity* for the change of population size

$$\frac{\Delta N / N(t)}{\Delta t} \longrightarrow \frac{1}{N(t)} \frac{dN(t)}{dt}$$

with the limit $\Delta t \to 0$.

As these arguments, to construct a reasonable continuous model for a population dynamics with the differential equation, it is necessary to distinguish the factor to change the population size in the population level from that in the individual level. The former factor in the population level must be related to the mean change per unit time $\Delta N / \Delta t$, while the latter in the individual level must be to the mean change per individual per unit time $\Delta N / N(t) / \Delta t$. From this standpoint, the former factor should be introduced in the model with a relation to the momental velocity dN/dt, while the latter should be done with a relation to $(1/N)dN/dt$, taking account of four elements to determine the change of population size as indicated by (3.24).

From the next chapter, we are going to focus on the reasonability in the continuous time modeling for a population dynamics, where the importance of such distinction about the momental velocity for the change of population size will be significant.

Answer to Exercise

Exercise 3.1 (p. 105)

For Ricker model in Sect. 2.1.2:

$$x_{n+1} = \mathscr{R}_0 x_n e^{-\gamma x_n},$$

we can define the function φ as

$$\varphi(x(t); h) = \left\{ \mathscr{R}_0(h) e^{-\gamma(h)x(t)} - 1 \right\} x(t),$$

where we suppose that $\mathscr{R}_0$ and γ are functions of h as before. Further, in the way similar to the case of Beverton-Holt model, we can assume from the constraint for the function φ that $\mathscr{R}_0 \to 1$ and $\gamma \to 0$ as $h \to 0$. Lastly, we can obtain the following logistic equation again from (3.10):

$$\frac{dx(t)}{dt} = \left\{ \mathscr{R}_0'(0) - \gamma'(0)x(t) \right\} x(t).$$

References

1. R. Pearl, *The Biology of Death* (Lippincott, Philadelphia, 1922)
2. R. Pearl, The growth of populations. Quarter. Rev. Biol. **2**, 532–548 (1927)
3. R. Pearl, *The Rate of Living* (Alfred A. Knopf, New York, 1928)
4. P.F. Verhulst, Notice sur la loi que la population suit dans son accroissement. Correspond. Math. Phys. **10**, 113–121 (1838)
5. P.F. Verhulst, Recherches mathématiques sur la loi d'accroissement de la population. Nouv. mém. de l'Academie Royale des Sci. et Belles-Lettres de Bruxelles **18**, 1–41 (1845)
6. P.F. Verhulst, Deuxi'eme mêmoire sur la loi d'accroissement de la population. Mém. de l'Academie Royale des Sci., des Lettres et des Beaux-Arts de Belgique **20**, 1–32 (1847)

Chapter 4
Continuous Time Modeling for Birth and Death Processes

Abstract In this chapter, we shall describe the fundamental modeling with the birth-death stochastic process. The idea and concept in such a modeling is very important to understand the meaning of modeling about the deterministic mathematical model which superficially seems not to have any relation to the stochasticity in the phenomenon. In almost all modeling for the population dynamics, the deterministic model could be actually regarded as an approximation to describe an important aspect about the phenomenon. To understand the reasonability of the deterministic structure introduced in the model, it is necessary and useful to know its relation to a stochastic process. For this reason, this chapter contains the idea and concept essential throughout the contents in this book. As the simplest and most important stochastic process, the Poisson process is introduced and used in some parts of this book. Chapter 15 of Part II serves to provide the mathematical fundamentals about it.

4.1 Yule-Furry Process

In this section, we shall consider the population dynamics only with a birth process, like a clonal or vegetative reproduction. Death is neglected here, and the population size necessarily increases as time passes. Such a birth process we shall consider here is called *Yule-Furry process* (simply, Yule, or Furry process) in the stochastic process theory [2, 3].

4.1.1 Probability Distribution for Population Size

Let consider the probability $P(n, t)$ that the number of individuals is n at time t in a population. We now assume the following probability that a new individual (newborn) is generated in it during the period $[t, t + \Delta t]$ when the number of

© The Author(s), under exclusive license to Springer Nature Singapore Pte Ltd. 2022   109
H. Seno, *A Primer on Population Dynamics Modeling*, Theoretical Biology,
https://doi.org/10.1007/978-981-19-6016-1_4

individuals is n at time t:

$$\binom{n}{1} \times \{\beta \Delta t + o(\Delta t)\} = n\beta \Delta t + o(\Delta t),$$

where the first factor $\binom{n}{1} = {}_nC_1 = n$ gives the ways of which present individual (parent) produces the newborn. Positive parameter β represents the likeliness of the reproduction, whereas it does not mean the probability by itself. This mathematical modeling assumes that every present individual has the probability $\beta \Delta t + o(\Delta t)$ to produce a newborn in $[t, t + \Delta t]$. The reader may image a population of cells in a clonal culture, in which the reproduction is caused by the cell division. The stochastic process of birth with the above probability is classified to the *non-homogeneous Poisson process* with intensity βn where the value of n temporally changes (refer to Sect. 15.1).

Let us consider the initial condition given by

$$P(n, 0) = \delta_{n,n_0} = \begin{cases} 1 & (n = n_0 > 0); \\ 0 & (n \neq n_0), \end{cases}$$

where δ_{n,n_0} is the Kronecker delta, and n_0 is the initial population size at $t = 0$. With the same procedure as that described in Sect. 15.2 about Poisson distribution, we can derive the following system of ordinary differential equations with respect to the temporal change of $P(n, t)$:

$$\frac{dP(n_0, t)}{dt} = -\beta n_0 P(n_0, t)$$

$$\frac{dP(n, t)}{dt} = -\beta n P(n, t) + \beta(n - 1)P(n - 1, t) \qquad (n > n_0), \tag{4.1}$$

where we remark that mathematically $P(n, t) = 0$ for any $n < n_0$ and $t > 0$ because the population size never decreases with the assumption of no death. We shall apply here the method of *probability-generating function* to get the solution of the above system of ordinary differential equations for the probability distribution $\{P(n, t)\}$ [1]. Let us define the following probability-generating function for $\{P(n, t)\}$:

$$F(s, t) = \sum_{n=n_0}^{\infty} P(n, t)s^n, \tag{4.2}$$

where s is a dummy variable. From the equations in (4.1), we can derive the partial differential equation with respect to $F(s, t)$:

$$\frac{\partial}{\partial t} F(s, t) = \beta s (s - 1) \frac{\partial}{\partial s} F(s, t), \tag{4.3}$$

with the initial condition that $F(s, 0) = s^{n_0}$, and the boundary condition that $F(1, t) = 1$. The boundary condition is derived from the fact that the sum of probability $P(n, t)$ becomes 1 for any time t.

The solution of the above partial differential equation (4.2) with those initial and boundary conditions is given as follows, while we skip here the detail of calculation to derive it:

$$F(s, t) = e^{-n_0 \beta t} s^{n_0} \left[1 - (1 - e^{-\beta t}) s \right]^{-n_0} \tag{4.4}$$

$$= \sum_{n=n_0}^{\infty} (-1)^{n-n_0} \binom{-n_0}{n - n_0} e^{-n_0 \beta t} (1 - e^{-\beta t})^{n-n_0} s^n$$

$$= \sum_{n=n_0}^{\infty} \binom{n - 1}{n - n_0} e^{-n_0 \beta t} (1 - e^{-\beta t})^{n-n_0} s^n, \tag{4.5}$$

where we used the series expansion

$$(1 + x)^{-n} = \sum_{k=0}^{\infty} \binom{-n}{k} x^k = \sum_{k=0}^{\infty} \binom{n + k - 1}{k} (-x)^k$$

and the relation to what is called negative binomial coefficient

$$\binom{-a}{k} = \frac{(-a)(-a - 1)(-a - 2) \cdots (-a - k + 1)}{k!} = (-1)^k \binom{a + k - 1}{k}.$$

From (4.2) and (4.5), we have

$$P(n, t) = \frac{(n - 1)!}{(n_0 - 1)!(n - n_0)!} e^{-n_0 \beta t} (1 - e^{-\beta t})^{n-n_0}, \tag{4.6}$$

which indicates a *negative binomial distribution* or *Pascal distribution*.

For example, the negative binomial distribution is mathematically equivalent to the probability distribution with respect to the least number of coin tosses by which the face appears n_0 times. The probability is for the number

(continued)

of the appearances of back, which is a stochastic variable. In the above expression (4.6), n is the least number of coin tosses, $n - n_0$ is the number of the appearances of back, when the probability of the appearance of face at each toss is given by $e^{-\beta t}$.

4.1.2 Expected Population Size

In this section, let us consider the expected population size at time t:

$$\langle n \rangle_t := \sum_{n=n_0}^{\infty} n P(n, t). \tag{4.7}$$

For the probability-generating function $F(s, t)$ defined by (4.2), we have

$$\frac{\partial F(s, t)}{\partial s}\bigg|_{s=1} = \sum_{n=n_0}^{\infty} n P(n, t).$$

Hence, from the partial derivative of (4.4) in terms of s, we can easily derive $\langle n \rangle_t = n_0 e^{\beta t}$. Therefore, the expected population size $\langle n \rangle_t$ is exponentially increasing as time passes.

This result can be obtained in a different way. We can derive the following ordinary differential equation with respect to $\langle n \rangle_t$ from the system (4.1) with (4.7) (Exercise 4.1):

$$\frac{d \langle n \rangle_t}{dt} = \beta \langle n \rangle_t. \tag{4.8}$$

Since the initial value is given by $\langle n \rangle_0 = n_0$, the solution becomes $\langle n \rangle_t = n_0 e^{\beta t}$.

Exercise 4.1 Derive the ordinary differential equation (4.8) from the system (4.1) with (4.7).

The result of exponential growth of the expected population size $\langle n \rangle_t$ indicates that it follows *Malthus growth* with the *malthusian coefficient* β (The more detail description about *Malthus model* from the viewpoint of modeling will be given in Sect. 5.1). In the next section, we shall see that Malthus growth of the expected population size $\langle n \rangle_t$ appears even when the population dynamics is accompanied by a death process, although we neglected the death in this section.

4.2   Malthus Growth with Death

In this section, we introduce the death process to Yule-Furry process described in the previous section. As before, we give the probability that the number of individuals becomes $n + 1$ in $[t, t + \Delta t]$ when it is n at time t by $n\beta \Delta t + \mathrm{o}(\Delta t)$. Positive parameter β has the same meaning as before. Now in addition, we introduce the probability that the number of individuals becomes $n - 1$ in $[t, t + \Delta t]$ when it is n at time t by $n\mu \Delta t + \mathrm{o}(\Delta t)$. Positive parameter μ represents the likeliness of the death. In this birth-and-death process, the parameter β may be called *birth rate*, *coefficient of birth*, or *proliferation rate*. Correspondingly μ may be called *death rate* or *coefficient of death*. Note again that neither μ nor β means the probability by itself.

Taking account of the possibility to lead the extinction due to the death process, we can derive the following recurrence relations about the probability $P(n, t)$ according to the change of population size in $[t, t + \Delta t]$:

$$P(0, t + \Delta t) = P(0, t) + \{\mu \Delta t + \mathrm{o}(\Delta t)\} P(1, t) + \mathrm{o}(\Delta t);$$

$$P(1, t + \Delta t) = [1 - \{\beta \Delta t + \mathrm{o}(\Delta t)\}][1 - \{\mu \Delta t + \mathrm{o}(\Delta t)\}] P(1, t)$$
$$+ [1 - \{\beta \Delta t + \mathrm{o}(\Delta t)\}]\{2\mu \Delta t + \mathrm{o}(\Delta t)\} P(2, t) + \mathrm{o}(\Delta t);$$

$$P(n, t + \Delta t) = [1 - \{\beta n \Delta t + \mathrm{o}(\Delta t)\}][1 - \{\mu n \Delta t + \mathrm{o}(\Delta t)\}] P(n, t)$$
$$+ [1 - \{\beta (n + 1)\Delta t + \mathrm{o}(\Delta t)\}]\{\mu (n + 1)\Delta t + \mathrm{o}(\Delta t)\} P(n + 1, t)$$
$$+ \{\beta (n - 1)\Delta t + \mathrm{o}(\Delta t)\}[1 - \{\mu (n - 1)\Delta t + \mathrm{o}(\Delta t)\}] P(n - 1, t)$$
$$+ \mathrm{o}(\Delta t) \qquad (n = 2, 3, \dots).$$

Taking the limit as $\Delta t \to 0$, we can derive the following system of ordinary equations:

$$\frac{dP(0, t)}{dt} = \mu P(1, t);$$

$$\frac{dP(1, t)}{dt} = -(\beta + \mu) P(1, t) + 2\mu P(2, t);$$

$$\frac{dP(n, t)}{dt} = -(\beta + \mu)n P(n, t) + (n + 1)\mu P(n + 1, t) + (n - 1)\beta P(n - 1, t)$$
$$(n = 2, 3, \dots).$$
$$(4.9)$$

In the same way as to derive (4.8) in the previous section, we can derive the following ordinary differential equation with respect to $\langle n \rangle_t$:

$$\frac{d\langle n \rangle_t}{dt} = (\beta - \mu)\langle n \rangle_t, \qquad (4.10)$$

where

$$\langle n \rangle_t := \sum_{n=0}^{\infty} n P(n, t). \tag{4.11}$$

Therefore, from (4.9), the temporal change of the expected population size $\langle n \rangle_t$ is regarded as Malthus growth with the malthusian coefficient $\beta - \mu$. Note that the population size may decrease now by the death process, so that the expected population size must be defined with $P(n, t)$ for all n as given by (4.11).

4.3 Death Process

In this section, we focus on a *cohort* which exists at $t = 0$ with n_0 individuals, and track the number of survived individuals in it as time passes. Although the cohort is usually defined as a collection of individuals with the same age, we define it here in a wider sense such that a collection of individuals who always has the same likeliness of death even if there would be a difference of age among them. The cohort necessarily becomes smaller due to the death process as time passes. We assume the probability that an individual dies in $[t, t + \Delta t]$ when the individual is alive at time t, given by $\mu \Delta t + o(\Delta t)$. The positive parameter μ is the death rate as before. Such a death process with the above probability is classified to the *homogeneous Poisson process* with intensity μ (refer to Sect. 15.1).

4.3.1 Survival Probability

Let us denote the probability that an individual survives during $[0, t]$ by $Q(t)$. The probability $Q(t + \Delta t)$ that an individual survives during $[0, t + \Delta t]$ must satisfy the following relation:

$$Q(t + \Delta t) = \left[1 - \{\mu \Delta t + o(\Delta t)\}\right] Q(t)$$

with the survival probability $1 - \{\mu \Delta t + o(\Delta t)\}$ during $[t, t + \Delta t]$. Hence we have

$$\frac{Q(t + \Delta t) - Q(t)}{\Delta t} = \left\{-\mu + \frac{o(\Delta t)}{\Delta t}\right\} Q(t),$$

and can get the following ordinary equation by taking the limit as $\Delta t \to 0$:

$$\frac{dQ(t)}{dt} = -\mu Q(t). \tag{4.12}$$

For the above assumption, every individual in the cohort is alive at $t = 0$, so that the probability that any individual survives at $t = 0$ must be 1, that is, $Q(0) = 1$. This gives the initial condition for the above ordinary equation. We can easily get the solution of (4.12) as $Q(t) = e^{-\mu t}$.

Exercise 4.2 When the death rate μ is a sufficiently smooth function of t, $\mu = \mu(t)$, find that the survival probability $Q(t)$ is given by $Q(t) = e^{-\langle\mu\rangle_t t}$ with the time-averaged death rate

$$\langle\mu\rangle_t := \frac{1}{t}\int_0^t \mu(\tau)d\tau.$$

4.3.2   Expected Life Span

Next let us consider the probability that an individual dies in $[t, t+\Delta t]$. It is given by the product of the survival probability $Q(t)$ during $[0, t]$ and the death probability $\mu\Delta t + o(\Delta t)$ that an individual dies during $[t, t + \Delta t]$ when the individual is alive at time t:

$$Q(t)\{\mu\Delta t + o(\Delta t)\} = e^{-\mu t}\mu\Delta t + o(\Delta t).$$

Therefore, the expected life span $\langle t \rangle$ for the cohort is mathematically calculated as follows:

$$\langle t \rangle = \int_0^\infty t \cdot e^{-\mu t}\mu\, dt = -\int_0^\infty t \frac{d}{dt}\{e^{-\mu t}\}\, dt$$

$$= \left[-t\,e^{-\mu t}\right]_0^\infty + \int_0^\infty e^{-\mu t}\, dt = \frac{1}{\mu}.$$

Consequently, the expected life span is equal to the inverse of the death rate μ.

In this mathematical argument, the function $f(t) = \mu e^{-\mu t}$ may be called *probability density function* for the life span, and we can define the *probability [cumulative] distribution function* by

$$F(t) = \int_0^t f(\tau)\, d\tau = 1 - e^{-\mu t}. \tag{4.13}$$

The function $F(t)$ means the probability that an individual dies until time t. The probability distribution function (4.13) indicates an exponential distribution. Therefore, the life span under the death process with a homogeneous Poisson process follows an exponential distribution.

The above probability distribution function $F(t)$ can be derived in a different way. From the modeling about the death process considered now, the probability $F(t)$ that an individual dies until time t must satisfy the following recurrence relation:

$$F(t + \Delta t) = F(t) + \{\mu \Delta t + o(\Delta t)\}\{1 - F(t)\}.$$

With the limit as $\Delta t \to 0$, we can derive the following ordinary differential equation:

$$\frac{dF(t)}{dt} = \mu\{1 - F(t)\}. \tag{4.14}$$

Since any individual is alive at $t = 0$, the initial condition is given by $F(0) = 0$. The solution of (4.14) with $F(0) = 0$ becomes (4.13).

When the death rate μ is a suffuciently smooth function of t, $\mu = \mu(t)$, we can derive the following equation with respect to the expected life span $\langle t \rangle$:

$$\langle t \rangle = \int_0^\infty t f(t) dt = \int_0^\infty t \cdot e^{-\langle \mu \rangle_t t} \mu(t) dt$$

$$= \int_0^\infty t \mu(t) \exp\left[-\int_0^t \mu(\tau) d\tau \right] dt = -\int_0^\infty t \frac{d}{dt}\left\{ \exp\left[-\int_0^t \mu(\tau) d\tau \right] \right\} dt$$

$$= \left[-t \exp\left[-\int_0^t \mu(\tau) d\tau \right] \right]_0^\infty + \int_0^\infty \exp\left[-\int_0^t \mu(\tau) d\tau \right] dt,$$

where we used the result of Exercise 4.2. If

$$\lim_{t \to \infty} t \exp\left[-\int_0^t \mu(\tau) d\tau \right] = 0,$$

then we have

$$\langle t \rangle = \int_0^\infty \exp\left[-\int_0^t \mu(\tau) d\tau \right] dt = \int_0^\infty e^{-\langle \mu \rangle_t t} dt. \tag{4.15}$$

Exercise 4.3 Derive the expected life span $\langle t \rangle$ for each of the following death rates $\mu = \mu(t)$:

(a) $\mu(t) = mt$ with a positive constant m.
(b) The following periodic function of t with period h:

$$\mu(t) = \begin{cases} \mu_1 & \text{for } t \in [kh, kh + \theta h); \\ \mu_2 & \text{for } t \in [kh + \theta h, (k+1)h) \end{cases} \quad (k = 0, 1, 2, \ldots),$$

where μ_i $(i = 1, 2)$, h and θ are positive constants with $\theta \leq 1$.

4.3.3 Probability Distribution for Cohort Size

Along the same framework of mathematical modeling, we introduce here the following probability that the death of an individual occurs in $[t, t + \Delta t]$ when the number of individuals in the cohort is n:

$$\binom{n}{1} \times \{\mu \Delta t + o(\Delta t)\} = n\mu \Delta t + o(\Delta t). \tag{4.16}$$

As for the probability $P(n, t)$ that the number of individuals in the cohort is n at time t, we can derive the following system of ordinary differential equations in the same way as for Yule-Furry process in Sect. 4.1.

$$\frac{dP(n_0, t)}{dt} = -\mu n_0 P(n_0, t);$$

$$\frac{dP(n, t)}{dt} = -\mu n P(n, t) + \mu(n+1)P(n+1, t) \quad (0 < n < n_0);$$

$$\frac{dP(0, t)}{dt} = \mu P(1, t),$$

$$\tag{4.17}$$

with the initial condition that $P(n, 0) = \delta_{n, n_0}$. Differently from Yule-Furry process, the system (4.17) consists of $n_0 + 1$ equations.

From the first equation of (4.17) and the initial condition, we can easily obtain the solution

$$P(n_0, t) = e^{-n_0 \mu t}.$$

In the same way of derivation as that for Poisson distribution in Sect. 15.2, we substitute

$$P(n_0 - 1, t) = u_{n_0-1}(t)e^{-\mu(n_0-1)t}$$

for the second equation of (4.17). Again with the initial condition, we can derive the following equation for $u_{n_0-1}(t)$:

$$\frac{du_{n_0-1}}{dt} = \mu n_0 e^{-\mu t},$$

and get $u_{n_0-1}(t) = n_0(1 - e^{-\mu t})$. Thus we find

$$P(n_0 - 1, t) = n_0(1 - e^{-\mu t})e^{-\mu(n_0-1)t}.$$

Applying the same procedure, we can find

$$\frac{du_{n_0-2}}{dt} = \mu n_0(n_0 - 1)(1 - e^{-\mu t})e^{-\mu t} = \frac{n_0(n_0 - 1)}{2}\frac{d}{dt}(1 - e^{-\mu t})^2,$$

and subsequently

$$P(n_0 - 2, t) = \frac{n_0(n_0 - 1)}{2}(1 - e^{-\mu t})^2 e^{-\mu(n_0-2)t}.$$

In this way, we can prove by the mathematical induction that

$$P(n, t) = \binom{n_0}{n}(1 - e^{-\mu t})^{n_0-n}e^{-n\mu t} \tag{4.18}$$

for $n = 1, 2, \ldots, n_0$.

Especially, from the third equation of (4.17), the initial condition, and the formula of $P(1, t)$, we can find that $P(0, t) = (1 - e^{-\mu t})^{n_0}$. Hence, with the conventional mathematical definitions $0! = 1$ and $\binom{n}{0} = 1$, the result (4.18) gives every $P(n, t)$ for $n = 0, 1, 2, \ldots, n_0$.

The probability distribution (4.18) of $\{P(n, t)\}$ is a *binomial distribution*. Especially when $n_0 = 1$, it can be called *Bernoulli distribution*. It is the probability distribution for the number of faces in n_0 coin tosses. For the case of the members in a cohort, each member has alternative status, alive or dead. For this reason, the binomial distribution could be acceptable.

From the Eq. (4.18) in the extremal case of $n_0 = 1$, the survival probability of an individual $Q(t)$ in Sect. 4.3.1 is easily obtained as $e^{-\mu t}$, since $Q(t) = P(1, t)$ from their definitions.

Exercise 4.4 Derive the probability distribution $\{P(n, t)\}$ for the time-dependent death rate $\mu = \mu(t)$.

4.3.4 Expected Population Size

We can easily derive the ordinary differential equation

$$\frac{d \langle n \rangle_t}{dt} = -\mu \langle n \rangle_t$$

with respect to the expected population size of the cohort at time t (see Excercise 4.1):

$$\langle n \rangle_t := \sum_{n=0}^{n_0} n P(n, t).$$

With the initial condition $\langle n \rangle_0 = n_0$, we find $\langle n \rangle_t = n_0 e^{-\mu t}$. Hence the cohort size is exponentially decreasing toward the extinction. It may be regarded as Malthus growth with a negative malthusian coefficient.

4.3.5 Average Life Span for Extinct Population

In this section, we track a collection of individuals, say, a population which has gone extinction with Malthus growth with the malthusian coefficient $-\nu$:

$$N(t) = N(0)e^{-\nu t}, \tag{4.19}$$

and derive the average [mean] life span $\bar{t}$ over the population to show that it is given by $1/\nu$.

Differently from the consideration on a cohort in the previous sections, we do not care of the physiological difference in the collection of individuals now. The members may be different in their nature about the death. We will consider the average of the life span $\bar{t}$ from a certain moment indexed by $t = 0$ over the collection of individuals which goes extinct in an exponential manner, neglecting what causes the death. For this reason, the argument in this section must be distinguished from that in the previous sections on the stochastic process with the death rate μ.

Since the population becomes extinct at the end after the monotonic decrease, it is determined when each member in it dies. That is, the deaths are distributed on the time line. In other words, the collection of the moments of the death leads to the distribution of life span in the population.

Now the number of deaths between t and $t + \Delta t$ is given by

$$N(t) - N(t + \Delta t) = N(0)\{e^{-vt} - e^{-v(t+\Delta t)}\}. \tag{4.20}$$

Thus, the ratio of deaths to the initial population size $N(0)$ becomes

$$\frac{N(t) - N(t + \Delta t)}{N(0)} = e^{-vt} - e^{-v(t+\Delta t)}, \tag{4.21}$$

which can be regarded as the probability that a randomly chosen member in the initial population has the life span between t and $t + \Delta t$.

Let us consider the frequency density distribution of life span $f(t)$ and the cumulative frequency distribution $F(t)$:

$$F(t) = \int_0^t f(\tau)d\tau. \tag{4.22}$$

$F(t)$ gives the frequency of members who have the life span not beyond t. It satisfies that $F(0) = 0$ and

$$\lim_{t \to \infty} F(t) = \int_0^\infty f(\tau)\, d\tau = 1.$$

This limit indicates that every member must die at a finite moment. The frequency density distribution $f(t)$ has the relation to $F(t)$ as given by (4.22), and the following relation at the same time:

$$f(t) = \frac{dF(t)}{dt}.$$

The frequency density distribution $f(t)$ and the cumulative frequency distribution $F(t)$ are mathematically equivalent to the probability density function and the probability [cumulative] distribution function defined in Sect. 4.3.2. However, the present argument is not based on any given probability distribution, but regard the emerging phenomenon as a stochastic process with a probability distribution. Malthus growth (4.19) is an emerging phenomenon here given a priori, and we will derive the distributions $f(t)$ and $F(t)$ from it.

From the definition of the cumulative frequency distribution, the difference $F(t + \Delta t) - F(t)$ means the frequency of members who have the life span between t and $t + \Delta t$. Thus, it is equal to (4.21), and we have

$$\frac{F(t + \Delta t) - F(t)}{\Delta t} = -\frac{e^{-\nu(t+\Delta t)} - e^{-\nu t}}{\Delta t}.$$

With the limit as $\Delta t \to 0$, we obtain the differential equation

$$\frac{dF(t)}{dt} = -\frac{d}{dt} e^{-\nu t},$$

which immediately gives

$$f(t) = \nu e^{-\nu t}. \tag{4.23}$$

It is easy to confirm that the improper integral of (4.23) over $[0, \infty)$ converges to 1. This means that the frequency density distribution is an exponential distribution. The cumulative frequency distribution $F(t)$ is obtained by integrating (4.23) as

$$F(t) = 1 - e^{-\nu t}. \tag{4.24}$$

The average life span $\bar{t}$ can be calculated with the frequency density distribution (4.23) by

$$\bar{t} := \int_0^\infty t f(t) \, dt,$$

and we find $\bar{t} = 1/\nu$.

Exercise 4.5 Derive the formula to give the average life span $\bar{t}$ when the population size N is decreasing to become extinct as

$$N(t) = N(0) e^{-\int_0^t \nu(\tau) d\tau} \tag{4.25}$$

with the malthusian coefficient $-\nu = -\nu(t)$ which temprally varies, keeping negative for any $t > 0$.

More generally, we may consider the case where the population size monotonically decreases toward the extinction as $N(t) = N(0)g(t)$. The function $g(t)$ of time t is now assumed to be monotonically decreasing in terms of t and satisfy that $g(0) = 1$ and $\lim\limits_{t\to\infty} g(t) = 0$. We further assume that $g(t)$ is smooth enough to be continuous and differentiable in terms of t. This assumption may be regarded as a reasonable mathematical simplification about the modeling for such an extinctive population.

Along the same arguments as given in the above, the number of deaths between t and $t + \Delta t$ is given by

$$N(t) - N(t + \Delta t) = N(0)\{g(t) - g(t + \Delta t)\},$$

and we can find

$$\frac{dF(t)}{dt} = -\frac{dg(t)}{dt}.$$

From the conditions $F(0) = 0$ and $g(0) = 1$, we obtain $F(t) = 1 - g(t)$. On the other hand, the above differential equation means that $f(t) = -dg(t)/dt$. Therefore, the average life span $\bar{t}$ must be mathematically defined by

$$\bar{t} := \int_0^\infty t f(t)\, dt = -\int_0^\infty t\, \frac{dg(t)}{dt}\, dt = \left[-tg(t) \right]_0^\infty + \int_0^\infty g(t)\, dt.$$

Depending on the nature of the function g, the above calculation may result in the positively infinite. In such a case, the average life span $\bar{t}$ is infinite. Some may hardly understand this situation, since the infinite life span means being immortal. From the viewpoint of mathematical modeling, the mathematical limit as $t \to \infty$ may not be regarded as the infinity of real time. Frequently it is regarded as corresponding to a mathematical approximation to the state after a sufficiently long time. In this sense, the infinity of average life span itself does not mean nothing, whereas it may be regarded as the theoretical implication that the average life span is much long, or alternatively it would not be decidable from the standard definition. Even in such a case of infinite life span, the reasonability of the modeling for the death process cannot be denied only from it, and can be acceptable.

4.3.6 Expected Extinction Time

Let us consider next the expected extinction time $\langle T_e \rangle$ at which a cohort goes extinct, that is, the expected life span of cohort. It is equivalent to the expected moment at which the last alive member of the cohort dies. Since the probability that a death occurs in $[t, t + \Delta t]$ with a sufficiently short interval Δt is assumed to be given by $\mu n \Delta t + o(\Delta t)$, the probability that more than one deaths occur in $[t, t + \Delta t]$ is given by $o(\Delta t)$. Hence, the probability that the cohort becomes extinct in $[t, t + \Delta t]$ is given by $\mu \Delta t + o(\Delta t)$ when the cohort size is 1 at time t, while it is given by $o(\Delta t)$ when the cohort size is more than 1 at time t. From these arguments, we can find the following mathematical expression of the probability that the cohort becomes extinct in $[t, t + \Delta t]$:

$$P(1, t) \cdot \{\mu \Delta t + o(\Delta t)\} + \sum_{k=2}^{\infty} P(k, t)o(\Delta t) = n_0(1 - e^{-\mu t})^{n_0 - 1} e^{-\mu t} \mu \Delta t + o(\Delta t).$$

For the initial cohort size n_0, the expected extinction time $\langle T_e \rangle_{n_0}$ can be defined as follows by taking the average in terms of time t with the above probability:

$$\langle T_e \rangle_{n_0} = \int_0^{\infty} t \cdot n_0 (1 - e^{-\mu t})^{n_0 - 1} e^{-\mu t} \mu \, dt \quad \left[= \int_0^{\infty} t \frac{d}{dt} (1 - e^{-\mu t})^{n_0} dt \right]$$

This integral can be actually calculated as follows:

$$\langle T_e \rangle_{n_0} = \int_0^{\infty} t \cdot n_0 \left[\sum_{k=0}^{n_0-1} \binom{n_0 - 1}{k} (-e^{-\mu t})^k \right] e^{-\mu t} \mu \, dt$$

$$= n_0 \mu \sum_{k=0}^{n_0-1} \binom{n_0 - 1}{k} (-1)^k \left\{ \int_0^{\infty} t e^{-(k+1)\mu t} dt \right\}$$

$$= n_0 \mu \sum_{k=0}^{n_0-1} \binom{n_0 - 1}{k} (-1)^k \frac{1}{(k + 1)^2 \mu^2}$$

$$= \frac{1}{\mu} \sum_{k=0}^{n_0-1} \binom{n_0}{k + 1} (-1)^k \frac{1}{k + 1} = \frac{1}{\mu} \sum_{k=1}^{n_0} \binom{n_0}{k} (-1)^{k+1} \frac{1}{k}. \tag{4.26}$$

This result can be proved to be equivalent to

$$\langle T_e \rangle_{n_0} = \frac{1}{\mu} \sum_{k=1}^{n_0} \frac{1}{k}. \tag{4.27}$$

Exercise 4.6 Prove the equivalence of (4.26) to (4.27).

The expression (4.27) can be derived in a different way which would have a wide applicability. Firstly let us consider the expected duration $\langle T_1 \rangle_{n_0}$ that the initial cohort size n_0 becomes $n_0 - 1$. This is the expected time at which the first death occurs. Suppose that the death does not occur until time t but does in $[t, t + \Delta t]$. The probability is given by

$$P(n_0, t) \cdot \left\{ \mu \Delta t \times \binom{n_0}{1} + \mathrm{o}(\Delta t) \right\} = \mathrm{e}^{-n_0 \mu t} n_0 \mu \Delta t + \mathrm{o}(\Delta t).$$

Hence the expected duration $\langle T_1 \rangle_{n_0}$ can be derived as follows:

$$\langle T_1 \rangle_{n_0} = \int_0^\infty t \, \mathrm{e}^{-n_0 \mu t} n_0 \mu \, dt = -\int_0^\infty t \frac{d}{dt} \mathrm{e}^{-n_0 \mu t} dt = \frac{1}{n_0 \mu}.$$

Next, the time from the moment that the cohort size becomes $n_0 - 1$ to the moment of extinction is the same as the expected extinction time $\langle T_e \rangle_{n_0 - 1}$ for the initial cohort size $n_0 - 1$. Since the duration until the first death occurs for the initial cohort with size n_0 and the time until the cohort extinction from the moment of the first death are independent of each other, we can find the following relation:

$$\langle T_e \rangle_{n_0} = \langle T_1 \rangle_{n_0} + \langle T_e \rangle_{n_0 - 1} = \frac{1}{n_0 \mu} + \langle T_e \rangle_{n_0 - 1}.$$

This can be regarded as the recurrence relation to determine the sequence $\{\langle T_e \rangle_1, \langle T_e \rangle_2, \ldots, \langle T_e \rangle_k, \ldots\}$. It is not difficult to derive (4.27) from this recurrence relation.

4.4 Net Reproduction Rate

In this section, we consider the net reproduction rate for the population with a Malthus growth of Sect. 4.2. Since the death and birth are assumed independent of each other, the net reproduction rate can be uniquely defined for any individual alive at each time t. We shall consider an individual at the initial time $t = 0$. From the assumption given in Sect. 4.2, the probability that an individual alive at $t = 0$ makes reproduction (e.g., cell division) to produce a newborn in $[t, t + \Delta t]$ is given by $\beta \Delta t + \mathrm{o}(\Delta t)$ with the birth rate β (refer also to Yule-Furry process in Sect. 4.1). It must be remarked that the probability of the reproduction to produce more than one newborns is now given by $o(\Delta t)$. On the other hand, from the arguments in Sect. 4.3, the probability that the individual dies in $[t, t + \Delta t]$ is given by $\mu \Delta t + o(\Delta t)$ with the death rate μ, and the probability to survive until time t by $Q(t) = e^{-\mu t}$.

Let $p(k, t)$ denote the probability that an individual alive at $t = 0$ survives and produces k offsprings until time t. In the way same as that for Malthus growth in Sect. 4.2, we can find the following relations:

$$p(0, t + \Delta t) = \left[1 - \{\beta \Delta t + o(\Delta t)\}\right]\left[1 - \{\mu \Delta t + o(\Delta t)\}\right]p(0, t) + o(\Delta t);$$

$$p(k + 1, t + \Delta t) = \{\beta \Delta t + o(\Delta t)\}\left[1 - \{\mu \Delta t + o(\Delta t)\}\right]p(k, t)$$
$$+ \left[1 - \{\beta \Delta t + o(\Delta t)\}\right]\left[1 - \{\mu \Delta t + o(\Delta t)\}\right]p(k + 1, t) + o(\Delta t)$$
$$(k = 0, 1, 2, \ldots).$$

With the limit of these relations as $\Delta t \to 0$, we can derive the following system of ordinary differential equations:

$$
\begin{aligned}
\frac{dp(0, t)}{dt} &= -(\beta + \mu)p(0, t); \\
\frac{dp(k + 1, t)}{dt} &= \beta p(k, t) - (\beta + \mu)p(k + 1, t) \qquad (k = 0, 1, 2, \ldots).
\end{aligned}
\tag{4.28}
$$

The initial condition is given by $p(k, 0) = \delta_{k,0}$. It is clear that $p(0, 0) = 1$ and $p(k, 0) = 0$ for any $k > 0$, since the individual is alive at $t = 0$ while there is no reproduction at the moment. This initial value problem can be easily solved (Exercise 4.7):

$$p(k, t) = \frac{(\beta t)^k}{k!} e^{-(\beta + \mu)t}, \tag{4.29}$$

where we used a conventional definition $0! = 1$.

From this result, we can find that

$$\sum_{k=0}^{\infty} p(k, t) = \sum_{k=0}^{\infty} \frac{(\beta t)^k}{k!} e^{-(\beta + \mu)t} = e^{\beta t} \cdot e^{-(\beta + \mu)t} = e^{-\mu t} = Q(t).$$

This indicates that the sum $\displaystyle\sum_{k=0}^{\infty} p(k, t)$ means the probability that the individual survives until time t, independently of how many offsprings it produces, which must be equal to $Q(t)$ from the definition.

Exercise 4.7 Derive (4.29).

With the argument same as that for the expected life span in Sect. 4.3.2, the probability that an individual dies in $[t, t + \Delta t]$ after producing k offsprings is given by

$$\{\mu \Delta t + o(\Delta t)\} p(k, t) = \mu p(k, t) \Delta t + o(\Delta t).$$

Therefore, we find the expected number of offsprings produced by an individual before the death, that is, the net reproduction rate $\mathscr{R}_0$ as follows:

$$\mathscr{R}_0 = \sum_{k=0}^{\infty} \int_0^{\infty} k \cdot \mu p(k, t) \, dt = \sum_{k=0}^{\infty} k \int_0^{\infty} \mu p(k, t) \, dt = \sum_{k=0}^{\infty} k \mathscr{P}(k), \qquad (4.30)$$

where

$$\mathscr{P}(k) := \int_0^{\infty} \mu p(k, t) \, dt$$

is the probability that an individual produces k offsprings before the death, independently of when it dies. From (4.29), we find the followings:

$$\mathscr{P}(0) = \int_0^{\infty} \mu \, e^{-(\beta+\mu)t} \, dt = \frac{\mu}{\beta + \mu}; \qquad \mathscr{P}(k) = \frac{\beta}{\beta + \mu} \mathscr{P}(k-1),$$

and consequently

$$\mathscr{P}(k) = \frac{\mu}{\beta + \mu} \left(\frac{\beta}{\beta + \mu} \right)^k. \qquad (4.31)$$

Finally, substituting (4.31) for (4.30), we can get the net reproduction rate

$$\mathscr{R}_0 = \sum_{k=0}^{\infty} k \frac{\mu}{\beta + \mu} \left(\frac{\beta}{\beta + \mu} \right)^k = \frac{\beta}{\mu}. \qquad (4.32)$$

This result can be understood as $\mathscr{R}_0 = \beta \langle t \rangle$ with the expected life span $\langle t \rangle = 1/\mu$ obtained in Sect. 4.3.2. In this sense, the net reproduction rate is given by the product of the birth rate that means the number of newborns per unit time per individual and the expected life span.

The relation that (the net reproduction rate) = (the number of newborns per unit time per individual) × (the expected life span) cannot be necessarily satisfied if the death rate depends on the reproduction. In the above arguments, it was an important assumption that the death rate is independent of time and the status of individual, and the death is independent of the reproduction.

The logic to derive the net reproduction rate in this section is the same as that for the *basic reproduction number* for the epidemic dynamics (refer to Sect. 9.2.3 and 9.3.4). In other words, the net reproduction rate in the epidemic dynamics may be called basic reproduction number. For the population dynamics of transmissible disease spread, the basic reproduction number is most generally defined as *the expected number of new cases of an infection caused by an infective individual, in a population consisting of susceptible contacts only.* Since the net reproduction rate is defined as the expected number of mature females produced by a mature female (p. 16 in Sect. 1.5), a mature female corresponds to an infective individual, and produced mature females does to new cases of infection in these definitions.

4.5 Logistic Equation

The logistic equation (2.15) can be derived for the density-dependent population growth by a Poisson process. Similarly with the arguments about Malthus growth with a Poisson process in Sect. 4.2, let us assume that the number of individuals in the population becomes $n + 1$ from n in $[t, t + \Delta t]$ with probability $n\beta \Delta t + o(\Delta t)$. Now, differently from Sect. 4.2, we assume that an individual dies in $[t, t + \Delta t]$ with probability $n\mu \Delta t + o(\Delta t)$. In this case, we can introduce the probability that the number of individuals becomes $n-1$ from n in $[t, t+\Delta t]$ as $n^2 \mu \Delta t + o(\Delta t)$. This modeling may be regarded as following the assumption that the individual death is affected by a density effect. Similarly with Malthus growth in Sect. 4.2, from the assumption that the individual death is independent of the death of any other, the death in a sufficiently short interval Δt occurs with probability given by the sum of the probability of the death for every individual, that is, by multiplying the number of alive individuals to the probability of the individual death.

In the similar way of the arguments about Malthus growth with a Poisson process in Sect. 4.2, we can get the following system of ordinary differential equations with respect to the probability $P(n, t)$ that the population size is n at time t:

$$\frac{dP(0, t)}{dt} = \mu P(1, t);$$

$$\frac{dP(1, t)}{dt} = -(\beta + \mu)P(1, t) + 4\mu P(2, t);$$

$$\frac{dP(n, t)}{dt} = -(\beta + n\mu)n P(n, t) + (n + 1)^2 \mu P(n + 1, t) + (n - 1)\beta P(n - 1, t)$$

$$(n = 2, 3, \dots).$$

Further, with respect to the expected number of individuals at time t,

$$\langle n \rangle_t = \sum_{k=0}^{\infty} k P(k, t),$$

it is not difficult to derive the following ordinary differential equation from the above system (refer to Exercise 4.1):

$$\frac{d\langle n \rangle_t}{dt} = \beta \langle n \rangle_t - \mu \langle n^2 \rangle_t = (\beta - \mu \langle n \rangle_t)\langle n \rangle_t - \mu \sigma_t^2,$$

where $\langle n^2 \rangle_t$ means the second moment which is the expected squared value of the population size at time t, $\langle n^2 \rangle_t = \sum_{k=0}^{\infty} k^2 P(k, t)$. Besides, $\sigma_t^2 = \langle n^2 \rangle_t - \langle n \rangle_t^2$ means the variance of the population size at time t. Therefore, if the variance σ_t^2 is small enough to make $\sigma_t^2 \approx 0$, the temporal change of the expected population size $\langle n \rangle_t$ can be approximated well by a logistic equation.

Answer to Exercise

Exercise 4.1 (p. 112)

By differentiate both sides of the definition (4.7) of the expected population size $\langle n \rangle_t$, we have

$$\frac{d\langle n \rangle_t}{dt} = \sum_{n=n_0}^{\infty} n \frac{dP(n, t)}{dt} = n_0 \frac{dP(n_0, t)}{dt} + \sum_{n=n_0+1}^{\infty} n \frac{dP(n, t)}{dt}.$$

Substitution of (4.1) for this equation gives

$$\frac{d\langle n \rangle_t}{dt} = -\beta n_0^2 P(n_0, t) + \sum_{n=n_0+1}^{\infty} n\{-\beta n P(n, t) + \beta(n - 1)P(n - 1, t)\}$$

$$= -\beta n_0^2 P(n_0, t) - \beta \sum_{n=n_0+1}^{\infty} n^2 P(n, t) + \beta \sum_{n=n_0+1}^{\infty} n(n - 1)P(n - 1, t)$$

$$= -\beta n_0^2 P(n_0, t) - \beta \sum_{n=n_0+1}^{\infty} n^2 P(n, t) + \beta \sum_{n=n_0}^{\infty} (n + 1)n P(n, t)$$

$$= -\beta \left\{ n_0^2 P(n_0, t) + \underbrace{\sum_{n=n_0+1}^{\infty} n^2 P(n, t)}_{} \right\} + \beta \sum_{n=n_0}^{\infty} n^2 P(n, t) + \beta \sum_{n=n_0}^{\infty} n P(n, t)$$

$$= \sum_{n=n_0}^{\infty} n^2 P(n,t)$$

$$= \beta \sum_{n=n_0}^{\infty} n P(n, t) = \beta \langle n \rangle_t.$$

Exercise 4.2 (p. 115)

We have $Q(t + \Delta t) = [1 - \{\mu(t)\Delta t + \mathrm{o}(\Delta t)\}] Q(t)$. By the same arguments as in the main text, we can derive the expression of $Q(t)$. When μ is constant independently of time, $\langle \mu \rangle_t$ is equal to the constant μ. As a temporally varying μ, we may consider an example such that the death rate depends on the season, or increases with the environmental degradation.

Exercise 4.3 (p. 117)

(a) From (4.15), we can easily get

$$\langle t \rangle = \left[-t \, \exp\left[-\int_0^t m\tau \, d\tau \right] \right]_0^{\infty} + \int_0^{\infty} \exp\left[-\int_0^t m\tau \, d\tau \right] dt$$

$$= \left[-t \, e^{-mt^2/2} \right]_0^{\infty} + \int_0^{\infty} e^{-mt^2/2} dt = \int_0^{\infty} e^{-mt^2/2} dt = \sqrt{\frac{\pi}{2m}}.$$

As a result about this case, the expected life span $\langle t \rangle$ is inversely proportional to the square root of parameter m. As an example, we may consider the temporal increase of death rate due to the physiological aging. As the other example, we may consider the environmental degradation due to the population's activities. The above result may be translated as an implication that, even if the velocity of the environmental degradation becomes double, the expected life span would become shorter while it is greater than half of the past.

(b) In this case, we can prove that

$$\lim_{t \to \infty} -t \, \exp\left[-\int_0^t \mu(\tau) \, d\tau \right] = 0,$$

since

$$t \exp\left[-\int_0^t \mu(\tau)\, d\tau\right] \le t\, e^{-\mu_m t}$$

with $\mu_m := \min\{\mu_1, \mu_2\}$, and $t\, e^{-\mu_m t} \to 0$ as $t \to \infty$. From (4.15), the expected life span $\langle t \rangle$ can be calculated as follows:

$$\langle t \rangle = \lim_{n\to\infty} \int_0^{nh} \exp\left[-\int_0^t \mu(\tau)\, d\tau\right] dt$$

$$= \lim_{n\to\infty} \sum_{k=0}^{n-1} \left\{ \int_{kh}^{kh+\theta h} \exp\left[-\int_0^t \mu(\tau)\, d\tau\right] dt + \int_{kh+\theta h}^{(k+1)h} \exp\left[-\int_0^t \mu(\tau)\, d\tau\right] dt \right\}$$

$$= \lim_{n\to\infty} \sum_{k=0}^{n-1} \left[\frac{1}{\mu_1}(1 - e^{-\mu_1\theta h})e^{-\mu_1 kh} + \frac{1}{\mu_2}e^{-\mu_2\theta h}\{1 - e^{-\mu_2(1-\theta)h}\}e^{-\mu_2 kh} \right]$$

$$= \frac{1 - e^{-\mu_1\theta h}}{1 - e^{-\mu_1 h}} \frac{1}{\mu_1} + \left(1 - \frac{1 - e^{-\mu_2\theta h}}{1 - e^{-\mu_2 h}}\right) \frac{1}{\mu_2}.$$

This result indicates that the expected life span $\langle t \rangle$ cannot be given by the inverse of a simple arithmetic mean of the death rates $\langle \mu \rangle := \theta\mu_1 + (1-\theta)\mu_2$. When the period h is sufficiently short, or when the death rates μ_1 and μ_2 are sufficiently small, we find that

$$\langle t \rangle \approx \theta\frac{1}{\mu_1} + (1 - \theta)\frac{1}{\mu_2},$$

which means that $\langle t \rangle$ can be approximated by the arithmetic mean of the inverse of death rates.

Exercise 4.4 (p. 119)

Again in the same way of derivation as that for Poisson distribution in Sect. 15.2, we can get the formula of probability $P(n, t)$ as

$$P(n, t) = \binom{n_0}{n}\left(1 - e^{-\langle\mu\rangle_t t}\right)^{n_0 - n} e^{-n\langle\mu\rangle_t t},$$

where

$$\langle \mu \rangle_t := \frac{1}{t} \int_0^t \mu(\tau)\, d\tau,$$

which means the long-term average of the death rate.

Exercise 4.5 (p. 121)

With the arguments same as in Sect. 4.3.5, we can derive the frequency density distribution function $f(t)$, correspondingly to (4.23), as

$$f(t) = v(t) \, \exp\left[-\int_0^t v(\tau)\, d\tau\right].$$

We must remark the general arguments given in p. 122 too. Hence we can get the cumulative frequency distribution function $F(t)$:

$$F(t) = 1 - \exp\left[-\int_0^t v(\tau)\, d\tau\right].$$

Therefore, as a result, the averaged life span $\bar{t}$ in this case can be given by

$$\bar{t} = \int_0^\infty t v(t) \exp\left[-\int_0^t v(\tau)\, d\tau\right] dt.$$

Let us consider the function $V(t) := \int_0^t v(\tau)\, d\tau$ which is a monotonically increasing function of t because $v(t)$ is positive for any $t > 0$. Therefore, with the transformation of variable for the integrand in the integral of $\bar{t}$, $\varsigma = V(t)$, we have

$$\bar{t} = \int_0^\infty V^{-1}(\varsigma)\, e^{-\varsigma} d\varsigma,$$

since $d\varsigma = V'(t)\, dt = v(t)\, dt$. V^{-1} is the inverse function of $V(t)$. We note that, from the Laplace transformation of $V^{-1}(\varsigma)$,

$$\mathcal{F}(s) = \mathcal{L}[V^{-1}(\varsigma)] := \int_0^\infty V^{-1}(\varsigma)\, e^{-\varsigma s} d\varsigma,$$

the averaged life span $\bar{t}$ can be given by $\bar{t} = \mathcal{F}(1)$ too.

Exercise 4.6 (p. 124)

From the binomial expansion

$$\sum_{k=1}^{n_0} \binom{n_0}{k} x^k = (1+x)^{n_0} - 1,$$

we can get the following relation:

$$\sum_{k=1}^{n_0} \binom{n_0}{k} x^{k-1} = \frac{(1+x)^{n_0} - 1}{x} = \frac{(1+x)^{n_0} - 1}{(1+x) - 1} = \sum_{k=0}^{n_0-1} (1+x)^k.$$

By the indefinite integral for the above equation, we can derive

$$\sum_{k=1}^{n_0} \binom{n_0}{k} \frac{x^k}{k} = \sum_{k=1}^{n_0-1} \frac{1}{k+1} (1+x)^{k+1} + x + C, \tag{4.33}$$

where C is an arbitrary constant. By substituting $x = 0$ for this equation, we find

$$C = -\sum_{k=2}^{n_0} \frac{1}{k}.$$

Thus, by substituting $x = -1$ for (4.33), we can get

$$\sum_{k=1}^{n_0} \binom{n_0}{k} \frac{(-1)^k}{k} = -1 - \sum_{k=2}^{n_0} \frac{1}{k} = -\sum_{k=1}^{n_0} \frac{1}{k}.$$

Applying this relation for (4.26), we can lastly derive (4.27).

Exercise 4.7 (p. 125)

From the first equation of the system (4.28) and the initial condition that $p(0, 0) = 1$, we find $p(0, t) = e^{-(\beta+\mu)t}$. Next let us substitute $p(k, t) = u_k(t) e^{-(\beta+\mu)t}$ for the second equation of (4.28). Then we obtain

$$\frac{du_{k+1}(t)}{dt} = \beta u_k(t).$$

Since $u_k(0) = 0$ from the initial condition $p(k, 0) = 0$, we can mathematically get the following equation from this differential equation:

$$u_{k+1}(t) = \beta \int_0^t u_k(\tau)\, d\tau. \tag{4.34}$$

Now, we have $u_0(t) = 1$ because of $p(0, t)$ obtained in the above, we find from (4.34) that $u_1(t) = \beta t$. Hence, by the mathematical induction with (4.34), we can prove that

$$u_k(t) = \frac{(\beta t)^k}{k!}.$$

Finally, we can derive (4.29).

References

1. W. Feller, *An Introduction to Probability Theory and Its Applications*, vol. 1, 3rd edn. (Wiley, New York, 1968)
2. J. Medhi, *Stochastic Models in Queueing Theory*, 2nd edn. (Academic Press, Amsterdam, 2002)
3. S.M. Ross, *Introduction to Probability Models*, vol. 12 (Academic Press, Amsterdam, 2019)

Chapter 5
Continuous Time Modeling for Single Species Population Dynamics

Abstract This chapter is on the continuous time modeling for the single-species population dynamics. The contents serves the readers with the essential idea to introduce the density effect in the mathematical modeling. So this chapter becomes basic to understand the modelings in the subsequent chapters. As a specific topic, we describe the modeling of what is called metapopulation dynamics model too.

5.1 Malthus Model

For a microorganism population in a laboratory culture, the death could be negligible in a sufficiently long period, when the energy supply and the space for the reproduction would be kept in a stationary condition. In such an environmental condition, the fertility could be assumed to be constant independently of the population growth. In other words, the expected number of offsprings produced by an individual per unit time could be a constant r independent of time. Hence from the arguments in Sect. 3.4, the population dynamics could be assumed to follow the relation

$$\frac{\Delta N / N(t)}{\Delta t} = r, \tag{5.1}$$

that is,

$$\Delta N = N(t + \Delta t) - N(t) = (r \Delta t) N(t). \tag{5.2}$$

This relation means that the expected number of newborns per individual during the time interval Δt is given by $r \Delta t$. Taking the limit as $\Delta t \to 0$ for (5.1)

© The Author(s), under exclusive license to Springer Nature Singapore Pte Ltd. 2022
H. Seno, *A Primer on Population Dynamics Modeling*, Theoretical Biology,
https://doi.org/10.1007/978-981-19-6016-1_5

as described in Sect. 3.4, we have the following continuous time model for the population dynamics:

$$\frac{1}{N(t)} \frac{dN(t)}{dt} = r, \tag{5.3}$$

which means that the *per capita momental velocity* of the change of population size, simply called *per capita growth rate* of population, is assumed constant. This model is what is called *Malthus model* which already appeared in Sects. 3.3.1, 4.1.2, and 4.2. The Eq. (5.3) shows Malthus model with the malthusian coefficient r. As clear from the above assumption, the malthusian coefficient r means the per capita growth rate of population.

In the case of laboratory culture mentioned above, the malthusian coefficient depends on the environmental condition provided for the culture. Assuming it constant, it can be regarded as the possibly highest per capita growth rate of population under the given environmental condition. In this sense, r is sometimes called *intrinsic (natural) growth rate* or *intrinsic rate of natural increase*. However, the malthusian coefficient does not necessarily mean the intrinsic growth rate, as described in the subsequent part of this section.

In a more general sense, as indicated by (3.24) in Sect. 3.4, the malthusian coefficient r in Malthus growth depends also on the death and migration of individuals in the population. It becomes clearer with substituting (5.2) for (3.24) that Malthus growth could appear, for example, when the increase due to the immigration is proportional to the population size with a constant per capita birth rate, death rate, and emigration rate. Especially for a closed population defined in Sect. 3.4, it could appear when the per capita birth and death rates are constant since the right side of (3.24) is determined only by them. In such a case of closed population, we have $r = \beta - \mu$ with the birth rate β and death rate μ which are positive constants. Hence for a closed population, $r > 0$ if $\beta > \mu$ while $r < 0$ if $\beta < \mu$.

The solution for the ordinary differential equation (5.3) becomes

$$N(t) = N(0)\,e^{rt}, \tag{5.4}$$

which shows the exponential growth of population size. As shown by the examples in Fig. 5.1, the temporal change of population size $N(t)$ is given by an unbounded increasing concave curve for $r > 0$, and by a decreasing concave curve asymptotically approaching zero for $r < 0$.

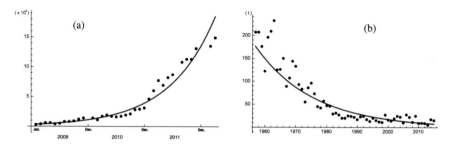

Fig. 5.1 Fitting of Malthus growth curve (5.4) (solid curve) to the real data (solid plots) with the least square method: (**a**) Monthly data of active users of Facebook in Japan (The Japan Ministry of Internal Affairs and Communications, 2015), $r = 0.103$ and $N(\text{Jan. 2009}) = 0.365 \times 10^6$ ($R^2 = 0.967$); (**b**) Annual data of catching glass eels in Japan (The Japan Fisheries Agency, 2015), $r = -0.054$ and $N(1957) = 176.52$ ($R^2 = 0.849$)

5.2 Gompertz Curve

Temporal change of the environment may affect the physiological condition so as to temporally change the per capita growth rate r: $r = r(t)$. Even in this case, the ordinary differential equation (5.3) can be solved by the method of variable separation (refer to Sect. 13.1.2):

$$N(t) = N(0) \exp\left[\int_0^t r(\tau)\, d\tau \right].\tag{5.5}$$

If $r(t) > 0$ for any $t > 0$, the right side of (5.5) is monotonically increasing as time passes, while, if $r(t) < 0$ for any $t > 0$, it is monotonically decreasing toward zero. When the sign of $r(t)$ is temporally changing, the population size at sufficiently large time depends on the nature of the temporal change of $r(t)$.

Let us consider the case where the malthusian coefficient $r(t)$ has a periodical change due to the seasonal variation of environmental condition:

$$r(t) = \bar{r} + \frac{\sigma}{2} \sin(\omega t + \phi).\tag{5.6}$$

The long-term average of the malthusian coefficient

$$\langle r \rangle := \lim_{T \to \infty} \frac{1}{T} \int_0^T r(t)\,dt$$

(continued)

can be easily obtained as $\langle r \rangle = \bar{r}$. The positive parameter σ means the amplitude in the temporal change of $r(t)$, and the positive parameter ω the angular frequency. Since

$$\int_0^t r(\tau)d\tau = \bar{r}t - \frac{\sigma}{2\omega}[\cos(\omega t + \phi) - \cos\phi], \qquad (5.7)$$

we can find that the principal trend of the temporal change of population size is determined by the long-term average of the malthusian coefficient $\bar{r}$. The second term in the right side of (5.7) is to introduce a periodical oscillation with a finite amplitude. Since the first term in the right side of (5.7) is proportional to time t, the value of (5.7) is mainly determined by the first term after sufficient long time. As shown in Fig. 5.2, the nature of the seasonal change in the malthusian coefficient does not contribute to the long-term trend of population growth.

As a special example, for $r(t) = r_0 e^{-t/\eta}$ with positive constants r_0 and η such that the malthusian coefficient in (5.5) temporally decreases toward zero, we have

$$N(t) = N(0) \exp\left[r_0\eta(1 - e^{-t/\eta})\right]. \qquad (5.8)$$

As seen in Fig. 5.3, the temporal change of population size shows an S-shape curve. This curve given by (5.8) has been empirically used to be fit to the experimental/observed data. It is called *Gompertz curve* or *Gompertz-Wright curve*. The naming is after the fitting to a data of adult death rate by Benjamin Gompertz (1779–1865) in 1825 [4]. Famous population geneticist Sewall G. Wright (1889–

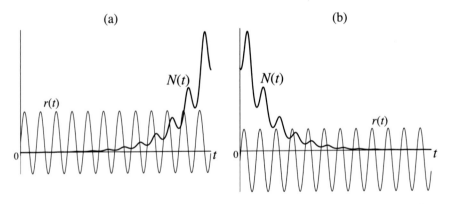

Fig. 5.2 Malthus growth (5.5) with a periodically changing malthusian coefficient $r(t)$ given by (5.6). (a) $\bar{r} > 0$; (b) $\bar{r} < 0$

Fig. 5.3 Gompertz
curve (5.8) which
corresponds to Malthus
growth (5.5) with the
malthusian coefficient
temporally decreasing from
positive to zero: $r(t) \propto e^{-t/\eta}$
with a positive constant η

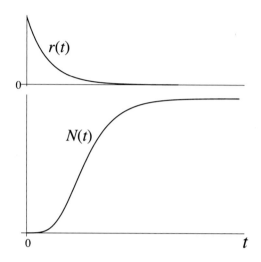

1988) used it as a mathematical model for the growth of individual body size in
1926 [16].

5.3 Logistic Equation

In this section, let us consider the case where the per capita growth rate becomes
smaller as the population density gets larger. This is the case, for example, that the
availability of necessary resource is limited, or that the environment is degraded
by the waste produced by the population itself. In the modeling for such a case,
the right side of (5.3) is given by a function decreasing in terms of the population
size N: $r = r(N)$. This means that the momental velocity of population size per
individual is determined by a negative density effect as in Sect. 2.1.

The simplest mathematical modeling for such a relation of the per capita growth
rate to the population size N is introduced by a linear function

$$r = r(N) = r_0 - \beta N. \qquad (5.9)$$

This is reasonable as what is called the zeroth approximation for such a negative
density effect. In this modeling, the parameter $r_0 = r(0)$ means the intrinsic growth
rate, since it gives the upper bound for the per capita growth rate. As the population
density gets sufficiently low, if the resource available for each individual becomes
large, or if the environmental degradation is moderated enough to be negligible by
the recovery process, the per capita growth rate could have the maximal fertility
which the species potentially has. It is now represented by r_0. The positive parameter
β characterizes the sensitivity of the per capita growth rate for the density effect.
The larger β introduces the severer density effect on the per capita growth rate.

Thus the parameter β may be regarded as the coefficient of density effect, while it is sometimes called *Verhulst coefficient* or *Verhulst-Pearl coefficient* (refer to p. 103 of Sect. 3.3.2).

With (5.9), the population dynamics model is given by

$$\frac{dN(t)}{dt} = \{r_0 - \beta N(t)\} N(t). \tag{5.10}$$

This is a *logistic equation*, already mentioned by (2.15) in Sect. 2.1.3, and also in Sects. 2.6.2, 3.3.2, 3.3.3, 3.3.4, and 4.5. Although the logistic equation (5.10) is a nonlinear ordinary differential equation, it can be explicitly solved by the method of variable separation (refer to Sect. 13.1.2):

$$N(t) = \frac{r_0/\beta}{1 - \left\{1 - \frac{r_0/\beta}{N(0)}\right\} e^{-r_0 t}}. \tag{5.11}$$

Even for the logistic equation with temporally varying r_0 and β,

$$\frac{dN(t)}{dt} = \{r_0(t) - \beta(t)N(t)\} N(t), \tag{5.12}$$

it is possible to get the following explicit solution:

$$N(t) = \left[\Xi(t) \int_0^t \frac{\beta(v)}{\Xi(v)} dv + \frac{\Xi(t)}{N(0)}\right]^{-1}, \tag{5.13}$$

where $\Xi(t) := \exp\left[-\int_0^t r_0(\tau)d\tau\right]$. Since the logistic equation (5.12) can be classified to what is called *Bernoulli equation* in the theory of ordinary differential equation, it can be mathematically translated to a linear ordinary differential equation with the variable transformation $y(t) = 1/N(t)$, as described in Sect. 13.1.4. By the way to solve the linear ordinary differential equation, we can derive the solution (5.13) (refer to Sect. 13). The temporal variation by (5.12) is significantly affected by the nature of time-dependence of r_0 and β as exemplified by a numerical calculation shown in Fig. 5.4.

As indicated by Figs. 2.11 (p. 47), 3.1 (p. 99), and 5.5a in this section, the temporal change of population size $N(t)$ by the logistic equation (5.10) shows a monotonic approach to the equilibrium value r_0/β from any positive initial size $N(0)$. However, it should be remarked that, as seen in Fig. 5.5a, the population growth curve of logistic equation (5.10) is not necessarily of S-shape, differently from the Gompertz curve in Sect. 5.2. When the positive initial value $N(0)$ is less than $r_0/2\beta$, the curve (5.11) becomes an S-shaped monotonically increasing

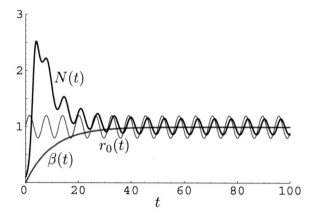

Fig. 5.4 A numerical example of the temporal variation by the non-autonomous logistic equation (5.12) with $r_0(t) = 1 + 0.2\sin(t)$ and $\beta(t) = 1 - e^{-0.1t}$; $N(0) = 0.1$. The population size approaches a periodical stationary oscillation with the mean value 1

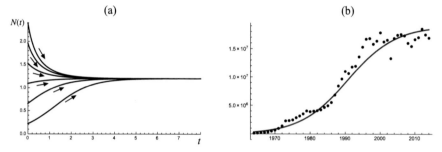

Fig. 5.5 The logistic growth (5.11). (**a**) Numerically drawn curves from different initial values with $r_0 = 1.2$ and $\beta = 1.0$. Every curve monotonically approaches the equilibrium value $r_0/\beta = 1.2$. (**b**) A fitting of (5.11) to the annual data about the number of Japanese tourists in abroad (Japan National Tourism Organization, 2015), by the least square method with the hypothesized carrying capacity of 1.9 million: $r_0 = 0.147$ and $N(1964) = 368000$ ($R^2 = 0.9233$). The actual data is however 128000 for 1964

one. In contrast, when $N(0)$ is greater than $r_0/2\beta$ and less than r_0/β, it does not become S-shape, but monotonically increasing convex. The critical value $r_0/2\beta$ is sometimes called *half saturation value*, since it is the half value of the equilibrium value r_0/β which the population size asymptotically approaches. Moreover, when $N(0)$ is greater than r_0/β, it becomes monotonically decreasing concave. In some literatures, an S-shaped curve would be called the logistic curve, or the curve of logistic growth (refer to Sect. 3.2). It may be incorrect and is misleading.

Exercise 5.1 Show that the velocity of the temporal change of population size by (5.10) takes maximum when the population size N is at the half saturation value $r_0/2\beta$.

For the population dynamics by the logistic equation (5.10), the equilibrium value r_0/β means the *carrying capacity* introduced at p. 34 of Sect. 2.1.1. It may be called *satuation density* too. Substituting the parameter of carrying capacity $K = r_0/\beta$ for (5.10), it can be rewritten in a mathematically equivalent form, the logistic equation (2.15) in p. 45 of Sect. 2.1.3, which is the most popular form used in many literatures as the logistic equation. Actually, both of the logistic equations (2.15) and (5.10) are easily proven to be mathematically equivalent to the following non-dimensionalized form:

$$\frac{dx(\tau)}{d\tau} = \{1 - x(\tau)\}x(\tau), \tag{5.14}$$

with appropriate variable and parameter transformations.

The ordinary differential equation which consists of only the dependent variable(s) and its derivative(s) is called *autonomous*. If the ordinary differential equation contains some functions of independent variable t in any part of it, it is called *non-autonomous*. Malthus model (5.3) and the logistic equation (5.10) are autonomous, while (5.6) and (5.12) are non-autonomous.

Rigorously, these two logistic equations (2.15) and (5.10) must be distinguished from each other in the sense of modeling. For (2.15), the per capita growth rate (5.3) is given by

$$r = r(N) = r_0\left(1 - \frac{N}{K}\right). \tag{5.15}$$

In the mathematical modeling, we find that r_0/K of (5.15) corresponds to β of (5.9). For the logistic equation (5.10) with (5.9), since the carrying capacity is given by r_0/β, the carrying capacity is proportional to the intrinsic growth rate r_0. In contrast, for (2.15) with (5.15), the carrying capacity is given by K, a parameter independent of r_0. This means that the carrying capacity is determined a priori as a characteristic of the habitat environment. In the case of (5.10), the carrying capacity is determined a posteriori as a consequence of the balance between the fertility and density effect.

For the logistic equation (2.15) with (5.15), the velocity of the population size change per individual is proportional to $(K - N)/K$. From this aspect

(continued)

of modeling, we may regard the velocity as determined by the proportion of the "rest of environment" to the carrying capacity K. In other words, when a given capacity K is "consumed" by the population size N, the velocity of the size change per individual is proportional to the unused amount of capacity, $K - N$. In this translation of the modeling, the intrinsic growth rate r_0 may be regarded as the coefficient of capacity consumption. This is a frequently used translation of modeling for the logistic equation (2.15), since it may be simple to understand the modeling. However, we must remark that the population dynamics with a logistic growth could not be necessarily shown by the logistic equation (2.15). Especially, if there is a temporal/seasonal change of environmental condition which affects the intrinsic growth rate, the density effect, or the carrying capacity itself, such an environmental influence must be reasonably introduced in the modeling. As shown by numerical examples in Fig. 5.6 about the logistic equation (2.15) with the time-dependent $r_0 = r(t)$ and $K = K(t)$, such an introduction of temporal change for those parameters can cause a significant difference in the characteristics which depends on how it is introduced in the model.

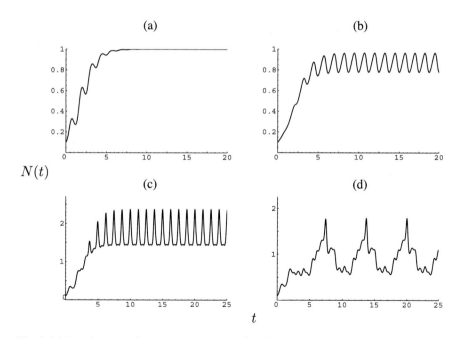

Fig. 5.6 Numerical examples of the temporal variation of population size by the non-autonomous logistic equation (2.15) with (5.15) where $r_0 = r_0(t) = 1 + (\sigma_r/2)\sin(\omega_r t + \phi_r)$ and $K = K(t) = 1 + (\sigma_K/2)\sin(\omega_K t + \phi_K)$. $(\sigma_r, \omega_r, \phi_r, \sigma_K, \omega_K, \phi_K) =$ (**a**) $(4.0, 5.0, 0, 0, 0, 0)$; (**b**) $(0, 0, 0, 1.0, 5.0, 0)$; (**c**) $(4.0, 5.0, 0, 1.0, 5.0, 5.6)$; (**d**) $(4.0, 5.0, 0, 1.0, 4.0, 0)$. Commonly $N(0) = 0.1$

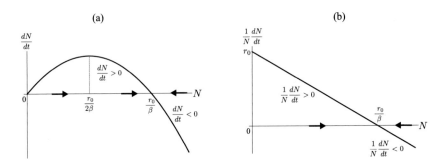

Fig. 5.7 The dynamical nature of logistic equation (5.10) in the one dimensional phase space. (**a**) The graph for the density dependence of the velocity of size change dN/dt; (**b**) The graph for that of per capita growth rate $(1/N)dN/dt$. The per capita growth rate is decreasing in terms of the population size N, and is negative for N greater than the equilibrium size r_0/β

From the aspect of dynamical system given by (5.10), the value of N represents the *state* of the population dynamics. In this sense, each value of N indicates a point on the number line which is generally called *one dimensional phase space*. The equilibrium value r_0/β is shown as a point $N = r_0/\beta$ in the phase space, the *equilibrium point*.

The dynamical nature of logistic equation (5.10) can be understood from the solution (5.11), while it can be investigated and understood from the equation (5.10) itself. This is not only a useful qualitative analysis on the dynamical nature, but also a meaningful approach to the dynamical nature in order to identify the meanings of modeling. As illustrated by Fig. 5.7, the direction of temporal change in the population size is determined by the sign of the velocity of population size change dN/dt or the per capita growth rate $(1/N)dN/dt$. On the number line, that is, in the one dimensional phase space, the population size increases if the sign is positive so that the point corresponding to the value of N moves rightward where the sign is positive. If the sign is negative, the population size decreases so that the point of N moves leftward on the number line. Hence, from Fig. 5.7, we can conclude that the value of N monotonically approaches the carrying capacity r_0/β from any positive initial value $N(0)$.

For modeling the logistic growth by (5.10), we introduced (5.9) as the simplest assumption for the per capita growth rate under the negative density effect, since the linear function is mathematically simplest. Now we may consider a more general function for the negative density effect like

$$r = r(N) = r_0 - \beta N^\alpha, \tag{5.16}$$

(continued)

where the additional positive parameter α means the sensitivity of growth rate to the density effect. For the smaller α, the decrease of per capita growth rate due to the density effect is steeper in the range of low population density while it is more moderate in the range of high population density. Then for the population dynamics model by (5.3) with the per capita growth rate (5.16), the carrying capacity is given by $[r_0/\beta]^{1/\alpha}$ which depends on α too. The model (5.3) with (5.16) was studied by Gilpin and Ayala [3]. Actually, the further past work by F.E. Smith [13], showed that the density dependence of the per capita growth for the population of *Daphnia magna* became a decreasing convex curve in the experimental data. This may be regarded as corresponding to the case of $\alpha < 1$ in (5.16).

Even with the per capita growth rate (5.16), the ordinary differential equation (5.3) is a Bernoulli equation (refer to Sect. 13.1.4). Indeed, with the variable transformation $y(t) = 1/[N(t)]^\alpha$, we can derive a linear ordinary differential equation of y, and consequently get the following solution:

$$N(t) = \left[\frac{r_0/\beta}{1 - \left\{ 1 - \frac{r_0/\beta}{[N(0)]^\alpha} \right\} e^{-\alpha r_0 t}} \right]^{1/\alpha}. \qquad (5.17)$$

We may understand from this solution that the qualitative nature of the temporal variation of N is the same as that of the standard logistic equation (5.10) or (2.15) with (5.15), while such a qualitative correspondence can be equivalently shown by the way to use the one-dimensional phase space as described above.

5.4 Verhulst Model

Even when the temporal change of a population size could be regarded as continuously occurring, the actual observation usually gives the data of a sequence of sizes with a certain time interval. Now let us consider such a sequence of population sizes when the population dynamics follows a logistic equation (5.10).

Like a series of snapshots, let us see the logistic growth given by (5.11) with a time step h. From (5.11), the population sizes $N_k = N(kh)$ at $t = kh$ and $N_{k+1} = N((k+1)h)$ at $t = (k+1)h$ are given by

$$N_k = \frac{r_0}{\beta - \left\{ \beta - \frac{r_0}{N(0)} \right\} e^{-r_0 kh}}; \quad N_{k+1} = \frac{r_0}{\beta - \left\{ \beta - \frac{r_0}{N(0)} \right\} e^{-r_0(k+1)h}}. \qquad (5.18)$$

Eliminating $e^{-r_0 kh}$ in these two equations, we can get the following recurrence relation [2]:

$$N_{k+1} = \frac{1}{1 + \phi_{r_0}(h)\beta N_k} \cdot e^{r_0 h} N_k, \tag{5.19}$$

where

$$\phi_{r_0}(h) = \frac{e^{r_0 h} - 1}{r_0}. \tag{5.20}$$

The recurrence relation (5.19) may be regarded as a discrete time model of population dynamics, and it is called *Verhulst model*. This naming is after the work by Pierre F. Verhulst (1804–1849) [14, 15] in which he used a mathematically equivalent discrete time model about the variation of populations in France and some other European countries (see [1]). It is clear that the recurrence relation (5.19) is mathematically equivalent to Beverton-Holt model which was described in Sect. 2.1.1. For this reason, Beverton-Holt model can be regarded reasonably in a modeling sense as the discrete time model corresponding to the logistic equation, as already mentioned in Sect. 2.1.3.

From the above derivation of Verhulst model (5.19), the temporal sequence of population sizes by (5.19) can track exactly, that is, with no error, the time continuous change of population size by the corresponding logistic equation (5.10). Further, it is easy to confirm that, at the limit as $h \to 0$, the recurrence relation (5.19) converges to (5.10).

We must note the form of the density effect in Verhulst model (5.19) in comparison to that in the logistic equation (5.10), since they have the exact correspondence. As a consequence, the density effect corresponding to (5.9) for the logistic equation (5.10) is introduced in the corresponding discrete time model by a rational function like (2.2) in Beverton-Holt model.

Such a correspondence with respect to the modeling of density effect between the time continuous and discrete models was intuitively applied for the modeling of the discrete time model such as Leslie-Gower model of the population dynamics of competing species described in Sect. 2.3.1.

We may consider the following more general ordinary differential equation for a population dynamics:

$$\frac{dN(t)}{dt} = \{r_0 - D(N(t))\} N(t), \tag{5.21}$$

(continued)

where $D(N)$ is a function introducing a density effect on the per capita growth rate. It can be shown that the recurrence relation

$$N(t+h) = \frac{1}{1 + \phi_{r_0}(h)D(N(t))} \cdot N(t)e^{r_0 h} \tag{5.22}$$

with $\phi_{r_0}(h)$ defined by (5.20) has the same qualitative nature of dynamics as (5.21) when the function $D(N)$ satisfied a certain general condition [12]. Such a correspondence about the qualitative nature of dynamics between (5.21) and (5.22) as well as (5.10) and (5.19) may be called *dynamical consistency*.

5.5   Logistic Equation to Logistic Map

The logistic map (2.13) or equivalently (12.4) in Sect. 12.1.3 is defined with a quadratic polynomial as well as the logistic equation (5.10) (equivalently (2.15) or (5.14)). Such a similarity between discrete and continuous time models may be mathematically understood with a time-discretization, sometimes called *simple discretization* which is well-known to be used in the *Euler method* for the numerical calculation of differential equation. It can give a piece-wise numerical approximation of the solution [8, 9].

Let us consider an approximation of the solution for the logistic equation (5.10) about a short interval $[t, t + \Delta t]$. We use here the following approximation of the derivative:

$$\frac{dN(t)}{dt} = \lim_{h \to 0} \frac{N(t+h) - N(t)}{h} \approx \frac{N(t + \Delta t) - N(t)}{\Delta t}. \tag{5.23}$$

This is the simplest approximation making use of the mathematical definition of derivative, called *simple discretization*. Applying the approximation (5.23) for the logistic equation (5.10), we can obtain the equation

$$\frac{\tilde{N}(t + \Delta t) - \tilde{N}(t)}{\Delta t} = \{r_0 - \beta \tilde{N}(t)\} \tilde{N}(t),$$

that is,

$$\tilde{N}(t + \Delta t) = \tilde{N}(t) + \{r_0 - \beta \tilde{N}(t)\} \tilde{N}(t) \Delta t \tag{5.24}$$

$$= (1 + r_0 \Delta t) \left\{ 1 - \frac{\beta \Delta t}{1 + r_0 \Delta t} \tilde{N}(t) \right\} \tilde{N}(t).$$

This recurrence relation can determine the value of $\widetilde{N}(t + \Delta t)$ at time $t + \Delta t$ by that of $\widetilde{N}(t)$ at time t. Thus, it can be used to give a sequence of values $\{\widetilde{N}(0), \widetilde{N}(\Delta t), \widetilde{N}(2\Delta t), \widetilde{N}(3\Delta t), \dots\}$ approximating the solution of the logistic equation (5.10), where $\widetilde{N}(0) = N(0)$. This numerical approximation may be called (forward) *Euler method*.

Substituting $\mathcal{R}_0 = 1 + r_0 \Delta t$ and $x_k = \beta \Delta t \widetilde{N}(k\Delta t)/(1 + r_0 \Delta t)$ $(k = 0, 1, 2, \dots)$ for (5.24), we can find the logistic map (12.4) in Sect. 12.1.3. Therefore, the logistic map may be regarded as an approximation with such a simple discretization for the logistic equation, as already mentioned at the end of Sect. 2.1.3. However, the approximation by (5.23) can be valid only for sufficiently small Δt.

The sequence $\{\widetilde{N}(0), \widetilde{N}(\Delta t), \widetilde{N}(2\Delta t), \widetilde{N}(3\Delta t), \dots\}$ generated by the recurrence relation (5.24) converges to that of values correspondingly determined by the solution of the logistic equation (5.10) at the limit as $\Delta t \to 0$. As long as $\Delta t > 0$, however, there must be an error between the value of N and the corresponding value of $\widetilde{N}$. Figure 5.8 numerically illustrates such a difference in the values, which appears bigger for the larger Δt. The time step size Δt must satisfy a certain condition for the sequence by the recurrence relation (5.24) to approximate well the solution of the logistic equation (5.10).

It can be easily found that the dynamics by (5.24) has the equilibria 0 and r_0/β, which are identical with those of the logistic equation (5.10). However, as seen in Sects. 2.1.3, 12.1.3, and 12.1.5, the logistic map (2.13) or (12.4) generates a periodic solution or a chaotic variation for a large value of $\mathcal{R}_0$, while the logistic equation (5.10) shows only a monotonic approach to the carrying capacity r_0/β. Further, although such a monotonic approach to r_0/β by the logistic equation (5.10) appears even for the initial value $N(0) > r_0/\beta$, the logistic map (2.13) may show a non-monotonic approach to it as in the upper second graph of Fig. 2.9 in p. 43. Moreover, the recurrence relation (5.24) generates negative values for the initial value $\widetilde{N}(0) > r_0/\beta + 1/(\beta\Delta t)$. For these reasons, the logistic map itself cannot be regarded as the discrete version of logistic equation, nor the discrete time model corresponding to the logistic equation, as well as the argument in the previous section from a different viewpoint.

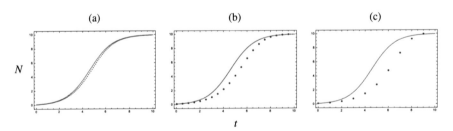

Fig. 5.8 Numerical plots of $\{\widetilde{N}(0), \widetilde{N}(\Delta t), \widetilde{N}(2\Delta t), \widetilde{N}(3\Delta t), \dots\}$ versus the curve of the solution (5.11) of the logistic equation (5.10). (**a**) $\Delta t = 0.1$; (**b**) $\Delta t = 0.5$; (**c**) $\Delta t = 1.0$. Commonly, $r_0 = 1.0$; $\beta = 0.1$; $N(0) = \widetilde{N}(0) = 0.1$

Exercise 5.2 Find the condition for Δt such that the recurrence relation (5.24) generates a monotonic sequence $\{\widetilde{N}(0), \widetilde{N}(\Delta t), \widetilde{N}(2\Delta t), \widetilde{N}(3\Delta t), \dots\}$ approaching r_0/β, respectively for $\widetilde{N}(0) < r_0/\beta$ and for $\widetilde{N}(0) > r_0/\beta$.

Now let us consider the other application of simple discretization for the logistic equation (5.10). First, the general population dynamics model

$$\frac{1}{N(t)} \frac{dN(t)}{dt} = r(N(t), t)$$

can be rewritten to

$$\frac{d \ln N(t)}{dt} = r(N(t), t).$$

Applying the simple discretization for the derivative in the left side, we can get

$$\frac{\ln \widetilde{N}(t + \Delta t) - \ln \widetilde{N}(t)}{\Delta t} = r(\widetilde{N}(t), t),$$

that is,

$$\ln \widetilde{N}(t + \Delta t) = \ln \widetilde{N}(t) + r(\widetilde{N}(t), t)\Delta t,$$

and subsequently,

$$\widetilde{N}(t + \Delta t) = \widetilde{N}(t) \cdot e^{r(\widetilde{N}(t), t)\Delta t}.$$

With the per capita growth rate (5.9) for the logistic equation, we arrive at the following recurrence relation:

$$\widetilde{N}(t + \Delta t) = \widetilde{N}(t) \cdot e^{\{r_0 - \beta \widetilde{N}(t)\}\Delta t}. \tag{5.25}$$

Lastly, with notations $N_k := \widetilde{N}(k\Delta t)$ ($k = 0, 1, 2, \dots$) and $\mathcal{R}_0 := e^{r_0 \Delta t}$, we can find the clear mathematical correspondence of the above recurrence relation to Ricker model (2.7) in Sect. 2.1.2. This relation of the logistic equation to Ricker model was mentioned in p. 46 of Sect. 2.1.3 too.

Exercise 5.3 Find the condition for Δt such that the recurrence relation (5.25) generates a monotonic sequence $\{\widetilde{N}(0), \widetilde{N}(\Delta t), \widetilde{N}(2\Delta t), \widetilde{N}(3\Delta t), \dots\}$ approaching r_0/β, respectively for $\widetilde{N}(0) < r_0/\beta$ and for $\widetilde{N}(0) > r_0/\beta$. Further, try to discuss which of (5.24) and (5.25) could be regarded as the better approximation for the logistic equation (5.10), from the results of Exercises 5.2 and 5.3.

5.6 Allee Effect

The per capita growth rate for the logistic equation is monotonically decreasing in terms of the density, in a linear manner as (5.9) or (5.15) defines (Fig. 5.7b). Therefore, from the definition of Allee effect described in Sect. 2.2.1, the population dynamics by the logistic equation does not involve the Allee effect. Although the net growth rate of population size for the logistic equation has a unimodal density dependence as seen in Fig. 5.7a, it does not mean any positive density effect for the per capita growth rate.

We may consider the following simple model with a positive density effect:

$$\frac{1}{N(t)} \frac{dN(t)}{dt} = \{aN(t) - v\}\{w - bN(t)\}, \tag{5.26}$$

where parameters a, b, v, and w are positive constants. The first factor in the right side means a positive density effect, and the second a negative density effect. In this modeling, the per capita growth rate is determined by the product of these two factors of density effect.

This model is the case of Fig. 5.9b when $v/a \neq w/b$. There is a threshold value $N_c = \min[v/a, w/b]$ for the population density N. When $N < N_c$, the population size monotonically decreases to become extinct, while, when $N > N_c$, it monotonically approaches the positive equilibrium value $N^* = \max[v/a, w/b]$ to become persistent. Thus, the population dynamics by (5.26) has the nature of

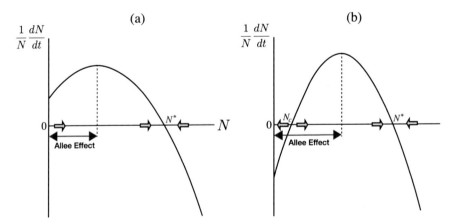

Fig. 5.9 The density effect for the per capita growth rate when the Allee effect exists. (**a**) weak Allee effect; (**b**) strong Allee effect. Refer to Sect. 2.2.3. In the case of (**a**), the population size necessarily approaches a positive equilibrium value from any positive initial size, while, in the case of (**b**), it monotonically decreases to become zero from sufficiently small initial size. Arrows on the horizontal axis indicates the direction of the temporal change in the population size (like Fig. 5.7 about the logistic equation). In both cases, the per capita growth rate becomes the maximum for a specific population density, that means the optimum density defined in Sect. 2.2.1

bistability, which appeared and discussed also for the discrete time model (2.17) in Sect. 2.2.1.

We should remark that the Allee effect does not necessarily bring the nature of bistability, as already mentioned in Sect. 2.2.3. For an illustrative example, the per capita growth rate like Fig. 5.9a involves the (weak) Allee effect, where it does not have the nature of bistability.

When $v/a \neq w/b$, the population dynamics by (5.26) has three equilibria, 0, N_c, and N^*. Although the ordinary differential equation (5.26) keeps the value N at N_c for any time $t > 0$ if the initial value is $N(0) = N_c$, the value N leaves away from N_c for the initial value $N(0) = N_c + \epsilon$ with any infinitesimal value $\epsilon \neq 0$. This is clear by the phase space argument (refer to the description in Sect. 5.3) with Fig. 5.9b for (5.26). This kind of equilibrium is called *unstable equilibrium* (refer to Sect. 14.1). If the population size stays at an unstable equilibrium, any perturbation in the population size causes the change of population size leaving from the equilibrium.

In contrast, the population size N asymptotically approaches the equilibrium N^* from the initial value $N(0) = N^* + \epsilon$ with any infinitesimal value $\epsilon \neq 0$. This means that, if the population size stays at the equilibrium value N^*, any sufficiently small perturbation in the population size cannot work to significantly change the size, and the population size tends to return to N^*. This kind of equilibrium is called *locally asymptotically stable equilibrium* (refer to Sect. 14.1).

In the exclusive case where $v/a = w/b$, it is easy to show that the right side of (5.26) is negative for any non-negative value of N except for $N_c^* = v/a = w/b$. Thus, the population size necessarily decreases as time passes when $N \neq N_c^*$. As seen by the phase space argument in this case, the population size monotonically decreases to become zero if $N(0) < N_c^*$, while it monotonically decreases to become N_c^* if $N(0) > N_c^*$. In this case, the population dynamics by (5.26) has two equilibria, 0 and N_c^*. Equilibrium 0 can be easily shown to be locally asymptotically stable, taking account of only positive perturbation because of the non-negative value for N from the modeling definition. The other equilibrium N_c^* is unstable, since any negative perturbation causes the change of population size leaving from it.

5.7   Metapopulation Model

5.7.1   Levins Model

Metapopulation means a collection of subpopulations separated in space which have some ecological interactions, for example, by migration between them. This concept has created by an ecological researcher Richard 'Dick' Levins (1930–2016) in his work of 1969 [10]. He expressed it as "a population of populations". Levins [10, 11] proposed also a new mathematical modeling for the population dynamics based on the concept. His idea of the modeling has been widely accepted and applied for a variety of theoretical researches in landscape ecology and conservation biology, especially in the ecological field treating the population dynamics for a large spatial scale.

The essence of Levins' idea proposed in [10, 11] is to binarize the state of local habitat, briefly *patch*, with respect to the trend of the subpopulation size in it. The possible alternative states are 'occupied' and 'empty' (or 'vacant'). This idea ignores the detail of subpopulation size in each patch, and focuses on the establishment of subpopulation at the patch in a certain ecological sense. Hence in other words, it may be regarded as a modeling of *coarse-graining* population dynamics. The metapopulation dynamics in this meaning is to describe the temporal change of the frequency of occupied patches in the whole habitat which consists of a collection of local habitats fragmented in space, that is, habitable patches.

The mathematical model proposed by Levins [10, 11] is as follows:

$$
\begin{aligned}
\frac{dE}{dt} &= eP - cPE; \\
\frac{dP}{dt} &= cPE - eP,
\end{aligned}
\tag{5.27}
$$

where E and P mean the frequencies of 'empty' and 'occupied' patches respectively. They may be regarded as the probabilities of such states at an arbitrarily chosen patch. The occupied patch is the local habitat established by a subpopulation in an ecological sense. The empty patch is the local habitat regarded as with no such an establishment, which may be vacant before any settlement, or may be emerged after the extinction of subpopulation previously inhabiting there. Such an extinction may be caused for example, by a stochastic disturbance in the environment with a climate change or human interruption. The parameter e is the coefficient of the occurrence of event to make the patch 'empty' (e.g., extinction or abandonment). The parameter c is the coefficient of the settlement to make the patch 'occupied'. The term cP introduces the likeliness of the occurrence of settlement by the immigration from occupied patches. This indicates that the probability of the settlement to an empty patch is proportional to the frequency of occupied patches in the whole habitat, whereas cP does not have any meaning of probability in itself (refer to the below arguments in the gray box). The model (5.27) is frequently called *Levins model*.

Let us suppose that there are N patches in the whole habitat for the metapopulation. Then $NP(t)$ and $NE(t)$ respectively define the (expected) numbers of occupied and empty patches at time t. Now, the probability that an empty patch is settled by immigrants in a sufficiently short period $[t, t+\Delta t]$ is assumed to be given by $\rho NP(t)\Delta t + o(\Delta t)$, where the positive parameter ρ is the coefficient reflecting the easiness/hardness of settlement in the empty patch. At the same time, the probability that an occupied patch becomes empty in $[t, t + \Delta t]$ is assumed to be given by $e\Delta t + o(\Delta t)$. Then the change in the frequency of empty patches during $[t, t + \Delta t]$, that is, $NE(t + \Delta t) - NE(t)$ satisfies the following relation:

$$NE(t + \Delta t) - NE(t) = -\{\rho NP(t)\Delta t + o(\Delta t)\}NE(t) + \{e\Delta t + o(\Delta t)\}NP(t).$$

Dividing both sides by $N\Delta t$ and taking the limit as $\Delta t \to 0$, we can get the first equation of (5.27) with $c = \rho N$.

From the modeling and definition, it is always satisfied that $0 \le E \le 1, 0 \le P \le 1$, and $E + P = 1$. Hence, the model (5.27) can be expressed by the following one dimensional ordinary differential equation:

$$\frac{dP}{dt} = cP(1 - P) - eP \tag{5.28}$$

$$= \{(c - e) - cP\}P. \tag{5.29}$$

The Eq. (5.28) is sometimes introduced as Levins model instead of (5.27). As seen from the Eq. (5.29), Levins model (5.28) is mathematically equivalent to the logistic equation (5.10). If $c \le e$, the frequency of occupied patches P monotonically decreases to asymptotically converge to 0, that is, the population goes extinct (Fig. 5.10a). If $c > e$, the frequency P asymptotically converges to equilibrium $1 - e/c$ (Fig. 5.10b). In this case, the eventual positive frequency of occupied patches means the persistence of population in the habitat. Consequently, it is implied that the population persistence requires a sufficient activity of migration between patches in such a fragmented habitat.

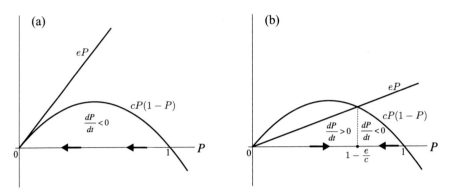

Fig. 5.10 The phase space consideration on the nature of Levins model (5.28). (**a**) For $c \leq e$, the population goes extinct; (**b**) For $c > e$, the population asymptotically approaches the persistent equilibrium

5.7.2   3-State Metapopulation Model

A straightforward extension of Levins model (5.27) is by introducing different states of patch in the modeling. In this section, we consider such a model proposed by Ilkka Hanski (1953–2016) in 1985 [5] (see also [6] and [7, p. 61]):

$$\frac{dE}{dt} = e_S S - cLE;$$
$$\frac{dS}{dt} = cLE + e_L L - e_S S - rS - mLS; \qquad (5.30)$$
$$\frac{dL}{dt} = rS + mLS - e_L L,$$

where E, S, and L are the frequencies of empty patches, occupied patches with a small population size, and occupied patches with a large population size respectively. The state of settled patch is now binarized into two different ones. This modeling provides a metapopulation dynamics with three states of patch. It is satisfied that $E(t) + S(t) + L(t) = 1$ for any time t, which can be reasonably embodied in (5.30) such that $d\{E(t) + S(t) + L(t)\}/dt = 0$ for any time t.

The parameter c is the coefficient of initial settlement by a small subpopulation immigrating in an empty patch. It must be noted that, in (5.30), the migration from patches of the state S is assumed negligible for the settlement. The parameter e_S is the coefficient of the transition to the state 'empty' E for the small subpopulation at the state S. The parameter r is the coefficient of the growth of small subpopulation to the large one with the reproduction in the patch, while e_L is the coefficient of the decline of large subpopulation to the small one with some causes to reduce the population size. The parameter m is the coefficient of the increase of subpopulation size at the state S by the immigration from the patches of the state L. The net immigration rate per small subpopulation is introduced by mL which is proportional to the frequency of large subpopulations. Hanski mentioned in his book [7] that the most important factor in this model is the immigration to small subpopulations from large subpopulations. The effect of such an immigration may be called *rescue effect* for a reason described in the following part.

Since $E(t) + S(t) + L(t) = 1$ for any time t, the system (5.30) is mathematically equivalent to the following two dimensional system:

$$
\begin{aligned}
\frac{dE}{dt} &= e_S(1 - E - L) - cLE; \\
\frac{dL}{dt} &= (r + mL)(1 - E - L) - e_L L.
\end{aligned}
\tag{5.31}
$$

The extinct equilibrium $(1, 0)$ always exists. By the local stability analysis (refer to Sects. 14.2 and 14.3), we can get the following result on the stability of the extinct equilibrium $(1, 0)$:

$$
\begin{cases}
\text{unstable (saddle)} & \text{if } \dfrac{c}{e_S} > \dfrac{e_L}{r}; \\[2ex]
\text{locally asymptotically stable (stable node)} & \text{if } \dfrac{c}{e_S} < \dfrac{e_L}{r}.
\end{cases}
\tag{5.32}
$$

First, let us consider the case of $m = 0$ for (5.31), when there is no contribution of the immigration from large subpopulations to the state transition of small subpopulation. For the system (5.31) with $m = 0$, we can easily find that the unique non-trivial equilibrium

$$
(E^*, L^*) = \left(E^\dagger, \frac{1 - E^\dagger}{1 + e_L/r} \right)
\tag{5.33}
$$

exists if and only if $E^\dagger := e_S e_L/(cr) < 1$. Further with the local stability analysis on this equilibrium, we can get the following result about its existence and stability (Exercise 5.4):

$$
\begin{cases}
\text{absent} & \text{if } \dfrac{c}{e_S} < \dfrac{e_L}{r}; \\[2ex]
\text{unique and locally asymptotically stable (stable node)} & \text{if } \dfrac{c}{e_S} > \dfrac{e_L}{r}.
\end{cases}
$$

Exercise 5.4 Derive the above result on the local stability of the non-trivial equilibrium (E^*, L^*) given by (5.33) when $m = 0$.

For the system (5.31) with $m > 0$, there is a bistable case with two non-trivial equilibria, E_u and E_s, as illustrated by Fig. 5.11. In such a bistable case, the population goes extinct if the frequency of empty patches is beyond a threshold

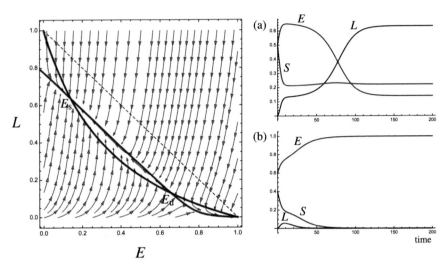

Fig. 5.11 Numerical example of a bistable case for the 3-state metapopulation model (5.30). Nullclines and vector flows in the (E, L)-phase plane (left) and the temporal changes of variables from different initial conditions (right). The dashed line in the (E, L)-phase plane indicates $E + L = 1$. In **(a)** of right figures, the system asymptotically approaches a persistent state from $(E(0), S(0), L(0)) = (0.5, 0.5, 0.0)$, while in **(b)**, it asymptotically approaches the extinct state from $(E(0), S(0), L(0)) = (0.6, 0.4, 0.0)$. Commonly, $e_S = 0.08$; $c = 0.2$; $e_L = 0.5$; $r = 0.02$; $m = 2.2$

at a certain moment, while it asymptotically approaches a persistent state if the frequency of empty patches is below the threshold.

The existence of such a bistable situation can be illustrated also by the bifurcation diagram as shown in Fig. 5.12. It is indicated that there are threshold values μ_c and ν_c for parameters m/r and c/e_S respectively such that

When $\dfrac{m}{r} \leq \mu_c$,
$$\begin{cases} \text{the population goes extinct if } \dfrac{c}{e_S} < \dfrac{e_L}{r}; \\[2ex] \text{the population approaches the persistent state } E_s \text{ if } \dfrac{c}{e_S} > \dfrac{e_L}{r}; \end{cases}$$

When $\dfrac{m}{r} > \mu_c$,
$$\begin{cases} \text{the population goes extinct if } \dfrac{c}{e_S} < \nu_c; \\[2ex] \text{a bistable situation appears if } \nu_c < \dfrac{c}{e_S} < \dfrac{e_L}{r}; \\[2ex] \text{the population approaches the persistent state } E_s \text{ if } \dfrac{c}{e_S} > \dfrac{e_L}{r}. \end{cases}$$

The threshold value μ_c is given by $e_L/r + 1$, while ν_c must depend on m/r and e_L/r as seen in Fig. 5.13.

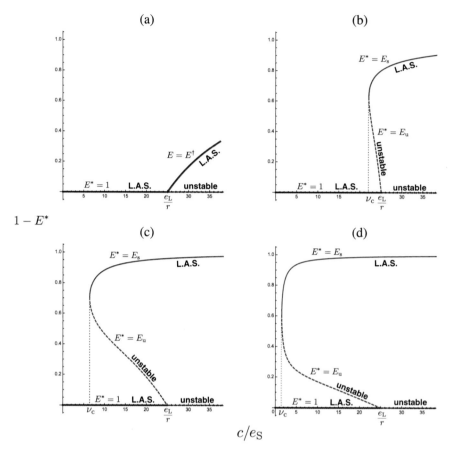

Fig. 5.12 Numerically drawn bifurcation diagrams about the value $1 - E^* = S^* + L^*$ at the equilibrium state for the 3-state metapopulation model (5.30). (**a**) $m = 0$; (**b**) $m/r = 30.0$; (**c**) $m/r = 50.0$; (**d**) $m/r = 110.0$. Commonly, $e_L/r = 25.0$. In figures, "L.A.S." means "locally asymptotically stable"

The bifurcation at $c/e_S = e_L/r$ when $m/r \leq \mu_c$ in Fig. 5.12a is classified in what is called *supercritical bifurcation* in the dynamical system theory, at which one equilibrium switches its stability from asymptotically stable to unstable while the other asymptotically stable equilibrium comes to exist at the same time. In contrast, the bifurcation at $c/e_S = e_L/r$ when $m/r > \mu_c$ in Fig. 5.12c–d is called *subcritical bifurcation*. The subcritical bifurcation has a structure called *hysteresis*. On the other hand, the bifurcation at $c/e_S = \nu_c$ is what is called *saddle-node bifurcation*.

Fig. 5.13 Parameter
dependence of the asymptotic
behavior for the 3-state
metapopulation model (5.30).
Numerical result for
$e_L/r = 25.0$. There are
possible three different
asymptotic behavior,
indicated by thick black solid
boundaries

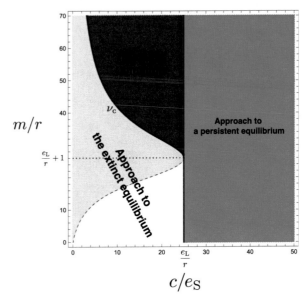

It is clearly indicated by Figs. 5.12a and 5.14a that, when there is no non-trivial equilibrium, the extinct equilibrium $(1, 0)$ is globally asymptotically stable (refer to Sect. 14.8). Even when a non-trivial equilibrium exists, the extinct equilibrium $(1, 0)$ is locally asymptotically stable if the parameter c/e_S is not so large such that $c/e_S < e_L/r$. When the parameter c/e_S is sufficiently large and a unique non-trivial equilibrium (E^*, L^*) exists with $E^* = E_s$, equilibrium (E^*, L^*) is globally asymptotically stable as seen in Fig. 5.14b.

Exercise 5.5 With the local stability analysis, show that the non-trivial equilibrium (E^*, L^*) is a locally asymptotically stable node (sink) or saddle when $m > 0$.

As a consequence, the population goes extinct when the immigration from the patches of large subpopulation to the empty patch is little active with small c, while it persists when it is sufficiently active with large c. However, even when the immigration from the patches of large subpopulation to the empty patch is little active, the population may persist with an initial condition if the immigration from the patches of large subpopulation to the patch of small subpopulation is sufficiently active. This is a bistable case as shown in Figs. 5.11 and 5.12b–d. The *rescue effect* called by Hanski [7] belongs to this case. In other words, the immigration from the patches of large subpopulation can work as a rescue effect to reduce the frequency of patches with small subpopulation, and subsequently it can decrease the probability of extinction.

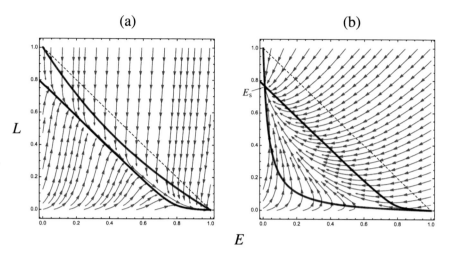

Fig. 5.14 Nullclines and vector flows in the (E, L)-phase plane. Numerically drawn for (**a**) $c = 0.05$ when there is no non-trivial equilibrium; (**b**) $c = 2.5$ when there is a unique non-trivial equilibrium. Commonly, $e_S = 0.08$; $e_L = 0.5$; $r = 0.02$; $m = 2.2$

Answer to Exercise

Exercise 5.1(p. 142)

The logistic equation (5.10) can be rewritten as

$$\frac{dN}{dt} = \left(\frac{r_0}{2\beta}\right)^2 - \beta\left(N - \frac{r_0}{2\beta}\right)^2.$$

The right side takes maximum when $N = r_0/2\beta$, that is, the net growth rate of population size dN/dt takes maximum $(r_0/2\beta)^2$ when the population size becomes the half saturation value $r_0/2\beta$.

Exercise 5.2 (p. 148)

As mentioned in Sect. 5.5, the recurrence relation (5.24) is mathematically equivalent to the logistic map (2.13) in Sect. 2.1.3 and (12.4) in Sect. 12.1.3. Thus, as described in Sects. 2.1.3 and 12.1.3, it is necessary for a monotonic approach to the equilibrium r_0/β that $1 < \mathcal{R}_0 = 1 + r_0\Delta t \leq 2$. Since $r_0 > 0$ and $\Delta t > 0$, this necessary condition leads to the following condition about the time step size Δt:

$$\Delta t \leq \frac{1}{r_0}. \tag{5.34}$$

We must remark that the recurrence relation (5.24) becomes nonsense as an approximation for the logistic equation for $\widetilde{N}(0) \geq r_0/\beta + 1/(\beta \Delta t)$, since the subsequent sequence $\{\widetilde{N}(\Delta t), \widetilde{N}(2\Delta t), \widetilde{N}(3\Delta t), \widetilde{N}(4\Delta t), \ldots\}$ becomes negative or all zero. Hence, we are going to consider the recurrence relation (5.24) only for the initial value $\widetilde{N}(0) < r_0/\beta + 1/(\beta \Delta t)$.

Next, when $1 < \mathcal{R}_0 \leq 2$, the cobwebbing method (refer to Sect. 12.1.2) for the logistic map (12.4) shows that the orbit $\{\widetilde{N}(0), \widetilde{N}(\Delta t), \widetilde{N}(2\Delta t), \widetilde{N}(3\Delta t), \ldots\}$ from the initial value $\widetilde{N}(0) < r_0/\beta$ necessarily approaches equilibrium r_0/β in a monotonically increasing manner, as seen from Fig. 12.2b in Sect. 12.1.3, because the derivative of the graph at the intersection with the line of $x_{k+1} = x_k$, that is, at equilibrium r_0/β, is non-negative and less than 1. Therefore, the condition (5.34) is sufficient for the sequence $\{\widetilde{N}(k\Delta t)\}$ to approach equilibrium r_0/β in a monotonically increasing manner when $\widetilde{N}(0) < r_0/\beta$.

For the case where $r_0/\beta < \widetilde{N}(0) < r_0/\beta + 1/(\beta \Delta t)$, the same arguments as the cobwebbing method on the logistic map (12.4) can be applied (see Fig. 12.2b). First, let us consider the case of $\mathcal{R}_0 = 1 + r_0 \Delta t = 2$, that is, when $\Delta t = 1/r_0$. In this case, the graph of the logistic map has the intersection with the line of $x_{k+1} = x_k$ at the extremal maximum of the graph, so that the derivative is zero at the intersection. It can be easily found from the recurrence relation (5.24) that $0 < \widetilde{N}(\Delta t) < r_0/\beta$ if $r_0/\beta < \widetilde{N}(0) < r_0/\beta + 1/(\beta \Delta t)$. Thus, the subsequent sequence $\{\widetilde{N}(\Delta t), \widetilde{N}(2\Delta t), \widetilde{N}(3\Delta t), \widetilde{N}(4\Delta t), \ldots\}$ must approach equilibrium r_0/β in a monotonically increasing manner, as already described in the above. As a result, when $\Delta t = 1/r_0$, we have $\widetilde{N}(0) > \widetilde{N}(\Delta t)$ and $\widetilde{N}(k\Delta t) < \widetilde{N}((k+1)\Delta t)$ for any $k > 0$. This means that the orbit from $\widetilde{N}(0) \in (r_0/\beta, r_0/\beta + 1/(\beta \Delta t))$ does not monotonically approach r_0/β but does in a manner like the upper second figure in Fig. 2.9 (p. 43) of Sect. 2.1.3.

Next let us consider the case of $\mathcal{R}_0 = 1 + r_0 \Delta t < 2$, that is, when $\Delta t < 1/r_0$. As exemplified in Fig. 12.2b, the graph of the logistic map has the intersection with the line of $x_{k+1} = x_k$ at a point in the left side of the extremal maximum of the graph. Thus, the cobwebbing method can show that $\widetilde{N}(k\Delta t) < \widetilde{N}((k-1)\Delta t)$ and $\widetilde{N}(k\Delta t) > r_0/\beta$ for any $k > 0$ if

$$\frac{r_0}{\beta} < \widetilde{N}(0) < \widetilde{N}_c := \frac{1}{\beta \Delta t}, \tag{5.35}$$

where $\widetilde{N}_c$ is derived from the root of the following equation of N such that $\widetilde{N}_c > r_0/\beta$:

$$N + (r_0 - \beta N)N\Delta t = \frac{r_0}{\beta}.$$

It is the larger intersection of the graph of the logistic map with the horizontal line of the value r_0/β. As a result, when $\Delta t < 1/r_0$, if the condition (5.35) is satisfied, the sequence $\{\widetilde{N}(k\Delta t)\}$ approaches r_0/β in a monotonically decreasing manner.

Further, it can be easily shown by the cobwebbing method that, when $\widetilde{N}(0) > \widetilde{N}_c$, we have $\widetilde{N}(\Delta t) < r_0/\beta$, so that the orbit $\{\widetilde{N}(k\Delta t)\}$ does not become monotonic but approaches r_0/β in a manner like the upper second figure in Fig. 2.9 (p. 43) of Sect. 2.1.3, as in the previous case mentioned above.

Lastly, let us consider the case of $\widetilde{N}(0) = \widetilde{N}_c$. In this case, we have $\widetilde{N}(\Delta t) = r_0/\beta$ and $\widetilde{N}(k\Delta t) = r_0/\beta$ for any $k > 0$. So the sequence $\{\widetilde{N}(k\Delta t)\}$ can be regarded as monotonically decreasing in a mathematically wide sense.

Consequently, for the initial value $\widetilde{N}(0) \in (r_0/\beta, r_0/\beta + 1/(\beta\Delta t))$, the condition (5.35) is necessary and sufficient for a monotonic approach of the sequence $\{\widetilde{N}(k\Delta t)\}$ to the equilibrium r_0/β in a decreasing manner. Hence, from the condition (5.35), we have

$$\Delta t \leq \frac{1}{\beta\widetilde{N}(0)}. \tag{5.36}$$

We note that $1/\{\beta\widetilde{N}(0)\} < 1/r_0$ in this case, since this is for $\widetilde{N}(0) > r_0/\beta$. Besides, when the condition (5.36) is satisfied, the initial value $\widetilde{N}(0)$ is necessarily less than $r_0/\beta + 1/(\beta\Delta t)$.

As a result from the conditions (5.34) and (5.36), if and only if

$$\Delta t \leq \min\left[\frac{1}{\beta\widetilde{N}(0)}, \frac{1}{r_0}\right] = \frac{1}{r_0}\min\left[\frac{1}{\beta\widetilde{N}(0)/r_0}, 1\right], \tag{5.37}$$

the sequence $\{\widetilde{N}(k\Delta t)\}$ generated by the recurrence relation (5.24) becomes monotonic to approach r_0/β.

Exercise 5.3 (p. 149)

As mentioned in Sect. 5.5, the recurrence relation (5.25) is mathematically equivalent to Ricker model (2.7) in Sect. 2.1.2. We shall now apply the logically same arguments as those in Exercise 5.2.

As described in Sect. 2.1.2, it is necessary for a monotonic approach to the equilibrium r_0/β that $1 < \mathcal{R}_0 = e^{r_0\Delta t} \leq e$. Since $r_0 > 0$ and $\Delta t > 0$, this necessary condition leads to the condition about the time step size Δt (5.34) same as for Exercise 5.2 about the recurrence relation (5.24).

Next, when $1 < \mathcal{R}_0 \leq e$, by the same arguments as in Exercise 5.2 with the cobwebbing method (refer to Sect. 12.1.2) for the logistic map (12.4), it can be shown that the orbit $\{\widetilde{N}(0), \widetilde{N}(\Delta t), \widetilde{N}(2\Delta t), \widetilde{N}(3\Delta t), \ldots\}$ from the initial value $\widetilde{N}(0) < r_0/\beta$ necessarily approaches equilibrium r_0/β in a monotonically increasing manner (refer to Fig. 5.15). Therefore, the condition (5.34) is sufficient for an approach of the sequence $\{\widetilde{N}(k\Delta t)\}$ from $\widetilde{N}(0) < r_0/\beta$ to equilibrium r_0/β in a monotonically increasing manner.

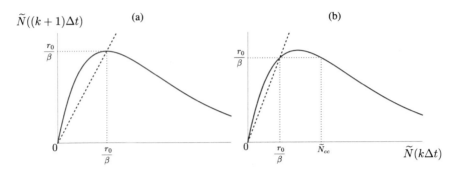

Fig. 5.15 The illustrative graph of Ricker model (5.25). (**a**) $\Delta t = 1/r_0$; (**b**) $\Delta t = 1/r_0$.

For $\widetilde{N}(0) > r_0/\beta$, let us first consider the case of $\mathscr{R}_0 = e^{r_0 \Delta t} = e$, that is, when $\Delta t = 1/r_0$, as in Exercise 5.2. In this case, as seen in Fig. 5.15a, the graph of Ricker model has the intersection with the line of $N_{k+1} = N_k$ at the extremal maximum of the graph, so that the derivative is zero at the intersection. It can be easily found from the recurrence relation (5.25) that $\widetilde{N}(\Delta t) < r_0/\beta$ if $\widetilde{N}(0) > r_0/\beta$. Thus, the subsequent sequence $\{\widetilde{N}(\Delta t), \widetilde{N}(2\Delta t), \widetilde{N}(3\Delta t), \widetilde{N}(4\Delta t), \ldots\}$ must approach equilibrium r_0/β in a monotonically increasing manner, as already described in the above. As a result, when $\Delta t = 1/r_0$, we have $\widetilde{N}(0) > \widetilde{N}(\Delta t)$ and $\widetilde{N}(k\Delta t) < \widetilde{N}((k+1)\Delta t)$ for any $k > 0$. This means that the orbit does not monotonically approach r_0/β from $\widetilde{N}(0) > r_0/\beta$.

Next let us consider the case of $\mathscr{R}_0 = e^{r_0 \Delta t} < e$, that is, when $\Delta t < 1/r_0$. In this case, the graph of Ricker model has the intersection with the line of $N_{k+1} = N_k$ at a point in the left side of the extremal maximum of the graph, as seen in Fig. 5.15b. Thus, the cobwebbing method can show that $\widetilde{N}(k\Delta t) < \widetilde{N}((k-1)\Delta t)$ and $\widetilde{N}(k\Delta t) > r_0/\beta$ for any $k > 0$ if

$$\frac{r_0}{\beta} < \widetilde{N}(0) < \widetilde{N}_{cc}, \tag{5.38}$$

where $\widetilde{N}_{cc}$ is the root of the following equation of N such that $\widetilde{N}_{cc} > r_0/\beta$:

$$N e^{(r_0 - \beta N)\Delta t} = \frac{r_0}{\beta}.$$

It is the larger intersection of the graph of Ricker model with the horizontal line of the value r_0/β. As a result, when $\Delta t < 1/r_0$, if the condition (5.38) is satisfied, the sequence $\{\widetilde{N}(k\Delta t)\}$ approaches r_0/β in a monotonically decreasing manner.

Further, it can be easily shown by the cobwebbing method that, when $\widetilde{N}(0) > \widetilde{N}_{cc}$, we have $\widetilde{N}(\Delta t) < r_0/\beta$, so that the orbit $\{\widetilde{N}(k\Delta t)\}$ does not become monotonic.

Lastly, let us consider the case of $\widetilde{N}(0) = \widetilde{N}_{cc}$. In this case, we have $\widetilde{N}(\Delta t) = r_0/\beta$ and $\widetilde{N}(k\Delta t) = r_0/\beta$ for any $k > 0$. So the sequence $\{\widetilde{N}(k\Delta t)\}$ can be regarded as monotonically decreasing in a mathematically wide sense.

Consequently, for the initial value $\widetilde{N}(0) > r_0/\beta$, the condition (5.38) is necessary and sufficient for a monotonic approach of the sequence $\{\widetilde{N}(k\Delta t)\}$ to equilibrium r_0/β in a decreasing manner. The condition (5.38) is mathematically equivalent to

$$\widetilde{N}(0)e^{\{r_0 - \beta\widetilde{N}(0)\}\Delta t} \geq \frac{r_0}{\beta}$$

as easily seen from Fig. 5.15b. That is, we find the following condition in terms of Δt:

$$\Delta t \leq \frac{1}{r_0} \frac{\ln[\beta\widetilde{N}(0)/r_0]}{\beta\widetilde{N}(0)/r_0 - 1}. \tag{5.39}$$

It can be easily shown that the right side of (5.39) is greater than $1/r_0$ for $\widetilde{N}(0) < r_0/\beta$, and less than $1/r_0$ for $\widetilde{N}(0) > r_0/\beta$, while it converges to $1/r_0$ as $\widetilde{N}(0) \to r_0/\beta$.

As a result from the conditions (5.34) and (5.39), if and only if

$$\Delta t \leq \frac{1}{r_0} \min\left[\frac{\ln[\beta\widetilde{N}(0)/r_0]}{\beta\widetilde{N}(0)/r_0 - 1}, 1\right], \tag{5.40}$$

the sequence $\{\widetilde{N}(k\Delta t)\}$ generated by the recurrence relation (5.25) becomes monotonic to approach r_0/β.

Now we shall compare the recurrence relation (5.25) with (5.24) with respect to which could be regarded as the better approximation for the logistic equation (5.10). Since the recurrence relation (5.24) can become an approximation for the logistic equation only for the initial value $\widetilde{N}(0) \in (0, r_0/\beta + 1/(\beta\Delta t))$, let us compare the recurrence relation (5.25) with (5.24) according to the condition for the monotonicity about the sequence generated by the recurrence relation only for $\widetilde{N}(0) \in (0, r_0/\beta + 1/(\beta\Delta t))$.

For $\widetilde{N}(0) \in (0, r_0/\beta]$, the obtained conditions (5.37) and (5.40) for the recurrence relations (5.24) and (5.25) respectively are identically given by (5.34). Hence for $\widetilde{N}(0) \in (0, r_0/\beta]$, they could be regarded as compatible with respect to the monotonicity of the sequence generated by them.

For $\widetilde{N}(0) > r_0/\beta$, we can prove that the following equality holds:

$$\frac{1}{\beta\widetilde{N}(0)/r_0} < \frac{\ln[\beta\widetilde{N}(0)/r_0]}{\beta\widetilde{N}(0)/r_0 - 1}.$$

Therefore, for $\widetilde{N}(0) \in (r_0/\beta, r_0/\beta + 1/(\beta\Delta t))$, the condition (5.40) is weaker than (5.37). Moreover, the condition (5.40) is applicable also for $\widetilde{N}(0) \geq r_0/\beta + 1/(\beta\Delta t)$. That is, the recurrence relation (5.25) can be used as an approximation for the logistic equation even for $\widetilde{N}(0) \geq r_0/\beta + 1/(\beta\Delta t)$ under the condition (5.40).

As a consequence from these arguments, it is implied that the recurrence relation (5.25) could be regarded as better than (5.24) according to the monotonicity of the sequence generated by the recurrence relation.

Exercise 5.4 (p. 155)

For the non-trivial equilibrium (E^*, L^*) of the system (5.31), we can derive the linearized system around (E^*, L^*) with the following Jacobian matrix A (refer to Sect. 14.2):

$$A = \begin{pmatrix} -a^* & -b^* \\ -c^* & -d^* \end{pmatrix}, \tag{5.41}$$

where a^*, b^*, c^*, and d^* are positive elements defined as

$$a^* := e_S + cL^*; \quad b^* := e_S + cE^*; \quad c^* := r + mL^*;$$

$$d^* := 2mL^* + r + e_L - m(1 - E^*) = r + mL^* + \frac{re_L}{r + mL^*}.$$

As described in Sect. 14.3, we can determine the local stability of (E^*, L^*) by the eigenvalues of A. The characteristic equation to determine the eigenvalue λ for A becomes

$$\lambda^2 + (a^* + d^*)\lambda + a^*d^* - b^*c^* = 0. \tag{5.42}$$

For the equilibrium (5.33) when $m = 0$, the characteristic equation (5.42) becomes

$$\lambda^2 + (e_S + cL^* + r + e_L)\lambda + cr(1 - E^*) = 0.$$

When the equilibrium (5.33) exists, we can easily find that the discriminant for this characteristic equation is necessarily positive, and two eigenvalues are different negative values, making use of the relation between the roots and coefficients. Therefore, the equilibrium (5.33) is locally asymptotically stable as a stable node, that is, a sink when it exists.

Exercise 5.5 (p. 158)

About Jacobian matrix A for the non-trivial equilibrium (E^*, L^*) of (5.31), given by (5.41) shown in the answer for Exercise 5.4, we have tr $A = a^* + d^* > 0$. Thus, by the relation between the roots and coefficients of the characteristic equation (5.42), if the eigenvalue is imaginary only when det $A = a^* d^* - b^* c^* > 0$, then the real part must be negative. If the eigenvalue is real, both eigenvalues are negative when det $A > 0$, while they have different signs when det $A < 0$. Therefore, when a non-trivial equilibrium (E^*, L^*) exists for (5.31), it is a locally asymptotically stable node (sink) if det $A > 0$, while it is unstable as a saddle if det $A < 0$. This result implies that, in a bistable case with two non-trivial equilibria, one is a sink and the other is a saddle.

References

1. N. Bacaër, A Short History of Mathematical Population Dynamics (Springer, London, 2011)
2. H. Fujita, S. Utida, The effect of population density on the growth of an animal population. Ecology **34**(3), 488–498 (1953)
3. M.E. Gilpin, F.J. Ayala, Global models of growth and competition. Proc. Natl. Acad. Sci. U.S.A. **70**, 3590–3593 (1973)
4. B. Gompertz, On the nature of the function expressive of the law of human mortality and on a new mode of determining life contingencies. Philos. Trans. Boy. Soc. Lond. A **115**, 513–585 (1825)
5. I. Hanski, Single-species spatial dynamics may contribute to long-term rarity and commonness. Ecology **66**, 335–343 (1985)
6. I. Hanski, Single-species metapopulation dynamics: concepts, models and observations. Biol. J. Linnean Soc. **42**, 17–38 (1991)
7. I. Hanski, Metapopulation Ecology. Oxford Series in Ecology and Evolution (Oxford University Press, Oxford, 1999)
8. F. John, Lectures on Advanced Numerical Analysis (Gordon and Breach, Science Publishers, New York, 1967)
9. E. Kreyszig, Advanced Engineering Mathematics, 10th edn. (Wiley, New York, 2011)
10. R. Levins, Some demographic and genetic consequences of environmental heterogeneity for biological control. Bull. Entomol. Soc. Am. **15**, 237–240 (1969)
11. R. Levins, Extinction, in: Some Mathematical Problems in Biology, ed. by M. Gerstenhaber, Lectures on Mathematics in the Life Sciences, vol. 2 (American Mathematical Society, Rhode Island, 1970), pp. 75–107
12. H. Seno, Some time-discrete models derived from ODE for single-species population dynamics: Leslie's idea revisited. Scientiae Mathematicae Japonicae **58**(2), 389–398 (2003)
13. F.E. Smith, Population dynamics in Daphnia magna and a new model for population growth. Ecology **44**, 651–663 (1963)
14. P.F. Verhulst, Notice sur la loi que la population suit dans son accroissement. Correspond. Math. Phys. **10**, 113–121 (1838)
15. P.F. Verhulst, Recherches mathématiques sur la loi d'accroissement de la population. Nouv. mém. de l'Academie Royale des Sci. et Belles-Lettres de Bruxelles **18**, 1–41 (1845)
16. S. Wright, Review of The Biology of Population Growth, by Raymond Pearl. J. Amer. Stat. Assoc. **21**, 493–497 (1926)

Chapter 6
Modeling of Interspecific Reaction

Abstract This chapter is devoted to the description of the idea to model the interspecific reaction for the continuous time population dynamics model. The contents could be rarely found in the other textbooks, though the topics are popular in mathematical biology.

6.1 Mass Action Type of Interaction

In this section, we shall describe an idea of mathematical modeling with the *mass action assumption* which is a basic modeling for the chemical kinetics. It provides some mathematical idea applied for the modeling of intra- and inter-specific reaction in biological population.

6.1.1 Mass Action Assumption

In the theory of chemical kinetics, the velocity of chemical reaction in a closed constant volume is defined by the velocity of the change in the concentration of a substrate or product by the reaction. Let us consider now the following simplest reaction as an example:

$$\alpha A + \beta B \to \gamma C, \tag{6.1}$$

where α, β, and γ are positive constants. The reaction of two substrates A and B makes the product C. For this chemical reaction, the velocity of the change in the concentration of chemical substances is mathematically expressed by $-d[A]/dt$, $-d[B]/dt$, and $d[C]/dt$ respectively, where [A], [B], and [C] mean

H. Seno, *A Primer on Population Dynamics Modeling*, Theoretical Biology,
https://doi.org/10.1007/978-981-19-6016-1_6

the corresponding concentrations. Since the definition of the velocity of the above chemical reaction $\mathcal{V}$ must be unique, it is defined as

$$\mathcal{V} = -\frac{1}{\alpha}\frac{d[A]}{dt} = -\frac{1}{\beta}\frac{d[B]}{dt} = \frac{1}{\gamma}\frac{d[C]}{dt}.$$

With this definition, the velocity is uniquely determined independently of which substrate or product is used to measure it.

Generally, the velocity of the chemical reaction is a function of the concentrations of all substances $c_1, c_2, \ldots, c_j$ involved in the reaction: $\mathcal{V} = \mathcal{V}(c_1, c_2, \ldots, c_j)$. The function of $\mathcal{V}$ is called *rate law* of the reaction in the theory of chemical kinetics [1]. In many cases, it is given by a power function like

$$\mathcal{V} = \mathcal{V}(c_1, c_2, \ldots, c_j) = \kappa\, c_1^{n_1} c_2^{n_2} c_3^{n_3} \cdots c_j^{n_j} \tag{6.2}$$

which may be an approximation for the rate law. The function significantly depends on the physical condition under which the chemical reaction occurs. The rate equation (6.2) is called *law of kinetic mass action* [1]. The constant κ is called *rate constant* for the chemical reaction, and each exponent n_k the *reaction order* or *kinetic order* for each substance in the chemical reaction. The sum of exponents, $n_1 + n_2 + \cdots + n_j$, may be used as the reaction order for the chemical reaction itself. In the theory of chemical kinetics, the *law of mass action* means that the chemical reaction *at the equilibrium state* satisfies that

$$c_1^{n_1} c_2^{n_2} c_3^{n_3} \cdots c_j^{n_j} = \text{constant}.$$

This law can be derived by the arguments of thermodynamics. From the dynamical system theory, this is the relation among chemical substances when the velocity of the chemical reaction reaches zero, which indicates the termination of reaction *or* the dynamical equilibrium where forward and backward reactions are balanced with each other in terms of the velocity.

As a fundamental modeling in mathematical biology, the power function like (6.2) has been applied for the velocity of the change in the population size (density). In such a modeling, a product of powered sizes (densities) appears in the formula for the velocity. It may be called the modeling with the *mass action assumption* for the intra- or the inter-specific reaction. It must be remarked that the mass action assumption in mathematical biology is not applied for the equilibrium state in the population dynamics. It may be regarded as corresponding to the law of kinetic mass action mentioned above with the rate law (6.2). For this reason, the mass action assumption in mathematical biology is different from the law of mass action in the chemical kinetics as the definition.

6.1.2 Lokta-Volterra Type of Interaction

We shall call here the mass action assumption with every reaction order of unity *Lotka-Volterra type of interaction* in the population dynamics modeling especially about the interspecific reaction. That is, according to the Lotka-Volterra type of interaction, the contribution of an interspecific reaction to the velocity of the change in the population size is introduced by the product of sizes (densities) of reacting populations (refer to Sects. 7.1 and 8.4). This may be the most conventional definition of the Lotka-Volterra type of interaction in mathematical biology.

This naming of the modeling is after an American scientist Alfred J. Lotka (1880–1949) and an Italian scientist Vito I. Volterra (1860–1940), especially after Lotka's works on the host-parasite system for the pest control in 1925 [9, 10] and Volterra's work on the fish prey-predator system of the Adriatic Sea in 1926 [16]. They indeed used the mass action assumption to introduce the interspecific reaction.

For such a modeling with the Lotka-Volterra type of interaction, some literatures describe it as follows:

> Consider an individual and the frequency of its encounters to the others. The frequency must become larger as the population density around the individual gets larger. Suppose that every individual is located at random in space, and moves at random. Further, assume that the spatial distribution of population density can be always approximated by a uniform distribution. This assumption may be called the assumption (or approximation) of *complete mixing* or *perfect mixing*. With this assumption, the frequency of encounters to the others can be assumed to be proportional to the population density.

The assumption in this description would appear natural since the *expected* number of the others in a fixed range around an arbitrarily chosen individual is proportional to the density under the complete mixing or under the uniform density distribution.

> If the population density is constant independently of time, the process of the random encounter to the others can be regarded as a Poisson process (refer to Sect. 15.1). The number of the encounters to the others follows a Poisson distribution (refer to Sect. 15.2). The modeling of the Lotka-Volterra type of interaction in the above corresponds to the case where the mean of the Poisson distribution is proportional to the population density.

Besides, as a mathematical simplification, let us assume that the duration for the interaction between individuals is ignored in comparison to the time scale of the change in the population size, so that such an interaction affects only the velocity of the change in the population size at each moment of the encounter. Following the temporal change in the population density under the complete mixing, the frequency of encounters to the others contributes to the momental velocity (rate) of interaction between individuals at each time.

Along this description of the Lotka-Volterra type of interaction, it is nonsense in general to introduce the contribution of an interaction among more than two individuals. Such an interaction among more than two individuals must occur by an encounter among them at the same moment because of the above assumption. Generally, the probability of such an encounter among more than two individuals at a moment could be regarded as a higher order of likelihood than that of an encounter between two individuals. Such an event with the higher order of likelihood hardly occurs in comparison to the event with the lower order of likelihood. Let us now denote the probability of an encounter between individuals A and B by p. The corresponding probability between individuals B and C is given also by p, since the encounter is assumed to be random. Even with this simple assumption, we may note that the probability of such an encounter among more than two individuals at a moment must have the order of p^2. More precisely in a modeling sense, such an encounter among more than two individuals at a moment would require some further conditions for its occurrence, and have the higher order. For this reason, the contribution of the interaction among more than two individuals to the velocity of the change in the population size is ignored about the mathematical modeling for population dynamics in general.

In actual population dynamics, the interaction between individuals takes a certain duration, and the spatial distribution of individuals could not be random or uniform. However, many mathematical models with the Lotka-Volterra type of interaction have been playing an important role to discuss a variety of actual population dynamics in the history of biological science. One of the reasons may be that the long-term or the large-scale characteristics of such population dynamics could match those of the mathematical model with the Lotka-Volterra type of interaction.

For example, even if each interaction between individuals does not have any spatio-temporal randomness or sufficiently short duration, the occurrence of the interaction could be approximated well by a random distribution over time and space. This is an approximation, what is called *mean field approximation*. Roughly saying, the non-randomness and non-instantaneousness of a population dynamics could cancel out in the mean over the population. This approximation would be applicable especially for a population with a sufficiently large size or high density. In contrast, for a population with a small population size or low density, the mean field approximation may be inappropriate for the modeling.

Although the Lotka-Volterra type of interaction is conventionally mentioned as a mathematical modeling for the interspecific reaction, the above-mentioned concept of its modeling is applicable even for the intraspecific reaction. When an interaction between individuals within a population has a contribution to the velocity of the change in the population size, the mass action assumption introduces the modeling by the square of the population density, as we shall describe in the next section.

6.1.3 Logistic Equation by Lotka-Volterra Type of Interaction

In this section, we shall see the modeling of a single species population with the mass action assumption to result in the logistic equation described in Sect. 5.3.

Now let us assume that the per capita growth rate of a population is given by a constant r_0. This means that the population grows exponentially with the malthusian coefficient r_0 (Sect. 5.1). We introduce an additional assumption that an interaction between individuals within the population causes a decrease in the population size. We may imagine a fatal battle or cannibalism for example. As mentioned in the previous section, we shall apply here the mass action assumption, that is, the Loka-Volterra type of interaction for the contribution of such an interaction to the population dynamics. Thus, we introduce it as the square of population density in the modeling for the velocity of the change in the population size. Hence, the population dynamics consists of Malthus growth and a process to decrease the population size by the Lotka-Volterra type of intraspecific reaction. Now we have the following population dynamics model:

$$\frac{dN}{dt} = r_0 N - \gamma N^2, \tag{6.3}$$

where the positive parameter γ indicates the strength of the influence of interaction on the population growth. As easily seen, this model is mathematically equivalent to the logistic equation (5.10) in Sect. 5.3.

It should be remarked that the above modeling does not include any density effect on the per capita growth rate, but assumes a process to decrease the population size itself. More clearly, the second term in the right side of (6.3) has no relation to the reproduction. For this reason, *the logistic equation* (6.3) *must be regarded as different in a modeling sense from the logistic equation* (5.10) with the per capita growth rate (5.9) under the density effect described in Sect. 5.3.

For this model, if we introduce a time-dependence in the malthusian coefficient, $r_0 = r_0(t)$, it means the temporal change of the intrinsic growth rate, which may be regarded as an influence of the environmental change on the physiological condition of individual. In contrast, if we introduce a time-dependence in the parameter γ, $\gamma = \gamma(t)$, it means a temporal change in the strength of the interaction between individuals. It may be regarded as an effect of the seasonal change of aggressiveness or mobility. In this way, the temporal change of γ must be independent of that of the fertility in the above modeling.

Even for the logistic equation (6.3), we can formally derive the formula of the "per capita growth rate" $(1/N)dN/dt$. However, it does not mean the per capita growth rate in the meaning of "growth" by the reproduction. It includes the negative factor *converted* from the reduction rate of the population size by the intraspecific reaction.

With the general mass action assumption, the interaction can be introduced by a power term with an exponent θ (> 0). We may get the following equation regarded as a generalization of (6.3):

$$\frac{dN}{dt} = r_0 N - \gamma N^{\theta}. \tag{6.4}$$

This ordinary differential equation can be solved as the logistic equation (5.10) in Sect. 5.3, and the solution becomes (5.17) given in Sect. 5.3 with $\alpha = \theta - 1$.

For $\theta > 1$, the nature of the population dynamics by (6.4) is qualitatively similar with that of the logistic equation and its extension with (5.16) in Sect. 5.3. In contrast, for $\theta \leq 1$, it has a great difference from them. When $\theta = 1$, the equation (6.4) becomes a Malthus model with the malthusian coefficient $r_0 - \gamma$. Hence, the population size exponentially grows if $r_0 > \gamma$, while it exponentially decreases to cause the extinction if $r_0 < \gamma$ (refer to Sect. 5.1).

Peculiarly when $\theta < 1$, the per capita growth rate $(1/N)dN/dt$ and the net growth rate dN/dt are negative for sufficiently small density N, and become positive for sufficiently large density N, as shown in Fig. 6.1. With the phase

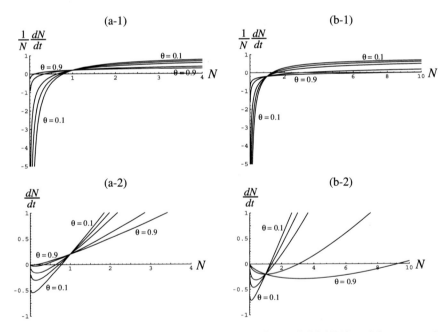

Fig. 6.1 The density dependence of the per capita growth rate $(1/N)dN/dt$ and the net growth rate dN/dt for (6.4) with $\theta < 1$. Numerically drawn with (**a**) $r_0 = 1$; $\gamma = 0.8$; $r_0/\gamma = 1.25 > 1$, (**b**) $r_0 = 0.8$; $\gamma = 1.0$; $r_0/\gamma = 0.8 < 1$. In each figure, the curves are for $\theta = 0.1, 0.3, 0.5, 0.8$, and 0.9

space argument described in Sect. 5.3, it is clear that the equilibrium $N = N^* = [r_0/\gamma]^{1/(\theta-1)}$ is unstable. For the initial value $N(0) < N^*$, the population size monotonically decreases toward zero, while, for $N(0) > N^*$, it monotonically increases unboundedly. In the latter case, since the term $r_0 N$ gets much larger than γN^θ for sufficiently large N, the population size is expected to asymptotically approach the exponential growth with the malthusian coefficient r_0.

In comparison to the logistic growth or the similar growth with $\theta > 1$, the population dynamics (6.4) with $\theta < 1$ does not approach any positive equilibrium like the carrying capacity, but approaches the extinctive state or increases unboundedly, depending on the initial value. Such a situation may be regarded as bistable in a wide sense. In the population dynamics (6.4), it is necessary for the interaction to *regulate* the population growth in order to lead the size to a carrying capacity. For the interaction term with the general mass action assumption, the case of $\theta < 1$ cannot work enough to regulate the growth by the reproduction with the intrinsic growth rate r_0. Moreover, when $\theta < 1$, it is satisfied that $N^\theta > N > N^2$ for sufficiently small $N < 1$. Thus, for sufficiently small population size, the interaction introduced by the term γN^θ is indeed stronger as θ gets smaller. This is the cause of the population extinction from sufficiently small initial value $N(0)$. In such a situation of sufficiently small population size, the interaction is so strong that the growth by the reproduction with the intrinsic growth rate r_0 cannot compensate the decrease due to the interaction. In contrast, when $\theta > 1$, the interaction can work enough to regulate the growth for large population size, and to become sufficiently weak for the growth under small population size.

6.1.4 Intraspecific Reaction and Density Effect

In the generalized model (6.4), there was no density effect on the reproduction itself, since it was constructed by introducing the intraspecific reaction to decrease the number of individuals itself. Now we shall consider the following model of population dynamics with the density effect on the reproduction in addition to the intraspecific reaction:

$$\frac{dN}{dt} = (r_0 - \beta N)N - \gamma N^\theta, \tag{6.5}$$

where the net growth rate is given by (5.9) in Sect. 5.3.

For this model, the population size does not diverge. When $\theta \geq 1$, the population size approaches a unique equilibrium in a monotonic manner for any positive initial value, similar as the logistic equation. Simply saying, when $\theta \geq 1$, the population dynamics has nature qualitatively similar with the logistic equation. In contrast, as illustratively shown in Fig. 6.2, the population dynamics with $\theta < 1$ shows characteristics different from the logistic growth.

Fig. 6.2 Numerical example of the temporal change in the population size by (6.5), with $r_0 = 1.2$; $\beta = 1.0$; $\gamma = 0.5$; $\theta = 0.5$. Curves of the temporal change from eight different initial values are drawn

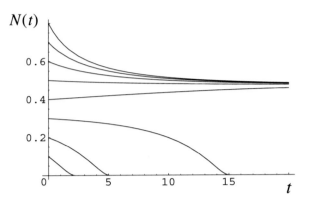

For the population dynamics with $\theta < 1$, there is the following positive critical value for the intrinsic growth rate r_0:

$$r_0^{\dagger} := \beta \left\{ \frac{\gamma(1 - \theta)}{\beta} \right\}^{1/(2-\theta)} + \gamma \left\{ \frac{\beta}{\gamma(1 - \theta)} \right\}^{(1-\theta)/(2-\theta)}. \tag{6.6}$$

The population goes extinct for any positive initial value if $r_0 < r_0^{\dagger}$. If $r_0 > r_0^{\dagger}$, there is a positive threshold value for the initial population size $N(0)$ such that the population monotonically goes extinct if $N(0)$ is less than it. If $N(0)$ is greater than it, the population size asymptotically approaches a positive equilibrium in a monotonic manner, as shown in Fig. 6.2. This may be regarded as a bistable case corresponding to that for the model in the previous section. Therefore, in this bistable case for the model (6.5) with $\theta < 1$, we find that the population dynamics involves indeed a strong Allee effect like Fig. 5.9b in p. 150 of Sect. 5.6. As seen from the bifurcation diagram in Fig. 6.3, a saddle-node bifurcation appears at $r_0 = r_0^{\dagger}$, which is similar to the 3-state metapopulation model (5.30) as shown in Fig. 5.12, whereas the whole structure of the bifurcation is different from each other.

For the critical case with $r_0 = r_0^{\dagger}$, the nature of population dynamics becomes qualitatively similar to the model (5.26) with $v/a = w/b$, which was described in the last part of Sect. 5.6. Mathematically, the unique positive equilibrium $\{\gamma(1 - \theta)/\beta\}^{1/(2-\theta)}$ is unstable, while the population size asymptotically approaches it from the initial value greater than it. The population size monotonically decreases toward zero from any positive initial value smaller than it.

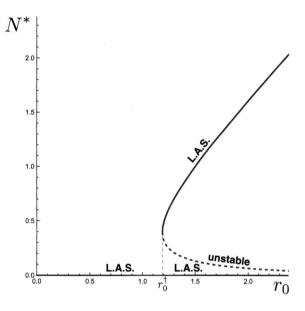

Fig. 6.3 Numerically drawn bifurcation diagrams in terms of the bifurcation parameter r_0 for the population dynamics model (6.5) with $\beta = 1.0; \gamma = 0.5; \theta = 0.5$, where $r_0^{\dagger} = 1.19055$. In figures, "L.A.S." means "locally asymptotically stable"

6.1.5   Consumer Population of Exhaustible Resource

In this section, we shall derive the logistic equation again with a modeling of the consumption of an exhaustible resource by a population, introducing the reaction between the resource and population by the Lotka-Volterra type of interaction.

Let $R(t)$ denote the available stock of a resource at time t, which is consumed by a population of size $N(t)$. Now, we introduce the velocity of the resource consumption by the Lotka-Volterra type of interaction between the resource and population, that is, with the product $N(t)R(t)$ at time t. Thus, we shall consider now the following model for their temporal change:

$$\begin{cases} \dfrac{dN(t)}{dt} = \gamma R(t)N(t); \\[2mm] \dfrac{dR(t)}{dt} = -\rho N(t)R(t), \end{cases} \tag{6.7}$$

where positive parameters γ and ρ mean respectively the per capita growth rate with a unit amount of resource and the coefficient of per capita resource consumption rate. We ignore the death and migration to cause the change of the population size. As already mentioned, the resource is exhaustible without any renewal process. Since the reproduction is possible to use the resource, the assumption implies that the population growth becomes dull as the resource gets consumed. For example, such a situation may be regarded as corresponding to a culture of microorganism in laboratory.

From (6.7), we can derive the following differential equation:

$$\frac{dR/dt}{dN/dt} = \frac{dR}{dN} = -\frac{\rho R N}{\gamma N R} = -\frac{\rho}{\gamma} \quad \text{(constant)}.$$

This equation indicates that dR/dN is constant independently of time. Integrating with N, we can easily solve it and get the following relation between N and R which is satisfied for any time t:

$$R(t) = -\frac{\rho}{\gamma} N(t) + C, \tag{6.8}$$

where C is the constant of integration, which must satisfy the following equation about the initial values $N(0)$ and $R(0)$ at time $t = 0$:

$$C = R(0) + \frac{\rho}{\gamma} N(0). \tag{6.9}$$

This result gives a *conservative quantity* $R(t) + \frac{\rho}{\gamma} N(t)$ for the system (6.7). The conservative quantity is uniquely determined by the initial values $N(0)$ and $R(0)$ from (6.8) and (6.9):

$$R(t) + \frac{\rho}{\gamma} N(t) = R(0) + \frac{\rho}{\gamma} N(0).$$

This equation describes the solution curve of $(N(t), R(t))$ in the (N, R)-phase plane too.

By substituting (6.8) and (6.9) for the right side of dN/dt in (6.7), we can obtain the following one dimensional ordinary differential equation about the temporal change of N:

$$\frac{dN(t)}{dt} = \{\gamma R(0) + \rho N(0) - \rho N(t)\} N(t). \tag{6.10}$$

The temporal change of the population size N follows the dynamics given by this ordinary differential equation which reflects the resource consumption in the background. It is clear that the equation (6.10) is mathematically equivalent to the logistic equation in Sect. 5.3. Thus, the population governed by the system (6.7) shows a logistic growth. The population size monotonically increases to approach an equilibrium, that is, a carrying capacity.

In the logistic growth, the carrying capacity is given by $N(0)+(\gamma/\rho)R(0)$, which depends on the initial values $N(0)$ and $R(0)$. This is significantly different from that for the logistic equation in Sect. 5.3 where the carrying capacity is determined independently of the initial value of the population size. The parameters in the logistic equation (5.10) in Sect. 5.3 are given as an innate nature for the population dynamics and the fixed environmental condition, so that the carrying capacity is determined by them as r_0/β. In constrast, the logistic growth by (6.10) is derived from the interaction between the resource and population, so that the carrying capacity is determined by the interaction itself.

Let us consider a modified model with a process of resource renewal, following the above modeling for (6.7):

$$\begin{cases} \dfrac{dN(t)}{dt} = \gamma R(t)N(t); \\ \dfrac{dR(t)}{dt} = -\rho N(t)R(t) + \epsilon R(t), \end{cases} \tag{6.11}$$

where the positive parameter ϵ is the resource renewal rate. Remark that this modeling of the resource renewal assumes a Malthus growth of the resource stock, so that it exponentially increases when the consumer population is absent, that is, when $N \equiv 0$. For the system (6.11), we can derive

$$\frac{dN}{dR} = \frac{\gamma N}{\epsilon - \rho N},$$

and find the conservative quantity

$$\epsilon \ln N(t) - \rho N(t) - \gamma R(t) = \epsilon \ln N(0) - \rho N(0) - \gamma R(0). \tag{6.12}$$

Hence, we can get the following equation to describe the temporal change of N in the same way:

$$\frac{dN(t)}{dt} = \left[\gamma R(0) + \epsilon \ln \frac{N(t)}{N(0)} - \rho \{N(t) - N(0)\} \right] N(t). \tag{6.13}$$

This is not the logistic equation. However, it can be easily shown that the population size N monotonically increases to approach an equilibrium like the logistic growth, while the resource becomes exhausted as well as in the previous model (6.7). Note that, in this modeling, the resource cannot be renewed once it is exhausted. For this reason, the resource may be, for example, a bioresource for the consumer, like a biological population as the food.

Exercise 6.1 Find how the population size N temporally varies by the following system of population-resource dynamics:

$$\begin{cases} \dfrac{dN(t)}{dt} = \gamma R(t)N(t); \\[2mm] \dfrac{dR(t)}{dt} = -\rho N(t)R(t) + A, \end{cases} \qquad (6.14)$$

where A is a positive constant. This is the model in which the resource has a constant inflow A from the outside of system. The resource stock linearly increases when the consumer population is absent.

6.2   Michaelis-Menten Type of Interaction

6.2.1   *Michaelis-Menten Reaction Velocity Equation*

In the theory of enzyme kinetics, there is a well-known reaction velocity equation called *Michaelis-Menten reaction velocity equation* or simply *Michaelis-Menten equation*. We shall describe the outline of its derivation with a specific mathematical treatment in this section, since the mathematical argument is useful and applicable for the modeling of population dynamics.

Enzyme is a macromolecule to work as a catalyst to effectively accelerate the chemical reaction even with relatively low concentration. It combines with a substrate and makes a chemical complex (coordination compound). Some enzymes work in most of biochemical reactions. The enzyme concentration is typically much low compared to the substrate concentration. So is the complex concentration in the enzyme kinetics. For this reason, in the theory of enzyme kinetics, it could be an appropriate approximation to suppose that the complex concentration is at a stationary state after a short-term state transition. We shall see here how this approximation works to derive an approximated reaction velocity equation in a simple enzyme kinetics.

Let us consider the following simplest enzyme kinetics with substrate S, enzyme E, complex X, and product P:

$$E + S \underset{k_{-1}}{\overset{k_1}{\rightleftharpoons}} X \underset{k_{-2}}{\overset{k_2}{\rightleftharpoons}} E + P. \qquad (6.15)$$

This enzyme kinetics is sometimes called *Michaelis-Menten structure*. The enzyme-substrate complex X may be called *Mihaelis-Menten complex* and sometimes mentioned as *ES complex*. The naming is after the chemists Leonor Michaelis (1875–1947) and Maud L. Menten (1879–1960) who theoretically studied the structure (6.15) and reasonably derived the reaction velocity equation (6.24) that we will derive by the following arguments.

This simple enzyme kinetics may be applied for a larger scale of biological reaction process. For example, we may consider a process such that a foreign substance like hormone, pathogen, or antigen operates on a cell of biological tissue, and the tissue is induced to produce another response matter like hormone, antigen, or antigen protein. In this process, E is the inactive receptor cell of the tissue, S the foreign substance, X the receptor cell activated by S, and P the response matter produced by the tissue. The parameter k_1 is the rate coefficient of the reaction between the foreign substance and receptor cell, and k_{-1} is the coefficient to index the probability of the dissociation of the foreign substance from the receptor cell. The receptor cell combined with the foreign substance produces the response matter with the production coefficient k_2. After the production of response matter, the receptor cell immediately loses its activity and returns to the inactive state. Since such a response matter could not combined with the receptor cell, it would be appropriate in this biological process to put $k_{-2} = 0$.

As well as in Sect. 6.1.1, let us denote the concentration of substance A by [A]. With the law of kinetic mass action in Sect. 6.1.1, we shall consider the following system of nonlinear ordinary differential equations as the dynamics to describe the temporal change of concentrations in the enzyme kinetics (6.15):

$$\frac{d[E]}{dt} = -k_1[E][S] - k_{-2}[E][P] + (k_{-1} + k_2)[X]; \tag{6.16}$$

$$\frac{d[X]}{dt} = k_1[E][S] + k_{-2}[E][P] - (k_{-1} + k_2)[X]; \tag{6.17}$$

$$\frac{d[S]}{dt} = -k_1[E][S] + k_{-1}[X]; \tag{6.18}$$

$$\frac{d[P]}{dt} = k_2[X] - k_{-2}[E][P]. \tag{6.19}$$

This is the system of an enzyme kinetics without any inflow or outflow of substances. Hence the above system satisfies the following equations for any time t as the conservation laws for it:

$$[E] + [X] = [E]_0; \tag{6.20}$$

$$[S] + [P] + [X] = [S]_0, \tag{6.21}$$

where $[E]_0$ and $[S]_0$ are respectively the initial concentrations of E and S before the reaction starts. At the same time, from (6.20) and (6.21), we have $d[E]/dt = -d[X]/dt$ and $-d[S]/dt = d[P]/dt + d[X]/dt$.

As mentioned before, we suppose that the enzyme concentration E is very low, and so is the complex concentration X, compared with the substrate and product concentrations, S and P: $[X] \ll [S]$ and $[X] \ll [P]$. For the conservation law (6.21), we can use the approximation

$$[S]_0 \approx [S] + [P]. \tag{6.22}$$

Further, we shall introduce an additional approximation that the temporal change in the complex concentration could be negligible, compared to that in substrate concentrations. This is what mentioned at the beginning of this section, and called *quasi-stationary state approximation (QSSA), quasi-steady state approximation (QSSA), stationary state approximation*, or *Briggs-Haldane approximation*. This approximation can be mathematically introduced as $d[X]/dt \approx 0$, that is, $d[E]/dt \approx 0$.

With (6.17) applied the QSSA and the conservation law (6.20), we can get the approximated expression of [E] and [X] by [S] and [P], and as a result, find the following approximated equation for the reaction velocity $\mathcal{V}$:

$$\mathcal{V} = -\frac{d[S]}{dt} \left(\approx \frac{d[P]}{dt} \right) \approx \frac{(V_m/K_m)[S] - (V_P/K_P)[P]}{1 + [S]/K_m + [P]/K_P}, \tag{6.23}$$

where

$$V_m = k_2[E]_0; \quad V_P = k_{-1}[E]_0; \quad K_m = \frac{k_{-1} + k_2}{k_1}; \quad K_P = \frac{k_{-1} + k_2}{k_{-2}}.$$

This equation is called *Michaelis-Menten reaction velocity equation* or *Michaelis-Menten equation* with *Michaelis constant* K_m. Those parameters K_m and K_P are related to the stationarity of the temporal change in the complex concentration.

When $k_{-2} = 0$ in the reaction system (6.15), for example, about a biological reaction process in a tissue, the Michaelis-Menten equation (6.23) becomes

$$\mathcal{V} = -\frac{d[S]}{dt} \left(\approx \frac{d[P]}{dt} \right) \approx \frac{(V_m/K_m)[S]}{1 + [S]/K_m}, = \frac{V_m}{1 + K_m/[S]}. \tag{6.24}$$

This reaction velocity has no dependence on the product concentration [P], and corresponds to the Eq. (6.23) with the limit as $K_P \to \infty$. In many cases, this Eq. (6.24) is called *Michaelis-Menten equation* too. Since the reaction velocity (6.24) depends only on the concentration of substrate S, we can find the dependence shown in Fig. 6.4. It is indicated that the velocity has an upper bound in terms of the concentration of substrate S, which may be regarded as a result of the limited

Fig. 6.4 The dependence of the reaction velocity of the Michaelis-Menten structure (6.24) on the concentration of substrate S

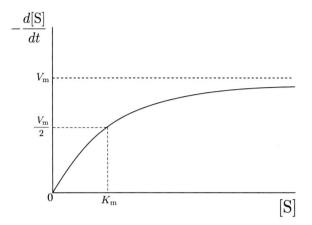

catalytic efficiency in the reaction. The similar nature can be found also for the general Michaelis-Menten equation (6.23).

> The Michaelis-Menten equation (6.24) mathematically coincides with *Holling's disc equation* argued later about the temporal change of prey population size in Sect. 8.5.1. It would be worth considering the reason of the coincidence in a modeling sense (refer to Sect. 8.5.1).

Now let us consider the initial reaction velocity for (6.23) just at the beginning of the reaction. Since no product has been produced at the initial, we have $[P] = 0$. From (6.23), we can obtain the initial reaction velocity

$$\mathcal{V}_0 \approx \frac{k_2[E]_0[S]}{[S] + (k_{-1} + k_2)/k_1} = \frac{V_m}{1 + K_m/[S]}. \tag{6.25}$$

Although this formula is the same as (6.24), their meanings are different from each other. Taking the inverse of (6.25), we can get

$$\mathcal{V}_0^{-1} \approx \frac{1}{V_m} + \frac{K_m}{V_m}[S]^{-1}, \tag{6.26}$$

which shows a linear relation between $\mathcal{V}_0^{-1}$ and $[S]^{-1}$. Therefore, if the data of $\mathcal{V}_0^{-1}$ and $[S]^{-1}$ are plotted for a Michaelis-Menten structure, they are expected to (approximately) show a linear correlation, with which the slope and intersection of the fitted line can be used to estimate the values of V_m and K_m. This approach has been successful in many cases, and today such a plot is called *Lineweaver-Burk plot* after the names of American chemists, Hans Lineweaver (1907–2009) and Dean Burk (1904–1988), in the theory of enzyme kinetics [8].

Exercise 6.2 In the same arguments, show that the Eq. (6.23) or (6.24) (further, (6.25) or (6.26)) is applicable for the following Michaelis-Menten structure with two different states of complex:

$$E + S \underset{k_{-1}}{\overset{k_1}{\rightleftharpoons}} X_1 \underset{k_{-2}}{\overset{k_2}{\rightleftharpoons}} X_2 \underset{k_{-3}}{\overset{k_3}{\rightleftharpoons}} E + P. \tag{6.27}$$

6.2.2 Reaction Velocity Equation with Inhibitor

In this section, we introduce an inhibitor I to the Michaelis-Menten structure (6.15) (or (6.27)). It is a substance to combine with the enzyme E, and works to reduce the concentration of active enzyme which can combine with the substrate:

$$E + I \underset{k_-}{\overset{k_+}{\rightleftharpoons}} Y, \tag{6.28}$$

where Y denotes the state at which I combines with E. Now with the law of kinetic mass action, let us assume that the temporal change of the concentrations [I] and [Y] is governed by

$$\frac{d[I]}{dt} = -k_+[E][I] + k_-[Y];$$

$$\frac{d[Y]}{dt} = k_+[E][I] - k_-[Y].$$

With the influence of introduced inhibitor, the temporal change of the enzyme concentration [E] is now governed by the following ordinary differential equation instead of (6.16):

$$\frac{d[E]}{dt} = -k_1[E][S] - k_{-2}[E][P] + (k_{-1} + k_2)[X] - k_+[E][I] + k_-[Y].$$

We apply the quasi-stationary state approximation (QSSA) for the reaction (6.28) between I and E, in addition to that for the complex in (6.15). That is, we use the QSSA such that $d[I]/dt \approx 0$. With the same arguments as in Sect. 6.2.1, we can obtain the reaction velocity $\mathcal{V}$ for the enzyme kinetics (6.15) with (6.28):

$$\mathcal{V} = -\frac{d[S]}{dt} \approx \frac{(V_m/K_m)[S] - (V_P/K_P)[P]}{1 + K_I[I] + [S]/K_m + [P]/K_P},$$

where $K_I = k_+/k_-$. The initial reaction velocity for this case becomes

$$\mathcal{V}_0 \approx \frac{V_m}{1 + (K_m/[S])(1 + K_I[I])}.$$

The arguments to arrive at these results can be regarded as a reasonable mathematical modeling to introduce the influence of the inhibitor in the formula of reaction velocity.

6.2.3 Application for Population Dynamics

The quasi-stationary state approximation (QSSA) is applicable for the modeling of biological population dynamics [2, 4, 5, 11–15]. In this section, we shall see such an application for the population dynamics.

Let us consider first the following interaction between two populations:

$$N_1 + N_2 \underset{\gamma_-}{\overset{\gamma_+}{\rightleftharpoons}} C \overset{\sigma}{\to} N_2(1 + \kappa), \tag{6.29}$$

where N_i $(i = 1, 2)$ indicates the individual free from the interaction, and C does the pair of individuals under the interaction. We consider only the interaction between individuals of different populations. The coefficient to enter in the interaction is γ_+. The coefficient γ_- is the coefficient for the end of interaction such that the interaction does not cause any influence (e.g., damage or benefit) on both individuals. After such an interaction without any influence, both individuals can return to the free state as before. In contrast, the coefficient σ is the coefficient for the end of interaction such that the individual of population 1 dies while the individual of population 2 makes reproduction with the rate κ. Hence, the reaction kinetics (6.29) can be regarded as an interspecific reaction like the predation or the parasitism by species 2 for species 1. More generally saying, it can be the process that species 2 exploits species 1 for the reproduction.

The kinetics (6.29) shows only the process of interaction between two populations. Now, taking account of the natural reproduction and death, let us consider the following reaction equations along the same way of its construction as in Sects. 6.2.1 and 6.2.2:

$$\frac{dN_1(t)}{dt} = g(N_1) - \gamma_+ N_2(t)N_1(t) + \gamma_- C(t); \tag{6.30}$$

$$\frac{dC(t)}{dt} = \gamma_+ N_2(t)N_1(t) - \mu_C(C) - \gamma_- C(t) - \sigma C(t); \tag{6.31}$$

$$\frac{dN_2(t)}{dt} = (1 + \kappa)\sigma C(t) - \mu_2(N_2) - \gamma_+ N_1(t)N_2(t). \tag{6.32}$$

We use here the italic symbol to indicate the densities of individual and pair, instead of bracket [ ] in the previous sections. The term $g(N_1)$ in (6.30) is the net growth rate of population 1 when population 2 is absent, that is, with no influence of the interaction. The term $\mu_2(N_2)$ in (6.32) is the death rate of population 2 by itself. The term $\mu_C(C)$ in (6.31) is the death rate of pair under the interaction, which causes the death of both individuals forming the pair. Remark that we give now $\mu_2(N_2)$ and $\mu_C(C)$ as general functions of N_2 and C respectively.

Now we apply the QSSA for the temporal change of the pair density C. For the QSSA, we assume that, in the time scale of the temporal change of the pair density C, the temporal change of the population densities N_1 and N_2 is negligible. In other mathematical words, the temporal change of the density of pairs C is assumed to be much faster than that of the population densities N_1 and N_2, so that the density of pairs C can be approximated well as taking a value sufficiently near the equilibrium (i.e., the quasi-stationary state) at each time t. Then we use the approximation $dC/dt \approx 0$ for (6.31), and get

$$\gamma_+ N_2(t)N_1(t) - \mu_C(C) - \gamma_- C(t) - \sigma C(t) \approx 0 \tag{6.33}$$

for each time t. Once we could get the expression of C in terms of N_1 and N_2 from (6.33), we can obtain a two dimensional closed system of N_1 and N_2 by substituting it for (6.30) and (6.32).

Let us consider a simplest modeling such that $\mu_C(C) = \delta_C C$ and $\mu_2(N_2) = \delta_2 N_2$ with the per capita natural death rates δ_C and δ_2 which are positive constants. Besides, let us introduce $\overline{N}_2(t) := N_2(t) + C(t)$ which means the total size of population 2, since it consists of free individuals and those under the interaction with individuals of population 1. In this modeling, from (6.33), we have

$$C(t) \approx \frac{N_1(t)}{k_h + N_1(t)} \overline{N}_2(t), \tag{6.34}$$

with $k_h := (\gamma_- + \delta_C + \sigma)/\gamma_+$.

Therefore, from (6.30)–(6.32) with the QSSA, we can obtain the following two dimensional closed system of N_1 and $\overline{N}_2$:

$$\begin{aligned}
\frac{dN_1(t)}{dt} &= g(N_1) - (\delta_C + \sigma)\frac{N_1(t)}{k_h + N_1(t)}\overline{N}_2(t); \\
\frac{d\overline{N}_2(t)}{dt} &= -\delta_2\overline{N}_2 + K(\delta_C + \sigma)\frac{N_1(t)}{k_h + N_1(t)}\overline{N}_2(t),
\end{aligned} \tag{6.35}$$

with $K := \{\sigma\kappa + \delta_2 - (\gamma_- + \delta_C)\}/(\delta_C + \sigma)$. Remark that N_1 means the density of individuals which can make the reproduction in population 1. As seen in (6.34) and subsequently in (6.35), the term of interaction between populations is given by a rational function of N_1 which is sometimes called *Michaelis-Menten type of interaction* in population dynamics.

As already mentioned in Sect. 6.2.1, the Michaelis-Menten type of interaction term mathematically coincides with *Holling's disc equation* argued later about the predator's functional response as (8.28) in Sect. 8.5.1. So the interaction term in (6.35) may be called *Holling type of interaction*, or *Holling's Type II response* in a sense of ecological modeling.

If $K \leq 0$, that is, if $\sigma\kappa + \delta_2 \leq \gamma_- + \delta_C$, then $N_2(t) \to 0$ as $t \to \infty$ that means the extinction of population 2, independently of the other parameters and the initial value $N_2(0) > 0$. This is the case where the death rate of pair under the interaction is sufficiently large, or where the interacting pair is much easy to dissolve. It is likely however that population 2 becomes extinct even when $K > 0$.

We must remark that the nature of population dynamics (6.35) does not necessarily correspond to that of (6.30–6.32), since such a correspondence could be expected only under the condition that the QSSA applied in the above arguments is valid as the appropriate modeling for the population dynamics. Although the system (6.35) generally shows the same qualitative behavior as (6.30–6.32) in most case even with the parameter values which cannot be regarded as corresponding to the condition for the application of the above QSSA, it must be mathematically inappropriate to regard the system (6.35) as equivalent to (6.30)–(6.32). Actually, as seen in the numerical example of Fig. 6.5 with $g(N_1) = (1 - N_1)N_1$, they may show different behaviors in some cases.

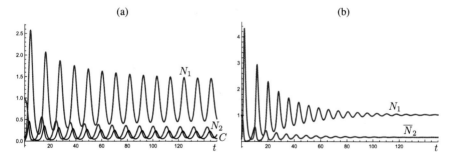

Fig. 6.5 Numerical example of the temporal change by the systems (6.30)–(6.32) and (6.35), with $g(N_1) = (1-N_1)N_1$; $\mu_C(C) = \delta_C C$; $\mu_2(N_2) = \delta_2 N_2$. Commonly, $\delta_C = 0.1$; $\delta_2 = 1.0$; $\gamma_+ = 1.0$; $\gamma_- = 0.1$; $\sigma = 1.0$; $\kappa = 7.0$; $(N_1(0), C(0), N_2(0)) = (1.0, 0.0, 0.1)$. In (**a**) for (6.30–6.32), the population dynamics asymptotically approaches a periodic solution, while, in (**b**) for (6.35), it asymptotically approaches a positive equilibrium with a damped oscillation

> The system (6.35) with $g(N_1) = (1 - N_1)N_1$, $\mu_C(C) = \delta_C C$, and $\mu_2(N_2) = \delta_2 N_2$ mathematically corresponds to what is called *Rosenzweig-MacArthur model* which will be discussed in Sect. 8.6. It is a model for the prey-predator population dynamics with the predator's functional response of Holling's Type II as mentioned also in the above. It will be described in Sect. 8.6 how the dynamics of Rosenzweig-MacArthur model is characterized by the existence of a periodic solution.

As the last subject in this section, let us consider a generalization of the modeling of (6.29) for a multi-species system. It becomes the following structure for the system with $m + \ell$ species ($i = 1, 2, \ldots, m; \ j = 1, 2, \ldots, \ell$):

$$
H_i + P_j \underset{\gamma_{ij}^-}{\overset{\gamma_{ij}^+}{\rightleftharpoons}} C_{ij} \overset{\sigma_{ij}}{\rightarrow} P_j(1 + \kappa_{ij}). \tag{6.36}
$$

This may be regarded as a model for the food web system of two trophic levels with m prey and ℓ predator species, where H_i is prey species i, and P_j predator species j. Applying the modeling same as that about (6.30)–(6.32), we shall consider here the following system:

$$
\frac{dH_i(t)}{dt} = g_i(H_i, t) - \sum_{j=1}^{\ell} \gamma_{ij}^+ H_i(t) P_j(t) + \sum_{j=1}^{\ell} \gamma_{ij}^- C_{ij}(t); \tag{6.37}
$$

$$
\frac{dC_{ij}(t)}{dt} = \gamma_{ij}^+ H_i(t) P_j(t) - D_{ij} C_{ij}(t) - \gamma_{ij}^- C_{ij}(t) - \sigma_{ij} C_{ij}(t); \tag{6.38}
$$

$$
\frac{dP_j(t)}{dt} = \sum_{i=1}^{m} (1 + \kappa_{ij}) \sigma_{ij} C_{ij}(t) - \delta_j P_j(t) - \sum_{i=1}^{m} \gamma_{ij}^+ H_i(t) P_j(t). \tag{6.39}
$$

The variables and parameters have their meanings corresponding to those in the system (6.30)–(6.32) considered in the earlier part of this section.

Again with the QSSA for the interaction between H_i and P_j for every pair of i and j, we use the approximation $dC_{ij}/dt \approx 0$ for any pair of i and j. Then, from (6.38), we have

$$
H_i(t) P_j(t) - k_{ij} C_{ij}(t) \approx 0 \tag{6.40}
$$

for any pair of i and j and any time t, where $k_{ij} := (\gamma_{ij}^- + D_{ij} + \sigma_{ij})/\gamma_{ij}^+$. Now let us introduce

$$\overline{P}_j(t) = P_j(t) + \sum_{i=1}^{m} C_{ij}(t), \tag{6.41}$$

which means the total population size of predator species j. From (6.40), we can obtain

$$C_{ij}(t) \approx \frac{H_i(t)/k_{ij}}{1 + \sum_{i=1}^{m} H_i(t)/k_{ij}} \overline{P}_j(t). \tag{6.42}$$

Applying these Eqs. (6.40)–(6.42) for (6.37)–(6.39), we can derive the following population dynamics model:

$$\frac{dH_i(t)}{dt} = g_i(H_i, t) - \sum_{j=1}^{\ell} (D_{ij} + \sigma_{ij}) \frac{H_i(t)/k_{ij}}{1 + \sum_{n=1}^{m} H_n(t)/k_{nj}} \overline{P}_j(t);$$

$$\frac{d\overline{P}_j(t)}{dt} = -\delta_j \overline{P}_j(t) + \sum_{i=1}^{m} K_{ji}(D_{ij} + \sigma_{ij}) \frac{H_i(t)/k_{ij}}{1 + \sum_{n=1}^{m} H_n(t)/k_{nj}} \overline{P}_j(t),$$

$$\tag{6.43}$$

where $K_{ji} := \{\sigma_{ij}\kappa_{ij} + \delta_j - (\gamma_{ij}^- + D_{ij})\}/(D_{ij} + \sigma_{ij})$. The term of interaction in (6.43) can be regarded as a reasonable form of the interaction for the interacting multi-species system, which has been derived as a generalization from the interacting two species system (6.35).

As a dynamical system, the nature of (6.43) could not necessarily match that of (6.37)–(6.39) even in a qualitative sense. The system (6.43) has been derived from (6.37)–(6.39) with the extremal approximation by the QSSA.

The theory of mathematical analysis on the nature of a dynamical system with two different time scales, that is, with the fast and slow processes, has been developed in applied mathematics. One of such mathematical analyses is what is called *two-timing method* [3, 6] in the *perturbation method* [7]. The QSSA is actually the zeroth approximation in the two-timing method for (6.37–6.39). In this sense, the dynamics (6.43) is regarded as the zeroth approximation for the dynamics (6.37–6.39).

The system (6.43) mathematically corresponds to the multi-species prey-predator dynamics model with (8.36) by Holling's disc equation in p. 235 of Sect. 8.5.2.

Answer to Exercise

Exercise 6.1 (p. 178)

For the system (6.14), we can easily derive the following equation:

$$\frac{d}{dt}\{\rho N(t) + \gamma R(t)\} = \gamma A \quad \text{(constant)}.$$

By integrating with t, we can get the following relation between N and R:

$$\rho N(t) + \gamma R(t) = \gamma At + \rho N(0) + \gamma R(0),$$

for the initial values $N(0)$ and $R(0)$. Substituting this relation for the first equation of (6.14), we can get

$$\frac{dN(t)}{dt} = \{\gamma At + \rho N(0) + \gamma R(0) - \rho N(t)\} N(t). \tag{6.44}$$

This corresponds to the logistic equation with the intrinsic growth rate which linearly depends on time. So using the solution (5.13) for the logistic equation (5.12) with temporally varying parameters in Sect. 5.3, we can mathematically get the solution $N(t)$ of the ordinary differential equation (6.44):

$$N(t) = \frac{\exp\left[(\gamma A/2)t^2 + \{\rho N(0) + \gamma R(0)\}t\right]}{\rho \displaystyle\int_0^t \exp\left[(\gamma A/2)\tau^2 + \{\rho N(0) + \gamma R(0)\}\tau\right]d\tau + 1/N(0)},$$

and find that

$$\lim_{t\to\infty} N(t) = \lim_{t\to\infty} \frac{\gamma At + \rho N(0) + \gamma R(0)}{\rho} = \infty, \tag{6.45}$$

making use of de l'Hôpital's law. Therefore, the population size unboundedly increases. This result can be implied by the modeling for the system (6.14).

First, from the first equation of (6.14), the population size N always increases as long as N and R are positive. From the modeling for the resource renewal, the resource cannot become exhausted in a finite time because there is always a resource inflow from the outside. This means that the resource stock R is always positive for any finite time, although we shall skip here the more mathematically rigorous proof of this nature. Thus, the population size N must be monotonically increasing for any finite time.

Next, from the nature of the logistic equation (5.10) to monotonically approach the carrying capacity r_0/β, the temporal change of N driven by (6.44) is expected to tend to come nearer to the value $N_\infty(t) := \{\gamma At + \rho N(0) + \gamma R(0)\}/\rho$ which can be regarded as corresponding to r_0/β for the logistic equation (5.10). This implies the expectation that the population size N approaches $N_\infty(t)$ as time passes, so that it comes to increase unboundedly in an approximately linear manner. This expectation is really reflected to the mathematical limit in (6.45).

Exercise 6.2 (p. 182)

With the same assumptions as in Sect. 6.2.1, we shall consider the following system to describe the temporal change in the substance concentrations in the enzyme kinetics:

$$\frac{d[E]}{dt} = -k_1[E][S] - k_{-3}[E][P] + k_3[X_2] + k_{-1}[X_1];$$

$$\frac{d[X_1]}{dt} = k_1[E][S] + k_{-2}[X_2] - (k_{-1} + k_2)[X_1];$$

$$\frac{d[X_2]}{dt} = k_2[X_1] + k_{-3}[E][P] - (k_{-2} + k_3)[X_2];$$

$$\frac{d[S]}{dt} = -k_1[E][S] + k_{-1}[X_1];$$

$$\frac{d[P]}{dt} = k_3[X_2] - k_{-3}[E][P],$$

with the conservation laws given by $[E]_0 = [E] + [X_1] + [X_2]$ and $[S]_0 = [S] + [P] + [X_1] + [X_2]$.

With the QSSA such that the temporal change in the complex concentration is negligible, we can use the approximations $d[X_1]/dt \approx 0$ and $d[X_1]/dt \approx 0$. Making use of the conservation laws, we can obtain the reaction velocity equation (6.23), or (6.24) with $k_{-3} = 0$. Further, we can obtain the equation for the initial reaction velocity $\mathcal{V}_0$, equivalent to (6.25) and (6.26). In these equations, the parameters are given for the enzyme kinetics (6.27) as follows:

$$K_m = \frac{k_2 k_3 + k_{-1}k_3 + k_{-1}k_{-2}}{k_1(k_{-2} + k_2 + k_3)}; \quad K_P = \frac{k_2 k_3 + k_{-1}k_3 + k_{-1}k_{-2}}{k_{-3}(k_{-2} + k_{-1} + k_2)};$$

$$V_m = \frac{k_2 k_3}{k_{-2} + k_2 + k_3}[E]_0; \quad V_P = \frac{k_{-2}k_{-1}}{k_{-2} + k_{-1} + k_2}[E]_0.$$

This arguments can be expanded for the case with more than two states of complex. This means that the Michaelis-Menten reaction velocity equation is applicable for the general Michaelis-Menten structure, independently of how many states the complex consist of. On the other hand, this result indicates that we could not determine the number of complex states even if the values V_m and K_m are estimated by the Lineweaver-Burk plot with the data of $\mathcal{V}_0^{-1}$ and $[S]^{-1}$.

References

1. I. Amdur, G.G. Hammes, *Chemical Kinetics: Principles and Selected Topics* (McGraw-Hill, New York, 1966)
2. J.A.M. Borghans, R.J. DeBoer, L.A. Segel, Extending the quasi-steady state approximation by changing variables. Bull. Math. Biol. **58**, 43–63 (1996)
3. N.F. Britton, *Reaction-Diffusion Equations and Their Applications to Biology* (Academic Press, London, 1986)
4. R.J. DeBoer, A.S. Perelson, Towards a general function describing T cell proliferation. J. Theor. Biol. **175**, 567–576 (1995)
5. G. Huisman, R.J. De Boer, A formal derivation of the "Beddington" functional response. J. Theor. Biol. **185**, 389–400 (1997)
6. D.W. Jordan, P. Smith, *Nonlinear Ordinary Differential Equations*. Oxford Applied Mathematics and Computing Science Series (Clarendon Press, Oxford, 1977)
7. P.B. Kahn, *Mathematical Methods for Scientists and Engineers: Linear and Nonlinear Systems* (Wiley, New York, 1990)
8. H. Lineweaver, D. Burk, The determination of enzyme dissociation constants. J. Am. Chem. Soc. **56**(3), 658–666 (1934)
9. A.J. Lotka, *Elements of Physical Biology* (Williams and Wilkins, Baltimore, 1925)
10. A.J. Lotka, *Elements of Mathematical Biology* (Dover, New York, 1956)
11. M.G. Pedersen, A.M. Bersani, E. Bersani, The total quasi-steady-state approximation for fully competitive enzyme reactions. Bull. Math. Biol. **69**, 433–457 (2007)
12. K.R. Schneider, T.Wilhelm, Model reduction by extended quasi-steady state approximation. J. Math. Biol. **40**, 443–450 (2000)
13. S. Schnell et al., The mechanism distinguishablility problem in biochemical kinetics: the single-enzyme, single-substrate reaction as a case study. C.R. Biol. **329**, 51–61 (2006)
14. L.A. Segel, M. Slemrod, The quasi steady-state assumption: a case study in perturbation. SIAM Rev. **31**, 446–477 (1989)
15. A.R. Tzafriri, E.R. Edelman, The total quasi-steady-state approximation is valid for reversible enzyme kinetics. J. Theor. Biol. **226**, 303–313 (2004)
16. V. Volterra, Variazione e fluttuazioni del numero d'individui in specie animali conviventi. Mem. Acad. Lincei. **6**, 30–113 (1926)

Chapter 7
Modeling for Competitive Relation

Abstract This chapter about the competition dynamics model that is very popular topics in ecology and mathematical biology. In this book, we shall see also some related classic theoretical topics which have been rarely mentioned in the modern textbooks. They may have a potential to provide cues for the readers to expand the idea for a new modeling about population dynamics.

7.1 Lotka-Volterra Competition Model

7.1.1 Influence of Competition

The following system of ordinary differential equations is a model of population dynamics for competing m species, called *Lotka-Volterra competition model*:

$$
\begin{cases}
\dfrac{dN_1(t)}{dt} = \left\{ r_1 - \beta_1 N_1(t) - \displaystyle\sum_{j=1;\, j\neq 1}^{m} \gamma_{1j} N_j(t) \right\} N_1(t); \\[2ex]
\dfrac{dN_2(t)}{dt} = \left\{ r_2 - \beta_2 N_2(t) - \displaystyle\sum_{j=1;\, j\neq 2}^{m} \gamma_{2j} N_j(t) \right\} N_2(t); \\[2ex]
\quad\vdots \\[1ex]
\dfrac{dN_k(t)}{dt} = \left\{ r_k - \beta_k N_k(t) - \displaystyle\sum_{j=1;\, j\neq k}^{m} \gamma_{kj} N_j(t) \right\} N_k(t); \\[2ex]
\quad\vdots \\[1ex]
\dfrac{dN_m(t)}{dt} = \left\{ r_m - \beta_m N_m(t) - \displaystyle\sum_{j=1;\, j\neq m}^{m} \gamma_{mj} N_j(t) \right\} N_m(t),
\end{cases}
\tag{7.1}
$$

where all parameters r_j, β_j, and γ_{ij} are nonnegative $(i, j = 1, 2, \ldots, m;\ i \neq j)$. As (5.9) for the logistic equation (5.10) in Sect. 5.3, the above Lotka-Volterra competition model is based on the modeling about the per capita growth rate for each species i as

$$\frac{1}{N_i(t)} \frac{dN_i(t)}{dt} = r_i - \beta_i N_i(t) - \sum_{j=1;\, j\neq i}^{m} \gamma_{ij} N_j(t), \qquad (7.2)$$

which assumes a linear density effect from populations of the other species. Besides, as described in Sect. 2.3, the interspecific competition is introduced as the negative density effect from the other species, which reduces the per capita growth rate. The parameter γ_{ij} indexes the strength of such an interspecific density effect from species j to species i, sometimes called (interspecific) *competition coefficient*. Correspondingly, the parameter β_i may be called *intraspecific competition coefficient* of species i. The parameter r_i indicates the upper bound for the per capita growth rate, and means the intrinsic growth rate of species i. In this modeling, the interspecific density effect is introduced in the per capita growth rate. As mentioned in Sect. 2.3, this could be reasonable for the *exploitative competition*.

An alternative modeling for the Lotka-Volterra competition model is to introduce the interspecific reaction to reduce the population size itself. This could be the case of *interference competition* (refer to Sect. 2.3). In this modeling, it would be more appropriate to express the system (7.1) in the following form ($k = 1, 2, \ldots, m$):

$$\frac{dN_k(t)}{dt} = \{r_k - \beta_k N_k(t)\} - \sum_{j=1;\, j\neq k}^{m} \gamma_{kj} N_j(t) N_k(t). \qquad (7.3)$$

This expression more clearly indicates that the interspecific reaction is introduced with the mass action assumption (Sect. 6.1.1). Especially, the effect of interspecific reaction on the population growth is introduced by the term proportional to the product of population densities of competing two species, which may be called *Lotka-Volterra type of interaction* (refer to Sect. 6.1.2).

Similarly as described for the logistic equations (5.10) and (6.3) in Sect. 6.1.3, the modeling of (7.3) must be regarded as different from that of (7.2). For both modelings, each population follows a logistic growth when no other species exist, that is, when there is no competition with the other species. The difference in the modeling is on the influence of the interspecific competition. For the competition model with (7.2), the interspecific competition affects the per capita growth rate as a density effect. For the competition model with (7.3), it is independent of the per capita growth rate, that is, it does not affect the reproduction rate but directly reduces the population size as an interference effect on the population growth.

It should be remarked that the competition model in which the interspecific reaction is introduced by the product of population densities of competing two species, that is, by the Lotka-Volterra type of interaction, may be called Lotka-Volterra competition model as a wide sense of its definition, independently of what modeling is applied for the intraspecific reproduction term.

For Lotka-Volterra competition models (7.2) and (7.3), it is satisfied in general for the competition coefficient that $\gamma_{ij} \neq \gamma_{ji} > 0$ $(i \neq j)$, since the influence of the interspecific competition must depend on the species' nature. Although we could mathematically consider the case where $\gamma_{ij} = 0$ and $\gamma_{ji} > 0$, such an interspecific relation between species i and j cannot be ecologically regarded as a competition. It may be regarded as an amensalism, since one species has a negative effect from the other while the other has nothing by the interspecific reaction. In contrast, we may consider Lotka-Volterra competition model (7.2) or (7.3) which includes $\gamma_{ij} = 0$ and $\gamma_{ji} = 0$ for some i and j. In such a model, there is no interspecific reaction between species i and j.

7.1.2 Competing Two Species System

Lotka-Volterra two species competition model of (7.1) for $m = 2$ has become popular in the history of population dynamics theory:

$$\begin{cases} \dfrac{dN_1(t)}{dt} = \{r_1 - \beta_1 N_1(t) - \gamma_{12} N_2(t)\} N_1(t); \\ \dfrac{dN_2(t)}{dt} = \{r_2 - \beta_2 N_2(t) - \gamma_{21} N_1(t)\} N_2(t), \end{cases} \tag{7.4}$$

with the initial condition such that $N_1(0) > 0$ and $N_2(0) > 0$, as a usual assumption for the model about competing two species population dynamics.

The behavior of $(N_1(t), N_2(t))$ can be classified in the following four cases (Fig. 7.1):

(a) When $r_1/\beta_1 < r_2/\gamma_{21}$ and $r_2/\beta_2 > r_1/\gamma_{12}$, species 1 goes extinct while species 2 survives as $t \to \infty$, independently of the initial condition.

(b) When $r_1/\beta_1 > r_2/\gamma_{21}$ and $r_2/\beta_2 > r_1/\gamma_{12}$, one of two species goes extinct while the other survives, depending on the initial condition. This is a bistable case.

(c) When $r_1/\beta_1 < r_2/\gamma_{21}$ and $r_2/\beta_2 < r_1/\gamma_{12}$, two species survive and coexist as $t \to \infty$, independently of the initial condition.

(d) When $r_1/\beta_1 > r_2/\gamma_{21}$ and $r_2/\beta_2 < r_1/\gamma_{12}$, species 2 goes extinct while species 1 survives as $t \to \infty$, independently of the initial condition.

Two species can coexist only in the case of (c). As a special case, when $r_1 = r_2$, the condition for (c) becomes such that $\beta_1 > \gamma_{21}$ and $\beta_2 > \gamma_{12}$. This condition means that the effect of intraspecific competition (i.e., the density effect within the population of same species) is stronger than that of interspecific competition.

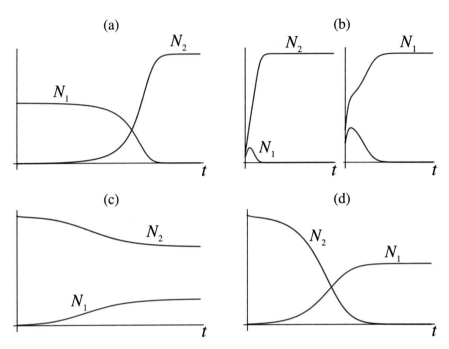

Fig. 7.1 Numerical examples of the temporal change of N_1 and N_2 by Lotka-Volterra two species competition model (7.4). (**a**) $(\gamma_{12}, \gamma_{21}) = (2.0, 1.5)$; (**b**) $(\gamma_{12}, \gamma_{21}) = (2.0, 2.5)$; (**c**) $(\gamma_{12}, \gamma_{21}) = (0.8, 1.2)$; (**d**) $(\gamma_{12}, \gamma_{21}) = (0.8, 2.5)$. Commonly, $r_1 = r_2 = 1.0$; $\beta_1 = 1.8$; $\beta_2 = 1.0$. In (**b**), one of two species goes extinct, depending on the initial condition (bistable case)

The equilibrium for Lotka-Volterra competition model (7.4) can be easily obtained from the parallel equations of $dN_1(t)/dt = 0$ and $dN_2(t)/dt = 0$. That is, the equilibrium $(N_1, N_2) = (N_1^*, N_2^*)$ must satisfy the following two equations:

$$(r_1 - \beta_1 N_1^* - \gamma_{12} N_2^*)N_1^* = 0; \quad (r_2 - \beta_2 N_2^* - \gamma_{21} N_1^*)N_2^* = 0.$$

Thus we can find possible four equilibria E_0, E_1, E_2, and E_3:

$$E_0(0, 0), \quad E_1\left(\frac{r_1}{\beta_1}, 0\right), \quad E_2\left(0, \frac{r_2}{\beta_2}\right), \quad E_3\left(\frac{r_1\beta_2 - r_2\gamma_{12}}{\beta_1\beta_2 - \gamma_{12}\gamma_{21}}, \frac{r_2\beta_1 - r_1\gamma_{21}}{\beta_1\beta_2 - \gamma_{12}\gamma_{21}}\right),$$

$$\tag{7.5}$$

in which equilibrium E_3 is feasible only when it consists of only positive values. If it contains a negative value, E_3 is nonsense, since it is not reachable from the positive initial condition. In such a case, we conventionally say that equilibrium E_3 *does not exist.*

It can be mathematically proved for the system (7.4) that N_1 and N_2 cannot become zero or negative for any time $t > 0$ from any positive initial condition, whereas we shall not describe here the proof. Such a positiveness of N_1 and N_2 is never *because* the variables N_1 and N_2 must be nonnegative from the meaning of population size, whereas it can be shown by the mathematical analysis on the system (7.4) that the mathematical nature of (7.4) matches such a biological requirement.

Exercise 7.1 Show that the necessary and sufficient condition for the existence of equilibrium E_3 is to satisfy one of the followings: (i) $\mathcal{R}_1 > 1$ and $\mathcal{R}_2 > 1$; (ii) $\mathcal{R}_1 < 1$ and $\mathcal{R}_2 < 1$ with respect to

$$\mathcal{R}_1 := \frac{\beta_1/r_1}{\gamma_{21}/r_2}; \quad \mathcal{R}_2 := \frac{\beta_2/r_2}{\gamma_{12}/r_1}.$$

The existence and stability of equilibria for Lotka-Volterra competition model (7.4) can be determined easily with the *isocline method* described in Sect. 14.7. For (7.4), the nullclines for N_1 in the (N_1, N_2)-phase plane are two lines, $N_1 = 0$ and $r_1 - \beta_1 N_1 - \gamma_{12} N_2 = 0$. Those for N_2 are $N_2 = 0$ and $r_2 - \beta_2 N_2 - \gamma_{21} N_1 = 0$. Thus, as shown in Fig. 7.2, we can easily find four cases according to their spatial configuration, which divides the (N_1, N_2)-phase plane into two or three regions with respect to the combination of signs of dN_1/dt and dN_2/dt.

Figure 7.2a or d shows the case where the trajectory of (N_1, N_2) eventually enters the middle region and asymptotically approaches E_1 or E_2 on an axis. In Fig. 7.2c, the trajectory asymptotically approaches E_3 in the first quadrant. In Fig. 7.2b, the trajectory asymptotically approaches E_1 or E_2, which indicates a bistable situation. Actually, as seen in Fig. 7.3, numerically drawn trajectories of (N_1, N_2) in the phase plane for Lotka-Volterra competition model (7.4) show a clear correspondence to the above result by the isocline method.

Exercise 7.2 Find the correspondence of each four cases shown in Fig. 7.2 to (a–d) in p. 193 respectively according to the behavior of $(N_1(t), N_2(t))$ by Lotka-Volterra competition model (7.4).

The local stability analysis on the equilibrium of Lotka-Volterra two species competition model (7.4) can derive the result corresponding to the above conclusion by the isocline method. Taking account of the condition for the existence of E_3 given

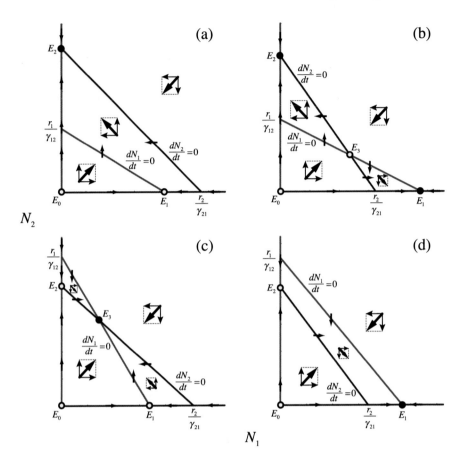

Fig. 7.2 Four cases of the spatial configuration of nullclines for Lotka-Volterra competition model (7.4). The equilibria E_1, E_2, and E_3 are given by (7.5)

in Exercise 7.1, we can obtain the following result by the local stability analysis described in Sect. 14.4 (p. 430):

$$\mathcal{R}_1 > 1 \text{ and } \mathcal{R}_2 > 1 \implies E_3 \text{ is locally asymptotically stable (node);}$$

$$\mathcal{R}_1 < 1 \text{ and } \mathcal{R}_2 < 1 \implies E_3 \text{ is unstable (saddle);} \tag{7.6}$$

$$\text{Otherwise} \implies E_3 \text{ does not exist,}$$

where $\mathcal{R}_1$ and $\mathcal{R}_2$ are defined in Exercise 7.1. Therefore, we can find the correspondence of the case (c) in p. 193 to the case where E_3 is locally asymptotically stable, and the case (b) to the case where E_3 is unstable. Refer to the cases (b) and (c) in Figs. 7.1, 7.2, and 7.3 too.

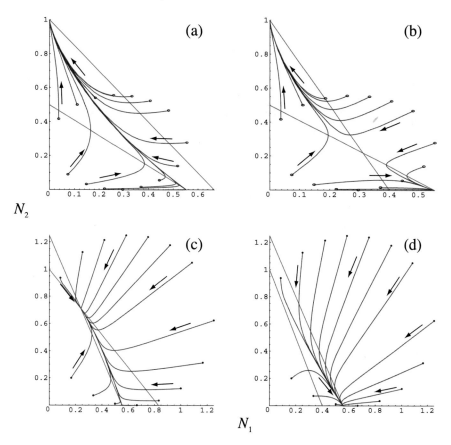

Fig. 7.3 Trajectories in the (N_1, N_2)-phase plane for Lotka-Volterra competition model (7.4). Numerically drawn for some different initial conditions commonly with $r_1 = r_2 = 1.0$; $\beta_1 = 1.8$; $\beta_2 = 1.0$. (**a**) $(\gamma_{12}, \gamma_{21}) = (2.0, 1.5)$; (**b**) $(\gamma_{12}, \gamma_{21}) = (2.0, 2.5)$; (**c**) $(\gamma_{12}, \gamma_{21}) = (0.8, 1.2)$; (**d**) $(\gamma_{12}, \gamma_{21}) = (0.8, 2.5)$. See also Figs. 7.1 and 7.2

The local stability analysis on the other equilibria, E_0, E_1, and E_2, is not difficult, and we can easily get the eigenvalues for each of them as follows:

E_0	E_1	E_2
r_1, r_2	$-r_1, r_2 - \gamma_{21}\dfrac{r_1}{\beta_1}$	$r_1 - \gamma_{12}\dfrac{r_2}{\beta_2}, -r_2$

Since every parameter is positive, the above result on the eigenvalues shows that equilibrium E_0 is always unstable. E_1 is locally asymptotically stable if $r_2/\gamma_{21} < r_1/\beta_1$, while it is unstable if $r_2/\gamma_{21} > r_1/\beta_1$. E_2 is locally asymptotically stable if $r_1/\gamma_{12} < r_2/\beta_2$, while it is unstable if $r_1/\gamma_{12} > r_2/\beta_2$. These results clearly coincide with (a–d) in p. 193, and in Figs. 7.1, 7.2, and 7.3.

As expected from the isocline method in Figs. 7.2c and 7.3c, the coexistent equilibrium E_3 is globally asymptotically stable if it is locally asymptotically stable. This can be proved with a Lyapunov function as described in Sect. 14.8 (see Exercise 14.2 and the subsequent part). Moreover, the Poincaré-Bendixson Trichotomy Theorem in Sect. 14.9 is applicable for the dynamical nature of Lotka-Volterra competition model (7.4), and we can show the global asymptotic stability of E_3 as well when it exists and is locally asymptotically stable.

As seen in Fig. 7.2b, the system (7.4) can show a bistable situation in which the trajectory asymptotically approaches alternatively equilibrium E_1 or E_2, depending on the initial condition. As numerically indicated by the thick dashed curve in Fig. 7.4, such a bistable situation divides the phase plane in two, respectively corresponding to the set of initial conditions from which the trajectory asymptotically approaches E_1 or E_2. The boundary for those two regions is called *separatrix* in the dynamical system theory. The existence of such a boundary curve in the phase plane is apparently implied also by numerical calculations of (b) in Fig. 7.3.

Fig. 7.4 Numerically drawn vector flows in the (N_1, N_2)-phase plane for Lotka-Volterra competition model (7.4) in a case corresponding to (**b**) of Fig. 7.1, 7.2, and 7.3 when a bistable situation appears with $r_1 = 1.0$; $r_2 = 0.8$; $\beta_1 = 1.8$; $\beta_2 = 1.0$; $\gamma_{12} = 2.0$; $\gamma_{21} = 2.5$. The thick dashed curve indicates the separatrix approximately drawn in this numerical calculation

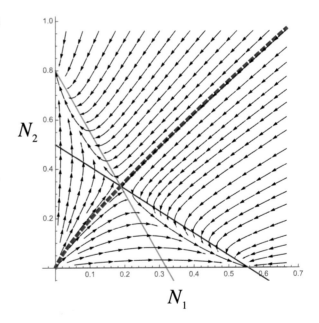

7.2 Competition for Resource

7.2.1 MacArthur's Modeling

In this section, we shall see a modeling used by Robert H. MacArthur (1930–1972) for the multi species population dynamics with the competition for a common resource. We describe here a generalized version of what he discussed in Chap. 2 of his famous book "Geographical Ecology: Patterns in the Distribution of Species" published in 1972 [1].

Let us consider different k resources used by n species of populations. Stock of the j th resource at time t is denoted by $R_j(t)$ ($j = 1, 2, \ldots, k$). Similarly with Sect. 6.1.5 for the single species population dynamics, let us assume that the per capita growth rate r_i of species i depends only on the resource use, and we give it now by the following linear relation between them ($i = 1, 2, \ldots, n$):

$$r_i = r_i(R_1, R_2, \ldots, R_{k-1}, R_k) = -R_{c,i} + \sum_{j=1}^{k} \alpha_{ij} R_j, \qquad (7.7)$$

where $R_{c,i}$ and α_{ij} are positive constants characterizing species i. Parameter α_{ij} indexes the efficiency for the use of resource j by species i. It may reflect a preference about the resource for species i. It can be regarded as the index about the degree of the dependence of species i on resource j. The larger α_{ij} means the stronger dependence on resource j. The parameter α_{ij} may be called (energy) *conversion coefficient* which appeared in Sect. 2.4.1 since it determines how a used resource is converted to the per capita growth rate (see also Sect. 8.4). Parameter $R_{c,i}$ is the threshold for the resource use of species i. The population growth is possible only when the total use of resources is beyond it. If the total use of resources is not beyond it, the population cannot grow and decreases its size. From the basic modeling described in Sect. 5.1, we consider the following population dynamics for species i with the per capita growth rate (7.7) ($i = 1, 2, \ldots, n$):

$$\frac{1}{N_i(t)} \frac{dN_i(t)}{dt} = -R_{c,i} + \sum_{j=1}^{k} \alpha_{ij} R_j(t). \qquad (7.8)$$

Next we consider the modeling for the temporal change of the stock $R_j(t)$ of resource j. As one of the simplest mathematical modelings, we assume that the velocity of resource consumption is proportional to the population size, and consider the following dynamics for resource j ($j = 1, 2, \ldots, k$):

$$\frac{dR_j(t)}{dt} = D_j(R_j(t)) - \sum_{i=1}^{n} \beta_{ji} N_i(t) R_j(t). \qquad (7.9)$$

The term $D_j(R_j(t))$ gives the recruitment dynamics of resource j when it is not used by any species, which is assumed to depend on the stock of resource j only. We do not assume here any interaction between different resources, although they could have it, for example, when they are bioresources. Positive parameter β_{ji} means the coefficient of consumption efficiency for resource j by species i. It characterizes the specific relation between species i and resource j. It is remarked that this modeling for the relation between the population and resource with respect to the temporal change of the resource stock is the same as in Sect. 6.1.5 with the Lotka-Volterra type of interaction between resource and population.

Let us introduce here the following simple recruitment term $D_j(R_j(t))$ for resource j:

$$D_j(R_j(t)) = \lambda_j R_j(t) - \gamma_j \{R_j(t)\}^2, \tag{7.10}$$

where a positive constant parameter λ_j is the recruitment coefficient for the stock of resource j. The negative term with a positive parameter γ_j means the natural decay of resource j, that is, the rate with which the stock of resource j becomes unavailable or non-usable. By the resource dynamics (7.9) with (7.10), the stock of resource j monotonically approaches λ_j/γ_j if it is not used by any species, since it is given by the equation mathematically equivalent to the logistic equation.

Getting (7.8), (7.9), and (7.10) together, we have the dynamics of n species populations and k resources, in which n species have a competitive relation with respect to the resources. The competitive relation is really an indirect competition (refer to Sect. 2.3). The use of a resource by a species works to reduce the stock, and subsequently the reduction in the stock works to reduce the per capita growth rate of every species.

Now we shall apply the quasi-stationary state approximation (QSSA; refer to Sects. 6.2.1 and 6.2.3) for the dynamics of (7.8)–(7.10). Let us suppose that the velocity of the temporal change of the resource stock is much faster than that of the population size. In the time scale of the temporal change of the resource stock, the temporal change of population sizes are approximately negligible. In the time scale of the temporal change of population sizes, the resource stock appears to instantaneously reflect the use by populations. In other words, the resource stock appears to be determined instantaneously by the population sizes at each moment. Subsequently, as we will see in the following arguments, the temporal change of each population size appears to be determined by the population sizes in the system at each moment, and could be observed as the result of interspecific direct reaction, that is, like a direct competition (refer to Sect. 2.3).

Hence, to consider the population dynamics, we introduce the QSSA about the resource dynamics with $dR_j/dt \approx 0$ for any time t, that is, from (7.9) with (7.10),

$$\lambda_j - \gamma_j R_j(t) - \sum_{i=1}^{n} \beta_{ji} N_i(t) \approx 0 \qquad (j = 1, 2, \ldots, k).$$

Then we use the following approximated relation of the resource stock to the population sizes at each time t:

$$R_j(t) \approx \frac{\lambda_j}{\gamma_j} - \sum_{i=1}^{n} \frac{\beta_{ji}}{\gamma_j} N_i(t) \qquad (j = 1, 2, \ldots, k). \tag{7.11}$$

We now consider only the case where $R_j > 0$. From (7.9) with (7.10), $R_j(t) = 0$ for any $t \geq t_1$ if $R_j(t) = 0$ for a moment $t = t_1$. That is, once a resource is exhausted, it cannot recover in the resource dynamics by (7.9) with (7.10).

By substituting (7.11) for (7.8), we obtain the following population dynamics model approximated by the QSSA from the system (7.8)–(7.10):

$$\frac{1}{N_i(t)} \frac{dN_i(t)}{dt} \approx -R_{c,i} + \sum_{j=1}^{k} \alpha_{ij} \left\{ \frac{\lambda_j}{\gamma_j} - \sum_{\ell=1}^{n} \frac{\beta_{j\ell}}{\gamma_j} N_\ell(t) \right\}$$

$$= \Lambda_i - \sum_{\ell=1}^{n} B_{i\ell} N_\ell(t) \qquad (i = 1, 2, \ldots, n), \tag{7.12}$$

where

$$\Lambda_i := -R_{c,i} + \sum_{j=1}^{k} \alpha_{ij} \frac{\lambda_j}{\gamma_j}; \quad B_{i\ell} := \sum_{j=1}^{k} \alpha_{ij} \frac{\beta_{j\ell}}{\gamma_j} \qquad (i, \ell = 1, 2, \ldots, n).$$

This system (7.12) is mathematically equivalent to Lotka-Volterra competition model (7.1). Consequently we have found that the resource-population dynamics model considered in this section could be approximated by the Lotka-Volterra competition model when the temporal change of the resource stock is much faster than that of population sizes in the dynamics of (7.8)–(7.10).

7.2.2 Tilman's Modeling

George David Tilman (1949–) proposed a mathematical modeling to theoretically discuss the dependence of interspecific competition dynamics on the common resource use in 1982 [2] (also see [3]). We shall describe in this section the idea of his modeling for two species system with the interspecific competition about two common resources.

Let us suppose that the resource stock has equilibrium $(\overline{R}_1^*, \overline{R}_2^*)$ when there is no consumption for those two resources. The stock of two resources at time t are now denoted by $R_1(t)$ and $R_2(t)$, the population size of species 1 by $N_1(t)$, and that of species 2 by $N_2(t)$. Tilman's model of two species and two resources is now given by

$$
\begin{cases}
\dfrac{dN_i(t)}{dt} = r_i(R_1(t), R_2(t))N_i(t) & (i = 1, 2); \\[4mm]
\dfrac{dR_j(t)}{dt} = D_j(R_j(t)) - \beta_{j1}N_1(t)R_j(t) - \beta_{j2}N_2(t)R_j(t) & (j = 1, 2).
\end{cases}
\tag{7.13}
$$

The population dynamics given by the former equation of (7.13) is based on the assumption same as that for the MacArthur's model (7.7) in Sect. 7.2.1: The per capita growth rate r_i is determined by the stock of two resources, so that it is given by a function of them, $r_i = r_i(R_1(t), R_2(t))$. The resource dynamics is the same as (7.9) in Sect. 7.2.1. The equilibrium value of the stock $\overline{R}_j^*$ without the consumption now satisfies that $D_j(\overline{R}_j^*) = 0$ ($j = 1, 2$). Let us assume now that, without the consumption, each resource asymptotically approaches the equilibrium in a monotonic manner. Therefore, in our modeling, we assume the following features of $D_j(R_j)$ ($j = 1, 2$): $D_j(R_j) > 0$ for $R_j < \overline{R}_j^*$, and $D_j(R_j) < 0$ for $R_j > \overline{R}_j^*$.

In this section to describe Tilman's way of the theoretical argument on the competition for the common resources, we shall set the following assumption for the per capita growth rate r_i:

$$
r_i(R_1, R_2)
\begin{cases}
= 0 & \text{if } R_1 = R_{i,1}^c \text{ and } R_2 \geq R_{i,2}^c; \\[2mm]
= 0 & \text{if } R_1 \geq R_{i,1}^c \text{ and } R_2 = R_{i,2}^c; \\[2mm]
< 0 & \text{if } R_1 < R_{i,1}^c \text{ or } R_2 < R_{i,2}^c; \\[2mm]
> 0 & \text{if } R_1 > R_{i,1}^c \text{ and } R_2 > R_{i,2}^c,
\end{cases}
\tag{7.14}
$$

where $R_{i,1}^c$ and $R_{i,2}^c$ are positive constants characterizing species i according to the dependence of the reproduction on two resources respectively. These constants define the least stock of each resource to make the per capita growth rate positive. Even when one of two resources has the stock beyond the threshold, the per capita growth rate becomes negative if the other has the stock below the threshold. In such a case, the resource of the stock below the threshold is called *limiting factor* for the population growth in ecology. With the above assumption, the per capita growth rate becomes positive when and only when every resource has the stock beyond the threshold.

In some real cases, the population size does not necessarily decrease even if the stock of a resource becomes below the threshold for the per capita growth rate. For example, some microorganisms become dormant in the environment with sufficiently poor stock of a resource required for the reproduction. In an ideal situation, the population size of such an organism may not decrease nor increase even when the resource stock is below the threshold.

Taking account of the death rate under such a dormant state, for example, by a disease or an effect of immune system for such a microorganism in vivo, the population size may decrease even under the dormant state. In such a case, the per capita growth rate must be assumed to be negative under the dormant state. In contrast, in the laboratory culture under an appropriately controlled condition, such a death would be negligible, and then the assumption (7.14) could be replaced by another such as $r_i(R_1, R_2) = 0$ for any $R_1 \leq R_{i,1}^c$ and $r_i(R_1, R_2) = 0$ for any $R_2 \leq R_{i,2}^c$.

From the assumption (7.14), since $r_i(R_{i,1}^c, R_2) = 0$ for any $R_2 \geq R_{i,2}^c$ and $r_i(R_1, R_{i,2}^c) = 0$ for any $R_1 \geq R_{i,1}^c$, we find that, if an equilibrium with $R_1 = R_1^* > 0$, $R_2 = R_2^* > 0$, $N_1 = N_1^* > 0$, and $N_2 = N_2^* > 0$ exists for the system (7.13), then it must satisfy that $R_1^* = R_{i,1}^c$ or $R_2^* = R_{i,2}^c$. Indeed, from the population dynamics of (7.13), it is clear that the population size of species i increases as long as $R_1 > R_{i,1}^c$ and $R_2 > R_{i,2}^c$. As the population size gets larger, the right side of the resource dynamics of (7.13) becomes smaller. Even when the right side of the resource dynamics of (7.13) reaches zero, the population size increases as long as $R_1 > R_{i,1}^c$ and $R_2 > R_{i,2}^c$, and subsequently the right side becomes negative so that the resource stock turns to decrease. Further, even when the resource stock decreases, the population size increases as long as $R_1 > R_{i,1}^c$ and $R_2 > R_{i,2}^c$, so that the resource stock cannot turn to increase but keeps decreasing. The temporal change follows such a process until it is satisfied that $R_1 = R_{i,1}^c$ or $R_2 = R_{i,2}^c$. At the moment that $R_1 = R_{i,1}^c$ or $R_2 = R_{i,2}^c$, it is satisfied that $dN_i/dt = 0$, and the population growth of species i stops.

These arguments do not imply that the population size of species i necessarily approaches a positive equilibrium. After the population growth of species i stops at a moment with $R_1 = R_{i,1}^c$, the population size of species i may decrease if the stock of resource 2 decreases to satisfy that $R_2 < R_{i,2}^c$. In such a case, the resource 2 is the limiting factor for species i, and species i would become extinct. Moreover, since the resource is used by the other species at the same time, even if $R_1 = R_{i,1}^c$ or $R_2 = R_{i,2}^c$ at a moment, the resource consumption by the other species may cause the decrease of the resource stock, and the subsequent decrease of the population size of species i occurs with $dN_i/dt < 0$ after the moment. In such a case, we can apply the above arguments for the other species as well.

Tilman [2] systematically discussed the relation of the resources to the coexistence of competing two species, making use of the *zero-net-growth-isocline* (ZNGI)

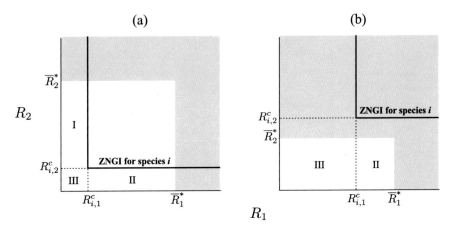

Fig. 7.5 Zero-net-growth-isocline (ZNGI) for species i, introduced by Tilman [2]. The assumption (7.14) is applied. (**a**) the ZNGI passes the region that $R_1 < \overline{R}_1^*$ and $R_2 < \overline{R}_2^*$; (**b**) the ZNGI does not pass the region that $R_1 < \overline{R}_1^*$ and $R_2 < \overline{R}_2^*$. Species i may be able to persist in the case of (**a**), while it cannot persist and goes extinct in the case of (**b**). For the resource stock of region I in the figure (**a**), the stock of resource 2 is beyond the threshold for species i, while that of resource 1 is below the threshold. So the population size of species i decreases. For the region II, resource 2 is the limiting factor to cause the decrease of population size. For the region III, both resources are below their thresholds, and the population size decreases

shown in Figs. 7.5 and 7.6. We describe the essence here, firstly focusing on one of two species.

In the case of Fig. 7.5b, when the ZNGI for species i does not pass the region where $R_1 < \overline{R}_1^*$ and $R_2 < \overline{R}_2^*$, species i cannot persist and goes extinct in such an environment. As indicated by the resource dynamics of (7.13), two resource stocks have the upper bounds $\overline{R}_1^*$ and $\overline{R}_2^*$ respectively, and each resource cannot become beyond the upper bound which is the natural assumption in an ecological sense. Thus, the resource stock must be always in the white region of Fig. 7.5. In the case of Fig. 7.5b, the white region does not intersect with the ZNGI. In the white region II, the stock of resource 1 is beyond the threshold for species i, while that of resource 2 is below the threshold. The resource 2 is the limiting factor in this case, and the population size of species i must decrease. In the white region III, the population size decreases as well, because of the shortage of both resource stocks. Therefore, in such a case, the per capita growth rate of species i must be negative, and the population goes extinct.

In contrast, in the case of Fig. 7.5a, the ZNGI for species i passes the white region, which indicates the possibility of the persistence of species i. For the white subregion above the ZNGI, the per capita growth rate of species i is positive. However, even in this case, the persistence of species i depends on the resource competition with the other species, and the extinction of species i may occur.

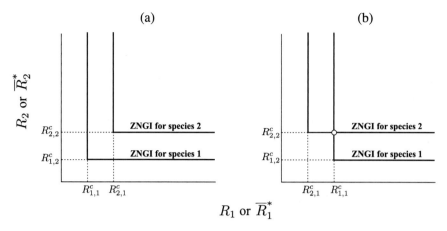

Fig. 7.6 Zero-net-growth-isoclines (ZNGIs) defined by Tilman [2]. The assumption (7.14) is applied. (**a**) the ZNGIs of two species do not have intersection; (**b**) the ZNGIs of two species have an intersection. In the case of (**a**), species 2 eventually goes extinct. In the case of (**b**), the coexistent equilibrium exists, which is indicated by the intersection, while its stability depends on the parameters for the population dynamics

Next, let us focus on the relation of the ZNGIs for two species. Figure 7.6a shows the case where two ZNGIs have no intersection, and the ZNGI for species 2 is located above the ZNGI for species 1. In this case, species 2 eventually goes extinct. When the resource stock is in the region above the ZNGI for species 2, the resource stock tends to approach the ZNGI for species 2, following the process described before. Since the resource stock on the ZNGI for species 2 is in the region above the ZNGI for species 1 in the case of Fig. 7.6a, the population size of species 1 must increase, so that the resource stock enters the region between the ZNGIs for two species, following the process described before. Hence, the resource stock finally reaches the ZNGI for species 1. When the resource stock is on the ZNGI for species 1, the per capita growth rate of species 2 is negative, and species 2 goes extinct. If the equilibrium reached by this process satisfies that $R_1 = R_{1,1}^c$ (on the vertical ZNGI for species 1), the resource dynamics of (7.13) gives the following equation with respect to the equilibrium population size N_1^* of species 1:

$$N_1^* = \frac{\gamma_1}{\beta_{11}}\left(\overline{R}_1^* - R_{1,1}^c\right) = \frac{\gamma_2}{\beta_{21}}\left(\overline{R}_2^* - R_2\right).$$

Subsequently, we can obtain the following equilibrium stock R_2^* of resource 2 at the equilibrium:

$$R_2^* = \overline{R}_2^* - \frac{\gamma_1\beta_{21}}{\gamma_2\beta_{11}}\left(\overline{R}_1^* - R_{1,1}^c\right). \tag{7.15}$$

In this argument, we find it necessary for the existence of such an equilibrium that $R_2^* \geq R_{1,2}^c$. This is because the resource stock $(R_{1,1}^c, R_2^*)$ is in the region below the ZNGI for species 1 if $R_2^* < R_{1,2}^c$, which is inconsistent with the positiveness of the equilibrium population size of species 1. Therefore, if $R_2^* < R_{1,2}^c$, the equilibrium such that $R_1 = R_{1,1}^c$ cannot exist. In such a case, it can be shown by the same argument that the equilibrium satisfying that $R_2 = R_{1,2}^c$ (on the horizontal ZNGI for species 1 in Fig. 7.6a) exists.

In the case of Fig. 7.6b when the ZNGIs for two species has an intersection, it is satisfied that $dN_1/dt = 0$ and $dN_2/dt = 0$ if $R_1 = R_{1,1}^c$ and $R_2 = R_{2,2}^c$. Thus, the intersection may be a coexistent equilibrium for two species. However, as shown by Tilman [2], such a coexistent equilibrium may not be reached from some initial condition. Depending on the upper bounds (R_1^*, R_2^*) for the resource stocks, it is likely that one species goes extinct. As an extremal case, the coexistent equilibrium may not be reached from any initial condition. As we expect, the coexistence of two species requires a certain condition related to (R_1^*, R_2^*).

We can see a numerical example in Fig. 7.7 about these arguments on the dynamics (7.13) with $D_j(R_j)$ given by (7.10) and the following per capita growth rate r_i $(i = 1, 2)$:

$$r_i = r_i(R_1, R_2)$$

$$= \begin{cases} 0.1(R_1 - R_{i,1}^c)(R_2 - R_{i,1}^c) & \text{if } R_1 > R_{i,1}^c \text{ and } R_2 > R_{i,2}^c; \\ 0 & \text{if } R_1 = R_{i,1}^c \text{ or } R_2 = R_{i,2}^c; \\ 0.1(R_1 - R_{i,1}^c) & \text{if } R_1 < R_{i,1}^c \text{ and } R_2 \geq R_{i,2}^c; \\ 0.1(R_2 - R_{i,2}^c) & \text{if } R_1 \geq R_{i,1}^c \text{ and } R_2 < R_{i,2}^c; \\ 0.1(R_1 - R_{i,1}^c) + 0.1(R_2 - R_{i,2}^c) & \text{if } _1 < R_{i,1}^c \text{ and } R_2 < R_{i,2}^c. \end{cases} \quad (7.16)$$

As a consequence, the coexistence of two species competing two common resources significantly depends on the nature of resource dynamics and the relation of species to it. In this theoretical argument, the resources may be regarded as representing the environmental condition for competing two species, The different resource dynamics could make the different fate of interspecific competition about the resource. Tilman's modeling provides a theoretical idea to discuss such a fate of interspecific competition.

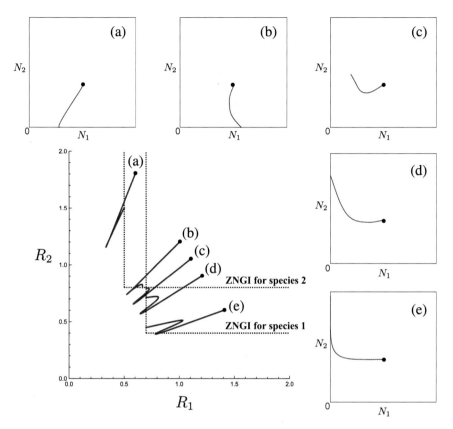

Fig. 7.7 Numerical example of the trajectories for the dynamics (7.13) with the per capita growth rate r_i $(i = 1, 2)$ given by (7.16) and the resource recruitment term $D_j(R_j)$ given by (7.10). (a) $(R_1^*, R_2^*) = (0.6, 1.8)$; (b) $(R_1^*, R_2^*) = (1.0, 1.2)$; (c) $(R_1^*, R_2^*) = (1.1, 1.05)$; (d) $(R_1^*, R_2^*) = (1.2, 0.9)$; (e) $(R_1^*, R_2^*) = (1.4, 0.6)$. The trajectories are for the initial condition given by $(R_1(0), R_2(0), N_1(0), N_2(0)) = (R_1^*, R_2^*, 14.5, 10.0)$, and commonly with $\lambda_j = 1.0$ $(j = 1, 2)$; $R_{1,1}^c = 0.5$; $R_{1,2}^c = 0.8$; $R_{2,1}^c = 0.7$; $R_{2,2}^c = 0.4$; $\beta_{11} = 0.02$; $\beta_{12} = 0.02$; $\beta_{21} = 0.02$; $\beta_{22} = 0.01$. Remark that, from $\lambda_j/\gamma_j = R_j^*$, γ_j is given by $1/R_j^*$ for each case in this numerical calculation

Answer to Exercise

Exercise 7.1 (p. 195)

The coexistent equilibrium E_3 given by (7.5) can be expressed with $\mathcal{R}_1$ and $\mathcal{R}_2$ as follows:

$$E_3\left(\frac{\mathcal{R}_2 - 1}{(\gamma_{21}/r_2)(\mathcal{R}_1\mathcal{R}_2 - 1)}, \frac{\mathcal{R}_1 - 1}{(\gamma_{12}/r_1)(\mathcal{R}_1\mathcal{R}_2 - 1)}\right).$$

Hence, the equilibrium values of E_3 become positive if and only if

$$(\mathcal{R}_1 - 1)(\mathcal{R}_1\mathcal{R}_2 - 1) > 0 \quad \text{and} \quad (\mathcal{R}_2 - 1)(\mathcal{R}_1\mathcal{R}_2 - 1) > 0.$$

This condition is satisfied when $\mathcal{R}_1 > 1$ and $\mathcal{R}_2 > 1$, or when $\mathcal{R}_1 < 1$ and $\mathcal{R}_2 < 1$. If neither of these two conditions is satisfied, E_3 contains non positive value. These two condition for the existence of E_3 correspond respectively to (b) and (c) shown in p. 193 about Lotka-Volterra two species competition system (7.4).

Exercise 7.2 (p. 195)

As shown in Fig. 7.2, the nullclines for Lotka-Volterra two species competition system (7.4) are lines. Therefore, their different spatial configurations in the first quadrant of the (N_1, N_2)-phase plane count $2 \times 2 = 4$, since they can be determined by the order of the intersections between internal nullclines and axes. The configurations correspond respectively to four cases shown in p. 193.

References

1. R.H. MacArthur, *Geographical Ecology: Patterns in the Distribution of Species* (Harper & Row, New York, 1972)
2. D. Tilman, *Resource Competition and Community Structure* (Princeton University Press, Princeton, 1982)
3. D. Tilman, Resources, competition and the dynamics of plant communities, in: *Plant Ecology*, ed. by M.J. Crawley (Blackwell, Oxford, 1986), pp. 51–75

Chapter 8
Modeling for Prey-Predator Relation

Abstract This chapter is about the prey-predator dynamics model that is very popular topics in ecology and mathematical biology. In this book, we shall see also some related classic theoretical topics which have been rarely mentioned in the modern textbooks. They may have a potential to provide cues for the readers to expand the idea for a new modeling about population dynamics.

8.1 Predator's Response

There are two aspects of the predator's response to the prey population through the predation: *numerical response* and *functional response* [16]. The numerical response means the change in the population size of predator by the predation, while the functional response does the change in the predation rate per predator according to the prey density. The functional response indicates the efficiency of predation, which reflects the change of predator's behavior/choice about the predation, and the numerical response does the reproduction with the predator's energy gain by the predation.

The following three types of functional response were proposed by Crawford S. Holling (1930–2019) in [5, 6], and have come to be conventionally used today in ecology (see Fig. 8.1):

Holling's Type I response   The predation rate per predator increasing (almost) proportionally to the prey density up to a certain critical value, beyond which it remains the maximal and insensitive to the prey density.

Holling's Type II response   The predation rate per predator increases as the prey density gets larger, while its increase becomes smaller at the same time. The predator rate has an increasing convex curve with respect to the prey density.

Holling's Type III response   The predation rate per predator increases as the prey density gets larger. Its increase becomes bigger up to a certain prey density, beyond which it becomes smaller. The predator rate has an S-shaped curve with respect to the prey density.

© The Author(s), under exclusive license to Springer Nature Singapore Pte Ltd. 2022   209
H. Seno, *A Primer on Population Dynamics Modeling*, Theoretical Biology,
https://doi.org/10.1007/978-981-19-6016-1_8

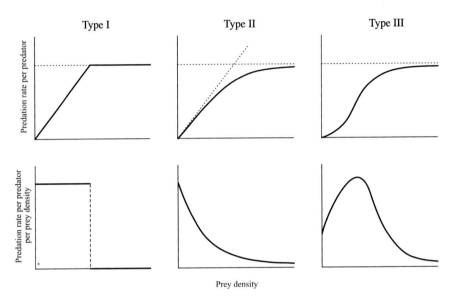

Fig. 8.1 Three types of the functional response in the predation by Holling [5, 6]

The functional response of predator necessarily affects the population dynamics between the predator and prey. As seen in Fig. 8.1, the predation rate per predator generally has an upper bound for sufficiently high prey density. Such a saturation of the predation rate is due to the *handling time* for the predation, for example, the time for searching, catching, or eating a prey individual. The existence of handling time for the predation leads to the supremum of the expected number of preys successfully caught by the predator per unit time. When the prey density is high, the expected time to find a prey by a predator would be short, whereas it must always take a certain time to handle the caught prey. For this reason, there must be the upper bound for the number of preys handled in unit time.

For Holling's Type III response, the predation rate per predator shows an S-shaped curve in terms of the prey density (Fig. 8.1). For example, when the prey density is low, the increase in the prey density causes a steep increase of the predation rate since the frequency of encounters between the prey and predator is induced to get larger. On the other hand, when the prey density is high, the handling time limits the predation efficiency, so that the increase in the prey density leads to only a gradual increase of the predation rate.

As the other factor of functional response, the predator may reduce the *searching effort* for the prey when the prey density is much low [4], since such a prey species appears less valuable for the predator due to the small

(continued)

expected energy gain from the prey in unit time with a small probability of successful find. In such a case, the predator would switch the choice of targeted prey species, if the other prey species is available for the predator. Such a predator is called *generalist* predator in the population dynamics (refer to Sect. 2.4.1). Then the change of targeted prey species is called *switching predation* which will be described later in Sect. 8.7.2. On the other hand, Holling [7] mentioned the other possibility that the capacity of predator's searching and catching becomes more effective as the prey density gets larger in a low range. Further, there may be the effect of limited space for the prey's refuge in the environment. When the prey density becomes beyond the limit, the predation success could gets easier since the preys' refuge gets less available.

8.2 Prey-Predator Population Dynamics

Let us denote the predator density by $P = P(t)$ and the prey density by $H = H(t)$ at time t. The predation rate per predator is now introduced as a function of P and H, $f(P, H)$. The total amount of preys successfully attacked by the predator population in a short period Δt, now denoted by $\Delta \mathcal{Y}$, is given by

$$\Delta \mathcal{Y} = f(P, H)P\Delta t + o(\Delta t). \tag{8.1}$$

As a reasonable modeling, the right side of (8.1) must be zero for any value of Δt as $H \to 0$, since the absence of prey causes no successful attack by the predator. Hence, we can assume that $f(P, 0) = 0$. Robert May mentioned the function f as the introduction of *functional response* of the predator in the prey-predator population dynamics [12]. The characteristics of predator's functional response can be included in the feature of function f as its modeling.

Since the functional response function $f(P, H)$ means the predation rate per predator, it may have the following nature in most cases:

$$\frac{\partial f}{\partial P} \leq 0; \qquad \frac{\partial f}{\partial H} \geq 0. \tag{8.2}$$

However, there may be some cases where the condition (8.2) does not hold. For example, if the predators tend to make a cooperative search for the prey, the searching efficiency may get higher as the predator density becomes

(continued)

larger, at least, in a certain range. Even in such a case, much high predator density would reduce the share of caught prey(s) for each predator, so that the value of $f(P, H)$ gets smaller. Then the predation rate $f(P, H)$ has a unimodal curve in terms of the predator density P.

Now let $H = H(t)$ denote the prey density at time t. The change of the prey density $H(t + \Delta t) - H(t)$ in a period Δt after time t is given by the decrease with the predation $\Delta \mathcal{Y}$ and the variation $\Delta \mathcal{G}$ due to the other processes including birth, death, and migration in the same period: $H(t + \Delta t) - H(t) = \Delta \mathcal{G} - \Delta \mathcal{Y}$. The variation $\Delta \mathcal{G}$ generally depends on the prey density itself, and must satisfy that $\Delta \mathcal{G} \to 0$ as $\Delta t \to 0$ since no period allows any variation (refer to the similar arguments in Sect. 3.3). Hence, from (8.1), we have

$$H(t + \Delta t) - H(t) = \Delta \mathcal{G} - f(P, H)P\Delta t + o(\Delta t). \qquad (8.3)$$

Dividing both sides of (8.3) by Δt and taking the limit as $\Delta t \to 0$, we can get the following ordinary differential equation to govern the population dynamics of prey:

$$\frac{dH(t)}{dt} = g(H) - f(P, H)P, \qquad (8.4)$$

where we define the limiting function g by

$$g(H) := \lim_{\Delta t \to 0} \frac{\Delta \mathcal{G}}{\Delta t},$$

which means the momental velocity of population size change for the prey when the predator is absent (refer to Sect. 3.4).

If the prey population is a closed population without any effect of migration on the size variation (refer to Sect. 3.4), the function g must satisfy that $g(0) = 0$, since no individual leads to no cause for any increase in the population size. For an open prey population, there may be an increase by the immigration, so that the population size could increase even when the size becomes zero at a moment. In such a case, the function g may satisfy that $g(0) \geq 0$.

Now let us introduce the population dynamics of predator by

$$\frac{dP(t)}{dt} = -m(P) + F(P, H)P, \qquad (8.5)$$

where the term $-m(P)$ means the momental velocity of population size change for the predator with the factors except for the predation. If the predator population is closed without any effect of migration on the size variation, this term $-m(P)$ may mean the natural death of predator to decrease the population size, and become negative. The term $F(P, H)P$ corresponds to the reproduction process of predator population by the predation. The function F means the per capita reproduction rate of predator. The factor represented by the function F is called *numerical response* of predator by Robert May [12] (refer to Sect. 8.1).

> The function F may be a function of the predation rate per predator $f(P, H)$, since the value of F must be determined by the energy gain for the reproduction with the predation. In general, it could not be proportional to that of f, while it may have a positive correlation with that of f. The amount of preys taken by a predator cannot produce the energy gain proportional to the amount due to the digestive efficiency etc. Thus, the dependence of F on f would be generally nonlinear. Besides, the value of F may depend on the physiological state depending on the environmental condition or the intraspecific density effect too. For these aspects about the numerical response, it is necessary to assume a reasonable function F in the model, with appropriate features corresponding to the nature of numerical response for the predator in the considered population dynamics.

8.3 Dynamics of Exhaustible Prey

In this section, we shall consider one of the simplest assumptions for the prey-predator relation. Let us assume first that there is no recruitment of prey population, that is, $\Delta G = 0$ for any time t. The prey is now an exhaustive resource for the predator. For example, this is the case of a culture of bacteria in vitro which a nutrient is given only once for the reproduction. There are many other examples of the predation in a similar situation in which the predation occurs only in a specific season when there is no reproduction of prey. We may regard the following modeling as for the population dynamics in such a season. Now from (8.3), we have the population dynamics of prey by

$$\frac{dH(t)}{dt} = -f(P, H)P. \tag{8.6}$$

When the population size of predator has already reached the equilibrium with the predation for the other preys, or when the recruitment of the predator population is out of the considered season with a negligible demographic change (e.g., due to the natural death) for the predator, we may assume that the population size of predator P can be regarded as constant independently of time. In such a case, the ordinary differential Eq. (8.6) can be mathematically solved by the method of variable separation (refer to Sect. 13.1.2):

$$\int_{H(0)}^{H(t)} \frac{dH}{f(P,H)} = -\int_0^t P d\tau = -Pt,$$

and we have

$$t = -\frac{1}{P} \int_{H(0)}^{H(t)} \frac{dH}{f(P,H)}. \tag{8.7}$$

This equation can be regarded as the inverse function of $H = H(t)$. Once the functional response function $f(P, H)$ is given, we can mathematically determine the temporal change of the prey density H by (8.7).

Let us consider here the following functional response function that satisfies the condition (8.2):

$$f(P,H) = \gamma \frac{H}{P}, \tag{8.8}$$

where γ is a positive constant to index the predation efficiency. In this case, we can easily calculate the integral in (8.7), and get

$$t = -\frac{1}{P} \cdot \frac{P}{\gamma} \{ \ln(H(t)) - \ln(H(0)) \},$$

that is,

$$H(t) = H(0)e^{-\gamma t}. \tag{8.9}$$

The prey density exponentially decreases independently of the predator density P.

When the functional response function f is given by (8.8), the total amount of preys successfully attacked by the predator population (8.1) in a short period Δt becomes $\Delta \mathcal{Y} = \gamma H \Delta t + o(\Delta t)$ which has the principal term independent of P. Therefore, even when the predator density varies, the total amount of preys successfully attacked by the predator population is always the same for any period Δt. This is because the predation rate per predator is now assumed to be inversely proportional to the predator density as given by (8.8).

Further, since the total amount of preys successfully attacked by the predator population is independent of the predator density, the total amount is determined only by the prey density. This would seem oversimplified, while such a situation may be approximately realized as a result of the decrease in the predation rate due to the intraspecific reaction in the predator population.

Next let us consider the total amount of preys successfully attacked by the predator population until time t:

$$\mathcal{Y}(t) = H(0) - H(t) = H(0)(1 - e^{-\gamma t}).$$

The amount per unit time is given by

$$\mathcal{Y}(t+1) - \mathcal{Y}(t) = H(0)e^{-\gamma t}(1 - e^{-\gamma}).$$

This is decreasing in terms of time t. Moreover it can be easily found that the time-averaged velocity of predation $\mathcal{Y}(t)/t$ is decreasing in terms of t as well.

We can define the following averaged predation rate per prey in unit time:

$$\frac{\mathcal{Y}(t+1) - \mathcal{Y}(t)}{H(t)} = 1 - e^{\gamma}.$$

This may be regarded as an index of the risk of predation for the prey. This predation rate per prey is constant independently of time t. If we consider the same averaged predation rate per prey in a short period Δt, then we can easily derive the following averaged predation rate per prey in unit time (refer to Sect. 3.4):

$$\frac{\mathcal{Y}(t + \Delta t) - \mathcal{Y}(t)}{H(t)\Delta t} \approx \gamma,$$

since

$$\mathcal{Y}(t + \Delta t) - \mathcal{Y}(t) = \frac{d\mathcal{Y}(t)}{dt}\Delta t + o(\Delta t) = H(0)\gamma e^{-\gamma t}\Delta t + o(\Delta t).$$

These results depend on the modeling of functional response function given by (8.8). So let us consider next the following function generalized from it:

$$f(P, H) = \gamma \frac{H^a}{P^b} \tag{8.10}$$

with positive parameters a and b, which satisfies (8.2) as well. The previous model can be regarded as for (8.10) with $a = 1$ and $b = 1$. In the following arguments, we will see that it was a very specific case.

The predator density P is assumed constant as before. From (8.7), we can derive

$$
t =
\begin{cases}
-\dfrac{1}{P} \cdot \dfrac{P^b}{\gamma} \{\ln[H(t)] - \ln[H(0)]\} & (a = 1); \\[2ex]
-\dfrac{1}{P} \cdot \dfrac{P^b}{\gamma} \dfrac{1}{1-a} \left\{ [H(t)]^{1-a} - [H(0)]^{1-a} \right\} & (a \neq 1),
\end{cases}
$$

and subsequently

$$
H(t) =
\begin{cases}
H(0)\mathrm{e}^{-\gamma P^{1-b} t} & (a = 1); \\[2ex]
\left\{ [H(0)]^{1-a} - (1-a)\gamma P^{1-b} t \right\}^{\frac{1}{1-a}} & (a \neq 1).
\end{cases}
\tag{8.11}
$$

With a specific modeling with $a = 1$ and $b = 0$, when the predation rate per predator f is proportional to the prey density, the total amount of preys successfully attacked by the predator population until time t is given by

$$
\mathcal{Y}(t) = H(0) - H(t) = H(0)(1 - \mathrm{e}^{-\gamma P t}).
\tag{8.12}
$$

This modeling may be called *Nicholson-Bailey type of predation process*. Indeed, this type of predation term has already appeared for the discrete time model called *Nicholson-Bailey model* in Sect. 2.4.2, whereas the modeling was different from what we are considering in this section. As the predator density gets larger, the total amount $\mathcal{Y}(t)$ becomes larger, while the amount per predator $\mathcal{Y}(t)/P$ becomes smaller. This nature is the same even for the model with $a = 1$ and $b < 1$, as seen from (8.11).

When the dynamics considered in this section is about the prey-predator dynamics in a season before their reproduction, we can derive the prey population size at the end of the predation season by $H(0) - \mathcal{Y}(T)$ with the season length T. We may construct a discrete time population dynamics for the annual variation of prey population size, as a modeling of population dynamics for the reproduction in the breeding season, which was described in Sect. 1.5 for the discrete time population dynamics. For the specific modeling with $a = 1$ and $b = 0$, it actually corresponds to Nicholson-Bailey model in Sect. 2.4.2, as mentioned in the above.

In contrast, for the model with $a > 1$, the formula (8.11) indicates that the prey density hyperbolically decreases toward zero as time passes. As the predator density gets larger, the velocity of decrease becomes more gradual when $b > 1$, and steeper when $b < 1$. Especially for the model with $b > 1$, this nature can be understood as the result of feature such that the predation rate per predator f becomes much smaller as the predator density gets larger.

The model with $a < 1$ has characteristics different from the above. As seen from (8.11), the prey density decreases until the following time t_c and becomes exhausted at $t = t_c$:

$$t_c = \frac{[H(0)]^{1-a}}{(1-a)\gamma P^{1-b}}.$$

This is a mathematically special case since the solution of (8.6) with (8.10) cannot be extended beyond $t = t_c$ in this case. When $b < 1$, the time of prey exhaustion t_c becomes earlier as the predator density gets larger. When $b > 1$, it becomes later.

8.4 Lotka-Volterra Prey-Predator Model

In this section, we shall consider again the specific modeling with $a = 1$ and $b = 0$ for the functional response function f given by (8.10), that is $f(P, H) = \gamma H$. Further we shall consider here the population dynamics of predator (8.5) with the simplest modeling of $F(P, H) = \kappa f(P, H)$, where κ is a positive constant frequently called (energy) *conversion coefficient*. The parameter κ indexes the efficiency of the conversion of the amount of preys caught by the predator into the energy for the reproduction. The per capita reproduction rate of predator F is now assumed to be proportional to the per capita predation rate for the prey f.

From (8.4) and (8.5), we have the following system of prey-predator population dynamics:

$$\begin{cases} \dfrac{dH(t)}{dt} = g(H) - \gamma H(t)P(t); \\ \dfrac{dP(t)}{dt} = -m(P) + \kappa\gamma H(t)P(t). \end{cases} \tag{8.13}$$

As a consequence of our modeling, the interaction between prey and predator appears as the terms given by the product of prey and predator densities. It may be regarded as the modeling with the *mass action assumption* for the interspecific reaction, that is, with the *Lotka-Volterra type of interaction* described before in Sect. 6.1.2. Thus in a wider sense, we may call the system (8.13) *Lotka-Volterra prey-predator model*.

The general Lotka-Volterra prey-predator model for the population dynamics of n predator species and ℓ prey species can be given by the following $n + \ell$ dimensional system:

$$
\begin{cases}
\dfrac{dH_i(t)}{dt} = g_i(H_i) - \displaystyle\sum_{k=1}^{n} \gamma_{ki} H_i(t) P_k(t) & (i = 1, 2, \ldots, \ell); \\[4mm]
\dfrac{dP_j(t)}{dt} = -m_j(P_j) + \displaystyle\sum_{k=1}^{\ell} \kappa_{jk} \gamma_{jk} H_k(t) P_j(t) & (j = 1, 2, \ldots, n).
\end{cases}
$$

$$(8.14)$$

Non-negative parameters γ and κ characterize the relation between prey and predator species according to the population dynamics.

The simplest Lotka-Volterra prey-predator model is derived by applying Malthus growth (refer to Sects. 3.3.1 and 5.1) for the recruitment term of prey population $g(H)$ and the death term of predator $-m(P)$ in (8.13), that is,

$$
g(H) = rH(t); \quad -m(P) = -\delta P(t).
$$

For the closed population dynamics (refer to Sect. 3.4), the positive parameters r and δ mean the intrinsic growth rate of prey and the natural death rate of predator respectively. Now we have the following simplest Lotka-Volterra prey-predator model:

$$
\begin{cases}
\dfrac{dH(t)}{dt} = rH(t) - \gamma H(t) P(t); \\[4mm]
\dfrac{dP(t)}{dt} = -\delta P(t) + \kappa \gamma H(t) P(t).
\end{cases}
$$

$$(8.15)$$

This is the pioneer prey-predator model considered by Lotka [10, 11] and Volterra [21, 22] in the early twenty century (see Sect. 6.1.2).

Lotka [10, 11] and Volterra [21, 22] applied the mass action assumption for the interspecific reaction between prey and predator, as the simplest approximation about it from the analogy with the chemical reaction kinetics (refer to Sect. 6.1). Hence their modeling was different from that with the functional response function in this section. However, the theoretical discussion of the population dynamics by the mathematical results on the

(continued)

model may be independent of such a difference in the modeling for (8.15). We shall remark here that the difference in the modeling could lead in general to some different theoretical discussion by the mathematical results on the model, whereas it is not the case for (8.15).

Exercise 8.1 For Lotka-Volterra prey-predator model (8.15), apply the isocline method described in Sect. 14.7. What feature can you find about the temporal change of H and P?

8.4.1 Trajectory in Phase Plane

The isocline method can give an information on the temporal change of H and P for (8.15) such that it may be oscillatory (Exercise 8.1 in the previous section). Actually as seen in Fig. 8.2, the temporal change by (8.15) necessarily becomes periodic with a finite period for any positive initial condition except for the coexistent equilibrium. More precisely, for any initial point $(H(0), P(0))$ in the (H, P)-phase plane except for the coexistent equilibrium, the point $(H(t), P(t))$ necessarily returns to the initial point $(H(0), P(0))$ with a finite period $T > 0$: $(H(T), P(T)) = (H(0), P(0))$. Hence the trajectory of $(H(t), P(t))$ in the (H, P)-phase plane becomes a closed curve, that is, a *periodic orbit*. Although it is usually hard to find the formula of such a periodic orbit, we can derive it for Lotka-Volterra prey-predator model (8.15) as described in the following.

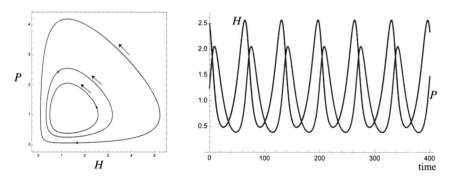

Fig. 8.2 Numerically drawn trajectories in the (H, P)-phase plane, and a temporal change for Lotka-Volterra prey-predator model (8.15) with $r = 0.1$; $\gamma = 0.1$; $\delta = 0.1$; $\kappa = 0.8$. In the (H, P)-phase plane, three trajectories from different initial conditions are drawn. Each closed curve for the periodic orbit in the (H, P)-phase plane is given by (8.17)

Let us denote the closed curve of the periodic orbit in the (H, P)-phase plane by an implicit function $Q(H, P) = 0$. Supposing that the curve is sufficiently smooth, we can assume the existence of a continuous and differential function φ from the implicit function theorem such that $P = \varphi(H)$ in the neighborhood of (h, p) on the curve $Q(H, P) = 0$, satisfying that

$$\frac{dP}{dH} = \frac{d\varphi(H)}{dH} = \varphi'(H) = -\frac{Q_H(H, \varphi(H))}{Q_P(H, \varphi(H))},$$

where $Q_H(H, P) = \partial Q(H, P)/\partial H$ and $Q_P(H, P) = \partial Q(H, P)/\partial P$. Since $P = P(t)$ and $H = H(t)$, we have

$$\frac{dP(t)}{dt} = \frac{d\varphi(H(t))}{dt} = \varphi'(H(t)) \frac{dH(t)}{dt}$$

by the derivative for a composite function. Thus the following relation holds:

$$\varphi'(H(t)) = \frac{dP(t)/dt}{dH(t)/dt}.$$

Substituting (8.15) for this equation, we can get

$$\varphi'(H) = \frac{d\varphi}{dH} = \frac{-\delta + \kappa\gamma H}{H} \cdot \frac{P}{r - \gamma P} = \frac{-\delta + \kappa\gamma H}{H} \cdot \frac{\varphi}{r - \gamma\varphi}. \tag{8.16}$$

This is the ordinary differential equation which $\varphi(H)$ must satisfy. The method of variable separation is applicable for (8.16) (refer to Sect. 13.1.2). Since

$$\int \frac{r - \gamma\varphi}{\varphi} d\varphi = \int \left(\frac{r}{\varphi} - \gamma\right) d\varphi = r \ln \varphi - \gamma\varphi + C_1;$$

$$\int \frac{-\delta + \kappa\gamma H}{H} dH = \int \left(-\frac{\delta}{H} + \kappa\gamma\right) dH = -\delta \ln H + \kappa\gamma H + C_2,$$

where C_1 and C_2 are undetermined constants, we can find from (8.16) that

$$r \ln \varphi - \gamma\varphi = -\delta \ln H + \kappa\gamma H + C_3,$$

with an undermined constant C_3. Therefore, from $P = \varphi(H)$, we can result that the following equation holds for any point (h, p) on the closed curve $Q(H, P) = 0$:

$$r \ln P - \gamma P + \delta \ln H - \kappa\gamma H = C_3.$$

Hence this equation holds for any time t. The undermined constant C_3 is uniquely determined if a point on the closed curve $Q(H, P) = 0$ is given. That is, the closed curve of the periodic orbit for the initial condition $(H(0), P(0))$ is given by

$$r \ln P - \gamma P + \delta \ln H - \kappa \gamma H = r \ln P(0) - \gamma P(0) + \delta \ln H(0) - \kappa \gamma H(0). \tag{8.17}$$

Since the Eq. (8.17) holds for any time t, we find that

$$V(t) := r \ln P(t) - \gamma P(t) + \delta \ln H(t) - \kappa \gamma H(t) \tag{8.18}$$

is a conserved quantity independent of time t for (8.16) (Exercise 8.2).

Exercise 8.2 By differentiating $V(t)$ in terms of t, check that the function $V(t)$ defined by (8.18) is a constant independent of time t.

8.4.2 Equilibrium and Averaged Population Size

It is easy to find that two equilibria always exist for Lotka-Volterra prey-predator model (8.15), $(0, 0)$ and $(\delta/(\kappa\gamma), r/\gamma)$. By the local stability analysis on each equilibrium (refer to Sects. 14.2 and 14.3), we can obtain explicitly the eigenvalues for it (Exercise 8.3).

Since the eigenvalues for equilibrium $(0, 0)$ are r and $-\delta$, it is a saddle point, and unstable (see Fig. 8.2 and Table 14.1 of Sect. 14.3). The eigenvalues for the coexistent equilibrium $(\delta/(\kappa\gamma), r/\gamma)$ are pure imaginaries $\pm i\sqrt{r\delta}$. Thus it is a center point, and mathematically Lyapunov stable (refer to Sect. 14.3). It has been already clear that the coexistent equilibrium $(\delta/(\kappa\gamma), r/\gamma)$ is not asymptotically stable (refer to Sect. 14.3), since the trajectory from the initial point different from $(\delta/(\kappa\gamma), r/\gamma)$ never returns to $(\delta/(\kappa\gamma), r/\gamma)$ as shown in the previous section.

Exercise 8.3 Linearizing Lotka-Volterra prey-predator model (8.15) around the equilibrium, derive the eigenvalues for each of equilibria $(0, 0)$ and $(\delta/(\kappa\gamma), r/\gamma)$.

Generally, even if the local stability analysis shows that an equilibrium is a center point, it may not be Lyapunov stable, as mentioned in Sect. 14.4. For Lotka-Volterra prey-predator model (8.15), the local stability analysis on equilibrium $(\delta/(\kappa\gamma), r/\gamma)$ clearly shows that it is a center point, whereas it is not possible to investigate only by the local stability analysis whether it is Lyapunov stable or not. For the center point, the stability is determined by the nature of original (nonlinear) dynamics. This is a reason why the stability of

(continued)

center point is sometimes called *neutral*. The linearized system can give some valuable information about the stability, while it is the extremal approximation for the dynamics just in the neighborhood of the equilibrium. We shall revisit this issue again in Sect. 8.4.3 on the structural stability.

Since the eigenvalues for the coexistent equilibrium $(\delta/(\kappa\gamma), r/\gamma)$ are pure imaginaries $\pm i\sqrt{r\delta}$, the linearized system of (8.15) around it has the solution in the form of (13.24) with $\rho = 0$ and $\omega = \sqrt{r\delta}$ in Sect. 13.2.4. Hence, the linearized system has a periodic solution with period $2\pi/\sqrt{r\delta}$. This implies that the periodic solution of (8.15) near equilibrium $(\delta/(\kappa\gamma), r/\gamma)$ approximately has the period $2\pi/\sqrt{r\delta}$.

As we have already seen, Lotka-Volterra prey-predator model (8.15) has a periodic solution with a finite period T. Since the periodic solution is uniquely determined by the initial condition, so is the period T. Now we can define the following averaged population sizes over the period, $\overline{H}$ and $\overline{P}$, for each initial condition:

$$\overline{H} := \frac{1}{T} \int_0^T H(t)\,dt; \quad \overline{P} := \frac{1}{T} \int_0^T P(t)\,dt. \tag{8.19}$$

From (8.15), we have

$$\frac{1}{H(t)} \frac{dH(t)}{dt} = r - \gamma P(t),$$

and, by the integral for both sides of this equation,

$$\int_0^T \frac{1}{H(t)} \frac{dH(t)}{dt}\,dt = \int_0^T r - \gamma P(t)\,dt,$$

we can derive the following equation:

$$\ln H(T) - \ln H(0) = 0 = rT - \gamma T \overline{P}.$$

Therefore, we find that $\overline{P} = r/\gamma$. In the same way, we can find that $\overline{H} = \delta/(\kappa\gamma)$ (Exercise 8.4). Consequently, the averaged population sizes over the period is necessarily equal to the values at the coexistent equilibrium $(\delta/(\kappa\gamma), r/\gamma)$ located inside of the closed curve of the periodic orbit in the (H, P)-phase plane, independently of the initial condition.

Exercise 8.4 Show that the averaged prey population size $\overline{H}$ over the period T defined by (8.19) for (8.15) becomes $\overline{H} = \delta/(\kappa\gamma)$. Moreover, show that the total

amount of preys successfully attacked by the predator population during the period
T,

$$\mathcal{Y}_T := \int_0^T \gamma H(t) P(t) \, dt$$

is given by $\gamma T \overline{H} \, \overline{P}$. This result indicates that the averaged predation rate over the
period T is given by $r\delta/(\kappa\gamma)$ independent of the initial condition.

8.4.3 Structural Stability

In this section, to understand further the nature of Lotka-Volterra prey-predator
model (8.15), we shall consider the following system that is mathematically added
sufficiently smooth two variable functions $\mathcal{F}_i$ ($i = 1, 2$) to (8.15):

$$\begin{cases} \dfrac{dH(t)}{dt} = rH(t) - \gamma H(t)P(t) + \varepsilon_1 \mathcal{F}_1(H(t), P(t)); \\[2mm] \dfrac{dP(t)}{dt} = -\delta P(t) + \kappa \gamma H(t)P(t) + \varepsilon_2 \mathcal{F}_2(H(t), P(t)), \end{cases} \tag{8.20}$$

where ε_1 and ε_2 are non-zero constants. We assume that the two variable functions
$\mathcal{F}_1$ and $\mathcal{F}_2$ satisfy that $\mathcal{F}_i = 0$ ($i = 1, 2$) for $(H, P) = (0, 0)$ and $(\delta/(\kappa\gamma), r/\gamma)$.
Hence the system (8.20) has the same equilibria $(0, 0)$ and $(\delta/(\kappa\gamma), r/\gamma)$ as Lotka-
Volterra prey-predator model (8.15). When $|\varepsilon_1|$ and $|\varepsilon_2|$ are sufficiently small,
the terms $\varepsilon_1 \mathcal{F}_1(H(t), P(t))$ and $\varepsilon_2 \mathcal{F}_2(H(t), P(t))$ may be called the *perturbation
terms* for (8.15).

By the linearization of (8.20) around equilibrium (H^*, P^*) (refer to Sect. 14.2),
we can derive the following linearized system for (8.20) around (H^*, P^*):

$$\begin{cases} \dfrac{d\widetilde{h}(t)}{dt} = r\widetilde{h}(t) - \gamma\{P^*\widetilde{h}(t) + H^*\widetilde{p}(t)\} + \varepsilon_1 \partial_H \mathcal{F}_1^* \, \widetilde{h}(t) + \varepsilon_1 \partial_P \mathcal{F}_1^* \, \widetilde{p}(t); \\[2mm] \dfrac{d\widetilde{p}(t)}{dt} = -\delta\widetilde{p}(t) + \kappa\gamma\{P^*\widetilde{h}(t) + H^*\widetilde{p}(t)\} + \varepsilon_2 \partial_H \mathcal{F}_2^* \, \widetilde{h}(t) + \varepsilon_2 \partial_P \mathcal{F}_2^* \, \widetilde{p}(t), \end{cases} \tag{8.21}$$

where

$$\partial_H \mathcal{F}_i^* := \left. \frac{\partial \mathcal{F}_i}{\partial H} \right|_{(H,P)=(H^*,P^*)} \quad ; \quad \partial_P \mathcal{F}_i^* := \left. \frac{\partial \mathcal{F}_i}{\partial P} \right|_{(H,P)=(H^*,P^*)} \qquad (i = 1, 2).$$

Hence we have the following Jacobian matrix A for equilibrium $(H^*, P^*) = (0, 0)$:

$$A = \begin{pmatrix} r + \varepsilon_1 \partial_H \mathscr{F}_1^* & \varepsilon_1 \partial_P \mathscr{F}_1^* \\ \varepsilon_2 \partial_H \mathscr{F}_2^* & -\delta + \varepsilon_2 \partial_P \mathscr{F}_2^* \end{pmatrix},$$

which characteristic equation $\det(A - \lambda E) = 0$ becomes

$$(\lambda - r)(\lambda + \delta) = (\varepsilon_1 \partial_H \mathscr{F}_1^* + \varepsilon_2 \partial_P \mathscr{F}_2^*)\lambda - r\varepsilon_2 \partial_P \mathscr{F}_2^* + \delta \varepsilon_1 \partial_H \mathscr{F}_1^*$$
$$+ (\varepsilon_1 \partial_P \mathscr{F}_1^*)(\varepsilon_2 \partial_H \mathscr{F}_2^*) - (\varepsilon_1 \partial_H \mathscr{F}_1^*)(\varepsilon_2 \partial_P \mathscr{F}_2^*).$$

Thus, for equilibrium $(0, 0)$, we can see that the eigenvalues are clearly different from r and $-\delta$ unless $\partial_H \mathscr{F}_i^* = \partial_P \mathscr{F}_i^* = 0$ $(i = 1, 2)$. However, as long as $|\varepsilon_1|$ and $|\varepsilon_2|$ are sufficiently small, they are a pair of positive and negative real numbers, and equilibrium $(0, 0)$ is a saddle point. Hence for sufficiently small $|\varepsilon_1|$ and $|\varepsilon_2|$, the stability of equilibrium $(0, 0)$ is the same as that for (8.15), which means that the stability is not affected by the perturbation terms.

For equilibrium $(H^*, P^*) = (\delta/(\kappa\gamma), r/\gamma)$, we have Jacobian matrix

$$A = \begin{pmatrix} \varepsilon_1 \partial_H \mathscr{F}_1^* & -\delta/\kappa + \varepsilon_1 \partial_P \mathscr{F}_1^* \\ r\kappa + \varepsilon_2 \partial_H \mathscr{F}_2^* & \varepsilon_2 \partial_P \mathscr{F}_2^* \end{pmatrix}$$

and the characteristic equation

$$(\lambda - \varepsilon_1 \partial_H \mathscr{F}_1^*)(\lambda - \varepsilon_2 \partial_P \mathscr{F}_2^*) + (r\kappa + \varepsilon_2 \partial_H \mathscr{F}_2^*)(\delta/\kappa - \varepsilon_1 \partial_P \mathscr{F}_1^*) = 0.$$

Clearly, unless the coefficient for the first order term of λ, $\varepsilon_1 \partial_H \mathscr{F}_1^* + \varepsilon_2 \partial_P \mathscr{F}_2^*$ is zero, the eigenvalue never becomes purely imaginary for any small $|\varepsilon_i|$ $(i = 1, 2)$. This means that any small perturbation term added to (8.15) changes the stability of equilibrium $(\delta/(\kappa\gamma), r/\gamma)$, which cannot be a center point. Thus, the nature of periodic solution appeared for (8.15) is lost by any small perturbation term.

For the system (8.20) with a sufficiently small perturbation terms which do not necessarily satisfy that $\mathscr{F}_i = 0$ $(i = 1, 2)$ for every equilibrium, these arguments are applicable, while every equilibrium may be changed slightly from that for (8.15). The equilibrium corresponding to $(\delta/(\kappa\gamma), r/\gamma)$ is in general not a center point. Therefore, we have found that an arbitrary small perturbation term makes a mathematically drastic difference in the nature of the solution for the system (8.15). We may say now that the system (8.15) lacks the *structural stability*, and the system (8.15) is called *structurally unstable*. Actually, Lotka-Volterra prey-predator model (8.15) is well-known as an example of structurally unstable system.

> For most dynamical systems with the equilibrium as a center point, the stability of center point is vulnerable when the system is modified with such a perturbation term as above. This is another reason why the stability of center point is sometimes called neutral, other than the reason mentioned before in this section.

Exercise 8.5 For the system (8.20) with the perturbation terms $\mathscr{F}_1(H, P) = H^2$ and $\mathscr{F}_2(H, P) \equiv 0$ (therefore, no perturbation term is added for the population dynamics of predator), find all equilibria and determine the local stability of each equilibrium.

8.4.4 Predator vs Prey with Logistic Growth

In this section, we assume a prey with the logistic growth (refer to Sects. 3.2 and 5.3), and consider Lotka-Volterra prey-predator model (8.13) with

$$g(H) = \{r - \beta H(t)\} H(t),$$

that is,

$$\begin{cases} \dfrac{dH(t)}{dt} = \{r - \beta H(t)\}H(t) - \gamma H(t)P(t); \\[2mm] \dfrac{dP(t)}{dt} = -\delta P(t) + \kappa\gamma H(t)P(t). \end{cases} \tag{8.22}$$

This prey-predator model is mathematically equivalent to the system (8.20) with $\varepsilon_1\mathscr{F}_1(H, P) = -\beta H^2$ and $\varepsilon_2\mathscr{F}_2(H, P) \equiv 0$. As already shown in Sect. 8.4.3, the nature of the solution for (8.22) must be different from that for (8.15).

First, let us apply the isocline method (refer to Sect. 14.7) for the system (8.22). The nullclines for H and P in the (H, P)-phase plane are given by $\{(H, P) \mid (r - \beta H - \gamma P)H = 0\}$ and $\{(H, P) \mid (-\delta + \kappa\gamma H)P = 0\}$. They are all lines. We can find the following three different cases with respect to their spatial configuration as shown in Fig. 8.3: (a) $\beta < \beta^* := r\kappa\gamma/\delta$; (b) $\beta = \beta^*$; (c) $\beta > \beta^*$. At the same time, we can find that the system (8.22) has the following three equilibria:

$$E_0(0, 0); \quad E_1\left(\frac{r}{\beta}, 0\right); \quad E_2\left(\frac{r}{\beta^*}, \frac{r}{\gamma}\left(1 - \frac{\beta}{\beta^*}\right)\right),$$

In the case of Fig. 8.3b, c, the application of isocline method indicates that no coexistent equilibrium exists, and any trajectory asymptotically approaches

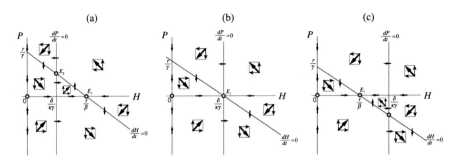

Fig. 8.3 Application of the isocline method for Lotka-Volterra prey-predator model (8.22). (a) $\beta < \beta^* := r\kappa\gamma/\delta$; (b) $\beta = \beta^*$; (c) $\beta > \beta^*$

equilibrium E_1 for any positive initial condition. Thus in such a case, the predator population goes extinct. In contrast, in the case of Fig. 8.3a with $\beta < \beta^*$, a coexistent equilibrium E_2 exists, while its stability cannot be determined only by the isocline method. The isocline method can imply just the possibility of oscillatory temporal change around equilibrium E_2, but cannot show any asymptotic behavior of the trajectory.

Next let us apply the local stability analysis on the equilibrium for (8.22). The linearized system around equilibrium (H^*, P^*) is given by

$$\begin{cases} \dfrac{d\widetilde{h}(t)}{dt} = (r - 2\beta H^*)\widetilde{h}(t) - \gamma\{P^*\widetilde{h}(t) + H^*\widetilde{p}(t)\}; \\ \dfrac{d\widetilde{p}(t)}{dt} = -\delta\widetilde{p}(t) + \kappa\gamma\{P^*\widetilde{h}(t) + H^*\widetilde{p}(t)\}. \end{cases}$$

Actually the local stability analysis for the system (8.22) was mathematically involved in Exercise 8.5 of the previous section (p. 225). From its results, we can find that the coexistent equilibrium E_2 is locally asymptotically stable whenever it exists as in Fig. 8.3a. Further, the local stability of equilibrium E_2 can be classified into the following two cases when it exists:

stable spiral when $0 < \beta < \beta^{**} := 2\kappa\gamma\left(-1 + \sqrt{1 + \dfrac{r}{\delta}}\right)$;

stable node when $\beta^{**} \leq \beta < \beta^* := \dfrac{\kappa\gamma r}{\delta}$.

Indeed, as shown by Fig. 8.4, the temporal change of prey and predator population sizes shows a damped oscillation toward the coexistent equilibrium when $0 < \beta < \beta^{**}$, while it does a monotonic approach toward it when $\beta^{**} \leq \beta < \beta^*$.

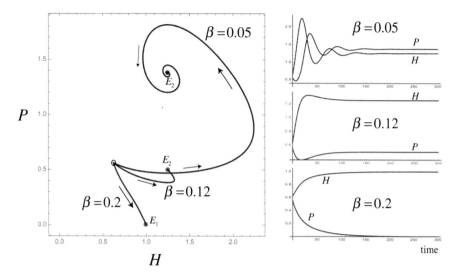

Fig. 8.4 Numerically drawn trajectories in the (H, P)-phase plane, and temporal changes for Lotka-Volterra prey-predator model (8.22) with different values of β and commonly, $r = 0.2$; $\gamma = 0.1$; $\delta = 0.1$; $\kappa = 0.8$; $(H(0), P(0)) = (0.625, 0.5625)$. For $\beta = 0.05$ ($< \beta^{**} = 0.117$), the coexistent equilibrium E_2 is a stable spiral, while, for $\beta = 0.12$ ($< \beta^* = 0.16$), E_2 is a stable node. For both cases, equilibrium E_1 is a saddle point. For $\beta = 0.2$, the coexistent equilibrium E_2 does not exist, when E_1 is a stable node

These results can be gotten together as the parameter dependence of the coexistence between prey and predator in Fig. 8.5. Especially, For the weak intraspecific density effect in the prey population (small β), the high predation efficiency (large γ), and the efficient predator reproduction (large $\kappa\gamma$), it is likely that the temporal change of prey and predator population sizes becomes oscillatory in a damped manner.

The description about the behavior of the solution for Lotka-Volterra prey-predator model (8.22) is based on the results obtained by the isocline method and the local stability analysis. However, they can be proved in a more precise mathematical sense, and we can get the further mathematical result about the behavior of the solution. As such a mathematical approach described in Sect. 14.8, an appropriately constructed *Lyapunov function* can be used to show the *global asymptotic stability* of the coexistent equilibrium E_2 for Lotka-Volterra prey-predator model (8.22). As the other mathematical approach, the Poincaré-Bendixson Trichotomy Theorem in Sect. 14.9 is applicable for the dynamical nature of Lotka-Volterra prey-predator model (8.22), and we can show the global asymptotic stability of E_2 as well when it exists.

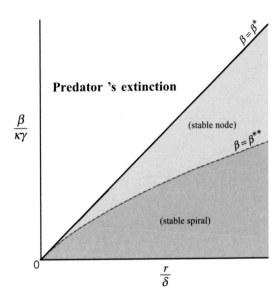

Fig. 8.5 Parameter
dependence of the
coexistence between prey and
predator for Lotka-Volterra
prey-predator model (8.22)

8.5 Holling's Disc Equation

In this section, we shall describe an idea of the modeling for the functional response function, proposed first by Crawford S. Holling in 1959 [5, 6] and discussed further by Tomoo Royama (1930–) [15], which is called today *Holling's disc equation*.

8.5.1 Disc Equation for a Single Prey Species

Let us assume the random predation such that the predator targets at every prey in the region of distance less than R from it. In other words, every prey in the disc area of radius R with the center of the predator can be attacked by the predator. Any prey out of the disc area is not attacked. This is a simple representation of the searching capacity of the predator. The movement of a predator is now traced in the two dimensional plane, drawing a band with the width of $2R$ as schematically shown in Fig. 8.6. The traced band may pass a part of area which was passed before. We now ignore any influence of such an overlapped visit to the same place by the predator, and assume that the predation success is independent of the past visit for any place in the plane. Any place is regarded as equivalent for the predator's visit. We are going to consider the amount of preys successfully attacked by the predator, making use of the cumulative total area traversed by the disc of radius R instead of the total area covered by the pattern generated by the trace of the disc movement. Moreover, we ignore any influence to each predator from the other predators' predation in the same plane.

Fig. 8.6 Trace of the disc area of radius R for the predation by a predator in plane

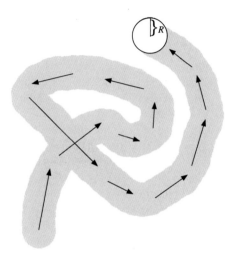

Let $V(t)$ denote the velocity of predator's movement at time t. Then the increment in the band area traversed by the predator's disc area in $[t, t + \Delta t]$ with sufficiently short period Δt is given by

$$2R \int_t^{t+\Delta t} V(z)\, dz = 2RV(t)\Delta t + \mathrm{o}(\Delta t).$$

Suppose that the number of predators in the same habitat (i.e., plane) is given by a constant P. Then the total increment of the area traversed by all predators' disc area in $[t, t + \Delta t]$ becomes $P\{2RV(t)\Delta t + \mathrm{o}(\Delta t)\}$. We assume that any overlapping of different disc areas at the same moment is negligible or does not occur, for example, due to a repulsive interaction between predators.

With these assumptions, we can define the amount $\Delta \mathcal{Y}$ of preys successfully attacked by P predators in $[t, t + \Delta t]$ by

$$\Delta \mathcal{Y} = P\big[\sigma\{2RV(t)H(t)\Delta t + \mathrm{o}(\Delta t)\}\big] = 2\sigma RV(t)PH(t)\Delta t + \mathrm{o}(\Delta t),$$

$$(8.23)$$

where $H(t)$ is the prey density at time t, and the positive parameter σ less than one means the probability of a predator's successful attack. The prey in the disc area can escape from the attack with probability $1 - \sigma$. Remark that the total number of preys located in the disc area traversed by a predator in $[t, t + \Delta t]$ is now given by $2RV(t)H(t)\Delta t + \mathrm{o}(\Delta t)$ in (8.23).

Further we assume that the spatial distribution of preys in the habitat can be regarded as random or statistically uniform (not necessarily regular in space) at any time t. Although the predators' attacks and predations could cause a spatial heterogeneity of the prey distribution, we assume a sufficiently fast diffusion of preys in the plane, and ignore such a spatial heterogeneity generated by the predation. For example, we may image a slowly swimming aquatic predator which catches preys floating around it. As the other example, we may image the immobile predator which catches preys coming near its location like spider, ant lion, and sea anemones etc., whereas, in this case, the movement of a disc area assumed here must be translated in another appropriate meaning in a modeling sense.

Let us denote the area of the habitat for the prey and predator considered now by S. The total amount of preys in the habitat is given by $SH(t)$ at time t. Now we assume that there is no recruitment of prey in the considered predation season. Thus the total amount of preys is monotonically decreasing as time passes. We are now focusing on the prey-predator dynamics in the predation season when there is no reproduction in the prey population. This assumption indicates that the considered prey is exhaustible like that in Sect. 8.3.

Under the above assumptions, the total amount of preys in the habitat $SH(t+\Delta t)$ must satisfy that $SH(t + \Delta t) = SH(t) - \Delta \mathcal{Y}(t)$, taking account of the predation in $[t, t + \Delta t]$. Thus from (8.23), we have

$$\frac{H(t + \Delta t) - H(t)}{\Delta t} = -\frac{\Delta \mathcal{Y}(t)}{S\Delta t} = -2\sigma R V(t) \cdot \frac{P}{S} \cdot H(t) + \frac{o(\Delta t)}{\Delta t}.$$

Finally, by the limit $\Delta t \to 0$, we can derive the following ordinary differential equation for the temporal change of prey density:

$$\frac{dH(t)}{dt} = -2\sigma R V(t) \cdot \frac{P}{S} \cdot H(t), \tag{8.24}$$

where P/S means the averaged predator density in the habitat.

In this equation, we note that the velocity of the decrease of prey density is proportional to the product of prey and predator densities. It corresponds to the Lotka-Volterra type of interaction in Sect. 6.1.2. This can be regarded as the result of assumptions given in this section which lead to the assumption mathematically equivalent to that of complete mixing for the Lotka-Volterra type of interaction.

Now, from (8.23) and (8.24), we can identify the functional response function f defined in Sect. 8.2, as $f = f(H, t) = 2\sigma R V(t)H(t)/S \propto H(t)$. Since the predation rate per predator is proportional to the prey density, the predation may be regarded as the Nicholson-Bailey type (refer to Sect. 8.3). Corresponding to the parameter γ in Sect. 8.3, the predation efficiency can be regarded now as given by the value of $2\sigma R V(t)/S$ at time t.

Actually, we can solve the ordinary differential Eq. (8.24), and get the solution

$$H(t) = H(0) \cdot \exp\left[-2\sigma R \int_0^t V(\tau)\frac{P}{S}d\tau \right],$$

and subsequently the total amount of preys successfully attacked by the predators until time t

$$\mathcal{Y}(t) = SH(0) - SH(t) = SH(0)\left\{1 - \exp\left[-2\sigma R \int_0^t V(\tau)\frac{P}{S}d\tau\right]\right\}.$$
(8.25)

This formula really indicates a Nicholson-Bailey type of predation process. Especially when the velocity of predator's movement is constant independently of time t, we have $\mathcal{Y}(t) = SH(0)(1 - e^{-2\sigma RV[P/S]t})$.

Introduction of Handling Time

As already mentioned in Sect. 8.1, the predation rate may not be proportional to the prey density, but be gradual increasing in terms of it toward a certain upper bound because of the handling time for the predation. Let us assume now the handling time h for the predation per prey as a constant independent of time. Then the predator cannot successfully get preys more than T/h in any period T.

Let Δy denote the amount of preys successfully attacked by a predator in $[t, t + \Delta\tau]$. Since the total handling time for the amount Δy is given by $h\Delta y$, the rest $\Delta\tau - h\Delta y$ can be regarded as the time for the predator's searching the prey in the period $\Delta\tau$. Remark that it must be satisfied that $\Delta y < \Delta\tau/h$, from the reasonability of modeling.

Regarding $\Delta\tau - h\Delta y$ as the time of predator's movement, from (8.23), we can now find the following equation with respect to $\Delta\mathcal{Y} = P\Delta y$:

$$\Delta\mathcal{Y} = P\left[\sigma\{2RV(t)H(t)(\Delta\tau - h\Delta y) + o(\Delta\tau)\}\right].$$

We can solve this equation in terms of $\Delta\mathcal{Y}$ and get

$$\Delta\mathcal{Y} = \frac{2\sigma RV(t)H(t)}{1 + h \cdot 2\sigma RV(t)H(t)}P\Delta\tau + o(\Delta\tau).$$

With the same argument to derive (8.24), we can obtain the following ordinary differential equation:

$$\frac{dH(t)}{dt} = -\frac{2\sigma RV(t)H(t)}{1 + h \cdot 2\sigma RV(t)H(t)} \cdot \frac{P}{S}.$$
(8.26)

Therefore, from (8.23) and (8.24), taking account of the handling time, we have now the following functional response function f:

$$f = f(H, t) = \frac{a(t)\sigma H}{1 + ha(t)\sigma H},$$
(8.27)

Fig. 8.7 Functional response by Holling's disc equation (8.28). The predation rate is bounded by $1/h$ which is the upper bound for the number of preys successfully attacked by a predator in unit time. This is a Holling's Type II response (refer to Sect. 8.1)

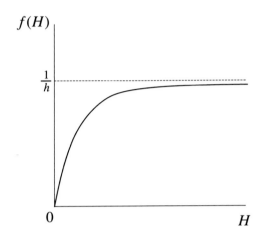

where $a(t) := 2RV(t)$. Generally, the functional response function, that is, the predation rate per predator in the following form is called *Holling's disc equation*:

$$f = f(H) \propto \frac{\alpha H}{1 + h\alpha H}, \tag{8.28}$$

where α is a positive constant. The above formula (8.27) can be regarded as a generalized type of Holling's disc equation. The functional response given by (8.28) is a *Holling's Type II response* defined in Sect. 8.1, as seen in Fig. 8.7.

As already mentioned in Sect. 6.2.1, Holling's disc equation is mathematically equivalent to the Michaelis-Menten equation (6.24). As argued in the above, Holling's disc equation (8.28) can be derived with the modeling taking account of the handling time which indicates a break in the search of prey. The requirement to have such a handling time suppresses the increase of the cumulative number of preys successfully attacked by the predator. In the Mihaelis-Menten structure (6.15) of Sect. 6.2.1, the creation of the enzyme-substrate complex X (Michaelis-Menten complex) works similarly to suppress the increase of the product concentration P. Such a similarity in the dynamical nature results in the same mathematical structure about the velocity of process about the predation and the chemical kinetics.

The idea of Holling [5, 6] to derive the disc Eq. (8.28) was conceptual. Let us consider randomly distributed points in a plane. Then we assume disc areas with radius R located randomly in a plane. If a randomly chosen point is located in a disc area, we regard it as the occurrence of predation such that a predator successfully attacks a prey. At the same time, we count the handling time h. He considered the repetition of this conceptual game-like stochastic process, and theoretically found the disc Eq. (8.28). Mathematically we must

(continued)

consider the expected number of points in a disc area over a large number of repetitions of the process. For more detail, see Royama [15] in which the author mathematically translated Holling's conceptual modeling and discussed its reasonability precisely. As easily expected, Holling's conceptual modeling was criticized for its application to the actual data about the parasitism or predation, while the disc equation was successful to be fit to the data.

8.5.2 Disc Equation for Multiple Species

In this section, we consider the modeling of the predation for m prey species by a generalist predator species. We take the same assumptions about the predation as in the previous section. Let us denote the density of prey species i by $H_i(t)$, the probability of a predator's successful attack for a prey of species i by σ_i, the handling time for it by h_i.

Assuming that the predation is at random (without any specific preference about the prey species), the expected number Δv_i of preys of species i in the total number of successfully attacked preys Δy in a short period $[t, t + \Delta\tau]$ is now given by

$$\Delta v_i = \frac{\sigma_i H_i(t)}{\sum_{j=1}^{m} \sigma_j H_j(t)} \Delta y + \mathrm{o}(\Delta\tau). \qquad (8.29)$$

This formula is based on the modeling that the effective density of prey species i for the predation is given by $\sigma_i H_i$ because the successful attack is determined by the probability σ_i. The total effective prey density is therefore given by $\sum_{j=1}^{m} \sigma_j H_j$ for the predator.

High density of a prey species is not necessarily beneficial for the predator if the attack for the prey species is little successful. For example, when $H_1 > H_2$ and $\sigma_1 < \sigma_2$, it may be satisfied that $\sigma_1 H_1 < \sigma_2 H_2$. Failure of the attack is nonsense for the predator, which may be regarded as the predation with no energy gain. From this point, it is theoretically reasonable to consider the above effective prey density for the predation.

On the other hand, the total time for the handling according to Δy is given by

$$\sum_{i=1}^{m} h_i \Delta v_i = \sum_{i=1}^{m} h_i \frac{\sigma_i H_i(t)}{\sum_{j=1}^{m} \sigma_j H_j(t)} \Delta y + \mathrm{o}(\Delta\tau),$$

since the successful attack for a prey of species i takes the handling time h_i. Therefore, by the same argument as in the previous section, we can define the searching time ΔT in the period $\Delta \tau$ by

$$\Delta T = \Delta \tau - \sum_{i=1}^{m} h_i \frac{\sigma_i H_i(t)}{\sum_{j=1}^{m} \sigma_j H_j(t)} \Delta y - \mathrm{o}(\Delta \tau). \qquad (8.30)$$

With the same assumptions for the movement of predator, the area traversed by a predator during ΔT of $[t, t + \Delta \tau]$ is given by $2RV(t)\Delta T + \mathrm{o}(\Delta \tau)$, so that the number of preys of species i successfully attacked by a predator in $\Delta \tau$ becomes $\sigma_i\{2RV(t)H_i(t)\Delta T + \mathrm{o}(\Delta \tau)\}$. Hence the total number Δy of preys successfully attacked by a predator in $[t, t + \Delta \tau]$ is given by

$$\Delta y = \sum_{i=1}^{m} 2RV(t)\sigma_i H_i(t)\Delta T + \mathrm{o}(\Delta \tau). \qquad (8.31)$$

Substituting (8.30) for (8.31) and solving it in terms of Δy, we can derive the following equation:

$$\Delta y = \frac{2RV(t) \sum_{i=1}^{m} \sigma_i H_i(t)}{1 + 2RV(t) \sum_{j=1}^{m} h_j \sigma_j H_j(t)} \Delta \tau + \mathrm{o}(\Delta \tau). \qquad (8.32)$$

The total amount of preys of species i in the habitat $SH_i(t + \Delta \tau)$ must satisfy that $SH_i(t + \Delta \tau) = SH(t) - P\Delta v_i$, taking account of the predation by P predators in $[t, t + \Delta \tau]$ as in the previous section. Thus, from (8.29) and (8.32), we can derive the following equation:

$$\frac{H_i(t + \Delta \tau) - H_i(t)}{\Delta \tau} = -\frac{2RV(t)\sigma_i H_i(t)}{1 + 2RV(t) \sum_{j=1}^{m} h_j \sigma_j H_j(t)} \cdot \frac{P}{S} + \frac{\mathrm{o}(\Delta \tau)}{\Delta \tau}.$$

With the limit as $\Delta \tau \to 0$, we have

$$\frac{dH_i(t)}{dt} = -\frac{2RV(t)\sigma_i H_i(t)}{1 + 2RV(t) \sum_{j=1}^{m} h_j \sigma_j H_j(t)} \cdot \frac{P}{S}. \qquad (8.33)$$

Therefore we can define the functional response for pray species i by the following function f_i:

$$f_i = f_i(H_1, H_2, \ldots, H_m, t) = \frac{a(t)\sigma_i H_i}{1 + a(t) \sum_{j=1}^{m} h_j \sigma_j H_j}, \qquad (8.34)$$

where $a(t) := 2RV(t)$.

The predation rate for prey species i, given by this function f_i, is clearly less than $1/h_i$. It is remarked that the predation rate significantly depends on the density of other prey species. Since the predator's attack to the other species must reduce the predation pressure for prey species i due to the handling time for them, this modeling can be regarded as a reasonable formulation to express such an indirect interspecific effect from the other prey species to prey species i through the predation by the common predator.

Now, let us introduce furthermore the predator's (expected) energy gain per prey of species i, denoted by e_i. This parameter e_i may be regarded as a weight according to the value of prey species i for the predator, since the difference of prey species must lead to the qualitative/quantitative difference as the bioresource for the predator. Let us denote by ΔE the energy gain per predator with the total number of successfully attacked preys Δy in a short period $[t, t + \Delta \tau]$. With (8.29), the energy gain ΔE in $[t, t + \Delta \tau]$ becomes

$$\Delta E = \sum_{i=1}^{m} e_i \Delta v_i = \sum_{i=1}^{m} e_i \frac{\sigma_i H_i(t)}{\sum_{j=1}^{m} \sigma_j H_j(t)} \Delta y + o(\Delta \tau).$$

From (8.32), we can derive

$$\Delta E = \sum_{i=1}^{m} e_i f_i(H_1, H_2, \ldots, H_m, t) \Delta \tau + o(\Delta \tau) \tag{8.35}$$

with the functional response function f_i given by (8.34). Since the predator's reproduction depends on the energy gain by the predation, the predator's numerical response function F of (8.5) in Sect. 8.2 must be related to the energy gain determined by the above equation. We shall consider the population dynamics model of one prey and one predator with Holling's disc equation and such a numerical response function in the next section.

As a simple expansion of the above modeling with Holling's disc equation, we can consider the predation by more than one predator species. Let us assume ℓ different species of predator for m prey species. The number of individuals of predator species j is now denoted by P_j. Predator species j is now characterized by the searching/attacking radius R_j, the moving velocity $V_j(t)$, the probability of successful predation for prey species i, σ_{ij}, and the handling time for prey species i, h_{ij}. Applying the above arguments for this case, we can easily derive

$$\frac{dH_i(t)}{dt} = -\sum_{k=1}^{\ell} \frac{2 R_k V_k(t) \sigma_{ik} H_i(t)}{1 + 2 R_k V_k(t) \sum_{j=1}^{m} h_{jk} \sigma_{jk} H_j(t)} \cdot \frac{P_k}{S}. \tag{8.36}$$

Consequently we can define the following functional response of predator species j for prey species i:

$$f_{ij} = f_{ij}(H_1, H_2, \ldots, H_m, t) = \frac{a_j(t)\sigma_{ij} H_i}{1 + a_j(t) \sum_{k=1}^{m} h_{kj}\sigma_{kj} H_k}, \qquad (8.37)$$

where $a_j(t) := 2R_j V_j(t)$. The interspecific relation of exploitative competition between predator species (refer to Sect. 2.3) is indirectly involved in the functional response function (8.37) through the dependence of prey densities affected by the other predators' predation.

As already mentioned in Sect. 6.2.3, the functional response function for the prey population dynamics of (6.43) is mathematically equivalent to (8.37) with time-independent a_j. Since the prey-predator population dynamics model (6.43) was derived with the quasi-stationary state approximation (QSSA) for the state of predator's handling the attacked prey, the functional response function for the prey population dynamics of (6.43) must contain a structure corresponding to the handling time. This would be the reasoning about the mathematical correspondence between the functional responses of (6.43) and (8.37).

8.6　Rosenzweig-MacArthur Model

In this section, we consider the following population dynamics model of one prey and one predator with the functional response by Holling's disc equation:

$$\begin{cases} \dfrac{dH(t)}{dt} = \{r - \beta H(t)\}H(t) - \dfrac{\gamma H(t)}{1 + h\gamma H(t)} P(t); \\[3mm] \dfrac{dP(t)}{dt} = -\delta P(t) + \kappa \dfrac{\gamma H(t)}{1 + h\gamma H(t)} P(t), \end{cases} \qquad (8.38)$$

where positive parameters r, β, δ, and κ have the same meanings respectively as those for Lotka-Volterra prey-predator model (8.22). Parameter γ is the coefficient of predation efficiency, and h is the handling time per prey. When the handling time is negligible, that is, when $h = 0$, the above system (8.38) coincides with Lotka-Volterra prey-predator model (8.22). After the study on the prey-predator population dynamics model (8.38) by American ecological scientists Michael L. Rosenzweig (1941–) and Robert H. MacArthur (1930–1972) in 1968 [14], it is sometimes called

Rosenzweig-MacArthur model today. Making use of the transformation of variables and parameters,

$$\widetilde{H}(t) := \frac{\beta H(t)}{r}; \quad \widetilde{P}(t) := \frac{\gamma P(t)}{r}; \quad \tau := rt; \quad \mu := \frac{\delta}{r}; \quad k := \frac{\kappa\gamma}{\beta}; \quad \eta := \frac{rh\gamma}{\beta},$$
(8.39)

we can get the following non-dimensionalized system for (8.38):

$$\begin{cases} \dfrac{d\widetilde{H}(\tau)}{d\tau} = \{1 - \widetilde{H}(\tau)\}\widetilde{H}(\tau) - \dfrac{\widetilde{H}(\tau)}{1 + \eta\widetilde{H}(\tau)} \widetilde{P}(\tau); \\[4mm] \dfrac{d\widetilde{P}(\tau)}{d\tau} = -\mu\widetilde{P}(\tau) + k\dfrac{\widetilde{H}(\tau)}{1 + \eta\widetilde{H}(\tau)} \widetilde{P}(\tau). \end{cases}$$
(8.40)

On the other hand, same as Lotka-Volterra prey-predator model (8.22), the prey population in the Rosenzweig-MacArthur model (8.38) follows the logistic growth with the carrying capacity r/β (refer to Sect. 5.3). Hence, for the ecological reasonability, we may assume that $0 < H(0) \le r/\beta$, that is, the initial population size of prey is not beyond the carrying capacity. The prey population size cannot beyond the carrying capacity r/β for any time t because its variation is governed by the reproduction with the logistic equation and the predation. Actually, since $dH/dt < 0$ for any $H > r/\beta$ and $P \ge 0$, we can mathematically prove that $0 < H(t) < r/\beta$ for any $t > 0$ with the initial condition such that $0 < H(0) \le r/\beta$ and $P(0) > 0$. Thus we consider the non-dimensionalized system (8.40) with the initial condition such that $0 < \widetilde{H}(0) < 1$ and $\widetilde{P}(0) > 0$, and we may assume that $0 < \widetilde{H}(\tau) < 1$ for any $\tau > 0$.

There are the following three equilibria for the system (8.40):

$$E_0(0,0); \quad E_1(1,0); \quad E_2(\widetilde{H}_+^*, (1 - \widetilde{H}_+^*)(1 + \eta\widetilde{H}_+^*)),$$
(8.41)

where $\widetilde{H}_+^* := 1/(k/\mu - \eta)$. The equilibrium E_2 exists when and only when $0 < \widetilde{H}_+^* < 1$, that is,

$$\frac{k}{\mu} - \eta > 1.$$
(8.42)

The local stability analysis (refer to Sect. 14.5) shows that the eigenvalues for E_0 are 1 and $-\mu$. Thus, equilibrium E_0 is always a saddle point. It is shown also that the eigenvalues for E_1 are -1 and $-\mu + k/(1 + \eta)$. Hence equilibrium E_1 can be classified as follows according to the stability:

$$\begin{cases} \text{stable node if } \dfrac{k}{\mu} - \eta < 1; \\[4mm] \text{saddle} \quad \text{if } \dfrac{k}{\mu} - \eta > 1. \end{cases}$$
(8.43)

As a result, we find that equilibrium E_1 is unstable as a saddle point when and only when equilibrium E_2 exists. As described in Sect. 14.5, we can obtain the following result on the stability of E_2 when it exists:

$$
\begin{cases}
\text{unstable (source)} & \text{if } \eta > 1 \text{ and } \dfrac{k}{\mu} - \eta > 2 + \dfrac{2}{\eta - 1}; \\[4ex]
\text{locally asymptotically stable (sink) if} & \begin{cases} \eta \leq 1 \\ \text{or} \\ \eta > 1 \text{ and } \dfrac{k}{\mu} - \eta < 2 + \dfrac{2}{\eta - 1}; \end{cases} \\[6ex]
\text{Lyapunov (neutrally) stable} & \text{if } \eta > 1 \text{ and } \dfrac{k}{\mu} - \eta = 2 + \dfrac{2}{\eta - 1}.
\end{cases}
\tag{8.44}
$$

This result is shown in Fig. 8.8.

It is indicated that there is no stable equilibrium in the first quadrant of $(\widetilde{H}, \widetilde{P})$-phase plane when equilibrium E_2 exists and is unstable, since the other equilibria E_0 and E_1 are both unstable as saddle points. In such a case, the solution of (8.40) asymptotically approaches a periodic orbit, that is, a closed curve in the $(\widetilde{H}, \widetilde{P})$-phase plane, called *limit cycle*, as shown in Fig. 8.9. The mathematical proof makes use of *Poincaré-Bendixson Theorem*, *Poincaré-Bendixson Trichotomy*, and the fact given in Exercise 14.3 of Sect. 14.9 (see [18] for more mathematical arguments).

In contrast, when the coexistent equilibrium E_2 does not exist, the above results by the local stability analysis and the isocline method can prove that the solution

Fig. 8.8 Parameter dependence of the existence and stability of equilibrium E_2 for (8.40). The solid line and curve indicate the boundary of parameter regions. The dashed lines are related to (8.42), (8.43), and (8.44)

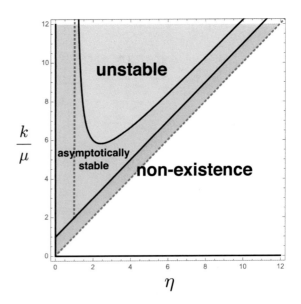

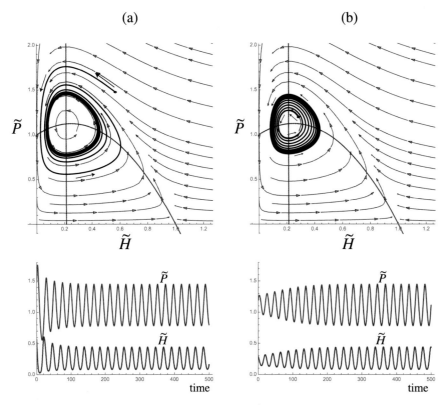

Fig. 8.9 A numerical example of the asymptotic approach of trajectory to a limit cycle in the $(\widetilde{H}, \widetilde{P})$-phase plane, and corresponding temporal change for the system (8.40). Nullclines are drawn too in the $(\widetilde{H}, \widetilde{P})$-phase plane. (**a**) $(\widetilde{H}(0), \widetilde{P}(0)) = (0.568, 1.476)$; (**b**) $(\widetilde{H}(0), \widetilde{P}(0)) = (0.285, 1.193)$. Commonly, $\mu = 0.15$; $k = 1.0$; $\eta = 2.0$

necessarily approaches the unique asymptotically stable equilibrium, alternatively E_0 or E_1.

At the critical case under the third condition in (8.44), equilibrium E_2 becomes unstable with purely imaginary eigenvalues, while it becomes unstable as an unstable spiral under the first condition in (8.44), and an asymptotically stable periodic orbit appears. This kind of bifurcation to a limit cycle at the critical case under the third condition in (8.44) is called (supercritical) *Hoph bifurcation* in the dynamical system theory.

The Rosenzweig-MacArthur model (8.38) with $h = 0$ coincides with Lotka-Volterra prey-predator model (8.22) for which no periodic solution appears. As seen in Fig. 8.10 about the parameter dependence of the prey-predator dynamics by the

Rosenzweig-MacArthur model (8.38), its nature is qualitatively the same as that of Lotka-Volterra prey-predator model (8.22) for sufficiently small h. Appearance of a limit cycle for the Rosenzweig-MacArthur model (8.38) requires a large h. This implies that the predator with a long handling time would cause a periodic variation when it coexists with the prey.

8.7 Regulation of Prey Use

As seen in the previous section on the Rosenzweig-MacArthur model, the functional response significantly affects the nature of prey-predator population dynamics. The coexistence between prey and predator would require some characteristics of the functional response. Actually, we can see from Fig. 8.10 about the Rosenzweig-MacArthur model (8.38) that the predator with an inappropriate functional response could not coexist with the prey. This implies that the coexistence between prey and predator could be understood as a result of the coevolution for the prey-predator relation between those species. The evolution of predation involves

- Which prey should be used;
- How much should be used.

The functional response can be regarded as related to the latter aspect.

The former aspect has been studied in what is called *diet selection theory* or *diet menu theory*. The latter has been studied in *foraging theory* with a relation to the ethology [3, 9, 17]. In such studies, the optimal strategy for the predator to maximize the *fitness* (defined for each aspect about the predator's behavior) has been discussed

Fig. 8.10 Parameter dependence of the asymptotic behavior of the solution for the Rosenzweig-MacArthur model (8.38). Solid line and curve indicate the boundary of parameter regions. See also Fig. 8.8

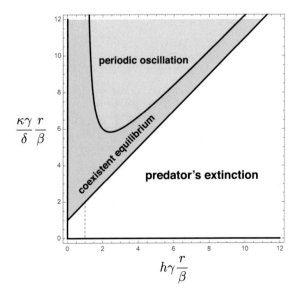

from the viewpoint of evolutionary biology. In this section, we consider some classic theories on such an aspect for a while. It will be seen that they are closely related to the reasonable modeling for the prey-predator population dynamics.

8.7.1 Diet Selection: Which Should Be Used?

Supposing that there are m prey species available for a predator, we shall consider the theoretical problem about which prey species should be used by the predator as the optimal predation. This problem has been studied with a variety of assumptions in the optimal diet selection theory.

Let us consider this problem here with the most classic assumption given by the following assumptions:

1. The predator searches a prey at random.
2. The density of each prey species is constant independently of the predation.
3. Any interaction between predators is negligible.
4. The frequency of a predator's encounters to the prey of species i per unit time is given by a species-specific constant λ_i $(i = 1, 2, \ldots, m - 1, m)$.
5. Each of the predator's attack to a prey is independent of the past experience of predation.
6. When the predator handles an attacked prey, it cannot search or attack any other prey.
7. When the predator attacks an individual of prey species i, the attack is successful with probability σ_i $(i = 1, 2, \ldots, m - 1, m)$.
8. When the attack to a prey of species i is successful, it takes the expected handling time h_i $(i = 1, 2, \ldots, m - 1, m)$.
9. If an attack fails, the handling time is negligible.
10. The expected energy gain by the predation of a prey of species i is given by e_i $(i = 1, 2, \ldots, m - 1, m)$.
11. The optimal diet selection maximizes the expected energy gain per unit time by the predation.

The last assumption in the above defines the optimality according to the diet selection in this theoretical consideration.

Although the predator does not necessarily attack an encountered prey, the higher frequency of encounters indicates the more easiness for the predator to find the prey. We now introduce the probability p_i $(i = 1, 2, \ldots, m - 1, m)$ that the predator attacks an individual of prey species i when the predator encounters it. The set of probabilities $\{p_1, p_2, \ldots, p_m\}$ represents the predator's behavior according to the diet selection. The frequency of a predator's encounters to the prey of any species per unit time is given by $\sum_{i=1}^{m} \lambda_i$ from the fourth assumption. Thus, the expected duration t_s until the predator encounters any prey individual is given by

$t_s = 1/\sum_{i=1}^{m} \lambda_i$. Then the probability q_i that, when the predator encounters a prey individual, it is of species i is given by $q_i = \lambda_i/\sum_{j=1}^{m} \lambda_j$.

First, let us derive the expected searching time T_s until the predator makes a successful attack to a prey after it starts the search for the prey. From the above modeling, even when the predator encounters a prey of species i, it does not attack it with probability $1 - p_i$. Besides, when the predator attacks a prey of species i, it is unsuccessful with probability $1 - \sigma_i$. Hence the probability that the predator does not attack an encountered prey or the attack fails is given by

$$\sum_{i=1}^{m} \{(1 - p_i)q_i + (1 - \sigma_i)p_i q_i\} = \sum_{i=1}^{m} (1 - \sigma_i p_i)q_i = 1 - \sum_{i=1}^{m} \sigma_i p_i q_i,$$

where $\sigma_i p_i q_i$ gives the probability that the predator encounters a prey of species i and succeeds in the attack for it. As a result, the formula

$$\sum_{i=1}^{m} \sigma_i p_i q_i \left\{1 - \sum_{i=1}^{m} \sigma_i p_i q_i\right\}^k$$

gives the probability that the predator makes the first successful attack to a prey encountered after k encounters without attack or with unsuccessful attack. Since the unsuccessful attack takes no handling time from the above ninth assumption, it takes no time if the predator does not attack an encountered prey or if the attack fails. Thus the successful attack after k encounters without attack or with unsuccessful attack takes the expected time $(k + 1)t_s$ after the predator starts the search for the prey. Lastly we can derive the expected searching time T_s as follows:

$$T_s = \sum_{k=0}^{\infty} (k + 1)t_s \left[\sum_{i=1}^{m} \sigma_i p_i q_i \left\{1 - \sum_{i=1}^{m} \sigma_i p_i q_i\right\}^k\right] = \frac{t_s}{\sum_{i=1}^{m} \sigma_i p_i q_i}. \quad (8.45)$$

Next, when the attack to a prey is successful, the expected handling time T_h for the successful attack to a prey is given by

$$T_h = \sum_{i=1}^{m} \frac{\sigma_i p_i q_i}{\sum_{j=1}^{m} \sigma_j p_j q_j} h_i, \quad (8.46)$$

where $\sigma_i p_i q_i / \sum_{j=1}^{m} \sigma_j p_j q_j$ gives the probability that, when the predator successfully attacks a prey, it is of species i. In the same way, we can give the expected energy gain E by the successful attack to a prey individual as follows:

$$E = \sum_{i=1}^{m} \frac{\sigma_i p_i q_i}{\sum_{j=1}^{m} \sigma_j p_j q_j} e_i, \quad (8.47)$$

With these modelings, the expected time necessary for the search and attack per prey is given by $T_s + T_h$. Hence, from (8.45), (8.46), and (8.47), the expected energy gain per unit time W can be defined by

$$W = \frac{E}{T_s + T_h} = \frac{\sum_{i=1}^{m} \sigma_i \lambda_i p_i e_i}{1 + \sum_{i=1}^{m} \sigma_i \lambda_i p_i h_i}. \tag{8.48}$$

As defined in the last assumption, the optimal diet selection must maximize the value W.

Now let us investigate the p_j-dependence of W. The partial derivative of W in terms of p_j,

$$\frac{\partial W}{\partial p_j} = \frac{\sigma_j \lambda_j e_j \left(1 + \sum_{i=1, i \neq j}^{m} \sigma_i \lambda_i p_i h_i\right) - \sigma_j \lambda_j h_j \sum_{i=1, i \neq j}^{m} \sigma_i \lambda_i p_i e_i}{\left(1 + \sum_{i=1}^{m} \sigma_i \lambda_i p_i h_i\right)^2}, \tag{8.49}$$

has the sign independent of p_j. If it is positive, the value of W increases as p_j gets larger, while if it is negative, it increases as p_j gets smaller. Hence p_j must take alternatively 0 or 1 for the optimal diet selection $(p_1, p_2, \ldots, p_m) = (p_1^*, p_2^*, \ldots, p_m^*)$, where p_j^* $(j = 1, 2, \ldots, m)$ is 0 or 1.

It is mathematically possible that $\partial W / \partial p_k = 0$ for a certain k with a specific set of values $\{p_1, p_2, \ldots, p_{k-1}, p_{k+1}, \ldots, p_m\}$. Then the value of p_k^* for the optimal diet selection is indefinite. However, this case must hold only with a specific relation among the parameters λ_i, h_i, and e_i $(i = 1, 2, \ldots, k-1, k+1, \ldots, m)$. Since the parameters are independent of each other according to the prey species, such a specific relation is regarded as hard to be established for any biological reason. Therefore, we ignore here such a specific case, and assume that the right side of (8.49) has the sign definitely positive or negative.

Now let us consider which p_i^* is zero. The right side of (8.49) can be transformed as follows:

$$\frac{\partial W}{\partial p_j} = \frac{\sigma_j \lambda_j h_j}{1 + \sum_{i=1}^{m} \sigma_i \lambda_i p_i h_i} \left(\frac{e_j}{h_j} - W\right). \tag{8.50}$$

Without any loss of generality, let us set the order of prey species as follows:

$$\frac{e_1}{h_1} \geq \frac{e_2}{h_2} \geq \frac{e_3}{h_3} \geq \cdots \geq \frac{e_{m-1}}{h_{m-1}} \geq \frac{e_m}{h_m}. \tag{8.51}$$

The value of e_i / h_i means the energy gain from the prey species i per handling time, which may be regarded as an index about the value of prey species i for the predator.

Suppose that the value of W takes its maximum W^* with the optimal diet selection $(p_1, p_2, \ldots, p_m) = (p_1^*, p_2^*, \ldots, p_m^*)$. In comparison with the above order of values e_i/h_i, there must exist the unique natural number k^* such that

$$\frac{e_1}{h_1} \geq \frac{e_2}{h_2} \geq \cdots \geq \frac{e_{k^*}}{h_{k^*}} > W^* > \frac{e_{k^*+1}}{h_{k^*+1}} \geq \cdots \geq \frac{e_m}{h_m},$$

where k^* may be 1 or m as mentioned later. Then the right side of (8.50) must be positive for $j = 1, 2, \ldots, k^*$ while it must be negative for $j = k^*+1, k^*+2, \cdots, m$. Therefore, it is concluded that $p_1^* = p_2^* = \cdots = p_{k^*}^* = 1$ and $p_{k^*+1}^* = p_{k^*+2}^* = \cdots = p_n^* = 0$ as the optimal diet selection. In the optimal diet selection, the predator necessarily attacks the encountered prey of species with the order higher than a critical value k^* about the index e_i/h_i, while it neglects the encountered prey of species with the order lower than it, that is, the species neglected by the predator is not the "prey" for it according to the optimal diet selection. From these theoretical arguments, we found that the optimal diet selection follows the all-or-none rule about the use of prey species.

From this result, we can derive the following condition to determine the unique natural number k^*:

$$\frac{e_{k^*}}{h_{k^*}} > W_{k^*} > \frac{e_{k^*+1}}{h_{k^*+1}}, \tag{8.52}$$

where we defined

$$W_k := \frac{\sum_{i=1}^{k} \sigma_i \lambda_i e_i}{1 + \sum_{i=1}^{k} \sigma_i \lambda_i h_i}$$

which means the expected energy gain per unit time when the predator uses the prey species only from the first to the kth in the above order for the value of prey species. Especially, since

$$W_1 = \frac{\sigma_1 \lambda_1 e_1}{1 + \sigma_1 \lambda_1 h_1} < \frac{e_1}{h_1}$$

for any positive λ_1, we can find that the optimal diet selection necessarily contains the prey species of the first in the above order. This may be taken natural and trivial from the above arguments, because, if the first species is not in the diet selection, no species is included in it so that the energy gain is zero, which is nonsense and contradictory for its maximization.

We remark here that the formula of W_k mathematically corresponds to the energy gain per unit time $\Delta E/\Delta \tau$ given by (8.35) with Holling's disc

(continued)

equation for more than one prey species in Sect. 8.5.2. As easily found, the number of preys within the searching area $a H_i$ in the arguments of Sect. 8.5.2 reasonably corresponds to the frequency of encounters λ_i in this section.

We can prove in addition the following nature about the optimal diet selection with a relatively large k^*:

$$\mathcal{W}_1 < \mathcal{W}_2 < \cdots < \mathcal{W}_{k^*-1} < \mathcal{W}_{k^*} > \mathcal{W}_{k^*+1} > \cdots > \mathcal{W}_{m-1} > \mathcal{W}_m.$$

It is clearly shown that the value of $\mathcal{W}_k$ takes the maximum for the unique k^*, and at the same time, the k-dependence of $\mathcal{W}_k$ is monotonically increasing for $k < k^*$ and decreasing for $k > k^*$. This indicates that the order of prey species defined by (8.51) is reasonable to index the value of prey species for the predator. From the above order with respect to the value of $\mathcal{W}_k$, the predator with the diet selection given by $k = k^*$ can expectedly gain the energy per unit time greater than any predator with that by $k \neq k^*$ can. Since the efficiency of energy gain by the predation is essential for the predator's reproduction, the larger $\mathcal{W}$ could be regarded as the greater fitness in the sense of evolutionary biology. Thus, the natural selection would favor the diet selection with $k = k^*$.

The above theoretical arguments significantly depend on the assumptions for the modeling. For instance, let us consider the ninth assumption that the handling time is negligible when the predator's attack fails. This assumption implies that the predator's attack itself would take sufficiently short time. This is not necessarily unrealistic. We may imagine the attack of bird which tries to catch a fish under water. In comparison, the predator's attack with sneaking and chasing, such as carnivorous fish, must take a significant time for an attack. Actually, replacing the above ninth assumption with that the handling time is the same independently of the attack's success when the predator attacks a prey, we can show that the optimal diet selection is different in general from that obtained in the above arguments (Exercise 8.6).

Exercise 8.6 If the above ninth assumption is replaced with that the handling time is the same independently of the attack's success when the predator attacks a prey, show that the optimal diet selection is different in general from the result obtained in this section.

Diet Selection for Two Prey Species

In this part, we focus on the special case of two prey species available for the predator. Without loss of generality, let us assume that $e_1/h_1 > e_2/h_2$. As already proved in the previous part, the predator must use the first prey for the optimal diet selection: $p_1^* = 1$. Thus we shall consider whether the predator uses the second species for the optimal diet selection. From the result with (8.52) in the previous part, if and only if $\mathcal{W}_1 < e_2/h_2$, the second species is included in the optimal diet menu.

We remark that this condition depends on λ_1 but does not on λ_2. The expected energy gain per unit time $\mathcal{W}_1$ is monotonically increasing in terms of λ_1. Hence, as the density of prey species 1 becomes lower by the predation, $\mathcal{W}_1$ must decrease due to the decline of the frequency to encounter the prey of species 1. Thus, if the predation begins with the condition that $\mathcal{W}_1 > e_2/h_2$, the density of prey species 1 decreases as the predation is going on, and then the condition may become unsatisfied. Once it comes to be satisfied that $\mathcal{W}_1 < e_2/h_2$, the expected energy gain per unit time becomes greater with the change of diet menu to use both of prey species.

As seen about the functional response with Holling's disc equation for more than one prey species in Sect. 8.5.2, the predation rate for prey species 1 is smaller when the predator uses the other prey species than when it uses only prey species 1. Hence the reproduction of prey species 1 may redeem the density enough to satisfy again the condition that $\mathcal{W}_1 > e_2/h_2$. These arguments imply the possibility that the optimal diet selection may repeat a change depending on the density change of prey by the predation. When such a repeated change of diet selection would converge to a fixed diet menu for the equilibrium densities of prey species 1 and 2 under the predation, the above arguments indicate that it must satisfy that $\mathcal{W}_1 = e_2/h_2$ and $\mathcal{W}_2 \geq \mathcal{W}_1$, where the latter condition is necessary since the use of only species 1 cannot make the equilibrium as mentioned in the above. At such an equilibrium, we find that, from the former condition,

$$\lambda_1 = \frac{1}{\sigma_1 h_1} \cdot \frac{e_2/h_2}{e_1/h_1 - e_2/h_2},$$

and from the latter condition, necessarily $\mathcal{W}_2 = \mathcal{W}_1$ independently of λ_2 as a result.

It must be remarked that the arguments in this section was based on the theory of optimal strategy from the viewpoint of evolutionary biology. Hence the arguments about the optimal diet selection in the first part was to be considered with the supposition of the equilibrium state for the population dynamics. Some readers may confuse this context, since the arguments would seem to consider the predator which could change the diet menu during the

(continued)

change of prey density with its predation. The arguments were to investigate the diet selection in order to determine which diet menu could make the energy gain per unit time maximal. In other words, they were to find a rule/algorithm to determine the optimal diet menu. It was a typical way of thinking the optimal strategy in evolutionary biology. In the last part of this section, we considered the change of prey density due to the predation. From the standpoint for the optimal strategy, it may be regarded as a shift of equilibrium prey density due to the predator's diet menu. In contrast, the way of consideration in the last part will be adapted for the switching predation in the next section as the other type of functional response in the population dynamics.

8.7.2 Switching Predation: How Much Should Be Used?

As seen in the previous section, the maximization of the energy gain for the predator depends on the diet selection. Further, as mentioned there, the diet selection depends on the densities of prey species available for the predator. This implies the possibility of a foraging strategy for the predator with the functional response to get the greater energy gain by the predation.

In the arguments on the optimal diet selection of the previous section, it is assumed that the prey density is constant independently of the predation, that is, the frequency of encounters to prey individuals is given as a constant λ_i for prey species i. In the context of population dynamics, it changes due to the predation. So it is worth considering the functional response to maximize the energy gain for the predator according to the prey-predator population dynamics in which the prey density changes by the predation. Such a functional response would be regarded as an optimal strategy with respect to the evolution of predation.

Let us consider m prey species available for the predator again. We shall introduce the concept of *predation effort* or *foraging effort* that reflects not only the energy but also time used for the predation/foraging. Let us denote here the total predation effort per unit time by C. It determines the energy gain per unit time by the predator how the predation effort is allocated to the foraging for each prey species.

We now introduce the allocation ratio of the predation effort for prey species i by θ_i ($0 \leq \theta_i \leq 1$; $i = 1, 2, \ldots, m$), which satisfies that $\sum_{i=1}^{m} \theta_i = 1$. Then we give the effort allocation per unit time to the foraging for prey species i by $c_i = \theta_i C$. Let us assume that the expected number of preys of species i successfully foraged per unit time is proportional to the effort allocation c_i. The net predation rate for the population of prey species i is given by $\alpha_i c_i H_i$ with a specific positive constant α_i and the population density H_i of prey species i. Parameter α_i could be regarded as the predation efficiency for prey species i. Hence the functional response for prey species i is given by $f_i = \alpha_i c_i$. We have now the expected energy gain E by the predation per unit time as follows:

$$E = \sum_{i=1}^{m} e_i f_i H_i = C \sum_{i=1}^{m} e_i \alpha_i \theta_i H_i, \tag{8.53}$$

where e_i is the expected energy gain for the predator to obtain by foraging a prey of species i.

Let us consider now the simplest case of two prey species when $m = 2$. Since $\theta_1 + \theta_2 = 1$ in this case, we have

$$\frac{\partial E}{\partial \theta_j} = C \left(e_j \alpha_j H_j - e_i \alpha_i H_i \right) \quad (i, j = 1, 2; i \neq j).$$

Hence, as long as $e_1 \alpha_1 H_1 > e_2 \alpha_2 H_2$, it is adaptive to increase the effort allocation θ_1 (decrease θ_2) in order to make the energy gain larger. Such an adaptive control of predation effort with changing the effort allocation to the foraging for each prey species is regarded as one of typical foraging behaviors called *switching predation*.

We remark that, along the present modeling, the allocation ratio of the predation effort θ_i must be a function of prey densities: $\theta_i = \theta_i(H_1, H_2)$. As θ_i gets larger, the predation pressure for prey species i becomes stronger while that for the other prey species j does weaker. The stronger predation pressure must make the density of prey species H_i smaller, and the weaker predation pressure must make H_j larger under the prey-predator population dynamics. Therefore, if an equilibrium exists for the prey-predator population dynamics with such a switching predation, it must be satisfied at the equilibrium that $\partial E/\partial \theta_1 = \partial E/\partial \theta_2 = 0$, that is, $e_1 \alpha_1 H_1 = e_2 \alpha_2 H_2$. Such an equilibrium may be regarded as a specific situation, what is called *ideal free distribution*. At the equilibrium, the energy gain per unit effort for one prey species is equal to that for the other. In this sense, the value of prey becomes the same independently of species at the equilibrium, according to the energy gain by the predation.

In an ethological context of evolutionary biology, the ideal free distribution is defined as a specific distribution of individuals in a spatially heterogeneous environmental condition. The high population density at a location means the smaller share of resource per individual there. When the spatial distribution of a resource is heterogeneous, if each individual can freely and ideally choose the settlement location to get the larger share of resource, we can define the equilibrium population density distribution in space by which every individual gets an even share of resource, following the spatially heterogeneous distribution of the resource. It is called the ideal free distribution [1, 2, 8, 20].

Ideal Switching Response

The adaptive control of predation effort with changing the effort allocation to the foraging for each prey species is given by the function $\theta_i = \theta_i(H_1, H_2)$ in the case of two prey species. It determines the functional response for prey species i at the same time, because $f_i = \alpha_i c_i = \alpha_i \theta_i C$. The following function can realize the ideal control of predation effort described in the previous section:

$$\theta_i = \theta_i(H_1, H_2) = \frac{(e_i \alpha_i H_i)^n}{(e_1 \alpha_1 H_1)^n + (e_2 \alpha_2 H_2)^n} \qquad (i = 1, 2), \qquad (8.54)$$

where the positive parameter n is introduced to index the responsiveness of the effort control to the prey density. The larger n indicates the higher responsiveness as seen in Fig. 8.11. In the exceptional case with $n = 0$, we have $\theta_1 = \theta_2 = 1/2$ independently of the prey density. This can be regarded as the case where the predator randomly forages the preys without any control of predation effort, since the effort allocation does not depend on the prey species.

With the effort allocation by (8.54), the expected energy gain per unit time E given by (8.53) becomes

$$E = C \frac{(e_1 \alpha_1 H_1)^{n+1} + (e_2 \alpha_2 H_2)^{n+1}}{(e_1 \alpha_1 H_1)^n + (e_2 \alpha_2 H_2)^n}. \qquad (8.55)$$

As indicated by Fig. 8.11, the switching predation with (8.54) ($n > 0$) can always provide E greater than the random predation ($n = 0$) can. For the larger n, the advantage of switching predation over the random predation becomes greater.

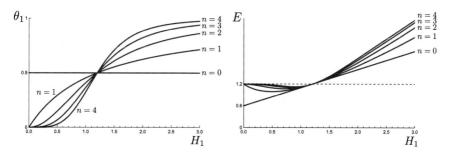

Fig. 8.11 The allocation ratio of predation effort given by (8.54) (the left figure). The function θ_1 is numerically drawn with $e_1 \alpha_1 = 1.0$; $e_2 \alpha_2 = 1.2$; $H_2 = 1.0$. On the right, drawn is the corresponding expected energy gain per unit time E given by (8.55) with $C = 1.0$

At the limit as $n \to \infty$ that is the case of extremal responsiveness for the effort control to the prey density, the allocation ratio (8.54) becomes

$$\theta_i = \theta_i(H_1, H_2) = \begin{cases} 0 & \text{for } e_i \alpha_i H_i < e_j \alpha_j H_j; \\ \frac{1}{2} & \text{for } e_i \alpha_i H_i = e_j \alpha_j H_j; \qquad (i = 1, 2; \ i \neq j). \\ 1 & \text{for } e_i \alpha_i H_i > e_j \alpha_j H_j; \end{cases}$$

This function provides what is called *bang-bang control* for the effort allocation, and the predator forages only one of two prey species unless $e_1 \alpha_1 H_1 = e_2 \alpha_2 H_2$. Thus, under the bang-bang control of predation, the predator appears to switch the diet menu, depending on the prey density.

As mentioned in the above, if there is the coexistent equilibrium for the prey-predator population dynamics, it must be satisfied that $e_1 \alpha_1 H_1 = e_2 \alpha_2 H_2$. At such an equilibrium, we have $\theta_1 = \theta_2 = 1/2$, so that the predator *apparently* takes a random predation. This is because the densities of prey species becomes such that the predator's effort allocation is even for each prey species. It is a result of the population dynamics with the predator's active control of prey densities by the switching predation toward the situation such that $e_1 \alpha_1 H_1 = e_2 \alpha_2 H_2$, that is, the ideal free distribution which is now given by $H_1 : H_2 = e_2 \alpha_2 : e_1 \alpha_1$.

Biased Switching Response

The effort allocation with (8.54) was ideal to increase the expected energy gain per unit time E. It has a close relation to the predation efficiency α_i and the energetic value e_i for prey species i. Now, let us consider the more general effort allocation by the following function:

$$\theta_i = \theta_i(H_1, H_2) = \frac{(\beta_i H_i)^n}{(\beta_1 H_1)^n + (\beta_2 H_2)^n} \qquad (i = 1, 2), \tag{8.56}$$

where the parameter β_i is regarded here as the index of predator's preference for prey species i. We assume that β_i is not necessarily related to the predation efficiency α_i and the energetic value e_i for prey species i. As β_1 gets larger and larger than β_2, the predation becomes strongly biased for prey species 1. Note that, although the effort allocation depends on the prey densities, it may not necessarily to increase the energy gain per unit time due to the existence of predator's preference for prey species.

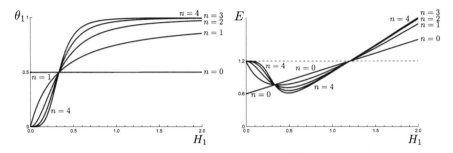

Fig. 8.12 The allocation ratio of predation effort given by (8.56). The function θ_1 is numerically drawn with $\beta_1 = 3.0$; $\beta_2 = 1.0$; $e_1\alpha_1 = 1.0$; $e_2\alpha_2 = 1.2$; $H_2 = 1.0$. On the right, drawn is the corresponding expected energy gain per unit time E given by (8.57) with $C = 1.0$

With the effort allocation by (8.56), the expected energy gain per unit time E given by (8.53) becomes

$$E = C\,\frac{e_1\alpha_1 H_1\,(\beta_1 H_1)^n + e_2\alpha_2 H_2\,(\beta_2 H_2)^n}{(\beta_1 H_1)^n + (\beta_2 H_2)^n}. \tag{8.57}$$

The switching predation with (8.56) is less adaptive than that with (8.54) due to the predator's preference for prey species. Actually, as seen in Fig. 8.12, the random predation ($n = 0$) can become more advantageous than the switching predation in some cases depending on the prey densities. The expected energy gain E given by (8.57) becomes the same as that by the random predation not only when $e_1\alpha_1 H_1 = e_2\alpha_2 H_2$ but also when $\beta_1 H_1 = \beta_2 H_2$. Moreover, we can easily show that, when

$$\min\left\{\frac{e_2\alpha_2}{e_1\alpha_1},\,\frac{\beta_2}{\beta_1}\right\}H_2 < H_1 < \max\left\{\frac{e_2\alpha_2}{e_1\alpha_1},\,\frac{\beta_2}{\beta_1}\right\}H_2,$$

the random predation is more advantageous than the switching predation with the effort allocation by (8.56).

Suppose again that there is the coexistent equilibrium for the prey-predator population dynamics. Then it must be satisfied that $\beta_1 H_1 = \beta_2 H_2$, because, if not, the effort allocation changes to cause a temporal variation of prey densities. Since the predation is apparently equivalent to the random one at the equilibrium, the expected energy gain per unit time E^* at the equilibrium for the switching predation with (8.56) can be expressed by the same formula as that with (8.54): $E^* = (C/2)(e_1\alpha_1 H_1 + e_2\alpha_2 H_2)$. However, we note that the equilibrium sizes of H_1 and H_2 are different from each other, depending on the prey-predator population dynamics in which the difference in the switching response of predator with the predator's preference for the prey species must result in the different equilibrium sizes.

Answer to Exercise

Exercise 8.1 (p. 219)

To apply the isocline method (refer to Sect. 14.7) for Lotka-Volterra prey-predator model (8.15), we need to find first the nullclines in the (H, P)-phase plane. From (8.15), the nullclines for H is given by $\{(H, P) \mid (r - \gamma P)H = 0\}$, and those for P is by $\{(H, P) \mid (-\delta + \kappa\gamma H)P = 0\}$. They are the axes, a horizontal line, and a vertical line in the (H, P)-phase plane. We can get the following figure by the isocline method to indicate the vector direction for the trajectory in each region bounded by the nullclines in the (H, P)-phase plane:

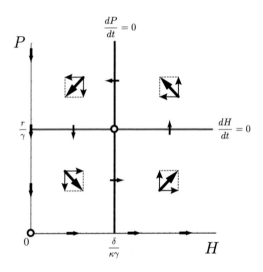

White circles indicate the equilibrium points. The above figure about the vector direction implies the possibility that the trajectory draws a curve moving around the coexistent equilibrium point $(\delta/(\kappa\gamma), r/\gamma)$, that is, the temporal change of H and P could show an oscillation around the equilibrium values $\delta/(\kappa\gamma)$ and r/γ respectively. However, we cannot get any information about the asymptotic behavior of the trajectory, whether it approaches one of equilibria or diverges.

Exercise 8.2 (p. 221)

By differentiating (8.18) in terms of t, we find that

$$\frac{dV(t)}{dt} = r\frac{1}{P(t)}\frac{dP(t)}{dt} - \gamma\frac{dP(t)}{dt} + \delta\frac{1}{H(t)}\frac{dH(t)}{dt} - \kappa\gamma\frac{dH(t)}{dt}$$

$$= \left\{\frac{r}{P(t)} - \gamma\right\}\frac{dP(t)}{dt} + \left\{\frac{\delta}{H(t)} - \kappa\gamma\right\}\frac{dH(t)}{dt}.$$

By substituting (8.15) for this equation, we find that

$$\frac{dV(t)}{dt} = \left\{\frac{r}{P(t)} - \gamma\right\}\{-\delta P(t) + \kappa\gamma H(t)P(t)\}$$

$$+ \left\{\frac{\delta}{H(t)} - \kappa\gamma\right\}\{rH(t) - \gamma H(t)P(t)\}$$

$$= -r\delta + r\kappa\gamma H(t) + \gamma\delta P(t) - \kappa\gamma^2 H(t)P(t)$$

$$+ \delta r - \delta\gamma P(t) - \kappa\gamma r H(t) + \kappa\gamma^2 H(t)P(t) = 0.$$

Therefore, $V(t)$ must be equal to a constant independent of time t.

Exercise 8.3 (p. 221)

By the linearization of Lotka-Volterra prey-predator model (8.15) around the equilibrium (H^*, P^*) (refer to Sect. 14.2), we can obtain the following linearized system of ordinary differential equations about the approximated perturbation $(\tilde{h}(t), \tilde{p}(t)) \approx (H(t) - H^*, P(t) - P^*)$ in the neighborhood of (H^*, P^*) (refer to Sect. 14.2):

$$\begin{cases} \dfrac{d\tilde{h}(t)}{dt} = r\tilde{h}(t) - \gamma\{P^*\tilde{h}(t) + H^*\tilde{p}(t)\}; \\ \dfrac{d\tilde{p}(t)}{dt} = -\delta\tilde{p}(t) + \kappa\gamma\{P^*\tilde{h}(t) + H^*\tilde{p}(t)\}. \end{cases} \tag{8.58}$$

Jacobian matrix for equilibrium (H^*, P^*) is given by

$$A = \begin{pmatrix} r - \gamma P^* & -\gamma H^* \\ \kappa\gamma P^* & -\delta + \kappa\gamma H^* \end{pmatrix}.$$

Hence, for the equilibrium $(H^*, P^*) = (0, 0)$, we have the diagonal matrix $A = \begin{pmatrix} r & 0 \\ 0 & -\delta \end{pmatrix}$, so that the eigenvalues of A, that is, those for equilibrium $(H^*, P^*) = (0, 0)$ are r and $-\delta$. For the coexistent equilibrium $(\delta/(\kappa\gamma), r/\gamma)$, we have the anti-diagonal matrix $A = \begin{pmatrix} 0 & -\delta/\kappa \\ r\kappa & 0 \end{pmatrix}$, so that the characteristic equation $\det(A - \lambda E) = 0$ becomes $\lambda^2 + r\delta = 0$, and we obtain the purely imaginary eigenvalues $\pm i\sqrt{r\delta}$.

Exercise 8.4 (p. 222)

From (8.15), we have

$$\frac{1}{P(t)}\frac{dP(t)}{dt} = -\delta + \kappa\gamma H(t),$$

and get the following equation by the definite integral for $t \in [0, T]$:

$$\ln P(T) - \ln P(0) = -\delta T + \kappa\gamma T\overline{H},$$

where we used the definition of $\overline{H}$ given by (8.19). Since $P(T) = P(0)$ from the periodicity, the left side is zero, so that we find $\overline{H} = \delta/(\kappa\gamma)$.

In the same way, the definite integral of the first equation in (8.15) for $t \in [0, T]$ results in the following equation:

$$H(T) - H(0) = rT\overline{H} - \mathcal{Y}_T.$$

Since $H(T) = H(0)$ from the periodicity, the left side is zero, we can get $\mathcal{Y}_T = rT\overline{H}$. On the other hand, we have already known that $\overline{P} = r/\gamma$, that is, $r = \gamma\overline{P}$. Hence we finally find that $\mathcal{Y}_T = \gamma T\overline{H}\,\overline{P}$.

Exercise 8.5 (p. 225)

In this case, the system (8.20) becomes

$$\begin{cases} \dfrac{dH(t)}{dt} = rH(t) - \gamma H(t)P(t) + \varepsilon_1\{H(t)\}^2; \\[2mm] \dfrac{dP(t)}{dt} = -\delta P(t) + \kappa\gamma H(t)P(t). \end{cases} \tag{8.59}$$

For this system, we can find that there could be three equilibria: $E_0(0,0)$; $E_2(\delta/(\kappa\gamma), r/\gamma + \varepsilon_1\delta/(\kappa\gamma^2))$; $E_1(-r/\varepsilon_1, 0)$. Although ε_1 must satisfy a condition for the reasonable equilibrium (H^*, P^*) with $H^* \geq 0$ and $P^* \geq 0$, we shall put aside the condition and investigate first the local stability for each of them.

The linearized system of (8.59) around equilibrium (H^*, P^*) is given by

$$
\begin{cases}
\dfrac{d\widetilde{h}(t)}{dt} = r\widetilde{h}(t) - \gamma\{P^*\widetilde{h}(t) + H^*\widetilde{p}(t)\} + 2\varepsilon_1 H^*\widetilde{h}(t); \\
\dfrac{d\widetilde{p}(t)}{dt} = -\delta\widetilde{p}(t) + \kappa\gamma\{P^*\widetilde{h}(t) + H^*\widetilde{p}(t)\},
\end{cases} \tag{8.60}
$$

where $(\widetilde{h}(t), \widetilde{p}(t))$ is the approximation for the perturbation $(h(t), p(t)) := (H(t) - H^*, P(t) - P^*)$ in the neighborhood of (H^*, P^*) (refer to Sect. 14.2). Jacobian matrix for (H^*, P^*) becomes

$$
A = \begin{pmatrix} r - \gamma P^* + 2\varepsilon_1 H^* & -\gamma H^* \\ \kappa\gamma P^* & -\delta + \kappa\gamma H^* \end{pmatrix}.
$$

Jacobian matrix for $E_0(0,0)$ becomes the diagonal matrix $A = \begin{pmatrix} r & 0 \\ 0 & -\delta \end{pmatrix}$, so that the eigenvalues are r and $-\delta$. These eigenvalues are the same as those for equilibrium $(0,0)$ about Lotka-Volterra prey-predator model (8.15). Equilibrium $E_0(0,0)$ is a saddle point, and the stability is not influenced by the perturbation term at all.

For equilibrium $E_2(\delta/(\kappa\gamma), r/\gamma + \varepsilon_1\delta/(\kappa\gamma^2))$, Jacobian matrix becomes

$$
A = \begin{pmatrix} \varepsilon_1\delta/(\kappa\gamma) & -\delta/\kappa \\ r\kappa + \varepsilon_1\delta/\gamma & 0 \end{pmatrix},
$$

so that we can get the characteristic equation

$$
\lambda^2 - \frac{\varepsilon_1\delta}{\kappa\gamma}\lambda + r\delta + \frac{\varepsilon_1\delta^2}{\kappa\gamma} = 0.
$$

As a result, we can find the following classification about the local stability of E_2, except for the case where the eigenvalues include zero (refer to Table 14.1 in p. 428 of Sect. 14.3):

Condition	Eigenvalues	Classification of equilibrium	Sign of equilibrium value
$\varepsilon_1 < \varepsilon^*$	Positive and negative	Saddle	$(+, -)$
$\varepsilon^* < \varepsilon_1 \leq \varepsilon_-^{**}$	Both negative	Stable node	$(+, +)$
$\varepsilon_-^{**} < \varepsilon_1 < 0$	Imaginary with negative real part	Stable spiral	
$\varepsilon_1 = 0$	Purely imaginary	Center	
$0 < \varepsilon_1 < \varepsilon_+^{**}$	Imaginary with positive real part	Unstable spiral	
$\varepsilon_1 \geq \varepsilon_+^{**}$	Both positive	Unstable node	

The critical value $\varepsilon^* := -r\kappa\gamma/\delta < 0$ is derived from the sign of the product of eigenvalues, and

$$\varepsilon_\pm^{**} := 2\kappa\gamma\left(1 \pm \sqrt{1 + \frac{r}{\delta}}\right)$$

are derived from the sign of the discriminant for the characteristic equation. For the excluded case where $\varepsilon_1 = \varepsilon^*$, the eigenvalues are 0 and $\varepsilon_1\delta/(\kappa\gamma) < 0$, so that the local stability analysis with the linearized system (8.60) cannot determine the stability of E_2. Moreover, since $(\delta/(\kappa\gamma), 0) = (-r/\varepsilon_1, 0)$ when $\varepsilon_1 = \varepsilon^*$, equilibrium E_2 merges with E_1. We will return to this case later.

Jacobian matrix for equilibrium E_1 becomes the triangular matrix

$$A = \begin{pmatrix} -r & r\gamma/\varepsilon_1 \\ 0 & -\delta - r\kappa\gamma/\varepsilon_1 \end{pmatrix},$$

so that the eigenvalues are $-r$ and $-\delta - r\kappa\gamma/\varepsilon_1$. Therefore, the local stability of E_1 can be classified as follows, except for the case where the eigenvalues include zero as already mentioned in the above:

Condition	Eigenvalues	Classification of equilibrium	Sign of equilibrium value
$\varepsilon_1 < \varepsilon^*$	Both negative	Stable node	$(+, 0)$
$\varepsilon^* < \varepsilon_1 < 0$	Positive and negative	Saddle	
$\varepsilon_1 > 0$	Both negative	Stable node	$(-, 0)$

The critical value $\varepsilon^* < 0$ is the same as before. When $\varepsilon_1 = \varepsilon^*$, the eigenvalues are $-r$ and 0, so that the local stability analysis with the linearized system (8.60) cannot determine the stability of E_1.

In the exclusive case of $\varepsilon_1 = \varepsilon^*$, the system (8.59) becomes

$$\begin{cases} \dfrac{dH(t)}{dt} = \dfrac{r\kappa\gamma}{\delta}\left\{\dfrac{\delta}{\kappa\gamma} - H(t)\right\}H(t) - \gamma H(t)P(t); \\ \dfrac{dP(t)}{dt} = \kappa\gamma\left\{-\dfrac{\delta}{\kappa\gamma} + H(t)\right\}P(t). \end{cases} \tag{8.61}$$

There are only two equilibria, $(0, 0)$ and $(\delta/(\kappa\gamma), 0)$. Let us apply the isocline method (refer to Sect. 14.7) for (8.61). Then we can get the following figure (a) about the vector direction in the (H, P)-phase plane:

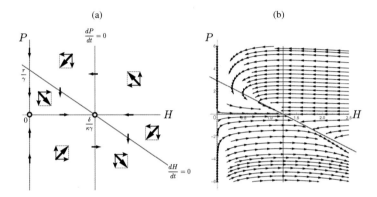

The isocline method implies that the trajectory from a positive initial point asymptotically approaches equilibrium $(\delta/(\kappa\gamma), 0)$. The numerically drawn vector flows shown in the above figure (b) clearly indicates this result. However, in a mathematical sense, equilibrium $(\delta/(\kappa\gamma), 0)$ is unstable, because both of the isocline method and the numerically drawn vector flows indicate the divergence of trajectory from an initial point such that $H(0) > 0$ and $P(0) < 0$.

Since the system (8.61) is not just a mathematical system of ordinary differential equations but a prey-predator population dynamics model, only the trajectory from a positive initial point is meaningful. Therefore, as our conclusion about the case of $\varepsilon_1 = \varepsilon^*$, $(\delta/(\kappa\gamma), 0)$ is globally asymptotically stable from the above result by the isocline method.

In a rigorous mathematical sense, the above conclusion with the isocline method requires the more mathematical proof, though we could clearly see the usefulness of the isocline method to get the mathematical nature of the trajectory.

(continued)

In the mathematical proof, it is necessary to show that the trajectory from a positive initial point cannot go out of the first quadrant of the (H, P)-phase plane in any finite time. Since the system (8.61) can be rewritten as

$$\begin{cases} \dfrac{1}{H(t)} \dfrac{dH(t)}{dt} = \mathscr{F}(H(t), P(t)) := \dfrac{r\kappa\gamma}{\delta}\left\{ \dfrac{\delta}{\kappa\gamma} - H(t) \right\} - \gamma P(t); \\[4mm] \dfrac{1}{P(t)} \dfrac{dP(t)}{dt} = \mathscr{G}(H(t), P(t)) := \kappa\gamma\left\{ -\dfrac{\delta}{\kappa\gamma} + H(t) \right\}, \end{cases} \tag{8.62}$$

we can mathematically get the following equations about $H(t)$ and $P(t)$ for the initial condition $(H(0), P(0)) = (H_0, P_0)$:

$$\begin{cases} H(t) = H_0 \exp\left[\displaystyle\int_0^t \mathscr{F}(H(s), P(s))\, ds \right]; \\[4mm] P(t) = P_0 \exp\left[\displaystyle\int_0^t \mathscr{G}(H(s), P(s))\, ds \right]. \end{cases} \tag{8.63}$$

These equations mathematically indicate that $H(t) > 0$ for any $t > 0$ for $H_0 > 0$, and $P(t) > 0$ for any $t > 0$ for $P_0 > 0$.

As the other way of mathematical argument on the same subject, we could carry out the proof making use of the contradiction of the supposition that the trajectory from a positive initial point intersects an axis at a finite time. In such a proof, we need to use the unique existence of the solution for the system (8.61). Since this book is not to focus on such a mathematical proof, we shall not go into it here. For the reader interested in such a mathematical aspect, some other text books focusing on the more mathematical aspect of the population dynamics model [19], or those of applied mathematics on the dynamical system [13] would be helpful.

As mentioned above, we are now interested only in the equilibrium inside the first quadrant of the (H, P)-phase plane. From this standpoint, we can get together the results obtained by the above analysis as follows:

Condition	E_0	E_2	E_1
$\varepsilon_1 < \varepsilon^*$		—	Stable node
$\varepsilon_1 = \varepsilon^*$		—	(Asymptotically stable)
$\varepsilon^* < \varepsilon_1 \leq \varepsilon_-^{**}$	Saddle	Stable node	Saddle
$\varepsilon_-^{**} < \varepsilon_1 < 0$		Stable spiral	
$\varepsilon_1 = 0$		Center	—
$0 < \varepsilon_1 < \varepsilon_+^{**}$		Unstable spiral	—
$\varepsilon_1 \geq \varepsilon_+^{**}$		Unstable node	—

For any positive ε_1, no asymptotically stable equilibrium exists, and the solution for the linear system (8.60) diverges. For any negative ε_1, the solution necessarily converges an asymptotically stable equilibrium. Consequently this result clearly indicates the structural instability of Lotka-Volterra prey-predator model (8.15) for the perturbation term.

Exercise 8.6 (p. 245)

When the handling time is the same independently of the attack's success, the attack necessarily takes a handling time specifically given for the attacked prey species even if it is unsuccessful. Thus, instead of T_s in the main text, we need to consider the expected duration until the first attack after the predator starts the search for the prey. Let us denote it by T_a now.

The probability that the predator does not attack an encountered prey is given by $\sum_{i=1}^{m}(1 - p_i)q_i = 1 - \sum_{i=1}^{m} p_i q_i$. Hence, the formula

$$\sum_{i=1}^{m} p_i q_i \left\{ 1 - \sum_{i=1}^{m} p_i q_i \right\}^k$$

gives the probability that the predator makes the first attack to a prey encountered after k encounters without attack. Same as in the main text, the attack after k encounters without attack takes the expected time $(k + 1)t_s$ after the predator starts the search for the prey. Lastly we can derive the expected time T_a as follows:

$$T_a = \sum_{k=0}^{\infty}(k + 1)t_s \left[\sum_{i=1}^{m} p_i q_i \left\{ 1 - \sum_{i=1}^{m} p_i q_i \right\}^k \right] = \frac{t_s}{\sum_{i=1}^{m} p_i q_i}. \tag{8.64}$$

Independently of whether the attack to a prey is successful or not, the expected handling time T_h' for the attack to a prey is now given by

$$T_h' = \sum_{i=1}^{m} \frac{p_i q_i}{\sum_{j=1}^{m} p_j q_j} h_i, \tag{8.65}$$

where $p_i q_i / \sum_{j=1}^{m} p_j q_j$ gives the probability that, when the predator attacks a prey, it is of species i. It is different from T_h by (8.46) in the main text. Further, the expected energy gain E' per attack is given by

$$E' = \sum_{i=1}^{m} \frac{p_i q_i}{\sum_{j=1}^{m} p_j q_j} \{\sigma_i e_i + (1 - \sigma_i) \cdot 0\} = \sum_{i=1}^{m} \frac{p_i q_i}{\sum_{j=1}^{m} p_j q_j} \sigma_i e_i. \tag{8.66}$$

This is because the energy gain is zero if the attack is unsuccessful.

Lastly, from (8.64), (8.65), and (8.66), the expected energy gain per unit time W' can be defined by

$$W' = \frac{E'}{T_a + T'_h} = \frac{\sum_{i=1}^{m} \sigma_i \lambda_i p_i e_i}{1 + \sum_{i=1}^{m} \lambda_i p_i h_i},$$

and we have

$$\frac{\partial W'}{\partial p_j} = \frac{\sigma_j \lambda_j e_j \left(1 + \sum_{i=1,i\neq j}^{m} \lambda_i p_i h_i\right) - \lambda_j h_j \sum_{i=1,i\neq j}^{m} \sigma_i \lambda_i p_i e_i}{\left(1 + \sum_{i=1}^{m} \lambda_i p_i h_i\right)^2},$$

which has the sign independent of p_j. Moreover, we have the other expression of this partial derivative as well as in the main text:

$$\frac{\partial W'}{\partial p_j} = \frac{\lambda_j h_j}{1 + \sum_{i=1}^{m} \lambda_i p_i h_i} \left(\frac{\sigma_j e_j}{h_j} - W'\right).$$

Hence, with the same arguments as in the main text, p_j must take alternatively 0 or 1 for the optimal diet selection $(p_1, p_2, \ldots, p_m) = (p_1^*, p_2^*, \ldots, p_m^*)$, and it can be expressed as $p_1^* = p_2^* = \cdots = p_{k^*}^* = 1$ and $p_{k^*+1}^* = p_{k^*+2}^* = \cdots = p_n^* = 0$ with the uniquely determined natural number k^*, when we set the order of prey species as follows:

$$\frac{\sigma_1 e_1}{h_1} \geq \frac{\sigma_2 e_2}{h_2} \geq \frac{\sigma_3 e_3}{h_3} \geq \cdots \geq \frac{\sigma_{m-1} e_{m-1}}{h_{m-1}} \geq \frac{\sigma_m e_m}{h_m}. \tag{8.67}$$

The specific natural number k^* is uniquely determined to satisfy that

$$\frac{\sigma_{k^*} e_{k^*}}{h_{k^*}} > \mathcal{W}'_{k^*} > \frac{\sigma_{k^*+1} e_{k^*+1}}{h_{k^*+1}},$$

where

$$\mathcal{W}'_k := \frac{\sum_{i=1}^{k} \sigma_i \lambda_i e_i}{1 + \sum_{i=1}^{k} \lambda_i h_i}.$$

In general, the order of prey species by (8.67) does not coincide with that by (8.51). This is because the order (8.67) depends on the probability distribution of successful attack $\{\sigma_i\}$, while (8.51) is independent of it. Hence, the species of high rank in (8.51) may be ordered as a low rank in (8.67) due to a small probability of successful attack, which means the difficulty for the predator to succeed in the attack for the prey species, for example, with the prey's protective behavior or resistant characteristics. Therefore, every prey species i with $p_i^* = 1$ in the above modeling does not coincide to a prey species j with $p_j^* = 1$ in the main text. That is, a prey species i with $p_i^* = 1$ in the above modeling may correspond to the species j with $p_j^* = 0$ in the main text.

References

1. M. Broom, J. Rychtář, *Game-Theoretical Models in Biology*. Chapman & Hall/CRC Mathematical and Computational Biology Series (CRC Press, Boca Raton, 2013)
2. M. Bulmer, *Theoretical Evolutionary Ecology* (Sinauer Associates Publishers, Sunderland, 1994)
3. J.M. Fryxell, P. Lundberg, *Individual Behavior and Community Dynamics*. Population and Community Biology Series, vol. 20 (Chapman & Hall, London, 1997)
4. M.P. Hassell, J.H. Lawton, R.M. May, Patterns of dynamical behaviour in single species populations. J. Anim. Ecol. **45**, 471–486 (1976)
5. C.S. Holling, Some charpe of predation and parasitism. Can. Entomol. **91**, 385–398 (1959)
6. C.S. Holling, The components of predation as revealed by a study of small mammals predation of the European pine sawfly. Can. Entomol. **91**, 292–320 (1959)
7. C.S. Holling, The functional response of predators to prey density and its role in mimicry and population regulation. Mem. Ent. Soc. Can. **45**, 43–60 (1965)
8. A.I. Houston, J.M. McNamara, *Models of Adaptive Behaviour: An Approach Based on State* (Cambridge University Press, Cambridge, 1999)
9. R.N. Hughes, *Diet Selection: An Interdisciplinary Approach to Foraging Behaviour* (Blackwell Scientific Publications, Oxford, 1993)
10. A.J. Lotka, *Elements of Physical Biology* (Williams and Wilkins, Baltimore, 1925)
11. A.J. Lotka, *Elements of Mathematical Biology* (Dover, New York, 1956)
12. R.M. May, Patterns in multispecies communities, in *Theoretical Ecology: Principles and Applications*, ed. by R.M. May (Blackwell, Oxford, 1981), pp. 197–227
13. R.C. Robinson, *An Introduction to Dynamical Systems: Continuous and Discrete*. Pure and Applied Undergraduate Texts, vol. 19, 2nd edn. (American Mathematical Society, Providence, 2012)
14. M. Rosenzweig, R. MacArthur, Graphical representation and stability conditions of predator-prey interaction. Am. Nat. **97**, 209–223 (1963)
15. T. Royama, A comparative study of models for predation and parasitism. Res. Popul. Ecol. **13**(S1), 1–91 (1971)
16. M.E. Solomon, The natural control of animal populations. J. Anim. Ecol. **2**, 235–248 (1949)
17. D.W. Stephens, J.R. Krebs, *Foraging Theory* (Princeton University Press, Princeton, 1986)
18. J. Sugie, Y. Saito, Uniqueness of limit cycles in a Rosenzweig-MacArthur model with prey immigration. SIAM J. Appl. Math. **72**(1), 299–316 (2012)
19. H.R. Thieme, *Mathematics in Population Biology* (Princeton University Press, Princeton, 2003)
20. T.L. Vincent, J.S. Brown, *Evolutionary Game Theory, Natural Selection, and Darwinian Dynamics* (Cambridge University Press, New York, 2005)
21. V. Volterra, Variazione e fluttuazioni del numero d'individui in specie animali conviventi. Mem. Acad. Lincei **6**, 30–113 (1926)
22. V. Volterra, Variations and fluctuations of the number of individuals in animal species living together, in *Animal Ecology*, ed. by R.N. Chapman (McGraw-Hill, New York, 1931), pp. 412–433

Chapter 9
Modeling with Class Structure

Abstract In this chapter, we shall focus on the modeling of the stage structured population dynamics especially in the context of the epidemic dynamics. It is the typical population structure for the epidemic dynamics to classify the individuals into susceptible, latent, infectious, recovered (immunized) ones, which are regarded as the physiological states with respect to the infection of a transmissible disease.

9.1 Structured Population

Problem on a population dynamics may need to take account of a structure within the population, like social structure, physiological structure, and ecological structure. The population with such an internal structure is called *structured population*. The social structure is based on the social behavior of individuals in the population, which follows, for example, *kin selection*, *sexual selection*, and *evolutionarily optimal strategy*. On the other hand, the physiological structure is based on the physiological state of individuals, for example, sex, age, size of a part of body, body weight, color, etc. As a specific example which will be mainly treated in this chapter, the physiological state with respect to the infection of a transmissible disease determines the structure of population under the epidemic dynamics. Such a physiological structure could have a close relation to the social structure. The ecological structure is based on the ecological interaction between individuals in the population. The group formation to compose subpopulations makes an example.

Now let us focus on the physiological structure which may be quantified by a certain measure. If an individual is characterized by a value x by the measure, we can identify the value of *state variable* for the individual by x. Each member of the population is distinguished in terms of the value of state variable, so that the population is characterized by the distribution of the state variable. It gives an expression of the physiological structure about the population. Generally a population characterized by a distribution of physiological state variable is called *physiologically structured population*.

When the measure is the age of individual, we can define the physiological structure by the age distribution for the population. Such a population with an

H. Seno, *A Primer on Population Dynamics Modeling*, Theoretical Biology,
https://doi.org/10.1007/978-981-19-6016-1_9

age distribution is sometimes called *age structured population* or *age classified population*. We will see the fundamental modeling for the age structured population dynamics in the subsequent Chap. 10.

The distribution of state variable may be discrete, when the state variable takes only a value belonging to a set consisting of countable number of values. The population with such a discrete distribution of state variable may be called *stage structured population* or *stage classified population*. If the age distribution in a population is by a number of age classes, the age structured population is a stage structured population.

In this chapter, we shall focus on the modeling of the stage structured population dynamics especially in the context of epidemic dynamics. It is the typical population structure for an epidemic dynamics to classify the individuals into susceptible, latent, infectious, recovered (immunized) ones, which are regarded as the physiological states with respect to the infection of a transmissible disease.

9.2   Spread of Transmissible Disease

The disease transmission process from an infective individual (we hereafter use "infective" as a single noun, following a convention in mathematical epidemiology) to a susceptible one (hereafter "susceptible" as a single noun) could be regarded as analogous to the interaction between predator and prey. The successful transmission of the disease causes the decrease in the subpopulation size of susceptibles while it does the increase in the subpopulation size of infectives. It can be regarded as corresponding to the process with the decrease in the prey population size by the predation and the increase in the predator population size by the reproduction with the energy gain by the predation.

In this section, we shall consider the modeling of the temporal change of the subpopulation sizes for the susceptible, infective, and if necessary, latent (incubated, encapsuled), and recovered (removed, immunized) classes. There could be the other classes, for example, the isolated (quarantined) and the died etc. The transmissible disease considered here may be regarded for example as panic or fear as a transmissible psychological state, or as addiction for drug, alcohol, gamble, gaming, etc. Moreover, we may consider the disease not only for human but also for plant or animal. Further, as an extended application from the clear similarity, we may consider the spread of mode, habit, rumor, fake news, innovative technology, or another information.

The disease transmission may have different aspects about the route: *horizontal transmission* and *vertical transmission*. We will assume only the horizontal transmission in this chapter. The horizontal transmission indicates the infection between individuals being at the same time and space. The vertical transmission means the infection of newborns at birth from the mother in a rigorous sense, while it may be referred as the transmission of a disease from generation to generation through the parent-child relationship.

In this chapter to describe the basic modelings, we shall consider the population dynamics about the spread of a transmissible disease by the horizontal transmission after its invasion in a population. We ignore any change of the population size in the time scale of the epidemic dynamics, that is, the spread of disease in the population. Thus we assume that the population size is constant independently of time, denoted by a positive constant N. Especially since we will describe only basic epidemic dynamics models, we assume no latent period for them in most of the subsequent part, so that the individual gets the infectivity to transmit the disease to the other immediately after the infection. Now let us denote the susceptible subpopulation size (density) by S_k at the beginning of the kth day after the initial invasion of a transmissible disease in the population. In the same way, I_k denotes the infective subpopulation size, which indicates the density of individuals who has the infectivity. R_k denotes the recovered subpopulation size, which indicates the density of individuals who has the immunity against the disease. From the assumption of constant population size, we have $S_k + I_k + R_k = N$ for any k. In Sect. 9.3.6, we will revisit these assumptions and give examples of modelings and models with modifying them.

In the other context of epidemic dynamics modeling, the class denoted by R may not be for individuals with the immunity. The use of the letter R is sometimes explained as the first letter of the word "Recovered" or "Recovery". However, according to the epidemic dynamics in a general sense, it would be better to be explained as the first letter of "Removed" or "Removal", since the individual of class R is generally assumed to have no contribution to the transmission of disease itself in the modeling. The class R consists of individuals removed from the epidemic dynamics. For example, if the disease is highly fatal, the class R indicates the subpopulation of dead, that is, the accumulation of dead by the disease. In contrast, if the disease is not fatal, it may indicate the subpopulation of isolated or hospitalized infectives after their detection. In this case, the individuals of class R is not at the recovered state, but they are out of the epidemic dynamics in a reasonable sense. In these cases with the different meaning of class R, the modeling described in this chapter would be reasonably modified, although we shall not consider such a modification any more in this book. The reader can easily find such modelings in the other literatures on the mathematical model for the epidemic dynamics [1, 2, 5, 6, 8, 15, 22].

9.2.1 Generic Discrete Time Model

Let us introduce the probability $P(j)$ that an individual contacts the route of disease transmission j times a day. The route of disease transmission means the chance to encounter the pathogen which could cause a transmissible disease by its successful infection to the host. For example, in the case of a sexually transmitted disease or a skin disease, the disease transmission requires a direct contact between individuals. So the probability $P(j)$ is determined by the frequency of contacts to the other individuals. In the case of malaria, dengue fever, or pine wilt disease, it requires the contact with a vector (mosquito or long-horned beetle) which carries the pathogen (microorganism or virus). Then $P(j)$ is determined by the frequency of contacts to the vector. In contrast, the transmission of influenza may be caused by the contact to droplets or aerosols containing the pathogen. Thus, in such a case, $P(j)$ is determined by the frequency of chances to stay in contaminated air or touch contaminated materials as potential routes for the disease transmission.

It must be remarked that $P(j)$ does not mean the probability of the infection. The infection may occur only by the contact to the infective individual, the carrier vector, or the contaminated air or materials. Any contact to the susceptible individual, non-carrier vector, or non-contaminated/purified/disinfected air/material cannot cause the infection. In the modeling for epidemic dynamics, it is important to take account of such a frequency/probability to contact the route of disease transmission, which depends on the characteristics of not only the disease but also the behavior of individuals and further the sanitary condition around them.

Today most of mathematical models for epidemic dynamics are given by the system of differential equations, and some of them may be treated as the typical models in many textbooks on mathematical biology and epidemiology. For such continuous time models, the logic of mathematical modeling described in this chapter for the discrete time model could be applied at least as an essential skeleton for its modeling.

Further, in most of the explanation on the modeling for the human epidemic dynamics, the infection with the contact between individuals is usually assumed. It may become misleading about the epidemic dynamics. We know that the essence of disease transmission is the contact not between individuals but between pathogen and individual. Even in a situation where no infective individual exists, there could be a possibility for a susceptible to get the infection, like the droplet-borne transmission through contaminated air/materials as mentioned in the above.

For example, as in the case of the COVID-19 or the infectious gastroenteritis, when the droplet would have a high probability to transmit the persistent virus to the others, the active pathogen density for the materials contaminated

(continued)

by the droplets significantly depends on the density of infectives in the population. However, it may not necessarily have significant relation to the frequency of contacts between individuals. For this reason, we understand that washing hands and clothes after getting chances to be in such situations would be effective to reduce the likeliness of the infection. Actually it is well-known that we need to make a highly cautious treatment about a vomit and excreta from the patient of an infectious gastroenteritis.

The description about the modeling in the context of the contact between individuals is straightforward and even appropriate for the disease with such a route of transmission. However, it is not necessarily applicable for many diseases. For this reason, we shall describe here the modeling in the context of the contact between pathogen and individual.

On the other hand, it is generally difficult to estimate such a frequency of contacts between pathogen and individual in the route of disease transmission. In contrast, today it may become possible to use a statistical analysis on a big data in order to get an estimation about the frequency of contacts between individuals. This modern situation may make it more feasible to consider the relation of the contacts between individuals and the epidemic dynamics.

With the probability distribution $\{P(j)\}$, we can define the expected number of contacts to the route of disease transmission per day by $\langle \pi \rangle = \sum_{j=0}^{\infty} j P(j)$. Let us assume that $\langle \pi \rangle$ is mathematically determined as a finite value. Next supposing that the ratio ϕ_k of j contacts in the kth day is expected for that to the pathogen, we give the following formula of the expected number of contacts to the route of disease transmission which has a possibility to cause the infection:

$$\sum_{j=0}^{\infty} \phi_k j P(j) = \phi_k \langle \pi \rangle.$$

ϕ_k can be regarded as the probability that a contact to the route of disease transmission is one which has a possibility to cause the infection, that is, which is contaminated by the pathogen.

In general, ϕ_k must depend on the density of infectives I_k in the population. As the density of infectives gets larger, the route of disease transmission is more likely to be contaminated by the pathogen. Besides the pathogen concentration at the route becomes higher. Hence ϕ_k could have a positive correlation with the number/density of infectives which produce the pathogen in the population. It may depends on the nature of the route of disease transmission, the behavior of individuals and the sanitary situation around them. For example, the spread of the custom to wear the mask at the social scene may work to weaken such a positive correlation.

Next we introduce the probability that a susceptible escapes from the infection with ℓ contacts to the contaminated route of disease transmission at the kth day, by $(1 - \beta_k)^\ell$ where β_k is the probability to get the infection by a contact to the contaminated route at the kth day $(0 < \beta_k < 1)$. On the other hand, when a susceptible contacts j times to the route of disease transmission at the kth day, the probability that ℓ of the contacts is to the contaminated route is given by the following from the assumption given above:

$$\binom{j}{\ell} \phi_k^\ell (1 - \phi_k)^{j-\ell} = \frac{j!}{\ell!(j-\ell)!} \phi_k^\ell (1 - \phi_k)^{j-\ell}.$$

Thus we have the following probability that a susceptible escapes from the infection with j contacts to the route of disease transmission at the kth day:

$$\sum_{\ell=0}^{j} (1 - \beta_k)^\ell \binom{j}{\ell} \phi_k^\ell (1 - \phi_k)^{j-\ell} = \sum_{\ell=0}^{j} \binom{j}{\ell} \left\{ \frac{\phi_k}{1 - \phi_k} (1 - \beta_k) \right\}^\ell (1 - \phi_k)^j$$

$$= \left\{ \frac{\phi_k}{1 - \phi_k} (1 - \beta_k) + 1 \right\}^j (1 - \phi_k)^j = (1 - \beta_k \phi_k)^j.$$

Therefore, the probability of infection at the kth day is correspondingly given by $1 - (1 - \beta_k \phi_k)^j$.

As a consequence of this modeling, we have the probability for a susceptible to escape from the infection at the kth day as $\sum_{j=0}^{\infty} (1 - \beta_k \phi_k)^j P(j)$, and the probability for a susceptible to get the infection at the kth day as $\sum_{j=0}^{\infty} \left\{ 1 - (1 - \beta_k \phi_k)^j \right\} P(j)$. The latter probability of the infection means what is called the *infection force* for the epidemic dynamics. The larger is the infection force, the disease spread is more severe.

With those assumptions derived with the modeling, we can consider the following simplest epidemic dynamics model to govern the daily change of the subpopulation sizes of susceptible S_k, infective I_k, and removed/recovered R_k (see Fig. 9.1):

$$S_{k+1} = \sum_{j=0}^{\infty} (1 - \beta_k \phi_k)^j P(j) S_k + (1 - m) q I_k + \theta R_k;$$

$$I_{k+1} = \sum_{j=0}^{\infty} \left\{ 1 - (1 - \beta_k \phi_k)^j \right\} P(j) S_k + (1 - q) I_k; \qquad (9.1)$$

$$R_{k+1} = mq I_k + (1 - \theta) R_k,$$

where it holds that $S_k + I_k + R_k = N$ for any k.

Here the parameter q means the probability that an infective recovers and loses the infectivity in one day $(0 < q \leq 1)$. Hence the expected duration of the infectivity

Fig. 9.1 State transitions for the epidemic dynamics model (9.1)

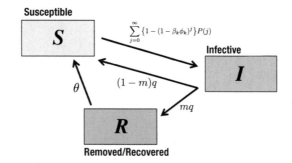

is now given by $1/q$ days as follows: Suppose an individual who is infected at a day. For the above model, such an individual becomes infective from the next day. So if the individual recovers k days later, the infective duration is regarded as k days since the individual is still infective at the day of the recovery. The probability of such a recovery after k days is now given by $(1 - q)^{k-1}q$. Thus the expected duration of the infectivity is mathematically derived by $\sum_{k=1}^{\infty} k(1-q)^{k-1}q = 1/q$ (refer to the last part of Sect. 1.5).

The parameter m means the probability that a recovered individual can get the immunity ($0 \leq m \leq 1$). When $m = 0$, the recovery cannot lead to the immunity at all. When $m = 1$, it can necessarily lead to the immunity. The individual with the immunity cannot get the infection, so that such an individual is removed from the epidemic dynamics. On the other hand, the recovered individual without getting the immunity is assumed to return to the susceptible state, that is, go back to the susceptible class, at which the individual can get the infection again like a susceptible who has not experienced the disease.

In the above model, we introduce the probability of the waning of the immunity per day by the parameter θ ($0 \leq \theta < 1$). For $\theta = 0$, the waning never occurs and the immune individual keeps the immunity at least in the time scale of the epidemic dynamics. By the same calculation as about the expected duration of the infectivity in the above, the expected duration of the immunity is now given by $1/\theta$.

9.2.2 Invasion Success of Transmissible Disease

In this section, we consider the condition for the success of the disease's invasion in the population under the epidemic dynamics given by (9.1). The *invasion success* is theoretically defined as the increase of infective population size at the initial of epidemic dynamics with a sufficiently small number of infective individuals in the population. Generally saying, the spread of a transmissible disease begins with its invasion success.

Now we take the initial condition for the infective population size as $I_0 \ll N$ at the zeroth day. Since the zeroth day is when the disease emerges in the population, we may reasonably suppose that no individual has the immunity, so that $R_0 = 0$. We do not consider now the other cases such that the vaccination is operated in advance or some immune individuals remain from the last season of epidemic dynamics. Thus we have $S_0 = N - I_0$. Further, from such a situation at the zeroth day, the probability ϕ_0 that a contact to the route of disease transmission has a possibility to cause the infection must be sufficiently small: $\phi_0 \ll 1$. So is the probability β_0 to get the infection by a contact to the contaminated route at the zeroth day: $\beta_0 \ll 1$. With these assumptions, let us consider the mathematical condition that $I_1 > I_0$, which now means the invasion success of the disease in the population by the above definition.

From the second equation of (9.1), we have

$$I_1 = \sum_{j=0}^{\infty} \left\{ 1 - (1 - \beta_0 \phi_0)^j \right\} P(j) S_0 + (1 - q) I_0$$

$$= \sum_{j=0}^{\infty} j \beta_0 \phi_0 P(j)(N - I_0) + o(\beta_0 \phi_0) + (1 - q) I_0, \qquad (9.2)$$

making use of Taylor expansion in terms of $\beta_0 \phi_0$. Since it is reasonable to assume that these probabilities given by ϕ_k and β_k are positively correlated with the infective population size I_k, let us assume further that the product $\beta_k \phi_k$ is given by an appropriately smooth increasing function of I_k: $\beta_k \phi_k = \mathcal{F}(I_k)$. Again by Taylor expansion, we have

$$\phi_0 \beta_0 = \mathcal{F}(I_0) = \mathcal{F}(0) + \mathcal{F}'(0) I_0 + o(I_0), \qquad (9.3)$$

where

$$\mathcal{F}'(0) = \left. \frac{d\mathcal{F}(I)}{dI} \right|_{I=0} > 0$$

because of the above-mentioned positive correlation to the infective population size I. In addition, we reasonably suppose that $\mathcal{F}(0) = 0$, assuming now that no infective (i.e., $I_k = 0$) means no pathogen in the route of disease transmission.

If the pathogen can keep active out of the host for a sufficiently long duration until its natural decay, it may become reasonable to assume that $\mathcal{F}(0) > 0$. In such a case, even when the infective is absent on a day, the active pathogen hiding in the environment may create a new infective individual. On the other hand, if the disease transmission is occurred by the vector as mentioned in the previous section, ϕ_k and β_k depend not directly on the infective population

(continued)

size but on the carrier vector population size. Also in such a case, even when the infective is absent on a day, the carrier vector may create a new infective individual. In these cases, the following arguments with $\mathcal{F}(0) = 0$ must be necessarily modified, although we shall not go into such a modeling any more in this book.

By substituting (9.3) for (9.2), we can get

$$
\begin{aligned}
I_1 &= \left\{ \sum_{j=0}^{\infty} j P(j) \mathcal{F}'(0) N + (1-q) \right\} I_0 + o(I_0) \\
&= \left\{ \langle \pi \rangle \mathcal{F}'(0) N + (1-q) \right\} I_0 + o(I_0),
\end{aligned}
\tag{9.4}
$$

where $\langle \pi \rangle$ is the expected number of contacts to the route of disease transmission per day, as defined in the previous section.

As a result, the mathematical condition for the invasion success to make $I_1 > I_0$ is given by $\langle \pi \rangle \mathcal{F}'(0) N + 1 - q > 1$, that is,

$$
\frac{\langle \pi \rangle \mathcal{F}'(0)}{q} N > 1.
\tag{9.5}
$$

If this condition is satisfied, the disease invasion is successful to make a spread in the population. The condition (9.5) implies that, when the duration of infectivity is sufficiently long (with a sufficiently small q), or when the individual mobility/activity is so high that the frequency of contacts to the route of disease transmission is sufficiently large (with a sufficiently large $\langle \pi \rangle$), the spread of the disease may occur.

9.2.3 Reproduction Number of Infectives

As an index about the tendency of the disease spread by an epidemic dynamics in a population, we may consider the expected number of new infectives produced by an infective during having the active infectivity. The expected number is not for a specific infective individual, but for an arbitrary infective individual, so that such an index could represent the tendency of the disease spread for the population.

We need some detail assumptions in order to make the conceptual definition about such an index more accurate, and give its mathematical definition [4, 6, 10, 19]. The typical conceptual definition is given as *the expected number of new infectives produced by an infective under the condition that the original infective has contacts only with susceptibles during having the active infectivity*. It is today well-known as the *basic reproduction number* for the disease spread.

It is actually impossible that an infective could have contacts only with susceptibles during having the active infectivity, even though it might occur stochastically with a certain probability. This is because the production of new infectives causes the possibility of contacts with them. Therefore, it is reasonable to understand the meaning of the basic reproduction number as the supremum, that is, the least upper bound for the expected number of new infectives produced by an infective. In this sense, the basic reproduction number can be defined as the expected number of new infectives produced by an infective during having the active infectivity under the condition that the original infective has the possibly highest efficiency to transmit the disease to susceptibles. In the actual disease transmission, the contacts only with susceptibles becomes harder as the infective produces new infectives around him/her.

> The basic reproduction number is defined with the same concept as the net reproduction rate given the definition in p. 16 of Sect. 1.5 and argued for the population dynamics with a Malthus growth in Sect. 4.4. The birth process of offsprings and the death process of parents are analogous to the production of new infectives by the disease transmission and the removal of infectives from the epidemic dynamics with the loss of infectivity. In such an analogy, it is characteristic for the epidemic dynamics that the new infectives can be produced from the susceptibles, so that the production must significantly depend on the number of susceptibles in the same population.

To derive the basic reproduction number for the epidemic dynamics (9.1), we must consider the supremum for the expected number of new infectives produced by an infective during having the infectivity. Now the number of new infectives produced at the kth day is given by the second equation of (9.1) as $\sum_{j=0}^{\infty} \left\{ 1 - (1 - \beta_k \phi_k)^j \right\} P(j) S_k$. In order to introduce the possibly highest efficiency to transmit the disease to susceptibles, when the infective has contacts only with susceptibles, we must suppose the situation that $I_k \ll N$ and $R_k = 0$. This supposed situation is the same as in Sect. 9.2.2. Hence, from (9.4), we have

$$\sum_{j=0}^{\infty} \left\{ 1 - (1 - \mathcal{F}(I_k))^j \right\} P(j) S_k = \langle \pi \rangle \mathcal{F}'(0) N I_k + \mathrm{o}(I_k)$$

with $\beta_k \phi_k = \mathcal{F}(I_k)$. This means that the expected number of new infectives produced by an infective in a day is given by $\langle \pi \rangle \mathcal{F}'(0) N$. We remark that this expected number is independent of k, since we used the above supposed situation to derive it.

As already mentioned in Sect. 9.2.1, the expected duration of the infectivity is given by $1/q$ for the epidemic dynamics (9.1). This is independent of k too, which indicates that any infective at any day has the expected duration of the infectivity after the day. Thus it is independent of when the individual is infected. In the other sense, all infectives are identical every day according to the epidemic dynamics.

Some readers may suspect that the infective who passes longer period after the infection could have a weaker infectivity and higher possibility to lose the infectivity and recover. Actually the immune response becomes more effective after the infection, and more likely to lead to the recovery. This indicates that the older infective would have lesser contribution to the disease transmission, that is, the efficiency of disease transmission for such an older infective would become lower. To introduce such a dependence of the infectivity on the time after the infection, it is necessary to take account of the *epidemic age* in the modeling for each infective individual after the infection. Then we must consider the distribution of epidemic age in the infective subpopulation, which temporally changes according to the epidemic dynamics. Even in such a modeling, however, the definition of the basic reproduction number is the same as given in the above.

In the actual epidemic dynamics of any transmissible disease, the duration of the infectivity depends on the infective individual's physiological condition (e.g., age and healthy state) too. For the epidemic dynamics model (9.1), the expected duration of the infectivity is given by $1/q$ while it does not mean that every infective has the same duration. Since the probability for the loss of infectivity per day is given by a constant q, we have already shown in Sect. 9.2.1 that the distribution of the duration of the infectivity follows the geometrical distribution defined as $(1 - q)^{k-1}q$ $(k = 1, 2, \ldots)$, although this is not the distribution of epidemic age.

Finally from the above arguments, we can define the basic reproduction number $\mathscr{R}_0$ for the epidemic dynamics model (9.1) by the product of the expected number of new infectives produced by an infective in a day, $\langle \pi \rangle \mathscr{F}'(0)N$, and the expected duration of the infectivity, $1/q$:

$$\mathscr{R}_0 = \langle \pi \rangle \mathscr{F}'(0)N \cdot \frac{1}{q}. \tag{9.6}$$

It must be remarked that the basic reproduction number $\mathscr{R}_0$ is independent of k, that is, of any day. This is taken natural because it is the index to express the supremum as mentioned above. The basic reproduction number is the index for the characteristics of the population under the epidemic dynamics. In other words, it indexes the degree of threat of the disease spread in the population.

If $\mathscr{R}_0 < 1$, the outbreak does not occur. It indicates that the supremum for the expected number of new infectives produced by an infective during having the infectivity is less than one. Thus, the number of infectives tends to decrease in the population. In contrast, when the outbreak occurs, it must be satisfied that $\mathscr{R}_0 > 1$.

This is because the outbreak means the increase of the number of infectives, so that the expected number of new infectives produced by an infective during having the infectivity must be greater than one.

As remarked above, the basic reproduction number $\mathcal{R}_0$ is defined as the supremum, the actual expected number of new infectives produced by an infective during having the infectivity must be less than it. Hence, even if $\mathcal{R}_0 > 1$, the actual expected number of new infectives produced by an infective during having the infectivity may be less than one on the way of epidemic dynamics, so that the number of infectives may decrease (we will argue such a behavior in the next section).

On the other hand, for the epidemic dynamics model (9.1), the condition that $\mathcal{R}_0 > 1$ coincides with that for the invasion success (9.5). Therefore, if $\mathcal{R}_0 > 1$, the disease succeeds in invading in the population and causes an increase of infectives toward the outbreak. For this reason, we can consider that, if $\mathcal{R}_0 > 1$, the outbreak occurs for (9.1).

Some readers may have expected the coincidence of those conditions for the basic reproduction number $\mathcal{R}_0$ and the invasion success of (9.5), from the corresponding supposition for their derivation. However, it must be remarked that they are different from each other according to the way of thought, that is, the modeling. The outbreak with the basic reproduction number $\mathcal{R}_0$ greater than one could be applied for the newly emergent disease in a population, or for the seasonally repetitive outbreak of an endemic disease. In contrast, since the introduction of an alien transmissible disease by the conqueror in a colony, the artificial diffusion of a harmful transmissible disease as a bioterrorism, or the spread of fear as a panic transmission through a population has a relatively large initial number of infectives, the possibility of outbreak cannot be discussed only by the basic reproduction number $\mathcal{R}_0$.

Similarly to the basic reproduction number, we can define an index to represent the epidemic situation at each day. As such an index, let us consider the supremum for the expected number of new infectives produced by an infective during having the infectivity, provided that the situation at the kth day is unchanged. It is sometimes called the *effective reproduction number*. For the epidemic dynamics model (9.1), the number of new infectives produced at the kth day is given by $\sum_{j=0}^{\infty} \left\{ 1 - (1 - \mathcal{F}(I_k))^j \right\} P(j) S_k$. Dividing this by the infective subpopulation size I_k gives the mean number of new infectives produced by an infective at the kth day. Hence we can define the effective reproduction number $\mathfrak{R}_k$ at the kth day for the epidemic dynamics model (9.1) as

$$\mathfrak{R}_k := \frac{1}{I_k} \sum_{j=0}^{\infty} \left\{ 1 - (1 - \mathcal{F}(I_k))^j \right\} P(j) S_k \cdot \frac{1}{q}, \tag{9.7}$$

since the expected duration of infectivity is given by $1/q$ for every day as explained before. As inferred from the derivation way of (9.7), it mathematically holds that $\mathfrak{R}_k \to \mathscr{R}_0$ as $(S_k, I_k) \to (N, 0)$.

As seen from (9.7), the effective reproduction number is defined for each day. Hence the difference in the initial condition makes it different. More generally saying, the effective reproduction number at a day depends on the history of epidemic dynamics. Although it must depend on the characteristics of the transmissible disease itself as well as $\mathscr{R}_0$, it indexes the situation at the kth day according to the undergoing epidemic dynamics. As the situation at the kth day is more serious, the increase in the number of infectives must be expected to become larger.

9.2.4 SIR, SIS, and SIRS Models

In this section, we are going to see some specific and simple models belonging to the epidemic dynamics model (9.1). Let us assume that the probability to get the infection by a contact to the contaminated route of disease transmission is constant for any day: $\beta_k = \beta$, and the probability that a contact is to the contaminated route of disease transmission ϕ_k is given by

$$\phi_k = \alpha \frac{I_k}{N}, \tag{9.8}$$

where α is a constant such that $0 < \alpha \leq 1$.

This modeling is based on the assumption that the possibility of the contamination for the route of disease transmission, that is, the risk of infection is proportional to the proportion of infectives in the population (with the proportional constant α). For example, a protective behavior like masking may suppress the dispersal of pathogens, and then the parameter α must be small.

In the modeling with the class R which means the state such that the infective is isolated or hospitalized, the infective belonging to the class R cannot contribute to the risk of infection. Then the above formula for ϕ_k may be changed to

$$\phi_k = \alpha \frac{I_k}{N - R_k}.$$

We shall not go into such a model any more in this book.

With the above assumptions, we have $\mathcal{F}'(0) = \beta\alpha/N$ for the epidemic dynamics model (9.1), so that, from (9.6), the basic reproduction number is given by

$$\mathcal{R}_0 = \frac{\beta\alpha\langle\pi\rangle}{q}. \tag{9.9}$$

Next, as did in Sect. 2.4.2, we shall introduce the following Poisson distribution for the probability $P(j)$ that an individual contacts the route of disease transmission j times a day:

$$P(j) = \frac{\gamma^j e^{-\gamma}}{j!} \quad (j = 0, 1, 2, \ldots) \tag{9.10}$$

with $P(0) = e^{-\gamma}$ (refer to Chap. 15). The expected number of contacts to the route of disease transmission per day $\langle\pi\rangle$ satisfies that $\langle\pi\rangle = \gamma$, and then from (9.9), the basic reproduction number is given by

$$\mathcal{R}_0 = \frac{\beta\alpha\gamma}{q}. \tag{9.11}$$

Further, since we have

$$\sum_{j=0}^{\infty}(1 - \beta_k\phi_k)^j P(j) = e^{-\gamma}\sum_{j=0}^{\infty}\frac{\{(1 - \beta_k\phi_k)\gamma\}^j}{j!}$$

$$= e^{-\gamma}\cdot e^{(1-\beta_k\phi_k)\gamma} = e^{-\beta_k\phi_k\gamma} = e^{-\beta\alpha\gamma I_k/N},$$

the infection force at the kth day is now given by $1 - e^{-\beta\alpha\gamma I_k/N}$, and the effective reproduction number $\mathfrak{R}_k$ defined by (9.7) becomes

$$\mathfrak{R}_k := \frac{S_k}{q I_k}(1 - e^{-\beta\alpha\gamma I_k/N}). \tag{9.12}$$

It can be easily proved that $\mathfrak{R}_k < \mathcal{R}_0$ for any $S_k > 0$ and $I_k > 0$.

SIR Model

In this part, we shall consider the model (9.1) with $m = 1$ and $\theta = 0$. Then every recovered individual can get the immunity, and the immunity is kept without its waning. Thus the state transition with respect to the disease is one-way as S $\rightarrow$ I $\rightarrow$ R. The epidemic dynamics model with such a one-way state transition is called *SIR model* (Fig. 9.2).

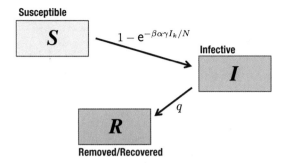

Fig. 9.2 State transitions for the SIR model (9.13)

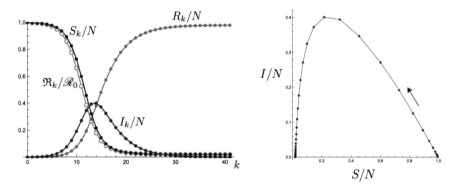

Fig. 9.3 A numerical calculation of the temporal change of variables for the SIR model (9.13) with $\beta\alpha\gamma = 1.0$; $q = 0.25$; $\mathscr{R}_0 = 4.0$; $(S_0/N, I_0/N, R_0/N) = (0.999, 0.001, 0.0)$. The temporal change of $\mathfrak{R}_k/\mathscr{R}_0$ is shown too

The SIR model for the epidemic dynamics model (9.1) with $m = 1$ and $\theta = 0$ becomes

$$S_{k+1} = S_k e^{-\beta\alpha\gamma I_k/N};$$

$$I_{k+1} = S_k\left(1 - e^{-\beta\alpha\gamma I_k/N}\right) + (1-q)I_k; \tag{9.13}$$

$$R_{k+1} = qI_k + R_k,$$

with the initial condition $(S_0, I_0, R_0) = (N - I_0, I_0, 0)$. For the SIR model (9.13), the infective population size eventually approaches zero as shown in Fig. 9.3, that is, $I_k \to 0$ as $k \to \infty$. Since the total population size is given as a constant N, the susceptible population size monotonically decreases as time passes, while the immune population size monotonically increases. Then the infective population size eventually decreases toward zero.

As seen in Fig. 9.3, there exists a positive value S_∞ for the SIR model (9.13) such that $S_k \to S_\infty > 0$ as $k \to \infty$. It indicates that there are susceptibles who have not experienced the infection until the end of epidemic dynamics when *disease-free equilibrium state* (DFE) is established, that is, when the disease is eliminated from the population. At the same time, the value $R_\infty = N - S_\infty$ gives the total number of individuals who have experienced the infection, which may be called the *final epidemic size*. In the following part, let us consider some detail nature of the epidemic dynamics by (9.13).

The above nature of SIR model is not necessarily applicable for the other SIR model with a temporally variable total population size. When the recruitment or the withdrawal of members by migration or temporal visit/trip must be taken into account for the epidemic dynamics, the high recruitment rate of susceptible members may maintain the production rate of new infectives, and subsequently the infective population remains for a long time in the population. Such a situation is called *endemic state*, and we say that the disease becomes endemic.

As a special case, when the time scale of epidemic dynamics is relatively large, that is, when the disease spread is relatively slow, the temporal change of total population size by birth and death could not be negligible. Then it is likely that the disease becomes endemic, like smallpox, tuberculosis, and measles. Such an endemic state could become unstable by the innovation of prevention and treatment for the disease, and transfer to the disease-free equilibrium state. The smallpox is one of historical examples about such diseases.

From the first and second equations of (9.13), we can find the following equality that holds for any $k \geq 0$:

$$\frac{S_{k+1}}{N} + \frac{I_{k+1}}{N} - \frac{1}{\mathscr{R}_0} \ln \frac{S_{k+1}}{N} = \frac{S_k}{N} + \frac{I_k}{N} - \frac{1}{\mathscr{R}_0} \ln \frac{S_k}{N}.$$

Therefore, for the initial condition $(S_0, I_0, R_0) = (N - I_0, I_0, 0)$, we have

$$\frac{S_k}{N} + \frac{I_k}{N} - \frac{1}{\mathscr{R}_0} \ln \frac{S_k}{N} = 1 - \frac{1}{\mathscr{R}_0} \ln \frac{S_0}{N} \tag{9.14}$$

for any $k \geq 0$. Since the right side of (9.14) is a constant independent of day (k), the left side defines a conserved quantity for the SIR model (9.13). Further, the Eq. (9.14) gives a relation between S_k/N and I_k/N for any $k \geq 0$, so that it can be regarded as the formula of a curve in the $(S/N, I/N)$-plane on which every point $(S_k/N, I_k/N)$ is located (see Fig. 9.4).

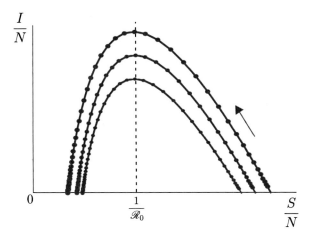

Fig. 9.4 Trajectories by the SIR model (9.13) in the $(S/N, I/N)$-plane for three different initial conditions $(S_0/N, I_0/N)$

From (9.14), we can find the following nature of the SIR model (9.13):

- The infective population size I_k is monotonically decreasing if $\mathscr{R}_0 S_0/N \leq 1$, while it increases in an early period and turns to monotonically decreasing if $\mathscr{R}_0 S_0/N > 1$.
- The maximal size of infective population when $\mathscr{R}_0 S_0/N > 1$ is below the following value $I_{\sup}$:

$$I_{\sup} := \left[1 - \frac{1}{\mathscr{R}_0}\left\{1 + \ln\left(\mathscr{R}_0 \frac{S_0}{N}\right)\right\}\right] N. \qquad (9.15)$$

The value $I_{\sup}$ gets larger as the initial susceptible population size S_0 is larger (i.e., for the smaller $I_0 = N - S_0$) (Fig. 9.4), and as the basic reproduction number $\mathscr{R}_0$ is greater (Fig. 9.5).

- The limiting value of $S_k \to S_\infty$ as $k \to \infty$ is given by the unique positive root less than N for the following equation:

$$\frac{S_\infty}{N} - \frac{1}{\mathscr{R}_0} \ln \frac{S_\infty}{N} = 1 - \frac{1}{\mathscr{R}_0} \ln \frac{S_0}{N}. \qquad (9.16)$$

From the first nature, we find that the infective population size is monotonically decreasing as time passes even for the basic reproduction number greater than one, if it satisfies that $1 < \mathscr{R}_0 \leq N/S_0$. Thus no outbreak occurs in such a case. Roughly saying, this is the case where the basic reproduction number $\mathscr{R}_0$ is slightly greater than one. Since $\mathscr{R}_0 > 1$, the disease invasion is successful in such a case, so that it

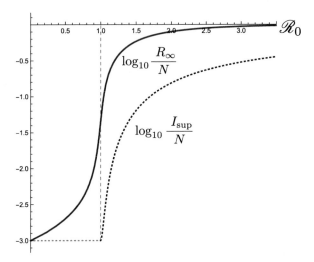

Fig. 9.5 Numerically drawn $\mathcal{R}_0$-dependence of I_{sup} and R_∞ by (9.15) and (9.17) for $I_0/N = 0.001$

must be a cautious situation from the viewpoint of public health about the epidemic size.

Next, let us consider the $\mathcal{R}_0$-dependence of the final epidemic size $R_\infty = N - S_\infty$. From (9.16), we can easily obtain the following equation to determine it:

$$\frac{R_\infty}{N} + \frac{1}{\mathcal{R}_0} \ln\left(1 - \frac{R_\infty}{N}\right) = \frac{1}{\mathcal{R}_0} \ln\left(1 - \frac{I_0}{N}\right). \tag{9.17}$$

As seen from Fig. 9.5, the final epidemic size R_∞ gets larger as the basic reproduction number $\mathcal{R}_0$ is greater. Thus the disease spread becomes more serious for the disease or community with the larger $\mathcal{R}_0$. Moreover, we note that the final epidemic size gets drastically larger as $\mathcal{R}_0$ is beyond one. Hence, it is implied that, even when the basic reproduction number $\mathcal{R}_0$ is slightly greater than one, the final epidemic size R_∞ could become much large, though the infective population size may appear to decrease as time passes.

Case of Uncertain Immunization

Let us consider next the model modified from (9.13), assuming that the recovery does not necessarily provide the effective immunity, that is, $0 < m < 1$. In this case, the recovered individual may get the infection again, so that we will need to

Fig. 9.6 State transitions for
the epidemic dynamics model
(9.18)

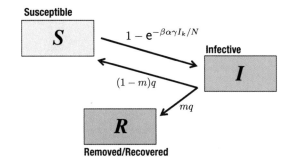

clarify the definition of the final epidemic size. The model is given as the following
system (Fig. 9.6):

$$S_{k+1} = S_k e^{-\beta \alpha \gamma I_k/N} + (1-m)q\,I_k;$$

$$I_{k+1} = S_k(1 - e^{-\beta \alpha \gamma I_k/N}) + (1-q)I_k; \qquad (9.18)$$

$$R_{k+1} = mq\,I_k + R_k.$$

The nature of the epidemic dynamics by this model is qualitatively similar with
that of the SIR model (9.13). However, as seen from (9.7), the immune population
size R_∞ at the end of the epidemic dynamics is smaller than that for the SIR model
(9.13), although this never indicates that the epidemic size was smaller than that by
the SIR model (9.13). Since the recovery necessarily provides the immunity for the
SIR model (9.13), we can regard the value of R_k as the number of individuals who
have experienced the infection. In contrast, for the model (9.18), we cannot do so,
because some recovered individuals can get the infection again. This means that the
susceptible subpopulation may contain the individuals who have experienced the
infection and failed to get the immunity after their recovery.

Hence, let us now introduce the variable C_k as the cumulative number of
individuals who have been infected and recovered until the kth day, which can
be determined by the recurrence relation $C_{k+1} = q\,I_k + C_k$ with $C_0 = 0$. As
shown by Fig. 9.7, such repetitive infections may result in the cumulative number
of infected individuals C_k greater than the total population size N. On the other
hand, the susceptible population size S_∞ at the end of the epidemic dynamics may
become greater than that for the SIR model (9.13), since the size S_∞ contains those
who have experienced the infection in past.

The uncertain immunization causes the recruitment of susceptible individuals,
so that the maximum of the infective population size tends to become larger, and
the infective population size decreases more slowly toward zero. Such repetitive
infections must lead to a heavier load for the medical services in the community.

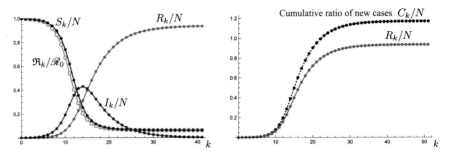

Fig. 9.7 A numerical calculation of the temporal change of variables for the epidemic dynamics model (9.18) with $m = 0.8$; $\beta\alpha\gamma = 1.0$; $q = 0.25$; $\mathscr{R}_0 = 4.0$; $(S_0/N, I_0/N, R_0/N) = (0.999, 0.001, 0.0)$

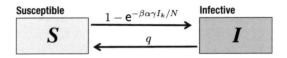

Fig. 9.8 State transitions for the SIS model (9.19)

SIS Model

If the recovery cannot bring the immunity at all, that is, when $m = 0$, the epidemic dynamics model (9.18) becomes the following system of difference equations:

$$S_{k+1} = S_k e^{-\beta\alpha\gamma I_k/N} + q I_k;$$
$$I_{k+1} = S_k (1 - e^{-\beta\alpha\gamma I_k/N}) + (1-q)I_k. \tag{9.19}$$

This is a kind of what is called *SIS model* (Fig. 9.8). The basic reproduction number $\mathscr{R}_0$ and the effective reproduction number $\mathfrak{R}_k$ are given again by (9.11) and (9.12) respectively.

When the pathogen is polymorphic or when there are a number of variants, the antigen generated with an infection may not be satisfactorily effective for the new infection. As the other example, the pathogen may hiddenly stay in the body of the recovered individual, and the repeated infection triggers the revival of the infectivity. In the other context, the addiction for drug, alcohol, gamble, or gaming could be regarded as the case.

Since $S_k + I_k = N$ for any k, the nature of the dynamical system (9.19) is essentially determined by the following one dimensional recurrence relation:

$$I_{k+1} = (N - I_k)(1 - e^{-\beta\alpha\gamma I_k/N}) + (1-q)I_k.$$

Hence we can apply the cobwebbing method (refer to Sect. 12.1.2), and find the following nature of the SIS model (9.19):

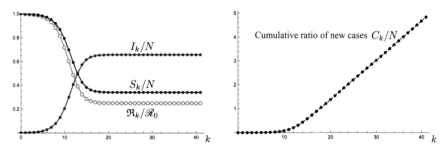

Fig. 9.9 A numerical calculation of the temporal change of variables for the SIS model (9.19) with $\beta\alpha\gamma = 1.0$; $q = 0.25$; $\mathcal{R}_0 = 4.0$; $(S_0/N, I_0/N, R_0/N) = (0.999, 0.001, 0.0)$

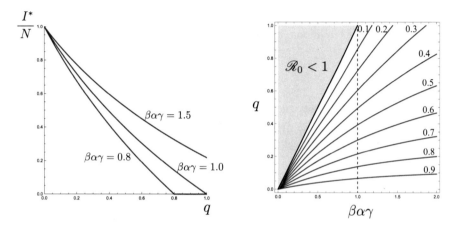

Fig. 9.10 Parameter dependence of the infective ratio I^*/N at the endemic equilibrium state for the SIS model (9.19). The right figure shows the contour map about I^*/N in the $(\beta\alpha\gamma, q)$-space

- When $\mathcal{R}_0 \leq 1$, the infective population size I_k monotonically decreases toward zero for any initial value $I_0 > 0$, and the population asymptotically approaches the disease-free equilibrium state.
- When $\mathcal{R}_0 > 1$, the infective population size I_k monotonically and asymptotically approaches a positive value I^* for any initial value $I_0 > 0$, and the population asymptotically approaches the endemic equilibrium state (Fig. 9.9). The endemic size I^* is given by the unique positive root for the following equation:

$$q\frac{I^*}{N} = \left(1 - \frac{I^*}{N}\right)(1 - e^{-\beta\alpha\gamma I^*/N}). \tag{9.20}$$

As indicated by Fig. 9.10, the infective population size I^* at the endemic equilibrium state gets larger as the duration of infectivity is longer with the smaller q. It gets larger as the basic reproduction number $\mathcal{R}_0 = \beta\alpha\gamma/q$ is greater.

Further, from the Eq. (9.20) to determine the value of I^* when $R_0 > 1$, we can find that the effective reproduction number $\mathfrak{R}_k$ defined by (9.12) is unity at the endemic equilibrium state. This means that the expected number of new infectives produced by an infective is one, which is consistent with the nature such that the infective population size I_k asymptotically approaches I^*, as shown by Fig. 9.9, and remains the same once it becomes I^*. Note that the epidemic dynamics always works at the endemic equilibrium state, keeping the infective population size I^* with the repetitive infection and recovery.

The endemic state is maintained by the balance in the processes that the susceptible becomes the infective by the infection and that the recovered becomes the susceptible. The repetitive infections keep occurring at the endemic state. Hence the cumulative number of individuals who have experienced the infection C_k is monotonically increasing as indicated by Fig. 9.9. Since new infectives are always produced at the endemic state, the medical services must undergo a heavy load for the treatment.

SIRS Model

If the immunity obtained by the recovery wanes, we have the following epidemic dynamics model from (9.1), which is a kind of what is called *SIRS model* (Fig. 9.11):

$$S_{k+1} = S_k e^{-\beta\alpha\gamma I_k/N} + \theta R_k;$$

$$I_{k+1} = S_k(1 - e^{-\beta\alpha\gamma I_k/N}) + (1 - q)I_k; \tag{9.21}$$

$$R_{k+1} = q I_k + (1 - \theta)R_k.$$

The basic reproduction number $\mathscr{R}_0$ and the effective reproduction number $\mathfrak{R}_k$ are given again by (9.11) and (9.12) respectively.

As seen in Fig. 9.12, the epidemic dynamics by (9.21) may approach an endemic equilibrium state, since the individuals who experienced the infection may return to

Fig. 9.11 State transitions for the SIRS model (9.21)

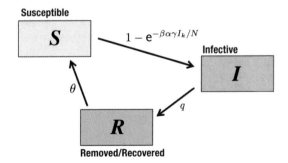

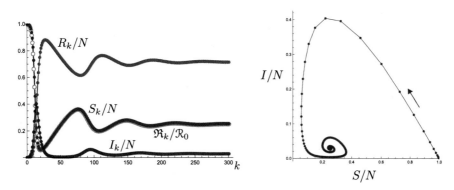

Fig. 9.12 A numerical calculation of the temporal change of variables for the SIRS model (9.21) with $\theta = 0.01$; $\beta\alpha\gamma = 1.0$; $q = 0.25$; $\mathscr{R}_0 = 4.0$; $(S_0/N, I_0/N, R_0/N) = (0.999, 0.001, 0.0)$

the susceptible class as well as the SIS model (9.19). The SIRS model (9.21) has the following nature:

- When $\mathscr{R}_0 \leq 1$, the infective population size I_k monotonically decreases toward zero for any initial value $I_0 > 0$, and the population asymptotically approaches the disease-free equilibrium state with $S_k \equiv N$, $I_k \equiv 0$ and $R_k \equiv 0$.
- When $\mathscr{R}_0 > 1$, the infective population size I_k asymptotically approaches a positive value I^* for any initial value $I_0 > 0$, and the population asymptotically approaches the endemic equilibrium state (Fig. 9.12). The endemic size I^* is given by the unique positive root for the following equation:

$$\frac{1}{q}\left(\frac{N}{I^*} - 1\right) = \frac{1}{\theta} + \frac{1}{1 - e^{-\beta\alpha\gamma I^*/N}}. \tag{9.22}$$

- The susceptible and infective population sizes S^* and I^* at the endemic equilibrium state satisfy the following equality:

$$\frac{I^*}{N} = \frac{\theta}{\theta + q}\left(1 - \frac{S^*}{N}\right). \tag{9.23}$$

Hence the infective population size I^* at the endemic equilibrium state is necessarily smaller than $\theta/(\theta + q)$.

- The ratio of the infective population size I^* to the immune population size R^* at the endemic equilibrium state is given by

$$I^* : R^* = \frac{1}{q} : \frac{1}{\theta}. \tag{9.24}$$

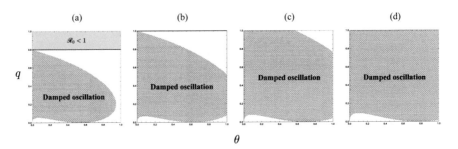

Fig. 9.13 Parameter dependence of the oscillatory behavior for the SIRS model (9.21). Numerically drawn with (**a**) $\beta\alpha\gamma = 0.8$; (**b**) $\beta\alpha\gamma = 1.0$; (**c**) $\beta\alpha\gamma = 1.2$; (**d**) $\beta\alpha\gamma = 1.5$

As well as the SIS model (9.19), it is determined by the basic reproduction number $\mathscr{R}_0$ whether the disease becomes endemic or not. The stability of the disease-free equilibrium state and the endemic equilibrium state can be analyzed by the local stability analysis (refer to Sect. 12.2). Further, it is easy to show that the effective reproduction number $\mathfrak{R}_k$ defined by (9.12) is unity at the endemic equilibrium state.

Differently from the SIS model (9.19), the SIRS model (9.21) may have a damped oscillation when the system approaches the endemic equilibrium state, as shown in Fig. 9.12. Even though the oscillation is damping, it could be observed as repeated outbreaks on the way of epidemic dynamics. Such a situation must be regarded as serious for the public health.

The SIRS model (9.21) with $\theta = 0$ coincides with the SIR model (9.13) for which no oscillation can appear in the temporal change. As indicated by Fig. 9.13, an oscillatory behavior can appear even with so small waning rate θ.

For the SIRS model (9.21), the expected duration of the immunity obtained by the recovery is given by $1/\theta$, making use of the same arguments to derive the expected duration of the infectivity $1/q$ as described before for the SIR model (9.13). Hence, as implied by Fig. 9.13, an oscillatory behavior is more likely to appear when the immunity is active for relatively long period with small waning rate and when the basic reproduction number $\mathscr{R}_0 = \beta\alpha\gamma/q$ is large. Inversely, when the immunity is short-term with large waning rate and when the basic reproduction number $\mathscr{R}_0$ is nearer to one, the temporal change of the infective population size is likely to be monotonic. Further, we can see that the oscillatory nature significantly depends on the value of parameter $\beta\alpha\gamma$. For a sufficiently large value of $\beta\alpha\gamma$, that is, when the infectivity is much strong (Fig. 9.13d), the oscillatory behavior may appear almost independently of the duration of infectivity and immunity. Then the value of the basic reproduction number $\mathscr{R}_0$ is much large too.

Now let us consider the characteristics of the endemic equilibrium state, making use of (9.22), (9.23), and (9.24). Figure 9.14 illustrates the θ-dependence and q-dependence of the population proportion $(S^*/N, I^*/N, R^*/N)$ at the endemic equilibrium state.

First, as the immunity is longer with smaller θ, the immune population size R^* at the endemic equilibrium state gets larger, while the infective population size I^*

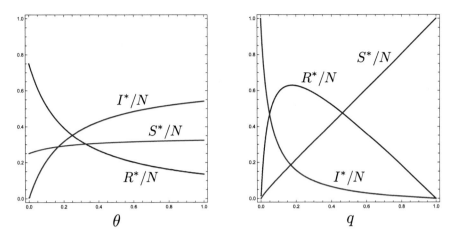

Fig. 9.14 θ-dependence with $q = 0.25$ ($\mathscr{R}_0 = 4.0$) and q-dependence with $\theta = 0.05$ of the population proportion (S^*/N, I^*/N, R^*/N) at the endemic equilibrium state for the SIRS model (9.21). Numerically drawn commonly with $\beta\alpha\gamma = 1.0$

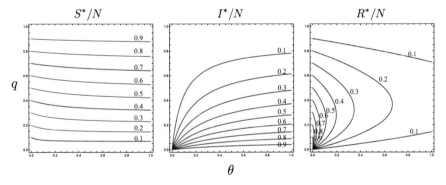

Fig. 9.15 Contour map of the population proportion (S^*/N, I^*/N, R^*/N) at the endemic equilibrium state for the SIRS model (9.21). Numerically drawn to show the (θ, q)-dependence with $\beta\alpha\gamma = 1.0$

does smaller. Inversely, as the immunity is shorter with larger θ, R^* gets smaller and I^* does larger. In this case, the susceptible population size S^* gets larger at the same time. This is because the immunity waning is more likely to occur, so that the recovered individuals transfers to the susceptible class faster to make the susceptible population size larger.

Next, as the infective can recover faster with larger q, the infective population size I^* gets smaller, while the susceptible population size S^* does larger. Interestingly the immune population size R^* does not have monotonic relation to the recovery rate. It takes the maximum at an intermediate value of q as indicated by (9.15). For faster or slower recovery, it gets smaller. This nature is clearly shown by Fig. 9.15 as well.

When the recovery is fast with large q, the effective reproduction number $\mathfrak{R}_k$ must be restricted to be small, because the duration of infectivity is short. The infective population size remains relatively small, so that the immune population size after the recovery is kept small.

On the other hand, when the recovery is slow with small q, the infectivity is long-term to cause the relatively large size of infective population. While the immune population size decreases by the immunity waning, the slow recovery tends to undercompensate such a decrease and leads to the relatively small size of immune population.

From these results on the SIRS model (9.21) with the immunity waning, we could get an implication that, even when the infective population size is large, the immune population size does not necessarily become large. They may have a non-monotonic relation.

There are a variety of problems for the epidemic dynamics. For example, the effect of vaccination on the epidemic size is one of typical problems. The vaccination can be regarded as the artificial immunization, which may have a possibility of waning. Then the epidemic dynamics with such a vaccination would have similar nature as the SIRS model in the above. On the other hand, the isolation affects directly the epidemic dynamics, as mentioned at the beginning of Sect. 9.2. Especially for the human epidemic dynamics, the behavioral change by the spread of information on the disease or the sanitary treatment policy for the route of disease transmission could alter the nature of the epidemic dynamics, which would have a relation in a feedback manner. Such a social response to the epidemic dynamics must have a relation to the cultural and social background too. The aspect of social science is very important for the research on the epidemic dynamics in the human community.

9.3 Kermack-McKendrick Model

The following system of ordinary differential equations as a continuous time epidemic dynamics model is the simplest version of the mathematical model studied by William O. Kermack (1898–1970) and Anderson G. McKendrick (1876–1943) in 1927 [9]:

$$\frac{dS(t)}{dt} = -\sigma I(t)S(t);$$

$$\frac{dI(t)}{dt} = \sigma I(t)S(t) - \rho I(t);$$

$$\frac{dR(t)}{dt} = \rho I(t).$$

$$(9.25)$$

This is popular today in mathematical epidemiology as *Kermack-McKendrick model* which is a continuous time SIR model with the susceptible population size $S(t)$, the infective population size $I(t)$, and the removed population size $R(t)$ at time t. Positive parameters σ and ρ are the coefficient of infection and the recovery rate respectively. As well as the discrete time SIR model (9.13) in Sect. 9.2.4, the total population size is constant independently of time t for Kermack-McKendrick model (9.25): $S(t)+I(t)+R(t) = N$ for any t. Actually, the system of ordinary differential Eq. (9.25) gives $d\{S(t) + I(t) + R(t)\}/dt = 0$, which mathematically means that the total population size $S(t) + I(t) + R(t)$ is constant independently of time t.

The continuous time SIR model (9.25) can be derived from the discrete time SIR model (9.13) by the time-step-zero limit described in Sect. 3.3. In the application of the time-step-zero limit, it is necessary to introduce the dependence of parameters γ and q on the time step size h such that $\gamma \rightarrow 0$ and $q \rightarrow 0$ as $h \rightarrow 0$ for the reasonable sake to take the limit as the time step size goes to zero (refer to Sect. 3.3). Then we can get the continuous time SIR model (9.25) with the correspondence such that $\sigma = \beta\alpha\gamma'(0)/N$ and $\rho = q'(0)$, where $\gamma'(0) := d\gamma(h)/dh\big|_{h\rightarrow 0}$ and $q'(0) := dq(h)/dh\big|_{h\rightarrow 0}$ which are now assumed finite and positive.

On the other hand, the modeling for Kermack-McKendrick model (9.25) is usually explained by the mass action assumption, and more precisely by the Lotka-Volterra type of interaction (refer to Sect. 6.1). As mentioned at the beginning of Sect. 9.2, the disease transmission process can be regarded as analogous to the interaction between predator and prey. Hence it could be reasonable to apply the other modeling on the prey-predator interaction for the disease transmission dynamics with some appropriate assumptions, like Holling's disc equation in Sect. 8.5.

9.3.1 Infection Force

Similarly to the modeling of the prey-predator interaction given in Sect. 8.2, we can introduce the infection force as the infection rate with a function of S and I ($R = N - S - I$), $\Lambda(S, I)$, and consider the change of susceptible population size in a short period Δt: $S(t+\Delta t) - S(t) = -\Lambda(S(t), I(t))S(t)\Delta t + o(\Delta t)$. Applying the way to derive the momental velocity of population size change in Sect. 3.4, we can get the ordinary differential equation to describe the temporal change of susceptible population size: $dS/dt = -\Lambda(S, I)S$. For Kermack-McKendrick model (9.25), the infection force is given by $\Lambda(S, I) = \sigma I$ that is proportional to the infective density in the population. The term of the interaction between susceptibles and infectives depends on what assumptions the infection rate, that is, the infection force would be modeled with. The simplest and least necessary assumptions are about the contact between individuals and the probability of an infectious contact, as argued already in Sect. 9.2.

Let us give the frequency of contacts per individual with the others in the interval $[t, t+\Delta t]$ by $\chi(t)\Delta t + \mathrm{o}(\Delta t)$ with a positive $\chi(t)$ that means the momental velocity for the number of contacts, that is, the contact rate between individuals. Besides, applying the *mean field approximation* (refer to Sect. 6.1.2), we assume that the probability that a contact by a susceptible is one with an infective is proportional to the proportion of infectives in the population $I(t)/N$. Then we assume that the probability that a susceptible gets infected in $[t, t + \Delta t]$ is proportional to

$$\frac{I(t)}{N}\chi(t)\Delta t + \mathrm{o}(\Delta t),$$

so that the infection force is proportional to $(I/N)\chi$. The detail of this modeling is determined by the function $\chi(t)$ which reflects the frequency of contacts between individuals in the population. Such a frequency of contacts must significantly depend on the route of disease transmission.

For a human sexually transmitted disease, the contact rate between individuals at the route of disease transmission, that is, the sexual contact may be regarded as a constant independent of time, since the frequency of such contacts would be determined by the established network of such sexual contacts in the population. Then χ is given as a positive constant. With such a modeling, the infection force is assumed to be proportional to the ratio of infectives in the population: $\Lambda \propto I/N$. It is sometimes called *ratio-dependent* type or *frequency-dependent* type of infection force.

In contrast, for a transmissible disease with the air-borne route of disease transmission, for example, by the droplet or the aerosol like the flue or COVID-19, we may assume that the contact rate is proportional to the population density: $\chi \propto N$, since the route of disease transmission is about the contacts between arbitrary individuals as already argued in Sect. 9.2.1. With such a modeling, the infection force is assumed to be proportional to the infective density in the population: $\Lambda \propto I$. Subsequently, the term of interaction between susceptibles and infectives is given by the product of their densities, that is, by the mass action assumption. Thus, such an infection force is sometimes called *mass action* type.

For the epidemic dynamics in which the temporal change of total population size is negligible and the total population size is assumed constant independently of time, the infection force is *mathematically* equivalent to each other of these two types. However, when the temporal change of total population size must be taken into account for the epidemic dynamics, the nature of epidemic dynamics model would be significantly different from each other for these two types of infection force. Moreover, even when the infection force is *mathematically* equivalent between them, the epidemiological arguments based on the mathematical results obtained by the analysis on the model

(continued)

would essentially depend on the assumption about the route of disease transmission or the nature of contact rate. For this reason, such a *mathematical equivalence* may not necessarily induce the same conclusion on the epidemic dynamics.

9.3.2 Invasion Success of Transmissible Disease

As considered for the discrete time epidemic dynamics model in Sect. 9.2.2, we can derive the condition for the invasion success of the transmissible disease about Kermack-McKendrick model (9.25). Let us assume the initial condition such that $I(0) = I_0 \ll N$ similarly to the argument in Sect. 9.2.2. This means the initial situation of the epidemic dynamics such that a sufficiently small number of infectives appear in the population. Since the invasion success causes the increase of the infective population size by such a few initial infectives, the following condition is necessary for it:

$$\frac{dI(t)}{dt}\bigg|_{t=0} = \{\sigma S(0) - \rho\} I(0) > 0.$$

Hence the condition for the invasion success can be given as

$$\frac{\sigma S(0)}{\rho} > 1. \tag{9.26}$$

Although the condition $I_0 \ll N$ is independent of this result, the concept of the invasion of a disease could be reasonable for such an initial condition with a sufficiently small number of infectives.

9.3.3 Final Epidemic Size

Since the right side of the first equation of (9.25) is always negative, the susceptible population size $S(t)$ is monotonically decreasing in terms of time t. If $S(t) > \rho/\sigma$ at time t, the right side of the second equation of (9.25) is positive, that is, the momental velocity of the temporal change of the infective population size is positive and the infective population size $I(t)$ increases at time t. If $S(t) < \rho/\sigma$, then $I(t)$ decreases. The right side of the third equation of (9.25) is always positive, so that $R(t)$ is monotonically increasing in terms of time t.

As seen in Fig. 9.16, the infective population size $I(t)$ asymptotically and exponentially approaches zero for sufficiently large t. This means that the disease

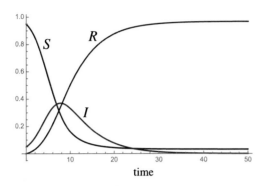
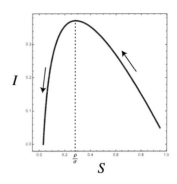

Fig. 9.16 Numerical example of the temporal change and trajectory in the (S, I)-phase plane by Kermack-McKendrick model (9.25) with $\sigma = 0.7$; $\rho = 0.2$; $(S(0), I(0), R(0)) = (0.95, 0.05, 0.0)$

eventually disappears in the population. In contrast, the susceptible population size does not approach zero, that is, there must a part of susceptibles who can escape from the infection until the end of epidemic dynamics. This is the same nature as the discrete time SIR model (9.13) in Sect. 9.2.4. Actually Kermack-McKendrick model (9.25) has the qualitatively same nature as the discrete time SIR model (9.13) as described below.

From the first and second equations of (9.25), we can find the following equation:

$$\frac{dS(t)}{dt} + \frac{dI(t)}{dt} = -\rho I(t) = \frac{\rho}{\sigma}\frac{1}{S(t)}\frac{dS(t)}{dt} = \frac{\rho}{\sigma}\frac{d}{dt}\{\ln S(t)\}.$$

By integrating both sides in terms of t, we have

$$S(t) + I(t) - \frac{\rho}{\sigma}\ln S(t) = S_0 + I_0 - \frac{\rho}{\sigma}\ln S_0 \qquad (9.27)$$

with $S(0) = S_0 > 0$ and $I(0) = I_0 > 0$. Similarly with the arguments for the discrete time SIR model (9.13) in Sect. 9.2.4, since the Eq. (9.27) holds for any time t, it can be regarded as the formula of a trajectory of the solution in the (S, I)-phase plane for (9.25) (Figs. 9.16 and 9.17).

From the Eq. (9.27), we can get

$$\frac{dI}{dS} = \frac{\rho}{\sigma}\frac{1}{S} - 1.$$

Thus, the trajectory can take the maximum for $S = \rho/\sigma$ (Figs. 9.16 and 9.17). Hence, at the moment that the infective population size takes the maximum on the way of the epidemic dynamics by (9.25), it holds that $S(t) = \rho/\sigma$. Mathematically this result coincides with that derived from the equation $dI(t)/dt = 0$ about the second equation of (9.25).

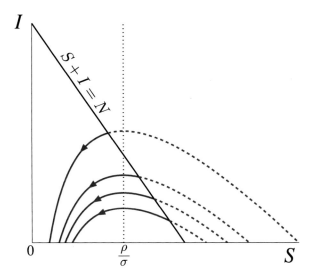

Fig. 9.17 Curves of (9.27) about the trajectory in the (S, I)-phase plane for Kermack-McKendrick model (9.25) for different four initial conditions $(S(0), I(0))$

In addition, we can easily prove that the curve of (9.27) in the (S, I)-phase plane necessarily has two positive intersection with the S-axis for any $S_0 > 0$ and $I_0 > 0$ (refer to Exercise 9.1 in the following part). This means the necessary existence of susceptibles who can escape from the infection until the end of epidemic dynamics.

Besides, since $S(t) + I(t) + R(t) = N$ for any t, $R(0) = 0$ and $S(0) + I(0) = N$, the right side of (9.27) is decreasing in terms of $S(0) = S_0 > 0$ as indicated by Fig. 9.17. Since the curve of (9.27) in the (S, I)-phase plane is uniquely determined for each initial condition $(S(0), I(0)) = (S_0, I_0)$, we can result that the curve for an initial condition is different from that for the other initial condition.

As the reasonable initial condition for almost all considerations with the epidemic dynamics by (9.25), it is assumed that $R(0) = 0$. Then we have $S(0) + I(0) = S_0 + I_0 = N$, and the initial point in the (S, I)-phase plane must be located on the line $S + I = N$. It is easily shown that the curve of (9.27) has the unique intersection with the line $S + I = N$, and the solution given by the curve for $t > 0$ is necessarily located in the region given by $S + I < N$. This is because $S + I = N - R$ is monotonically decreasing in terms of t due to the transition of individuals to the removed class R. Therefore, as seen from Fig. 9.17, the curve for an initial value of S_0 is different from that for the other initial value of S_0.

As a result, we find that, for the initial susceptible population size $S(0)$ larger than ρ/σ, the infective population size increases in the earlier period, takes a maximal at a moment, and subsequently decreases monotonically toward zero. For the initial susceptible population size $S(0)$ not beyond ρ/σ, the infective population size monotonically decreases toward zero (see Fig. 9.17).

The susceptibles who can escape from the infection until the end of epidemic dynamics is now mathematically denoted as $S_\infty := \lim_{t\to\infty} S(t)$. The size S_∞ may be called the final size of the susceptible population for the epidemic dynamics by (9.25).

Since $I_\infty := \lim_{t\to\infty} I(t) = 0$, we can get the following equation about the relation between the initial value S_0 and the final size S_∞:

$$S_\infty - \frac{\rho}{\sigma} \ln S_\infty = S_0 + I_0 - \frac{\rho}{\sigma} \ln S_0. \tag{9.28}$$

This equation uniquely determines the final size S_∞ for any given $S_0 > 0$ (Exercise 9.1). From (9.28), the final epidemic size $R_\infty = N - S_\infty$ for Kermack-McKendrick model (9.25) satisfies that

$$\mathscr{R}_0 = \frac{1}{R_\infty/N} \ln \frac{1 - I_0/N}{1 - R_\infty/N}, \tag{9.29}$$

where $\mathscr{R}_0 := \sigma N/\rho$, and we used the relations $S_0 + I_0 = N$, $I_\infty = 0$, and $S_\infty + R_\infty = N$. Since the final size S_∞ is uniquely determined by the Eq. (9.28), the final epidemic size R_∞ must be uniquely determined as well. As shown in Fig. 9.18, the final epidemic size R_∞ is monotonically increasing in terms of σ/ρ, and tends to drastically increase for $\mathscr{R}_0 := \sigma N/\rho > 1$. We will revisit this nature in the next section.

Exercise 9.1 Prove that the Eq. (9.28) uniquely determines a positive value of the final size $S_\infty < 1$ for any given $S_0 > 0$.

9.3.4   Reproduction Number of Infectives

As described in Sect. 9.2.3, the basic reproduction number $\mathscr{R}_0$ is defined as the expected number of new infectives produced by an infective under the condition that the original infective has contacts only with susceptibles during having the active infectivity, and is corresponding to the supremum for the expected number of new infectives produced by an infective. Besides, similarly to the definition for the discrete time epidemic dynamics in Sect. 9.2.3, we can define the effective reproduction number $\mathfrak{R}(t)$ as the supremum for the expected number of new infectives produced by an infective during having the infectivity, provided that the situation at time t is unchanged.

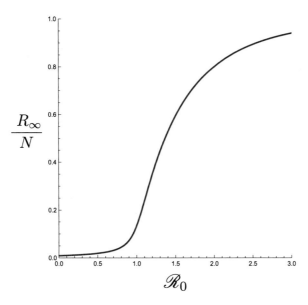

Fig. 9.18 $\mathscr{R}_0$-dependence of the final epidemic size R_∞ for Kermack-McKendrick model (9.25), determined by (9.29) with $I_0/N = 0.01$. $\mathscr{R}_0 := \sigma N/\rho$

For Kermack-McKendrick model (9.25), we have the expected number of new infectives produced by an infective during a small interval $(t, t + \Delta t)$ as

$$\frac{S(t) - S(t + \Delta t)}{I(t)} = -\frac{1}{I(t)}\frac{dS(t)}{dt}\Delta t + o(\Delta t) = \sigma S(t)\Delta t + o(\Delta t).$$

From the arguments in Sect. 4.3.2 or 4.3.5, the expected duration of infectivity for Kermack-McKendrick model (9.25) is given by $1/\rho$. Therefore we can find the effective reproduction number for Kermack-McKendrick model (9.25) as $\mathfrak{R}(t) = \sigma S(t)/\rho$. Hence we can define the basic reproduction number now as

$$\mathscr{R}_0 = \sup_S \mathfrak{R}(t) = \frac{\sigma N}{\rho}. \tag{9.30}$$

As already argued in Sect. 9.2.3, the outbreak does not occur if $\mathscr{R}_0 < 1$, while $\mathscr{R}_0 > 1$ if the outbreak occurs. Indeed we have gotten the corresponding nature of the epidemic dynamics by (9.25) in Sect. 9.3.3: If $S(t) > \rho/\sigma$ at time t, the infective population size $I(t)$ increases at time t, while, if $S(t) < \rho/\sigma$ at time t, it decreases at time t. This means that the infective population size $I(t)$ increases if $\mathfrak{R}(t) > 1$, while it decreases if $\mathfrak{R}(t) < 1$. Since $\mathfrak{R}(t) \leq \mathscr{R}_0$ for any t, the above proposition holds with respect to $\mathscr{R}_0$. This nature of the epidemic dynamics by (9.25) is demonstrated well by the $\mathscr{R}_0$-dependence of the final epidemic size R_∞ shown in Fig. 9.18.

The condition for the invasion success (9.26) derived in Sect. 9.3.2 can be expressed here as the condition for the effective reproduction number such that $\Re(0) > 1$. The invasion success and outbreak are more likely to occur for Kermack-McKendrick model (9.25) as the total population size N gets larger, since the former is proportional to $S(0)$ which becomes larger for larger N as well as the latter is. Thus, for the epidemic dynamics by (9.25), the larger population is more vulnerable to the spread of a transmissible disease.

9.3.5 Extension to SIS and SIRS Models

As well as the generic discrete time model (9.1) in Sect. 9.2.1, we can introduce the probability that the recovery is successful to get the immunity, m ($0 \leq m \leq 1$), and the waning rate of the immunity, ν, in Kermack-McKendrick model. Then we have the following extended model:

$$
\begin{aligned}
\frac{dS(t)}{dt} &= -\sigma I(t)S(t) + (1-m)\rho I(t) + \nu R(t); \\
\frac{dI(t)}{dt} &= \sigma I(t)S(t) - \rho I(t); \\
\frac{dR(t)}{dt} &= m\rho I(t) - \nu R(t).
\end{aligned}
\tag{9.31}
$$

From the arguments in Sect. 4.3.2 or 4.3.5, the expected duration of immunity for this model (9.31) is given by $1/\nu$. Note that the basic reproduction number $\mathscr{R}_0$ and the effective reproduction number $\Re(t)$ are defined the same as given in Sect. 9.3.4 for Kermack-McKendrick model (9.25), since the same derivation can be applied independently of the above extension: $\Re(t) = \sigma S(t)/\rho$; $\mathscr{R}_0 = \sigma N/\rho$.

SIS Model

If the immunity is ineffective as supposed for the discrete time SIS model (9.19) in Sect. 9.2.4, we have the following SIS model from (9.31) with $m = 0$:

$$
\begin{aligned}
\frac{dS(t)}{dt} &= -\sigma S(t)I(t) + \rho I(t); \\
\frac{dI(t)}{dt} &= \sigma S(t)I(t) - \rho I(t).
\end{aligned}
\tag{9.32}
$$

For this model, no immune individual exists. Since $S(t) + I(t) = N$ for any t, the nature of the epidemic dynamics by (9.32) is essentially determined by the following one dimensional ordinary differential equation:

$$\frac{dI(t)}{dt} = \sigma\{N - I(t)\}I(t) - \rho I(t) = \sigma\left\{N - \frac{\rho}{\sigma} - I(t)\right\}I(t).$$

This is mathematically equivalent to the logistic Eq. (5.10) in Sect. 5.3. Hence the temporal change of the infective population size is always monotonic for any initial value $I(0) > 0$, and

- when $N - \rho/\sigma \leq 0$, that is, when $\mathscr{R}_0 \leq 1$, the infective population size is monotonically decreasing toward zero;
- when $N - \rho/\sigma > 0$, that is, when $\mathscr{R}_0 > 1$, the infective population size monotonically approaches the equilibrium value given by $N - \rho/\sigma$.

The latter is the case where the population approaches an endemic equilibrium state.

Case of Uncertain Immunization

Next, similarly with the discrete time model (9.18) in Sect. 9.2.4, if the immunity never wanes once it is obtained while the recovery does not necessarily lead to the immunity, the model (9.31) becomes as follows with $\nu = 0$ and $m > 0$:

$$\frac{dS(t)}{dt} = -\sigma I(t)S(t) + (1 - m)\rho I(t);$$

$$\frac{dI(t)}{dt} = \sigma I(t)S(t) - \rho I(t); \tag{9.33}$$

$$\frac{dR(t)}{dt} = m\rho I(t).$$

Since the previous two equations are closed with respect to the variables S and I, the nature of this three dimensional system is essentially determined by the two dimensional system of those two equations. Thus, the isocline method is applicable (refer to Sect. 14.7).

As indicated by Fig. 9.19, we can find that the outbreak occurs only when $S(0) > \rho/\sigma$ as for Kermack-McKendrick model (9.25). When $S(0) \leq \rho/\sigma$, the infective population size is monotonically decreases toward zero, while the susceptible population size approaches a certain positive value as the epidemic dynamics is going to the end. More precisely with respect to the temporal change of the susceptible population size $S(t)$ and its final size $S_\infty := \lim_{t\to\infty} S(t)$, we can get the following nature:

- For $S(0)/N > (1 - m)/\mathscr{R}_0$, $S(t)/N$ monotonically decreases toward $S_\infty/N > (1 - m)/\mathscr{R}_0$.

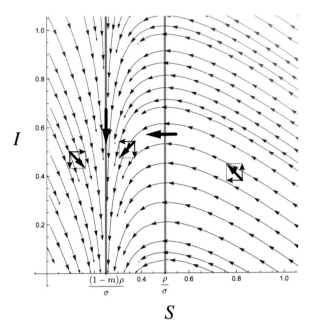

I

S

Fig. 9.19 Numerically drawn vector flows in the (S, I)-phase plane for the epidemic dynamics model (9.33) with $\sigma = 0.2$; $\rho = 0.1$; $m = 0.5$. Two vertical lines in the first quadrant are nullclines

- For $S(0)/N < (1 - m)/\mathscr{R}_0$, $S(t)/N$ monotonically increases toward $S_\infty/N < (1 - m)/\mathscr{R}_0$.

Since the susceptible class contains individuals who have experienced the infection and lost the immunity, the final size S_∞ does not mean the susceptibles who have escaped from the infection until the end of the epidemic dynamics even though it contains such individuals.

From (9.33), we have

$$\frac{dS}{dt} + \frac{dI}{dt} = -m\rho I. \tag{9.34}$$

Besides from the first equation of (9.33), we can derive

$$I = -\frac{1}{\sigma}\frac{d}{dt}\ln\left| -\sigma S + (1 - m)\rho \right|. \tag{9.35}$$

Substituting (9.35) for (9.34) and integrating it in terms of t, we can get the following equation which gives the curve of the trajectory in the (S, I)-phase plane:

$$S(t) + I(t) - \frac{m\rho}{\sigma} \ln \left| -\sigma S(t) + (1-m)\rho \right|$$

$$= S(0) + I(0) - \frac{m\rho}{\sigma} \ln \left| -\sigma S(0) + (1-m)\rho \right|. \tag{9.36}$$

Therefore we can obtain the following equation to determine the final size S_∞:

$$\frac{S_\infty}{N} - \frac{m}{\mathscr{R}_0} \ln \left| -\frac{S_\infty}{N} + \frac{1-m}{\mathscr{R}_0} \right| = 1 - \frac{m}{\mathscr{R}_0} \ln \left| -\frac{S_0}{N} + \frac{1-m}{\mathscr{R}_0} \right| \tag{9.37}$$

for the initial condition given by $(S(0), I(0), R(0)) = (N - I_0, I_0, 0)$ with $I_0 > 0$. Since $I(t) \to 0$ as $t \to \infty$, we can get the final value R_∞ from the equation $S_\infty + R_\infty = N$. However, it must be remarked that R_∞ does NOT mean the final epidemic size for the epidemic dynamics by (9.33). This is because the epidemic dynamics by (9.33) contains the re-infection after the recovery from the infective state as already mentioned about the meaning of S_∞.

To investigate the final epidemic size which is now defined as the cumulative number of individuals who have experienced the infection until time t, $C(t)$, we must integrate the flux of recovered individuals over time:

$$C(t) := \int_0^t \rho I(\tau) \, d\tau.$$

This is the conceptually same definition as C_k for the discrete time model (9.18) in Sect. 9.2.4.

The final epidemic size can be now given by $C_\infty := \lim_{t\to\infty} C(t)$. From the third equation of (9.33), we find that

$$C_\infty = \int_0^\infty \rho I(\tau) \, d\tau = \int_0^\infty \frac{1}{m} \frac{dR(\tau)}{d\tau} \, d\tau = \frac{R(\infty) - R(0)}{m} = \frac{R_\infty}{m} \tag{9.38}$$

for $R(0) = 0$. Consequently from (9.37), we can derive the following equation to determine the final epidemic size C_∞:

$$\frac{C_\infty}{N} + \frac{1}{\mathscr{R}_0} \ln \left| m\frac{C_\infty}{N} - 1 + \frac{1-m}{\mathscr{R}_0} \right| = \frac{1}{\mathscr{R}_0} \ln \left| \frac{I_0}{N} - 1 + \frac{1-m}{\mathscr{R}_0} \right|. \tag{9.39}$$

From the nature of the temporal change of the susceptible population size $S(t)$ and its final size S_∞ obtained by the isocline method at the beginning of this part, we can find the following nature of $C(t)$ and the final epidemic size C_∞:

- For $I(0)/N < 1 - (1 - m)/\mathscr{R}_0$, $C(t)/N$ monotonically increases toward $C_\infty/N < 1/m - (1 - m)/(m\mathscr{R}_0)$.
- For $I(0)/N > 1 - (1 - m)/\mathscr{R}_0$, $C(t)/N$ monotonically increases toward $C_\infty/N > 1/m - (1 - m)/(m\mathscr{R}_0)$.

Making use of this nature, we can get the following equation equivalent to (9.39):

$$m = \frac{I_0/N + (1 - 1/\mathscr{R}_0)(e^{\mathscr{R}_0 C_\infty/N} - 1)}{(C_\infty/N - 1/\mathscr{R}_0)e^{\mathscr{R}_0 C_\infty/N} + 1/\mathscr{R}_0}. \tag{9.40}$$

As seen from Fig. 9.20, the final epidemic size C_∞ is monotonically decreasing in terms of m. That is, as the probability to succeed in getting the immunity on the recovery is larger, the final epidemic size gets smaller.

When $m = 0$ for the model (9.33), it becomes equivalent to the SIS model (9.32). For the SIS model, if $\mathscr{R}_0 > 1$, the epidemic dynamics approaches an endemic equilibrium state, at which the cumulative number of individuals who have experience the infection $C(t)$ is monotonically increasing as time passes. Thus, as shown in Fig. 9.20, the final epidemic size C_∞ positively diverges as $m \to 0$. On the other hand, at the other extremum with $m = 1$, the model (9.33) is equivalent to the Kermack-McKendrick model (9.25), and the final epidemic size C_∞ becomes equal to R_∞ (< 1) discussed in Sect. 9.3.3 for it.

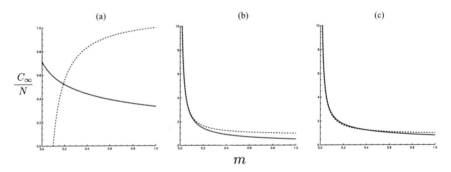

Fig. 9.20 The m-dependence of the final epidemic size C_∞ defined by (9.38) for the epidemic dynamics model (9.33) with (**a**) $\mathscr{R}_0 = 0.9$ and $I_0/N = 0.1$; (**b**) $\mathscr{R}_0 = 1.2$ and $I_0/N = 0.1$; (**c**) $\mathscr{R}_0 = 1.2$ and $I_0/N = 0.5$. The broken curve indicates $1/m - (1 - m)/(m\mathscr{R}_0)$

SIRS Model

When the recovery assures the immunity though the immunity may want to be lost, the epidemic dynamics model (9.31) becomes the following SIRS model:

$$\frac{dS(t)}{dt} = -\sigma I(t)S(t) + vR(t);$$

$$\frac{dI(t)}{dt} = \sigma I(t)S(t) - \rho I(t); \tag{9.41}$$

$$\frac{dR(t)}{dt} = \rho I(t) - vR(t).$$

It is easy to find that there are at most two equilibria $E_0(N, 0, 0)$ and

$$E_+\left(S^*, \frac{v}{\rho + v}(\mathscr{R}_0 - 1)S^*, \frac{\rho}{\rho + v}(\mathscr{R}_0 - 1)S^*\right) \quad \text{with } S^* := \frac{\rho}{\sigma}.$$

The former is the disease-free equilibrium state, and the latter the endemic equilibrium state. The endemic equilibrium state exists if only if $\mathscr{R}_0 > 1$.

From $S(t) + I(t) + R(t) = N$ for any t, the system (9.41) is mathematically equivalent to the two dimensional system of ordinary differential equations with respect to S and I, derived from the first and second equations of (9.41) with substituting $R(t) = N - S(t) - I(t)$ for the first. Hence we can use the isocline method, and find the following condition for the global stability of the disease-free equilibrium state E_0 (see Fig. 9.21a):

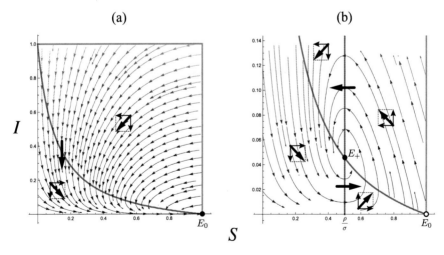

Fig. 9.21 Numerically drawn vector flows in the (S, I)-phase plane for the epidemic dynamics model (9.41) with $N = 1.0$; $\rho = 0.1$; $v = 0.01$; (**a**) $\sigma = 0.09$ ($\mathscr{R}_0 = 0.9$); (**b**) $\sigma = 2.0$ ($\mathscr{R}_0 = 2.0$). Nullclines are drawn too

- The disease-free equilibrium state E_0 is globally asymptotically stable if and only if $\mathscr{R}_0 \leq 1$.

It is shown too that, when $\mathscr{R}_0 \leq 1$, the infective population size monotonically decreases toward zero as time passes. That is, when $\mathscr{R}_0 \leq 1$, the epidemic dynamics monotonically approaches the disease-free equilibrium state. As for the stability of the endemic state E_+, however, it is impossible to get the result only by the isocline method (see Fig. 9.21b).

The local stability analysis (Sects. 14.2 and 14.3) can show that the endemic state E_+ is locally asymptotically stable whenever it exists with $\mathscr{R}_0 > 1$. Indeed, from the characteristic equation for E_+, $\lambda^2 + (\sigma I^* + v)\lambda + (\rho + v)\sigma I^* = 0$, it is easily seen that every eigenvalue has a negative real part: Re $\lambda < 0$. Since the system (9.41) is mathematically equivalent to the two dimensional system of ordinary differential equations as mentioned above, we can apply Poincaré-Bendixson Theorem (Sect. 14.9) for it. Then from Poincaré-Bendixson Theorem, we can get the following result for the case of $\mathscr{R}_0 > 1$:

- The endemic state E_+ is globally asymptotically stable if and only if $\mathscr{R}_0 > 1$.

The global stability of the endemic state E_+ can be proved also by the Lyapunov function (Sect. 14.8). Indeed the following function of S and I is a Lyapunov function about the endemic state E_+ for the system (9.41):

$$V(S, I) := \left\{(S - S^*) + (I - I^*)\right\}^2 + 2S^*\left\{(I - I^*) - I^* \ln \frac{I}{I^*}\right\}. \tag{9.42}$$

The function $V(S, I)$ is positive for any (S, I) such that $0 < S \neq S^*$ and $0 < I \neq I^*$, while it becomes zero only for $(S, I) = (S^*, I^*)$ (see Fig. 9.22). We can derive

$$\frac{dV}{dt} = -2v\left\{(S - S^*) + (I - I^*)\right\}^2 - 2\sigma S^*(I - I^*)^2,$$

which is negative for any (S, I) such that $S \neq S^*$ and $I \neq I^*$, while it becomes zero only for $(S, I) = (S^*, I^*)$. These features indicate that the function $V(S, I)$ is a strict Lyapunov function about the endemic state E_+ for the system (9.41). From the existence of this Lyapunov function, the endemic state E_+ is shown to be globally asymptotically stable when it exists.

As mentioned for the discrete time SIRS model in Sect. 9.2.4, the temporal change to approach the endemic equilibrium state depends on the parameters. Numerical example in Fig. 9.23 indicates that a damped oscillation to approach the endemic equilibrium state is likely to occur. Such an oscillation could be regarded

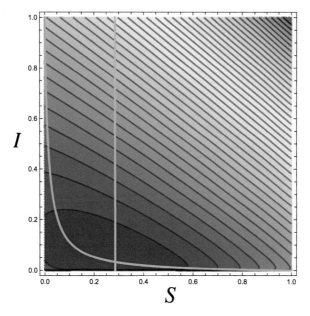

Fig. 9.22 Numerically drawn contour map of the value of Lyapunov function (9.42) in the (S, I)-phase plane about the system (9.41) with $N = 1.0$; $\sigma = 0.7$; $\rho = 0.2$; $\nu = 0.01$; $\mathcal{R}_0 = 3.5$. Nullclines for the system is shown too

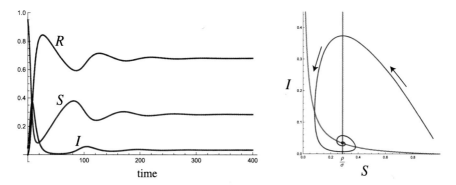

Fig. 9.23 Numerical example of the temporal change and the trajectory in the (S, I)-phase plane by the SIRS model (9.41) with $N = 1.0$; $\sigma = 0.7$; $\rho = 0.2$; $\nu = 0.01$; $\mathcal{R}_0 = 3.5$; $(S(0), I(0), R(0)) = (0.95, 0.05, 0.0)$

as a repetition of epidemic outbreaks which may cause a heavy load for the medical services and a strong fear of the disease in the community. If the eigenvalue for E_+ is imaginary, the temporal change to approach the endemic state E_+ becomes a damped oscillation. If it is real, such a damped oscillation does not occur.

By investigating the eigenvalue for E_+, we can get the following result about the temporal change to approach the endemic state E_+:

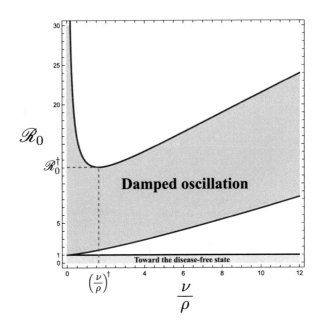

Fig. 9.24 Parameter dependence of the oscillatory behavior for the SIRS model (9.41). The region of the damped oscillation is given by the condition (9.43). For the other region with $\mathscr{R}_0 > 1$, the epidemic dynamics approaches the endemic equilibrium state without any lasting damped oscillation. $(\nu/\rho)^\dagger = (1 + \sqrt{5})/2 \approx 1.61803$; $\mathscr{R}_0^\dagger = (13 + 5\sqrt{5})/2 \approx 12.0902$

- When the endemic state E_+ exists with $\mathscr{R}_0 > 1$, the temporal change follows a damped oscillation to approach it if and only if

$$\left(\frac{\rho}{\nu} + 1\right)\left(\sqrt{1 + \frac{\nu}{\rho}} - 1\right)^2 < \mathscr{R}_0 - 1 < \left(\frac{\rho}{\nu} + 1\right)\left(\sqrt{1 + \frac{\nu}{\rho}} + 1\right)^2. \tag{9.43}$$

Otherwise, it approaches E_+ monotonically for sufficiently large t.

As seen in Fig. 9.24, a damped oscillation is likely to appear even for sufficiently small ν, which means the case where the waning of immunity hardly occurs and the expected duration of immunity is sufficiently long. This result on the occurrence of a damping oscillation is the same as the discrete time SIRS model in Sect. 9.2.4. Besides, similarly with it, we note that a damped oscillation is likely to occur for an intermediate range of the basic reproduction number $\mathscr{R}_0$. The epidemic dynamics with $\mathscr{R}_0$ near or much far from 1 does not show such a damped oscillation, but does a non-oscillatory approach to the endemic equilibrium state. Since $\mathscr{R}_0$ for the SIRS model (9.41) is proportional to the population size N, this result can be regarded as an implication that a damped oscillation is likely to occur for a population with an intermediate size.

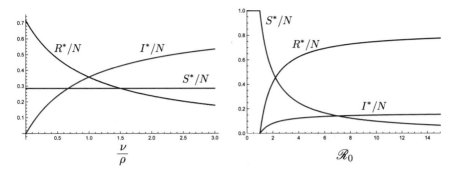

Fig. 9.25 Parameter dependence of the population proportion $(S^*/N, I^*/N, R^*/N)$ at the endemic equilibrium state for the SIRS model (9.41). Numerically drawn on the ν/ρ-dependence with $\mathscr{R}_0 = 3.5$ and the $\mathscr{R}_0$-dependence with $\nu/\rho = 0.2$

On the other hand, as shown in Fig. 9.25, the epidemic size I^* at the endemic equilibrium state is positively correlated to both values of ν/ρ and $\mathscr{R}_0$. As the duration of immunity gets shorter with smaller ν or as the infectivity gets stronger with larger $\mathscr{R}_0$, the epidemic size I^* becomes greater. The susceptible population S^* at the endemic equilibrium state is independent of the value of ν/ρ because $S^*/N = 1/\mathscr{R}_0$ for the SIRS model (9.41). Hence, it is independent of ν, that is, of how long the immunity is effective.

It must be remarked that, even at the endemic equilibrium state, the epidemic dynamics is going on with a cycle of state transition $S \to I \to R \to S$. Therefore, as mentioned for the discrete time SIS and SIRS models in Sect. 9.2.4, the endemic state gives a continuous load to the medical services. Since

$$\lim_{\nu/\rho \to \infty} \frac{I^*}{N} = 1 - \frac{1}{\mathscr{R}_0}; \quad \lim_{\mathscr{R}_0 \to \infty} \frac{I^*}{N} = \frac{\nu/\rho}{1 + \nu/\rho},$$

it is implied that the desirable medical services must have the capacity enough to provide the effective treatment even at the epidemic situation corresponding to the above limit.

9.3.6 Modeling with the Other Factors for Epidemic Dynamics

Latent Period

In general, the transmissible disease has a *latent (or incubation) period* during which the disease does not cause any clear symptom. The infected individual in such a latent period may not be distinguishable from the susceptible. The infected individual in the latent period, who appears like a susceptible, may have infectivity like one with the human immunodeficiency virus (HIV) to cause the acquired immunodeficiency syndrome (AIDS). In contrast, there are not a few transmissible diseases which have the latent period to cause little infectivity.

To introduce the contribution of a latent period to the epidemic dynamics, we need to add a class for infected individuals in the latent period, denoted here by E as a most popular symbol which is said to be after the word "encapsulation". With the introduction of class E, the class I means the subpopulation of infected individuals with infectious symptoms of the disease. Let us now consider a discrete time modeling of what is called *SEIR model* in which the state transition with respect to the disease is one-way as S → E → I → R. As a simplest assumption, we now assume the followings in addition to or replacing those for the discrete time SIR model (9.13) in Sect. 9.2.4:

(i) A susceptible gets the infection by a contact with an infected individual in the latent period with probability β.
(ii) The infected individual in the latent period cannot recover during it but necessarily develops symptoms after it;
(iii) The infected individual in the latent period develops symptoms with probability ω per day;
(iv) An infective with symptoms after the latent period is immediately isolated under a treatment, and stay away from any contact with susceptibles until the recovery to be discharged from the isolation.

The assumption (i) is more generally with respect to the susceptible's contact with the pathogen through the contact to the contaminated route of disease transmission, as described in Sect. 9.2.1. Because of the assumption (iv), the assumption (i) means that the pathogen contamination of the route of disease transmission would have a positive correlation principally with the density of infected individuals in the latent period. This is the essence of assumption (i) from the viewpoint of the modeling, while we describe here the modeling in the most frequently used context of the contact between individuals as expressed in the above assumption (i).

From the assumption (iv), since the infective with the symptom cannot contact any susceptible, the susceptible can have contacts only with susceptible, recovered or infected individuals in the latent period, which population size in total is now given by $N - I_k$ at the kth day. Hence, making use of the mean field approximation again as in Sect. 9.3.1, the probability ϕ_k that a contact of a susceptible is with an infected individual in the latent period at the kth day is now given by $\phi_k = \alpha E_k / (N - I_k)$ with a constant α such that $0 < \alpha \leq 1$.

With the same arguments in Sects. 9.2.1 and 9.2.4 applying Poisson distribution with the intensity γ for the number of contacts with the others, we can get the following discrete time SEIR model:

$$S_{k+1} = S_k e^{-\beta\alpha\gamma E_k/(N-I_k)};$$

$$E_{k+1} = S_k\left\{1 - e^{-\beta\alpha\gamma E_k/(N-I_k)}\right\} + (1-\omega)E_k;$$

$$I_{k+1} = \omega E_k + (1-q)I_k;$$

$$R_{k+1} = qI_k + R_k,$$

(9.44)

where E_k denotes the population size of infected individuals in the latent period at the kth day. With the same arguments as in Sects. 9.2.3 and 9.2.4, it is easy to show that the basic reproduction number for this SEIR model (9.44) is given by $\mathcal{R}_0 = \beta\alpha\gamma/\omega$. As seen from Fig. 9.26a, the epidemic dynamics by (9.44) has a peak for the population size of infected individuals in the latent period before a peak for that of infectives. Since it is assumed that the infected individual can be detected after developing symptoms, the number of infectives can provide the information of the disease spread necessarily with a time delay.

The corresponding continuous time SEIR model can be derived from the above (9.44), for example, by the time-step-zero limit described in Sect. 3.3. Same as the arguments at the beginning of Sect. 9.3 for Kermack-McKendrick model, we can

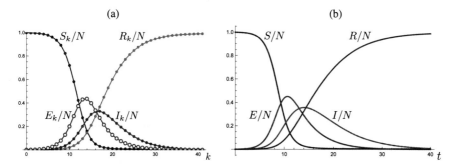

Fig. 9.26 Numerical calculation of the temporal change of variables for (**a**) the discrete time SEIR model (9.44) with $\beta\alpha\gamma = 1.0$; $\omega = 0.25$; $q = 0.25$, and (**b**) the continuous time SEIR model (9.45) with $b = 1.0$; $\kappa = 0.25$; $\rho = 0.2$. Commonly, $\mathcal{R}_0 = 4.0$; $(S_0/N, E_0/N, I_0/N, R_0/N) = (0.999, 0.001, 0.0, 0.0)$

derive the following system of ordinary differential equations as a continuous time
SEIR model corresponding to (9.44):

$$\frac{dS(t)}{dt} = -b\frac{E(t)}{N - I(t)}S(t);$$

$$\frac{dE(t)}{dt} = b\frac{E(t)}{N - I(t)}S(t) - \kappa E(t);$$

$$\frac{dI(t)}{dt} = \kappa E(t) - \rho I(t);$$

$$\frac{dR(t)}{dt} = \rho I(t),$$

(9.45)

where b is the infection coefficient, κ the rate of symptom development, and ρ
the recovery rate. The infection force Λ is now given by $\Lambda = bE/(N - I)$
which may be regarded as a ratio-dependent type (refer to Sect. 9.3.1). The basic
reproduction number $\mathscr{R}_0$ for the epidemic dynamics by (9.45) can be derived by the
same argument as in Sect. 9.3.4:

$$\mathscr{R}_0 = \sup_{(S,I)} \frac{1}{\kappa} \cdot b \frac{S}{N - I} = \frac{b}{\kappa},$$

(9.46)

since the expected duration of latent period for the SEIR model (9.45) is given by
$1/\kappa$. As seen from Fig. 9.26, the epidemic dynamics by (9.44) and (9.45) have the
qualitatively same nature.

On the other hand, the SEIR model (9.45) may be regarded as specific in a
sense because of the assumption (iv) on the perfect isolation of infectives. As a
simplest extension of Kermack-McKendrick model (9.25) with the introduction of
latent class, we may consider the following system, omitting the assumption (iv) on
the perfect isolation of infectives:

$$\frac{dS(t)}{dt} = -bE(t)S(t) - \sigma I(t)S(t);$$

$$\frac{dE(t)}{dt} = bE(t)S(t) + \sigma I(t)S(t) - \kappa E(t);$$

$$\frac{dI(t)}{dt} = \kappa E(t) - \rho I(t);$$

$$\frac{dR(t)}{dt} = \rho I(t),$$

(9.47)

where b and σ are infection coefficients about the individuals in the latent period
and with the symptom respectively. In general, the infectivity depends on the
physiological state of the infected individual, and those parameters b and σ are

different from each other. The basic reproduction number $\mathcal{R}_0$ for the epidemic dynamics by (9.47) becomes

$$\mathcal{R}_0 = \sup_S \left(\frac{1}{\kappa} \cdot bS \right) + \sup_S \left(\frac{1}{\rho} \cdot \sigma S \right) = N \left(\frac{b}{\kappa} + \frac{\sigma}{\rho} \right), \qquad (9.48)$$

since an infected individual can transmit the disease during the latent period of the expected duration $1/\kappa$ and during the period of symptoms of the expected duration $1/\rho$.

As seen by the comparison of the basic reproduction number (9.48) for (9.47) to (9.46) for (9.45), the basic reproduction number (9.48) is independent of the total population size N differently from (9.46). This is typical with respect to the difference between the epidemic dynamics with the ratio-dependent type of infection force and with the density-proportional type, that is, the infection force by the mass action assumption. As argued in Sect. 9.3.1, such a difference is caused by the nature of the route of disease transmission.

Further, as well as Kermack-McKendrick model (9.25), the SEIR model (9.47) has the following time-independent conserved quantity:

$$\frac{b}{\kappa} \{ S(t) + E(t) \} + \frac{\sigma}{\rho} \{ S(t) + E(t) + I(t) \} - \ln S(t) = C_0, \qquad (9.49)$$

where the constant C_0 is determined by the initial condition $(S(0), E(0), I(0))$ similarly as (9.27) and (9.36). For the initial condition given by $(S(0), E(0), I(0), R(0)) = (N - E_0, E_0, 0, 0)$ with $E_0 > 0$, we have $C_0 = \mathcal{R}_0 - \ln(N - E_0)$ with the basic reproduction number $\mathcal{R}_0$ defined by (9.48).

Since $E(t) \to 0$ and $I(t) \to 0$ as $t \to \infty$, we have the following equation to determine the final susceptible population size S_∞ at the end of epidemic dynamics by (9.47):

$$\mathcal{R}_0 \frac{S_\infty}{N} - \ln S_\infty = \mathcal{R}_0 - \ln(N - E_0).$$

Since $S_\infty + R_\infty = N$, we get the following equation to determine the final epidemic size R_∞ (Fig. 9.27):

$$\mathcal{R}_0 \frac{R_\infty}{N} + \ln \left(1 - \frac{R_\infty}{N} \right) = \ln \left(1 - \frac{E_0}{N} \right). \qquad (9.50)$$

From this Eq. (9.50), we can easily find the lower bound for the final epidemic size, $\inf R_\infty$, determined only by $\mathcal{R}_0$ and independent of the initial value E_0. It is given by the unique positive root less than one for the following equation of x: $\mathcal{R}_0 x/N +$

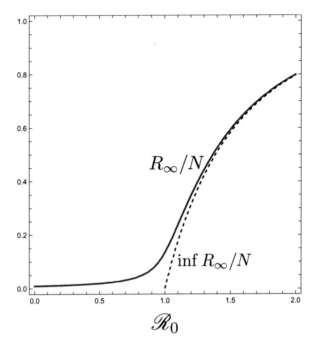

Fig. 9.27 $\mathcal{R}_0$-dependence of the final epidemic size R_∞ for the SEIR model (9.47), determined by (9.50) with $E(0)/N = 0.01$

$\ln(1 - x/N) = 0$ with $\mathcal{R}_0 > 1$. For $\mathcal{R}_0 \leq 1$, we mathematically define it as $\inf R_\infty = 0$. As seen in Fig. 9.27, the final epidemic size R_∞ tends to drastically increase for $\mathcal{R}_0 > 1$. This can be regarded as a manifestation of the importance of the threshold value 1 for the basic reproduction number $\mathcal{R}_0$ as well as the result discussed in Sect. 9.3.4 for Kermack-McKendrick model (9.25).

Vector-Borne Disease

Like malaria, dengue fever, and Zika virus infection by mosquito, or pine wilt disease by long-horned beetle, some transmissible diseases rely on specific animals as the *vector* which carries the pathogen to the host. The epidemic dynamics for such a *vector-borne disease* needs a specific modeling which contains the population dynamics about the vector.

We shall consider here a simplest modeling for such an epidemic dynamics of vector-borne disease, modifying Kermack-McKendrick model (9.25). We assume that the disease transmission occurs only by the contact between host and vector. Introducing the interaction between vector and host by the Lotka-Volterra type of

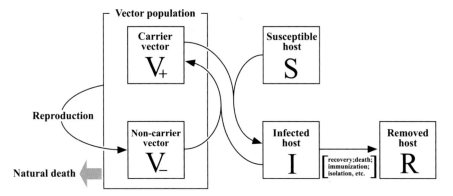

Fig. 9.28 Scheme of the epidemic dynamics by the model (9.51) for a vector-borne disease

interaction (Sect. 6.1.2), we can derive the following epidemic dynamics model for a vector-borne disease (Fig. 9.28):

$$\frac{dS(t)}{dt} = -a\beta S(t)V_+(t);$$

$$\frac{dI(t)}{dt} = a\beta S(t)V_+(t) - \rho I(t);$$

$$\frac{dR(t)}{dt} = \rho I(t); \tag{9.51}$$

$$\frac{dV_+(t)}{dt} = b\beta I(t)V_-(t) - \mu V_+(t);$$

$$\frac{dV_-(t)}{dt} = -b\beta I(t)V_-(t) - \mu V_-(t) + g(V(t)),$$

where $V_+(t)$ and $V_-(t)$ are the population sizes of vector carrying the pathogen and non-carrier vector respectively. The total population size of vector is denoted as $V(t) = V_+(t) + V_-(t)$. The function $g(V)$ indicates the net reproduction rate for the vector population, which is now assumed to depend only on the total population size of vector V. This means that we assume that the per capita growth rate of vector does not depend on whether the vector individual is the carrier or non-carrier of the pathogen. The coeffcients $a\beta$ and $b\beta$ are respectively for the infection of susceptible host from the carrier vector and for the transition of non-carrier vector to carrier by the pathogen transmission through the contact with the infected host. Parameter μ is the natural death rate of vector. Although the state transition of the host with respect to the disease is the type of SIR, the disease transmission is only via the contact with the vector.

Now, from (9.51), we can derive the following equation to govern the temporal change of the total population size of vector, $V(t)$:

$$\frac{dV(t)}{dt} = g(V(t)) - \mu V(t). \tag{9.52}$$

Since the population dynamics by (9.52) is independent of the epidemic dynamics, we shall consider the epidemic dynamics in the situation that the vector population reaches (or can be approximated to reach) the equilibrium state. This means that we shall suppose that the total population size of vector V is a constant V^* for any time t, which satisfies the equality $g(V^*) = \mu V^*$.

With this assumption about the vector population size, the system (9.51) can be reduced to the mathematically equivalent one with the lower dimension as follows:

$$\frac{dS(t)}{dt} = -a\beta S(t)V_+(t);$$

$$\frac{dI(t)}{dt} = a\beta S(t)V_+(t) - \rho I(t); \tag{9.53}$$

$$\frac{dV_+(t)}{dt} = b\beta I(t)\{V^* - V_+(t)\} - \mu V_+(t).$$

As seen in Fig. 9.29, the infected host necessarily decreases toward zero as time passes, since the total population size of host N is constant independently of time and the state transition of host is one-way as S $\to$ I $\to$ R. Hence, the carrier vector necessarily tends to disappear as well.

We assumed that $V(t) \to V^*$ as $t \to \infty$ for the vector population dynamics (9.52). Since the population dynamics (9.52) is now assumed to be independent of the epidemic dynamics, so is the convergence of V to V^*. However, in general, as long as considering the epidemic dynamics (9.51), the vector population size temporally varies with a relation to the densities V_+ and V_-. Therefore, the temporal changes of variables by the epidemic dynamics (9.51) and (9.53) could be generally different more or less from each other.

As mentioned in the above, since we assume that the total vector population size approaches an equilibrium as time passes, the epidemic dynamics (9.51) is expected to come to have the qualitatively same nature as (9.53) as times passes, whereas it requires a mathematical proof [15, 18]. The system (9.53) may be called the *limiting system* for (9.51), and the system (9.51) may be called *asymptotically autonomous* since the limiting system (9.53) is autonomous while the system (9.51) can be regarded as *non-autonomous* because it includes a function $V = V(t)$ of time.

(continued)

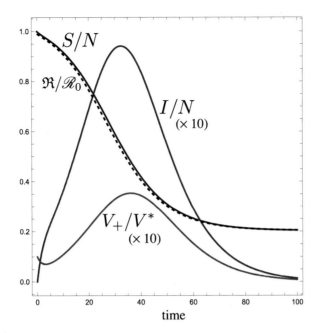

Fig. 9.29 Numerical calculations of the temporal change of variables for the epidemic dynamics model of vector-borne disease (9.53) with $a\beta V^* = 1.0$; $b\beta N = 0.1$; $\rho = 0.2$; $\mu = 0.25$; $\mathscr{R}_0 = 2.0$; $(S(0)/N, I(0)/N, V_+(0)/V^*) = (1.0, 0.0, 0.01)$. Temporal change of the effective reproduction number $\mathfrak{R}(t)$ defined by (9.54) is shown by the broken curve

> From the standpoint of modeling, it could be reasonable to consider the system (9.51) with the assumption $V(t) = V^*$, because it can be regarded as the epidemic dynamics model in the situation that the vector population has reached the equilibrium state.

To consider the reproduction number about the disease spread, we need to take account of the route of pathogen transmission. Let us focus on an infected host. The expected duration of host's "infectivity" for the vector is given by $1/\rho$. Now we can consider the effective reproduction number $\mathfrak{R}(t)$ for the host population at time t (refer to Sect. 9.2.3). An infected host is expected to produce $b\beta V_-(t)$ of carrier vectors per unit time. The expected life span of vector is given by $1/\mu$ (refer to Sects. 4.3.2 and 4.3.5). On the other hand, a carrier vector is expected to produce $a\beta S(t)$ of infected hosts per unit time. Therefore, we can define the effective reproduction number as

$$\mathfrak{R}(t) := \underbrace{\frac{1}{\mu} \cdot a\beta S(t)}_{\substack{\text{production of infected} \\ \text{hosts per carrier vector}}} \times \underbrace{\frac{1}{\rho} \cdot b\beta V_-(t)}_{\substack{\text{production of carrier} \\ \text{vectors per infected host}}} . \qquad (9.54)$$

Hence the basic reproduction number $\mathscr{R}_0$ can be defined as

$$\mathscr{R}_0 = \sup_{(S,V_-)} \mathfrak{R}(t) = \frac{1}{\mu} \cdot a\beta N \times \frac{1}{\rho} \cdot b\beta V^*. \qquad (9.55)$$

The latter factor $b\beta V^*/\rho$ means the supremum for the expected number of non-carrier vectors which become carrier by an infected host until the recovery, while the former factor $a\beta N/\rho$ means the supremum for the expected number of susceptible hosts infected by a carrier vector during its survival. Therefore, their product means the supremum for the expected number of infected hosts produced by an infected host until the recovery.

There are different mathematical ways to derive the basic reproduction number $\mathscr{R}_0$ for an epidemic dynamics model. One is to use the condition for the local stability of the disease-free equilibrium state. It would seem to match the necessarily satisfied nature of disease-free equilibrium state argued in Sect. 9.2.3. That is, if $\mathscr{R}_0 < 1$, the disease fails to invade in the population and becomes eliminated, while if $\mathscr{R}_0 > 1$, the disease succeeds in invading in the population and causes an increase of infectives and an outbreak. Hence, the disease-free equilibrium state is unstable if $\mathscr{R}_0 > 1$. However, this could not the case for some epidemic dynamics models as mentioned at the end of this section.

The other way well-known today uses what is called *next generation matrix*. We shall not touch its mathematical theory itself here but describe only the mathematical process to derive $\mathscr{R}_0$ given by (9.55) with the next generation matrix for the model (9.51). Readers interested in the mathematical theory can easily find the description in not a few literatures and textbooks on mathematical epidemiology (for example, see [3, 6, 7, 10, 15, 19, 22] or the original articles by van den Driessche and Watmough [20, 21]).

(continued)

Firstly, from the system (9.51), we pick out only equations to describe the route of disease transmission, that is, those especially related to the recruitment of infected hosts:

$$\frac{dI(t)}{dt} = a\beta S(t)V_+(t) - \rho I(t);$$
$$\frac{dV_+(t)}{dt} = b\beta I(t)V_-(t) - \mu V_+(t). \qquad (9.56)$$

Now we decompose the dynamical terms into two sets in which one shows the recruitment of infected hosts, and the other does the supplementary processes:

$$\frac{d\boldsymbol{x}}{dt} = \mathcal{F}(S, V_+) - \mathcal{V}(I, V_+, V_-),$$

where $\boldsymbol{x} := {}^\mathsf{T}\!\big(I(t), V_+(t)\big)$;

$$\mathcal{F}(S, V_+) := \begin{pmatrix} a\beta SV_+ \\ 0 \end{pmatrix}; \quad -\mathcal{V}(I, V_+, V_-) := \begin{pmatrix} -\rho I \\ b\beta IV_- - \mu V_+ \end{pmatrix}. \qquad (9.57)$$

The vector $\mathcal{F}$ is for the terms of the recruitment of infected hosts, while $-\mathcal{V}$ is for the other. Jacobian matrices of $\mathcal{F}$ and $\mathcal{V}$ about the disease-free equilibrium state $(S, I, R, V_+, V_-) = (N, 0, 0, 0, V^*)$ are given by

$$F := D\mathcal{F}\big|_{\mathrm{DFE}} = \begin{pmatrix} 0 & a\beta N \\ 0 & 0 \end{pmatrix}; \quad V := D\mathcal{V}\big|_{\mathrm{DFE}} = \begin{pmatrix} \rho & 0 \\ -b\beta V^* & \mu \end{pmatrix}.$$

The *next generation matrix* $\mathcal{K}$ is defined by FV^{-1}, that is,

$$\mathcal{K} = FV^{-1} = \begin{pmatrix} \dfrac{a\beta N \cdot b\beta V^*}{\rho\mu} & \dfrac{a\beta N}{\mu} \\ 0 & 0 \end{pmatrix}. \qquad (9.58)$$

The theory says that the maximum absolute value of the eigenvalue for $\mathcal{K}$ gives the basic reproduction number $\mathcal{R}_0$, which is defined as the *spectral radius* of $\mathcal{K}$ in mathematics. Therefore, from (9.58), we can get the basic reproduction number $\mathcal{R}_0$ given by (9.55). It must be remarked that, as mentioned in [7, Chap. 9] and [15, Chap. 5] (and references therein), there could be the other way of the decomposition different from (9.57), which

(continued)

results in a different formula of $\mathcal{R}_0$, typically corresponding to the square root of (9.55) (Exercise 9.2).

Exercise 9.2 Drive the basic reproduction number $\mathcal{R}_0$ for the epidemic dynamics model (9.51), making use of the next generation matrix with the following decomposition instead of (9.57) for (9.56):

$$\mathscr{F}(S, I, V_+, V_-) := \begin{pmatrix} a\beta S V_+ \\ b\beta I V_- \end{pmatrix}; \quad -\mathscr{V}(I, V_+) := \begin{pmatrix} -\rho I \\ -\mu V_+ \end{pmatrix}.$$

Long-Term Epidemic Dynamics

We have considered some models for the epidemic dynamics model until now under the assumption that the host population size is a constant independent of time. This assumption means that the epidemic dynamics is short-term enough to make the demographic change in the host population negligible with respect to the spread of disease in the population. Alternatively the model may be regarded as fundamental in order to theoretically consider the epidemic dynamics in which the population size of host is temporally varying, while the researches on such fundamental models have been useful to get valuable insights for a variety of problems in population dynamics as proved by the history in mathematical sciences.

We shall introduce here the demographic change in the host population under an epidemic dynamics. Such an epidemic dynamics must have a time scale comparable to that for the temporal change of the population size of host. For this reason, such a time scale must be relatively large, and the epidemic dynamics is sufficiently long-term.

Let us consider the following model for a vector-borne epidemic dynamics with the demographic change in the host population, modified from (9.51):

$$
\begin{aligned}
\frac{dS(t)}{dt} &= -a\beta S(t)V_+(t) - \delta S(t) + \Lambda(S(t), I(t), R(t)); \\[4pt]
\frac{dI(t)}{dt} &= a\beta S(t)V_+(t) - \delta I(t) - \rho I(t); \\[4pt]
\frac{dR(t)}{dt} &= \rho I(t) - \delta R(t); \\[4pt]
\frac{dV_+(t)}{dt} &= b\beta I(t)V_-(t) - \mu V_+(t); \\[4pt]
\frac{dV_-(t)}{dt} &= -b\beta I(t)V_-(t) - \mu V_-(t) + g(V(t)),
\end{aligned}
\qquad (9.59)
$$

where Λ and δ are the net recruitment rate and per capita death rate respectively for the host population. We assume now that the disease has no effect on the survival, that is, is non-fatal. From the former three equations about the host population dynamics, we have

$$\frac{dN(t)}{dt} = \Lambda(S(t), I(t), R(t)) - \delta N(t) \tag{9.60}$$

about the temporal change of the total population size $N(t) := S(t) + I(t) + R(t)$. Since we assume that the disease is non-fatal, the population dynamics with respect to the total population size of host can be assumed to have no or negligible relation to the disease. Hence, with (9.60), let us consider the situation that the total population of host has reached the equilibrium. In other words, the total population size of host is assumed to be at the equilibrium state with the value N^*: $N(t) = N^*$ for any time t. This makes it hold that $\Lambda(S(t), I(t), R(t)) - \delta N^* = 0$ for any time t, because the population dynamics with respect to the total population size of host is now assumed to have no relation to the epidemic dynamics.

Some readers may think that this assumption about the host population dynamics would be related to the quasi-stationary state approximation (QSSA) described in Sect. 6.2. However, it is not the case. As mentioned in the above, we are now considering a long-term epidemic dynamics for a non-fatal disease, which is the reason why we take account of the demographic change in the host population dynamics. Actually, the demographic change in the host population must have a certain influence on the epidemic dynamics, since it could provide a recruitment or reduction of susceptible hosts in the population. The assumption introduced above about the host population dynamics means that we are going to focus on the epidemic dynamics for the host population which has reached the equilibrium state according to the total population size, and we shall not consider the epidemic dynamics through the transient state of the host population dynamics about the total size. Thus, we do not assume any difference about the time scale between the population dynamics of host and vector. Therefore, we do not apply the QSSA for the population dynamics considered here.

Further, let us assume that the total population size of vector V is a constant V^* for any time t, so that we have $g(V) \equiv g(V^*) = \mu V^*$ as before. From the original model (9.59) with these assumptions, we can derive the following epidemic dynamics model with respect to the population frequencies defined as $f_S := S/N^*$, $f_I := I/N^*$, $f_R := R/N^*$, $v_+ := V_+/V^*$, and $v_- := V_-/V^*$:

$$\frac{df_S(t)}{dt} = -a\beta V^* f_S(t)v_+(t) - \delta f_S(t) + \delta;$$

$$\frac{df_I(t)}{dt} = a\beta V^* f_S(t)v_+(t) - \rho f_I(t) - \delta f_I(t);$$

$$\frac{df_R(t)}{dt} = \rho f_I(t) - \delta f_R(t);$$

$$\frac{dv_+(t)}{dt} = b\beta N^* f_I(t)v_-(t) - \mu v_+(t);$$

$$\frac{dv_-(t)}{dt} = -b\beta N^* f_I(t)v_-(t) - \mu v_-(t) + \mu,$$

and the mathematically equivalent closed system with the reduced dimension:

$$\frac{df_S(t)}{dt} = -a\beta V^* f_S(t)v_+(t) - \delta f_S(t) + \delta;$$

$$\frac{df_I(t)}{dt} = a\beta V^* f_S(t)v_+(t) - (\rho + \delta) f_I(t); \qquad (9.61)$$

$$\frac{dv_+(t)}{dt} = b\beta N^* f_I(t)\{1 - v_+(t)\} - \mu v_+(t);$$

since $f_S(t) + f_I(t) + f_R(t) = 1$ and $v_+(t) + v_-(t) = 1$ for any time t. As easily seen, this model (9.61) becomes equivalent to the previous one (9.53) when $\delta = 0$.

> The system (9.61) may be regarded as the limiting system for (9.59), since we assumed that $N(t) \to N^*$ and $V(t) \to V^*$ as $t \to \infty$ for (9.59). However, the arguments to lead to (9.61) were from the standpoint of modeling about the epidemic dynamics.

The basic reproduction number $\mathcal{R}_0$ for the epidemic dynamics by (9.59), that is, by (9.61) becomes

$$\mathcal{R}_0 = \frac{1}{\mu} \cdot a\beta N^* \times \frac{1}{\rho + \delta} \cdot b\beta V^*,$$

which can be derived by the same arguments as for (9.55). As for the existence of endemic equilibrium state and the local stability of equilibria for the system (9.61), we can get the following results:

- The disease-free equilibrium state (DFE) $(f_S, f_I, v_+) = (1, 0, 0)$ is locally asymptotically stable if $\mathcal{R}_0 < 1$, while it is unstable if $\mathcal{R}_0 > 1$.

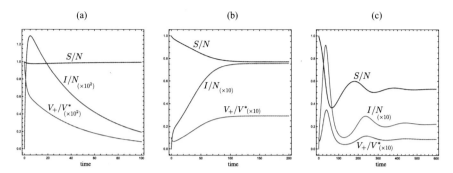

Fig. 9.30 Numerical calculations of the temporal change of variables for the epidemic dynamics model of vector-borne disease (9.61) with $a\beta V^* = 1.0$; $b\beta N^* = 0.1$; $\rho = 0.2$; $\mu = 0.25$; $(S(0)/N^*, I(0)/N^*, V_+(0)/V^*) = (1.0, 0.0, 0.01)$; (**a**) $\delta = 0.25$ ($\mathscr{R}_0 = 0.888$); (**b**) $\delta = 0.1$ ($\mathscr{R}_0 = 1.333$); (**c**) $\delta = 0.01$ ($\mathscr{R}_0 = 1.905$)

- The endemic equilibrium state $(f_S, f_I, v_+) = (f_S^*, f_I^*, v_+^*)$ uniquely exists if and only if $\mathscr{R}_0 > 1$, where

$$f_S^* = \frac{\delta/(a\beta N^*)}{\delta/(a\beta N^*) + v_+^*}; \quad f_I^* = \frac{\mu}{b\beta V^*}\frac{v_+^*}{1 - v_+^*}; \quad v_+^* = \frac{\delta/(a\beta N^*)}{\delta/(a\beta N^*) + 1/\mathscr{R}_0}\left(1 - \frac{1}{\mathscr{R}_0}\right),$$

  satisfying that $f_S^* + f_I^* = 1 - f_R^* < 1$ and $v_+^* < 1$.
- The endemic equilibrium state is locally asymptotically stable whenever it exists.

We can investigate the local stability of the DFE by the standard local stability analysis with the linearized system (refer to Sect. 14.2), whereas we can prove that of the endemic equilibrium state by making use of the *Routh-Hurwitz Criterion* with a little cumbersome calculation (refer to Sect. 14.6).

The endemic equilibrium state is globally asymptotically stable when it exists, though we shall not go into the mathematical proof here (see [23]). As numerically demonstrated in Fig. 9.30, the temporal change may be accompanied by a damped oscillation toward the endemic equilibrium state.

Sir Ronald Ross (1857–1932) presented the following epidemic dynamics model for the Malaria spread [16, 17]:

$$\begin{cases} \dfrac{df_I(t)}{dt} = a\beta V^*\{1 - f_I(t)\}v_+(t) - \delta f_I(t); \\[2mm] \dfrac{dv_+(t)}{dt} = b\beta N^* f_I(t)\{1 - v_+(t)\} - \mu v_+(t), \end{cases} \tag{9.62}$$

(continued)

which was studied further by George U. Macdonald (1925–1997) about the modification and estimation of parameter values [11–14]. This model is sometimes called *Ross-Macdonald model* today.

The model (9.62) can be now regarded as a simpler version of (9.61) with $\rho = 0$ and $f_S(t) + f_I(t) = 1$. Hence there are not the recovery or the removed class. Then we must assume that the infected host keeps its "infectivity" for the vector until its death. As seen from the similarity of two formulas in (9.62), this model can be regarded as the dynamics of pathogen transshipment between two different populations. It may be the simplest reasonable model for a vector-born "disease" with negligible fatality.

Since the model (9.62) is two dimensional, for example, the analysis with the isocline method (Sect. 14.7) can easily show that the DFE is globally asymptotically stable if $\mathcal{R}_0 := (a\beta N^*/\mu)(b\beta V^*/\delta) \leq 1$, while the endemic equilibrium state is globally asymptotically stable if $\mathcal{R}_0 > 1$. Further, it can be shown also that there is no oscillatory behavior in the temporal change by the model (9.62), differently from (9.61).

Answer to Exercise

Exercise 9.1 (p. 294)

Let us define $F(x) := x - (\rho/\sigma) \ln x$ for $x > 0$. Then the Eq. (9.28) is expressed as $F(S_\infty) = F(S_0) + I_0$. The curve $y = F(x)$ is concave with the extremal minimum at $x = \rho/\sigma$, satisfying that $\lim_{x \to 0+} F(x) = \infty$ and $\lim_{x \to \infty} F(x) = \infty$, as illustrated in the below figures:

(a)

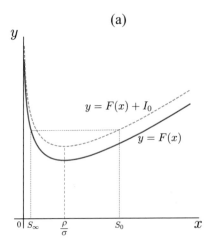

(b)

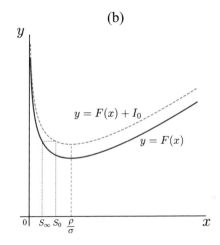

These are in the case of $F(\rho/\sigma) > 0$. The curve of $y = F(x) + I_0$ is equivalent to the parallel displacement of $y = F(x)$ by I_0 in the direction of y-axis.

Since $S(t)$ is monotonically decreasing as time passes, we must have $S_\infty < S_0$. From the above left graph (a) when $S_0 > \rho/\sigma$, the root S_∞ of the Eq. (9.28) must be determined as indicated in it. In the similar way, when $S_0 < \rho/\sigma$, it must be determined as indicated in (b). Hence it is clear that the root S_∞ of the Eq. (9.28) is uniquely determined.

Although these arguments are for the case of $F(\rho/\sigma) > 0$, it can be applied for the case of $F(\rho/\sigma) \leq 0$ too. Lastly these arguments show the unique existence of the root S_∞ of the Eq. (9.28).

Exercise 9.2 (p. 316)

For the decomposition given in this exercise, we have the following Jacobian matrices of $\mathcal{F}$ and $\mathcal{V}$ about the disease-free equilibrium state $(S, I, R, V_+, V_-) = (N, 0, 0, 0, V^*)$:

$$\mathcal{F} := D\mathcal{F}\big|_{\text{DFE}} = \begin{pmatrix} 0 & \alpha\beta N \\ b\beta V^* & 0 \end{pmatrix}; \quad \mathcal{V} := D\mathcal{V}\big|_{\text{DFE}} = \begin{pmatrix} \rho & 0 \\ 0 & \mu \end{pmatrix}.$$

Then the next generation matrix $\mathcal{K} = \mathcal{F}V^{-1}$ becomes

$$\mathcal{K} = \mathcal{F}V^{-1} = \begin{pmatrix} 0 & \dfrac{\alpha\beta N}{\mu} \\ \dfrac{b\beta V^*}{\rho} & 0 \end{pmatrix}.$$

The eigenvalues for $\mathcal{K}$ are purely imaginary $\pm i\sqrt{(\alpha\beta N/\mu)(b\beta V^*/\rho)}$, and therefore the basic reproduction number $\mathcal{R}_0$ is obtained as $\mathcal{R}_0 = \sqrt{(\alpha\beta N/\mu)(b\beta V^*/\rho)}$. This is the square root of that given by (9.55).

> Although those formulas for the basic reproduction number $\mathcal{R}_0$ are different, the condition for $\mathcal{R}_0 \gtrless 1$ is mathematically identical for both of them. However, for matching the definition of the basic reproduction number as the supremum of the expected number of new cases in Sect. 9.2.3, the above formula with the square root would not be reasonable as long as we consider the expected number of new infectives in the host population.

(continued)

In some literatures (for example, see [15]), the above basic reproduction number is explained as the geometric mean of the expected number of secondary infections by the host and that by the vector, since the initial infective individual could be a host or a vector. However, this explanation clearly show that it is not the expected number of new infective hosts regarded as those produced by an infected host, or that of new infective vectors as those produced by an infected vector. In general, for a vector-borne disease, we consider the vector as a carrier of pathogen for the host population. Therefore it is the main subject to be investigated how many new infective hosts are produced in the host population. In this sense, the above result with the square root would not be reasonable.

References

1. R.M. Anderson, R.M. May, Population biology of infectious diseases: I and II. Nature **280**, 351–367, 455–461 (1979)
2. R.M. Anderson, R.M. May, *Infectious Diseases of Humans: Dynamics and Control* (Oxford University Press, Oxford, 1991)
3. F. Brauer, C. Castillo-Chávez, *Mathematical Models in Population Biology and Epidemiology.* Texts in Applied Mathematics, vol. 40, 2nd edn. (Springer, New York, 2012)
4. P.L. Delamater et al., Complexity of the basic reproduction number (R). Emerg. Infect. Dis. **25**(1), 1–4 (2019)
5. O. Diekmann, J.A.P. Heesterbeek, *Mathematical Epidemiology of Infectious Diseases: Model Building, Analysis and Interpretation.* Wiley Series in Mathematical and Computational Biology (John Wiley & Son, Chichester, 2000)
6. O. Diekmann, H. Heesterbeek, T. Britton, *Mathematical Tools for Understanding Infectious Disease Dynamics.* Princeton Series in Theoretical and Computational Biology (Princeton University Press, Princeton, 2013)
7. H. Inaba, *Age-Structured Population Dynamics in Demography and Epidemiology* (Springer, New York, 2017)
8. M.J. Keeling, P. Rohani, *Modeling INfectious Diseases in Humans and Animals* (Princeton University Press, Princeton, 2008)
9. W.O. Kermack, A.G. McKendrick, A contribution to the mathematical theory of epidemics. Proc. R. Soc. A **115**, 700–721 (1927)
10. M.A. Lewis, Z. Shuai, P. van den Driessche, A general theory for target reproduction numbers with applications to ecology and epidemiology. J. Math. Biol. **78**, 2317–2339 (2019)
11. G. Macdonald, The analysis of infection rates in diseases in which superinfection occurs. Trop. Dis. Bull. **47**(10), 907–915 (1950)
12. G. Macdonald, The measurement of malaria transmission. Proc. R. Soc. Med. **48**, 295–301 (1955)
13. G. Macdonald, Epidemiological basis of malaria control. Bull. W.H.O. **15**, 613–626 (1956)
14. G. Macdonald, *The Epidemiology and Control of Malaria* (Oxford University Press, London, 1957)
15. M. Martcheva, *An Introduction to Mathematical Epidemiology.* Texts in Applied Mathematics (Springer, Boston, 2015)

16. R. Ross, An application of the theory of probabilities to the study of a priori pathometry—Part I. Proc. R. Soc. Lond. Ser A **92**, 204–230 (1916)
17. R. Ross, *Memoirs, With a Full Account of the Great Malaria Problem and Its Solution* (John Murray, London, 1923)
18. H.R. Thieme, *Mathematics in Population Biology* (Princeton University Press, Princeton, 2003)
19. P. van den Driessche, Reproduction numbers of infectious disease models. Infect. Dis. Model. **2**, 288–303 (2017)
20. P. van den Driessche, J. Watmough, Reproduction numbers and subthreshold endemic equilibria for compartmental models of disease transmission. Math. Biosci. **180**, 29–48 (2002)
21. P. van den Driessche, J. Watmough, Further notes on the basic reproduction number, in *Mathematical Epidemiology Lecture Notes in Mathematics*, vol. 1945, ed. by F. Brauer, P. van den Driessche, J. Wu. Berlin (Springer, Berlin, 2008), pp. 159–178
22. E. Vynnycky, R.G. White, *An Introduction to Infectious Disease Modelling* (Oxford University Press, Oxford, 2010)
23. H. Yang, H. Wei, Li X, Global stability of an epidemic model for vector-borne disease. J. Syst. Sci. Complex **23**, 279–292 (2010)

Chapter 10
Modeling for Age Structure

Abstract As described in Sect. 9.1, the biological population can have the structure according to a measure of the "age". In biology, the age distribution may be treated as a set of age classes. In such a case, the age distribution follows a discrete variable to index each age class. In the widest sense, the age can be regarded as the time lapse after the birth. Then the age is expressed by a continuous variable with real number. In this chapter, we shall describe the fundamentals on the modeling for the population dynamics with an age structure, discrete or continuous, and give the essential introduction about some biological concepts related to such theoretical approach with the mathematical modeling for the age structured population dynamics.

The biological population can have the structure according to a measure of the "age" as described in Sect. 9.1. In biology, the age distribution may be treated as a set of age classes. In such a case, the age distribution follows a discrete variable to index each age class. In a widest sense, the age can be regarded as the time lapse after the birth. Then the age is expressed by a continuous variable with real number.

In this chapter, we shall describe the fundamentals on the modeling for the population dynamics with an age structure, discrete or continuous, and at the same time give the essential introduction of some biological concepts related to such a theoretical approach with the mathematical modeling for the age structured population dynamics.

10.1 Discrete Time Model

In this section, we shall consider the fundamental mathematical modeling for the discrete time population dynamics with a number of age classes. The mathematical model can be expressed with a matrix. Let the number of age classes be m, and the subpopulation size at age class i be $n_{i,k}$ at time step k. As introduced in Chap. 1, the time step may correspond to the generation or season to define a cycle of reproduction. The most popular time step would be the year for the biological population dynamics. Here let us call it "generation" in a general sense, since we are going to consider the subpopulation sizes of all age classes, $\{n_{i,k} \mid i = 1, 2, \ldots, m\}$. The last age class m can be regarded as corresponding to the ecological life span (may correspond to the physiological life span under a specific environment controlled in laboratory) (refer to Sect. 1.5).

We have the following expression of the *age distribution* with the m dimensional vector n_k:

$$n_k \equiv \begin{pmatrix} n_{1,k} \\ n_{2,k} \\ \vdots \\ n_{m,k} \end{pmatrix}.$$

The generational change of age distribution can be mathematically expressed as

$$n_{k+1} = A_k n_k \tag{10.1}$$

with an $m \times m$ matrix A_k, which elements may depend on the age distribution itself as the general assumption: $A_k = A_k(n_k)$. The matrix A_k is called *transition matrix* or *projection matrix* for the discrete age structured population dynamics. The discrete time model with such a transition matrix is called *matrix model* in general.

> More generally, the transition matrix A_k may depend also on the age distribution older than the last generation: $A_k = A_k(n_k, n_{k-1}, n_{k-2}, \ldots)$. Since the transition matrix A_k does not necessarily contain only constant elements, the matrix model may have a nonlinearity in the mathematical expression. However, in some cases, the matrix model would indicate a population dynamics with a matrix with only constant elements, which leads to a linear model as appeared in Sect. 1.4.

10.1.1   Leslie Matrix Model

Let us consider the case where the "age" must increase by one when time proceeds by a generation, that is, by a time step. This assumption of aging is the most popular

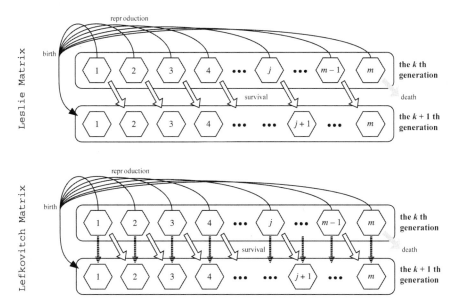

Fig. 10.1 The age/state transition for Leslie matrix model and for Lefkovitch matrix model

as the discrete time "age", like the chronological age. In such a case, the transition matrix A_k must have the following structure (see Fig. 10.1):

$$A_k \equiv \begin{pmatrix} b_1 & b_2 & b_3 & b_4 & b_5 & \cdots & b_m \\ a_1 & 0 & 0 & 0 & 0 & \cdots & 0 \\ 0 & a_2 & 0 & 0 & 0 & \cdots & 0 \\ \vdots & \ddots & \ddots & \ddots & & \vdots & \vdots \\ 0 & \cdots & 0 & a_j & 0 & \cdots & 0 \\ \vdots & & \vdots & \ddots & \ddots & \ddots & \vdots \\ 0 & 0 & 0 & \cdots & 0 & a_{m-1} & 0 \end{pmatrix}, \tag{10.2}$$

where zero are elements except for $(1, j)$-elements and $(j, j-1)$-elements ($j = 1, 2, \ldots, m$). This transition matrix is called *Leslie matrix* after the works on the age structured population dynamics with the transition matrix (10.2) by Patrick H. Leslie (1900–1972) in 1945 and 1948 [10, 11].

For a closed population dynamics (refer to Sect. 3.4) with Leslie matrix (10.2), the $(j+1, j)$-element a_j ($j = 1, 2, \ldots, m-1$) means the proportion of individuals which are of age class j and survive until the next generation to become of age class $j + 1$. In other words, $1 - a_j$ means the proportion of individuals which are alive at age class j and die before the next generation. Hence it must satisfy that $0 \le a_j \le 1$ ($j = 1, 2, \ldots, m-1$). From a viewpoint of modeling with a stochastic process, a_j

means the survival probability that an individual of age class j survives to reach the next generation, and $1 - a_j$ correspondingly does the death probability.

The element b_i ($i = 1, 2, \ldots, m$) indicates the recruitment with the reproduction by individuals of age class i. More precisely, b_i means the per capita growth rate (number of newborns, seeds, etc.) for the individual of age class i. Hence it is non-negative. If the reproduction is possible only after a age class J ($< m$) because of an immature period, we have $b_1 = b_2 = \cdots = b_J = 0$ in the Leslie matrix (10.2).

10.1.2 Lefkovitch Matrix Model

Differently from the previous case of "age" increasing generation by generation, we may consider the case where the transition to the next "age" class does not necessarily occur. Such "age" may be better to be called "state" more generally, like the stages of seed, immature, and mature (seedling) in the plant population. Such a structured population dynamics may be called *stage structured population* as already mentioned in Sect. 9.1. The epidemic dynamics with a structure with the state classes, susceptible, latent, infective (infectious), and recovered (immune) could be analogously regarded as such a stage structured population dynamics, which was discussed in Sects. 9.2 and 9.3.

Since some individuals may stay at the same age class even after the generation passes (see Fig. 10.1), the transition matrix becomes

$$A_k(\boldsymbol{n}_k) \equiv \begin{pmatrix} b_1 & b_2 & b_3 & b_4 & b_5 & b_6 & \cdots & b_m \\ a_1 & c_2 & 0 & 0 & 0 & 0 & \cdots & 0 \\ 0 & a_2 & c_3 & 0 & 0 & 0 & \cdots & 0 \\ \vdots & \ddots & \ddots & \ddots & \ddots & & \vdots & \vdots \\ 0 & \cdots & 0 & a_j & c_{j+1} & 0 & \cdots & 0 \\ \vdots & & \vdots & 0 & \ddots & \ddots & \ddots & \vdots \\ \vdots & & \vdots & \vdots & \ddots & \ddots & \ddots & 0 \\ 0 & \cdots & 0 & 0 & \cdots & 0 & a_{m-1} & c_m \end{pmatrix}. \tag{10.3}$$

This is sometimes called *Lefkovitch matrix*, after the works on the stage structured population dynamics by Leonard Lefkovitch (1929–2010) (especially after [9]).

For the stage structured dynamics of a closed population with (10.3), the principal diagonal element c_j ($j = 2, 3, \ldots, m$) means the proportion of individuals which remain alive at the same age class j until the next generation. Hence it must satisfy that $0 \le c_j \le 1$ ($j = 2, 3, \ldots, m$). Therefore $1 - c_j$ means the proportion of individuals which go out of age class j before the next generation, that is, those which alternatively die or survive to transfer to age class $j + 1$ until the next generation. It must be remarked that $1 - c_j$ does not necessarily mean the death probability. Further, from the meaning of the $(j + 1, j)$-element

a_j ($j = 1, 2, \ldots, m - 1$) as described in the previous section for Leslie matrix (10.2), it is necessary to satisfy that $0 \leq a_j + c_j \leq 1$ ($j = 2, 3, \ldots, m - 1$). Thus $1 - a_j - c_j$ means the proportion of individuals which are alive at age class j and die before the next generation.

For an *open* population with the migration of individuals to and from it, it is generally necessary to take account of the immigration to and the emigration from each age class. Hence, from the argument of Sect. 3.4 in a general context, the generational change of the subpopulation size of age class j becomes as follows:

$\big($subpopulation of age class j at the $k + 1$th generation$\big) =$

$\quad + \big($individuals remaining at age class j from the kth generation$\big)$

$\quad + \big($individuals transferring from age class $j - 1$ at the kth generation$\big)$

$\quad + ($migrants from and to age class j between the k th and

$\qquad k + 1$th generations$)$.

For the Lefkovitch matrix model with (10.3), the recurrence relation to give the subpopulation size $n_{j,k+1}$ of age class j at the kth generation generally becomes

$$n_{j,k+1} = a_{j-1}n_{j-1,k} + c_j n_{j,k} + \epsilon_{j,k}$$

with the number of immigrants $\epsilon_{j,k}$ (≥ 0) to age class j between the kth and $k + 1$th generations, so that the recurrence relation to govern the generational change of age distribution is described as $n_{k+1} = A_k n_k + \epsilon_k$ with $\epsilon_k :=$ $^T(\epsilon_{1,k}, \epsilon_{2,k}, \ldots, \epsilon_{m,k})$.

It must be remarked for this modeling that the emigration is included in the term $c_j n_{j,k}$. The number of emigrants cannot become beyond the subpopulation size of corresponding age class, while the number of immigrants may be independent of it. Hence, in this modeling, the parameter c_j means the proportion of individuals which do not emigrate from and remains alive at the same age class j until the next generation. Therefore $1 - c_j$ means the proportion of individuals which go out of age class j before the next generation, which consist of not only those which die or survive to transfer to age class $j + 1$ but also those which emigrate from the population until the next generation.

In a wider sense, the Lefkovitch matrix model may be defined with the transition matrix which has some positive values in the principal diagonal elements except for $(1, 1)$-element. The transition matrix (10.3) is the simplest one extended from the Leslie matrix model. Readers interested in the matrix model for the age structured population dynamics can find the introductory description, for example, in [3, 15, 16]. Books by Hal Caswell (1949–) [1, 2] describe the further discussion and application of the matrix model for the biological population dynamics.

10.1.3   Stable Age Distribution

For the age structured population dynamics by (10.1) with a transition matrix A_k, we can mathematically have the following expression of the age distribution at the kth generation:

$$\boldsymbol{n}_k = A_{k-1} A_{k-2} \cdots A_1 A_0 \boldsymbol{n}_0,$$

where $\boldsymbol{n}_0$ is the initial age distribution. When the transition matrix A_k is a constant matrix, $A_k = A$, it becomes

$$\boldsymbol{n}_k = A^k \boldsymbol{n}_0. \tag{10.4}$$

For the matrix model with such a constant transition matrix A, if there are an age distribution $\boldsymbol{n}^*$ and a positive constant λ such that

$$A\boldsymbol{n}^* = \lambda \boldsymbol{n}^*, \tag{10.5}$$

then $\boldsymbol{n}^*$ defines what is called the *stable age distribution*, more generally saying, *stable state distribution* for the age/state structured population dynamics with the transition matrix A. In mathematical words, the stable age distribution is given by the right eigenvector for a specific eigenvalue λ (> 0) of the matrix A.

If the population reaches a stable age distribution, the subpopulation size of every age class grows as a geometric progression with the common ratio λ. The population size increases if and only if $\lambda > 1$, while it decreases if and only if $\lambda < 1$, as described in Chap. 1. Now we can define the age frequency distribution by the vector

$$\boldsymbol{f}_k = \frac{1}{\sum_{i=1}^m n_{i,k}} \boldsymbol{n}_k,$$

where the element $f_{i,k}$ means the frequency/proportion of individuals in age class i at the kth generation. It can be proved from (10.5) that the age frequency distribution is generationally unchanged for the stable age distribution $\boldsymbol{n}^*$ (Exercise 10.1). This is the reason to use the word "stable" for the specific age distribution.

Exercise 10.1 Prove that the age frequency distribution is generationally unchanged for the stable age distribution n^*.

Suppose that the transition matrix A has distinct m eigenvalues λ_i ($i = 1, 2, \ldots, m$) such that $|\lambda_1| \leq |\lambda_2| \leq \cdots \leq |\lambda_m|$. Let us denote the right eigenvector for the eigenvalue λ_i by $u_i = {}^T(u_{i,1}, u_{i,2}, \ldots, u_{i,m})$ ($i = 1, 2, \ldots, m$) such that $Au_i = \lambda_i u_i$. Since the eigenvectors $\{u_i\}$ are linearly independent of each other, there is a unique set of constants c_i ($i = 1, 2, \ldots, m$) for arbitrarily given initial distribution n_0 such that $n_0 = c_1 u_1 + c_2 u_2 + \cdots + c_m u_m$. Let us now consider the case of $c_m \neq 0$. From (10.4), we have

$$n_k = A^k n_0 = A^k \sum_{i=1}^{m} c_i u_i = \sum_{i=1}^{m} c_i A^k u_i = \sum_{i=1}^{m} c_i \lambda_i^k u_i$$

$$= \lambda_m^k \left\{ c_1 u_1 \left(\frac{\lambda_1}{\lambda_m}\right)^k + c_2 u_2 \left(\frac{\lambda_2}{\lambda_m}\right)^k + \cdots + c_{m-1} u_{m-1} \left(\frac{\lambda_{m-1}}{\lambda_m}\right)^k + c_m u_m \right\},$$

where $|\lambda_i/\lambda_m| \leq 1$ for any $i < m$. Especially if $|\lambda_{m-1}| < |\lambda_m|$, we can find that

$$n_k \approx \lambda_m^k c_m u_m \tag{10.6}$$

for sufficiently large k, because $|\lambda_1/\lambda_m| \leq |\lambda_2/\lambda_m| \leq \cdots \leq |\lambda_{m-1}/\lambda_m| < 1$. In this case, the generational change of the age distribution eventually approaches a geometric progression with the common ratio λ_m. Hence the population goes extinct if $|\lambda_m| < 1$, while it geometrically grows if $|\lambda_m| > 1$.

Further from this result (10.6) about the asymptotic generational change of the age distribution, we can easily find that the age frequency distribution f_k becomes

$$f_k \approx \frac{1}{\sum_{i=1}^{m} u_{m,i}} u_m$$

for sufficiently large k. This means that the age frequency distribution f_k approaches a specific stable age frequency distribution f^* defined by the right eigenvector u_m for the eigenvalue λ_m of the largest absolute value as

$$f^* = \frac{1}{\sum_{i=1}^{m} u_{m,i}} u_m.$$

Although the above arguments are only for the case of $c_m \neq 0$, they are generally acceptable as the reasonable modeling. The initial age distribution n_0 to make $c_m = 0$ is specific, and is not worth considering from the standpoint of theoretical consideration on the biological population dynamics.

(continued)

This is because a little difference in the initial age distribution can make $c_m \neq 0$. Taking into account the environmental fluctuation in the biological system, the above arguments on the stable age distribution would be generally regarded as satisfactory in the theoretical consideration on the biological population dynamics.

Let us consider next the stable age distribution n^* for the Leslie matrix model with the transition matrix A given by (10.2). The characteristic equation $\left| A - \lambda E \right| = 0$ for A becomes the following polynomial equation of degree m:

$$\lambda^m - b_1 \lambda^{m-1} - a_1 b_2 \lambda^{m-2} - a_1 a_2 b_3 \lambda^{m-3} - \cdots$$

$$\cdots - \left\{ \prod_{j=1}^{m-2} a_j \right\} b_{m-1} \lambda - \left\{ \prod_{j=1}^{m-1} a_j \right\} b_m = 0. \tag{10.7}$$

Since the Leslie matrix is one of matrices that have only non-negative elements, Perron-Frobenius theorem says that the above characteristic Eq. (10.7) for A has a unique positive root λ_+ such that any other root has the absolute value less than λ_+ (for example, see [14, 15]). That is, λ_+ corresponds to λ_m in the above arguments which satisfies that $|\lambda_{m-1}| < \lambda_m$. The root λ_+ is called *principal eigenvalue*, *principal root*, or *dominant eigenvalue* for the characteristic equation or for the matrix A.

From (10.5), we can get the following expression of the stable age distribution n^* for the Leslie matrix model with (10.2) (Exercise 10.2):

$$n^* \equiv \begin{pmatrix} n_1^* \\ n_2^* \\ \vdots \\ n_j^* \\ \vdots \\ n_{m-1}^* \\ n_m^* \end{pmatrix} = n_m^* \cdot \begin{pmatrix} \lambda_+^{m-1}/(a_1 a_2 \cdots a_{m-1}) \\ \lambda_+^{m-2}/(a_2 a_3 \cdots a_{m-1}) \\ \vdots \\ \lambda_+^{m-j}/(a_j a_{j+1} \cdots a_{m-1}) \\ \vdots \\ \lambda_+/a_{m-1} \\ 1 \end{pmatrix}. \tag{10.8}$$

Although the above expression of the stable age distribution does not explicitly contain parameter b_j ($j = 1, 2, \ldots, m$), the principal root λ_+ depends on it. As already described before, the population with the stable age distribution shows a growth of geometric progression with the common ratio λ_+.

Exercise 10.2 Derive (10.8).

10.1.4 Reproductive Value

In this section, we are going to consider the *reproductive value* for a closed age structured population following the stable age distribution n^* according to the Leslie matrix model with (10.2). The reproductive value is defined as the expected total number of survived mature female offsprings born from a mother of an age onward. Roughly saying, it is an index of the "value" of a reproductive female at an age according to how its offsprings contribute to the future population size. This implies that the reproductive vale of a female gets smaller as the future population size becomes larger. As easily understood from the definition, the reproductive value becomes smaller for the older reproductive female, that is, the female with the larger age. This is because the expected total number of survived mature female offsprings produced by a female of an age onward gets smaller as the female becomes older. The reproductive value was defined first by Sir Ronald A. Fisher (1890–1962) in his book published in 1930 [5]. Today it has a close relation to the *fitness* in evolutionary biology. In this section, we are going to describe the mathematical modeling for the reproductive value about a closed age structured population of Leslie matrix model, for the purpose of clarifying the logical meaning of reproductive value.

> The reproductive value must be distinguished from the net reproduction rate defined in Sect. 1.5, while they are closely related to each other. Since the net reproduction rate is defined as the expected total number of reproductive females produced by a reproductive female, it is the supremum for the reproductive value of a reproductive female in terms of the age. The net reproduction rate is derived as the expected value for the whole reproductive period (age span) for a mature female, taking account of the survival probability in the period. Therefore it is an index of the reproductive potential according to the population. In contrast, the reproductive value is defined for a reproductive female of each age as seen in the following description of this section. Thus it is derived as the expected value for the rest of reproductive period at each age, taking account of the survival probability in the rest period for the reproduction.

Let us denote the reproductive value of an individual in age class j at a generation by v_j. The individual becomes age $j + 1$ and has the reproductive value v_{j+1} at the next generation if it survives until the next generation. Since the survival probability to the next generation is now given by a_j for the individual of age j, the individual cannot contribute to the future population size if it dies with probability $1 - a_j$. Since we assume that the population follows the stable age distribution, the population size geometrically grows with the growth rate λ_+ per generation as shown in Sect. 10.1.3. That is, the population size becomes larger if $|\lambda_+| > 1$ or smaller if $|\lambda_+| < 1$ respectively by λ_+ times at the next generation. Hence the reproductive value v_{j+1} becomes discounted by $1/\lambda_+$ times at the next generation in comparison with v_j at the present generation. On the other hand, the offsprings born from an individual of age j, which are given by b_j, have the total reproductive value $b_j v_1$ at the next generation, since they are of age class 1 at the next generation. This total reproductive value of offsprings is included to determine the reproductive value v_j as described below. Exceptionally, the reproductive value of an individual in the terminal age class m is determined only by the reproductive value of offsprings born from it, since such an individual cannot survive until the next generation.

Now we have the following equations corresponding to these arguments:

$$v_j = b_j \frac{v_1}{\lambda_+} + \left\{ a_j \frac{v_{j+1}}{\lambda_+} + (1 - a_j) \cdot 0 \right\} \qquad (1 \le j < m);$$

$$(10.9)$$

$$v_m = b_m \frac{v_1}{\lambda_+},$$

that is,

$$b_j v_1 + a_j v_{j+1} = \lambda_+ v_j \quad (1 \le j < m);$$

$$b_m v_1 \qquad\qquad = \lambda_+ v_m.$$

These equations can be expressed with vector and matrix as follows:

$$\begin{pmatrix} b_1 & a_1 & 0 & 0 & 0 & \cdots & 0 \\ b_2 & 0 & a_2 & 0 & 0 & \cdots & 0 \\ \vdots & \vdots & \ddots & \ddots & \vdots & & \vdots \\ b_j & 0 & \cdots & 0 & a_j & \cdots & 0 \\ \vdots & \vdots & & & \ddots & \ddots & \vdots \\ b_{m-1} & 0 & 0 & \cdots & 0 & 0 & a_{m-1} \\ b_m & 0 & 0 & \cdots & 0 & 0 & 0 \end{pmatrix} \begin{pmatrix} v_1 \\ v_2 \\ \vdots \\ v_j \\ \vdots \\ v_{m-1} \\ v_m \end{pmatrix} = \lambda_+ \begin{pmatrix} v_1 \\ v_2 \\ \vdots \\ v_j \\ \vdots \\ v_{m-1} \\ v_m \end{pmatrix}.$$

This $m \times m$ matrix is the transposed matrix of the Leslie matrix (10.2), that is, $^{\mathsf{T}}A$. Hence, with $v := {}^{\mathsf{T}}(v_1, v_2, \ldots, v_m)$, the above equation can be rewritten as

$$^{\mathsf{T}}vA = \lambda_+ {}^{\mathsf{T}}v, \qquad (10.10)$$

where the m dimensional vector $^{\mathsf{T}}v$ is the row vector given by the transposition of the column vector v, and we used a mathematical relation that $^{\mathsf{T}}(^{\mathsf{T}}Av) = {}^{\mathsf{T}}vA$. The Eq. (10.10) indicates that the row vector $^{\mathsf{T}}v$ is the left eigenvector for the principal eigenvalue λ_+ of matrix A.

Since any vector proportional to an eigenvector becomes the eigenvector for the same eigenvalue, the standard mathematical definition of the reproductive value is to use the value relative to the reproductive value at birth, that is, the reproductive value for the individual of age class 1. Mathematically it corresponds to determining the vector v such that $v_1 = 1$. Then the element v_j ($2 \leq j \leq m$) means the relative reproductive value for the individual of age class j. From (10.9), we can easily get the following formula about the general vector v:

$$\frac{v_j}{v_1} = \frac{\lambda_+^{j-1}}{\prod_{i=1}^{j-1} a_i} \sum_{k=j}^{m} \frac{\left(\prod_{i=1}^{k-1} a_i\right) b_k}{\lambda_+^k} \quad (2 \leq j \leq m), \tag{10.11}$$

where the right side gives the formula of the relative reproductive value for the individual of age class j.

Sir Ronald A. Fisher [5] defined the reproductive value V_x for the individual of age x about the age structured population for which the age is defined as a continuous non-negative value:

$$\frac{V_x}{V_0} = \frac{e^{rx}}{l_x} \int_x^{\infty} e^{-rt} l_t m_t \, dt, \tag{10.12}$$

where l_t is the mean survival rate of the individual with age t, m_t the expected number of offsprings produced by the individual with age t onward, and r the momental natural growth rate of the population size. V_0 means the reproductive value at birth. The right side of (10.12) gives the formula of the relative reproductive value for the individual of age class j. The reproductive value for the individual older than the reproductive age must be zero, so that the value of m_t is given by zero for such an old individual. For the growing population with a positive value of r, the above formula (10.12) indicates that the present offsprings have the higher value than the future ones have. For the declining population, it is inverse. The formula of reproductive value (10.11) for the Leslie matrix model with (10.2) has a clear mathematical correspondence to (10.12), and they has the same nature in a biological sense.

10.1.5 Sensitivity Analysis

About a matrix model, the *sensitivity analysis* generally means the analysis on the contribution of each element of the matrix to the nature of the temporal change of population size. A change of environmental condition may influence some factors governing the population dynamics, while the nature of population dynamics could be qualitatively maintained. For such a persistent population dynamics, it is biologically interesting which factor would be most influential in determining the nature of population dynamics. Mathematically according to the population dynamics model, it corresponds to how closely each parameter is related to the nature of dynamical system as the model.

Whereas the phrase "sensitivity analysis" may be sometimes used today in a much wider sense as the analysis on the parameter dependence according to a dynamical system, we shall describe in this section only the essential to mathematically estimate the sensitivity for the general matrix model with a constant transition matrix A (for further detail, see [1–3, 16]). Readers may be able to find the other aspect of such a structured population dynamics with a transition matrix.

Let us suppose that the transition matrix A has m distinct eigenvalues λ_i ($i = 1, 2, \ldots, m$), and denote the right eigenvector by $\boldsymbol{u}_i$ and the left eigenvector by $\boldsymbol{v}_i$ as column vectors for each eigenvalue λ_i. We have now $A\boldsymbol{u}_i = \lambda_i\boldsymbol{u}_i$ and $\boldsymbol{v}_i^* A = \lambda_i\boldsymbol{v}_i^*$, where, from the mathematical definition, the left eigenvector $\boldsymbol{v}_i$ satisfies the latter equation as the complex conjugate transposed vector of $\boldsymbol{v}_i$, $\boldsymbol{v}_i^* := {}^{\mathsf{T}}\overline{\boldsymbol{v}}_i$.

Generally every eigenvalue λ_i, right eigenvector $\boldsymbol{u}_i$ and left eigenvector $\boldsymbol{v}_i$ can be treated as continuous and differentiable functions of elements a_{ij} ($i, j = 1, 2, \ldots, m$). Thus the change in the values of elements of the transition matrix could alter them. From this viewpoint, we take the differential of the equation $A\boldsymbol{u}_i = \lambda_i\boldsymbol{u}_i$, and obtain the following relation of differentials:

$$(dA)\boldsymbol{u}_i + A\,d\boldsymbol{u}_i = (d\lambda_i)\boldsymbol{u}_i + \lambda_i d\boldsymbol{u}_i, \qquad (10.13)$$

where dA denotes the $m \times m$ matrix which elements are given by the differential of elements of A, da_{ij}, and $d\boldsymbol{u}_i$ does the column vector with elements of the differential $du_{i,j}$ ($j = 1, 2, \ldots, m$) about those of vector $\boldsymbol{u}_i$. Multiplying $\boldsymbol{v}_i^*$ from the left for both sides, we have

$$\boldsymbol{v}_i^*(dA)\boldsymbol{u}_i + \boldsymbol{v}_i^* A\,d\boldsymbol{u}_i = (d\lambda_i)\boldsymbol{v}_i^*\boldsymbol{u}_i + \lambda_i\boldsymbol{v}_i^* d\boldsymbol{u}_i,$$

where the second terms of both sides are the same as each other since $\boldsymbol{v}_i^* A = \lambda_i\boldsymbol{v}_i^*$. Hence we find the relation such that

$$d\lambda_i = \frac{\boldsymbol{v}_i^*(dA)\boldsymbol{u}_i}{\boldsymbol{v}_i^*\boldsymbol{u}_i}, \qquad (10.14)$$

where we must remark that the term $\boldsymbol{v}_i^*\boldsymbol{u}_i$ is the scalar product that leads to a non-zero scalar value when the matrix A has m distinct eigenvalues (Exercise 10.3).

Exercise 10.3 When the $m \times m$ transition matrix A has m distinct eigenvalues, show that the scalar product of the right eigenvalue $\boldsymbol{u}_i$ and the left eigenvalue $\boldsymbol{v}_i$, $\boldsymbol{v}_i^*\boldsymbol{u}_i$, does not become zero for every eigenvalue λ_i.

In the above description, we used the *differential* used in mathematics, though the arguments can be carried out in the following way essentially equivalent to it, which may be more straightforward to grasp the meaning from the viewpoint of modeling about the population dynamics.

Assume that each element a_{ij} of the transition matrix A changes by a sufficiently small perturbation δa_{ij}, that is, the element a_{ij} is replaced by $a_{ij} + \delta a_{ij}$. Following the change of elements, the eigenvalue and eigenvectors are changed from λ_i to $\lambda_i + \delta \lambda_i$ and from u_i to $u_i + \delta u_i$ respectively. Thus we have the following relation:

$$(A + \delta A)(u_i + \delta u_i) = (\lambda_i + \delta \lambda_i)(u_i + \delta u_i),$$

where δA denotes the $m \times m$ matrix with elements δa_{ij}. Neglecting all terms with the second order of perturbations δa_{ij}, $\delta \lambda_i$, and δu_i, we can get the following approximated relation from the above one:

$$(\delta A)u_i + A\delta u_i \approx (\delta \lambda_i)u_i + \lambda_i \delta u_i,$$

which clearly corresponds to (10.13). With this relation, we can carry out the same arguments as the above to derive (10.14).

From the above general result, we shall now consider the dependence only on a specific element $a_{k\ell}$ of A, fixing the other elements. In this case, any element of the matrix dA except for $da_{k\ell}$ is zero with $da_{k\ell} \neq 0$. From (10.14), we have

$$d\lambda_i = \frac{\overline{v}_{i,k}(da_{k\ell})u_{i,\ell}}{v_i^* u_i},$$

and consequently obtain the following equation about the partial derivative of the eigenvalue λ_i in terms of $a_{k\ell}$:

$$\frac{\partial \lambda_i}{\partial a_{k\ell}} = \frac{\overline{v}_{i,k}u_{i,\ell}}{v_i^* u_i}. \tag{10.15}$$

This partial derivative is translated as the *sensitivity* of the eigenvalue λ_i to the element $a_{k\ell}$ of A. The Eq. (10.15) indicates that it is proportional to the product of the kth element of left eigenvector and the ℓth element of right eigenvector, since the denominator $v_i^* u_i$ is independent of the choice of k and ℓ. As it has the larger absolute value, the eigenvalue λ_i has the higher sensitivity to the element $a_{k\ell}$. In other words, for the perturbation of an element $a_{k\ell}$ with the higher sensitivity, the eigenvalue λ_i is more significantly affected to change by the larger amount.

We can define the $m \times m$ matrix S_i which (k, ℓ)-element is given by (10.15) $(k, \ell = 1, 2, \ldots, m)$, called *sensitivity matrix* for the eigenvalue λ_i. From (10.15), it can be mathematically defined by

$$S_i := \frac{^\mathsf{T}(\boldsymbol{u}_i \boldsymbol{v}_i^*)}{\boldsymbol{v}_i^* \boldsymbol{u}_i}.$$

With the definition of sensitivity by (10.15), we may be able to compare the dependence of each eigenvalue on elements of A, though we cannot compare the dependence of eigenvalues on each element of A. For such a purpose, the following *elasiticity* is defined for the eigenvalue λ_i in terms of the element $a_{k\ell}$ of A:

$$e_{i,k\ell} := \frac{a_{k\ell}}{\lambda_i} \frac{\partial \lambda_i}{\partial a_{k\ell}} = \frac{\partial (\ln \lambda_i)}{\partial (\ln a_{k\ell})}.$$

From this definition, the elasticity means the relative change of λ_i, that is, $\delta\lambda_i/\lambda_i$ for the relative change of $a_{k\ell}$, given by $\delta a_{k\ell}/a_{k\ell}$ with a sufficiently small perturbation $\delta a_{k\ell}$. It mathematically satisfies that $\sum_{k,\ell=1}^{m} e_{i,k\ell} = 1$ (Exercise 10.4). Hence the elasticity $e_{i,k\ell}$ can be regarded as an index about the relative contribution of element $a_{k\ell}$ to the eigenvalue λ_i. Caswell [1, Section 9.2] advises against its misleading use, and suggests its logical use with clarifying the meaning and definition.

Exercise 10.4 Show that $\sum_{k,\ell=1}^{m} e_{i,k\ell} = 1$.

10.2 Continuous Time Model

In this section, we shall describe the modeling of the age structured population dynamics with the age defined as a continuous variable according to time. The age continuously varies as time passes.

10.2.1 Age Distribution Function

For the age continuously varying as time passes, the age of an individual a is given by a linear function of time t: $a = a(t) = t - t_0$ with the moment of its birth at $t = t_0$. Now let us denote the subpopulation size of individuals who have the age not beyond X at time t by $U(X, t)$. The function $U(X, t)$ is called *age distribution function.* We assume that it is a sufficiently smooth function of X and t. From the definition, it must be monotonically increasing non-negative function in terms of X.

The subpopulation size in a range of age $[X, X + \delta X]$ at time t is given by $\delta U(X, \delta X, t) = U(X + \delta X, t) - U(X, t)$. Let us focus on this subpopulation, that is, a *cohort* of the same age class in a wide sense (as for the definition of cohort, refer to Sect. 4.3). After a period Δt, the cohort transfers to that with the size $\delta U(X + \Delta t, \delta X, t + \Delta t)$ for a range of age $[X + \Delta t, X + \Delta t + \delta X)$ at time $t + \Delta t$, because age a at time t becomes age $a + \Delta t$ at time $t + \Delta t$. Thus, the difference

$$\delta Q(X, \delta X, t, \Delta t) = \delta U(X + \Delta t, \delta X, t + \Delta t) - \delta U(X, \delta X, t) \tag{10.16}$$

gives the variation of cohort size focused on now. Such a variation is caused by the death in the population and migration from/to it.

> As seen from the definition, $\delta U(X, 0, t) = U(X, t) - U(X, t) = 0$ for any X and t. Although $\delta U(X, \delta X, t)$ gives the cohort size in the age range $[X, X + \delta X)$, $\delta U(X, 0, t)$ does not mean the cohort size of any age X. This is a general mathematical nature of the distribution function for a continuous variable. Since X is the variable for the continuous age more than zero, mathematically $U(0, t) = 0$ for any t for the same reason.

Making use of Taylor expansion for a multi-variable function, we have

$$\delta U(X, \delta X, t) = U(X + \delta X, t) - U(X, t) = \frac{\partial U(X, t)}{\partial X} \delta X + o(\delta X), \tag{10.17}$$

and hence,

$$\delta U(X + \Delta t, \delta X, t + \Delta t)$$

$$= \frac{\partial U(x, t + \Delta t)}{\partial x} \bigg|_{x = X + \Delta t} \delta X + o(\delta X)$$

$$= \frac{\partial U(X, t + \Delta t)}{\partial X} \delta X + \frac{\partial^2 U(X, t)}{\partial X^2} \Delta t \delta X + o(\Delta t)\delta X + o(\delta X). \tag{10.18}$$

Therefore, from (10.16–10.18), we can derive

$\delta Q(X, \delta X, t, \Delta t)$

$$= \left[\frac{\partial U(X, t + \Delta t)}{\partial X} - \frac{\partial U(X, t)}{\partial X}\right]\delta X + \frac{\partial^2 U(X, t)}{\partial X^2}\Delta t \delta X + o(\Delta t)\delta X + o(\delta X)$$

$$= \left[\frac{\partial^2 U(X, t)}{\partial t \partial X}\Delta t + o(\Delta t)\right]\delta X + \frac{\partial^2 U(X, t)}{\partial X^2}\Delta t \delta X + o(\Delta t)\delta X + o(\delta X)$$

$$= \left[\frac{\partial^2 U(X, t)}{\partial t \partial X} + \frac{\partial^2 U(X, t)}{\partial X^2}\right]\Delta t \delta X + o(\Delta t)\delta X + o(\delta X). \qquad (10.19)$$

On the other hand, for a sufficiently smooth function $U(X, t)$, the variation of cohort size $\delta Q(X, \delta X, t, \Delta t)$ in the period Δt can be supposed to be a sufficiently smooth function of δX and Δt too. Now $\delta Q(X, \delta X, t, \Delta t)$ satisfies the following features as a function of δX and Δt:

(i) $\delta Q(X, \delta X, t, 0) = 0$ for any (X, t) and δX;
(ii) $\delta Q(X, 0, t, \Delta t) = 0$ for any (X, t) and Δt.

Since no passage of time causes no change of the cohort size, the feature (i) holds, similarly as the arguments in Sect. 3.3. The feature (ii) comes from the definition of $\delta Q(X, \delta X, t, \Delta t)$, since $\delta U(X, 0, t) = 0$ for any (X, t).

From the feature (i), we now note that, for $\Delta t \to 0$, the term $o(\delta X)$ of (10.19) must be zero, which was defined as the difference of terms $o(\delta X)$ in (10.17) and (10.18). Actually the term $o(\delta X)$ of (10.18) contains Δt which order we did not take care of. As easily seen from (10.18), it must become equivalent to the term $o(\delta X)$ of (10.17) for $\Delta t \to 0$. The term $o(\delta X)$ of (10.19) must be zero as well.

Now let us introduce the following equation about $\delta Q(X, \delta X, t, \Delta t)$:

$$\delta Q(X, \delta X, t, \Delta t) = M(X, \delta X, t)\delta U(X, \delta X, t)\Delta t + o(\delta X, \Delta t) \qquad (10.20)$$

with a sufficiently smooth function $M(X, \delta X, t)$ that means the momental variation rate of cohort size at time t (refer to Sect. 3.4), and the residue term $o(\delta X, \Delta t)$ such that

$$\lim_{\delta X \to 0}\frac{o(\delta X, \Delta t)}{\delta X} = 0; \qquad \lim_{\Delta t \to 0}\frac{o(\delta X, \Delta t)}{\Delta t} = 0.$$

The right side of (10.20) must satisfy the above features (i) and (ii). Making use of Taylor expansion of $\delta U(X, \delta X, t)$ and $M(X, \delta X, t)$, the Eq. (10.20) becomes

$$
\begin{aligned}
\delta Q(X, \delta X, t, \Delta t) &= M(X, \delta X, t)\left[\frac{\partial U(X, t)}{\partial X}\delta X + o(\delta X)\right]\Delta t + o(\delta X, \Delta t) \\
&= \left[M(X, 0, t) + \left.\frac{\partial M(X, x, t)}{\partial x}\right|_{x=0}\delta X + o(\delta X)\right] \\
&\quad \times \left[\frac{\partial U(X, t)}{\partial X}\delta X + o(\delta X)\right]\Delta t + o(\delta X, \Delta t) \\
&= M(X, 0, t)\frac{\partial U(X, t)}{\partial X}\delta X\Delta t + o(\delta X)\Delta t + o(\delta X, \Delta t).
\end{aligned}
$$
$$(10.21)$$

Consequently from the right sides of (10.19) and (10.21), we can find the following equation:

$$
-\mu(X, t)\frac{\partial U(X, t)}{\partial X} = \frac{\partial}{\partial X}\left\{\frac{\partial U(X, t)}{\partial X}\right\} + \frac{\partial}{\partial t}\left\{\frac{\partial U(X, t)}{\partial X}\right\}, \tag{10.22}
$$

where $\mu(X, t) := -M(X, 0, t)$. This partial differential equation governs the temporal change of age distribution given by $U(X, t)$.

10.2.2 von Foerster Equation

As the frequency density distribution of life span $f(t)$ in Sect. 4.3.5, we can define the age density distribution $u(a, t)$ as

$$
U(X, t) = \int_0^X u(a, t)\, da, \tag{10.23}
$$

and we have

$$
u(a, t) = \frac{\partial U(a, t)}{\partial a}. \tag{10.24}
$$

Since $U(X, t)$ is monotonically increasing in terms of X as seen from the definition given in the previous section, the Eq. (10.24) indicates that $u(a, t)$ must be non-negative for any a and t. Now, from (10.22), we can obtain the following partial differential equation with respect to the age density distribution $u(a, t)$:

$$
-\mu(a, t)u(a, t) = \frac{\partial}{\partial a}u(a, t) + \frac{\partial}{\partial t}u(a, t). \tag{10.25}
$$

This partial differential equation about the temporal change of age density distribution for an age structured population is today called *von Foerster equation* after the work by Heinz von Foerster (1911–2002) on the growth of microorganism population in 1959 [19]. Actually, in 1926, Anderson G. McKendrick (1876–1943) presented the same equation on the epidemic dynamics [12]. For this reason, the Eq. (10.25) may be called *McKendrick-von Foerster equation*. In mathematical biology, the Eq. (10.25) has been studied as models in a variety of contexts, for example, on the age distribution of proliferating cell population (for example, see [13, 17, 18]).

The derivation of (10.22) and (10.25) described in this section can be applied for the more general physiologically structured population (Sect. 9.1). Suppose that the physiological state of an individual $x = x(t)$ at time t follows the differential equation

$$\frac{dx(t)}{dt} = g(x, t), \tag{10.26}$$

where $g(x, t)$ is a sufficiently smooth function of x and t. The partial differential equation governing the temporal change of the state density distribution $u(x, t)$ can be derived as

$$-\mu(x, t)u(x, t) = \frac{\partial}{\partial x}\{g(x, t)u(x, t)\} + \frac{\partial}{\partial t}u(x, t) \tag{10.27}$$

along the same way for (10.25). For readers interested in the more mathematical detail about the derivation, refer to [13].

For an age structured closed population, every cohort of any age range must temporally decreases due to the death. In this case, $\delta Q(X, \delta X, t, \Delta t)$ defined by (10.16) is necessarily negative, so that $M(X, \delta X, t)\delta U(X, \delta X, t)$ in (10.20) and $M(X, 0, t)$ in (10.21) become negative. Then the function $\mu(x, t)$ in (10.25) is non-negative, which means the momental (per capita natural) death rate for the individual of age x at time t. For the open population with a migration process from/to the population, the left side of (10.25) must include the term about it.

We must remark that the value of $u(a, t)$ does not mean the cohort size of age a at time t. As argued in Sect. 10.2.1, the subpopulation size in a range of age $[X, X + \delta X)$ at time t is given by

$$\delta U(X, \delta X, t) = U(X + \delta X, t) - U(X, t) = \int_X^{X+\delta X} u(a, t)\, da.$$

It may seem that the cohort size of individuals with the same age X could be given by the limit as $\delta X \to 0$, whereas $\delta U(X, \delta X, t) \to 0$ as $\delta X \to 0$. From this mathematical feature, some readers might think that it means that there is no individual with age X. But it is not valid. As long as

$$\int_{X-\delta X/2}^{X+\delta X/2} u(a, t)\, da > 0$$

for arbitrary small $\delta X > 0$, such individuals with age X exist. From this argument, we can note that the positiveness of the value $u(a, t)$ could be regarded as indicating the existence of individuals of age a, while the value of $u(a, t)$ cannot mean the cohort size of individuals with the same age a. Mathematically, we have $\delta U(X, \delta X, t) = u(X, t)\delta X + o(\delta X)$ from the above equation. Also from this equation, we can find that $u(a, t)$ cannot have the meaning of population size by itself, since it must be meant not by $u(X, t)$ but by the product $u(X, t)\delta X$. We must not confuse the value of $u(a, t)$ with the cohort size of age a. Precisely the subpopulation size in an age structured population with the continuously variable age can be defined by the age distribution function $U(X, t)$ as argued in Sect. 10.2.1.

10.2.3 Population Renewal Process

As mentioned in the previous section, the death process can be introduced by the term $-\mu(a, t)u(a, t)$ in von Foerster equation (10.25). To complete the modeling for the age structured population dynamics, we must give the modeling for the recruitment or renewal process with the birth by the reproduction in the population.

From the definition of the distribution function $U(X, t)$ in the previous section, we can mathematical identify the subpopulation born in $[t, t + \Delta t]$ as

$$\Delta U(t) = U(\Delta t, t + \Delta t) - U(0, t) = U(\Delta t, t + \Delta t),$$

since $U(0, t) \equiv 0$ for any t. By Taylor expansion, we have

$$\Delta U(t) = \left[\frac{\partial U(x, t)}{\partial x} + \frac{\partial U(x, t)}{\partial t}\right]_{(x,t)=(0,t)} \Delta t + o(\Delta t). \tag{10.28}$$

On the other hand, from the definition of partial derivative, we find that

$$\left.\frac{\partial U(x, t)}{\partial t}\right|_{(x,t)=(0,t)} = \lim_{h \to 0} \frac{U(0, t+h) - U(0, t)}{h} = 0$$

for any t, because $U(0, t) \equiv 0$ for any t by the mathematical nature of $U(x, t)$ as mentioned in Sect. 10.2.1. From (10.28), we have

$$\frac{\Delta U(t)}{\Delta t} = \left.\frac{\partial U(x, t)}{\partial x}\right|_{(x,t)=(0,t)} + \frac{o(\Delta t)}{\Delta t} = u(0, t) + \frac{o(\Delta t)}{\Delta t}. \tag{10.29}$$

with (10.24). This equation indicates that the value of $u(0, t)$ means the momental reproduction rate (velocity) at time t, $\lim_{\Delta t \to 0} \Delta U(t)/\Delta t$ (refer to Sect. 3.4).

Let us assume that the *reproductive* cohort of $\delta U(X, \delta X, t)$ produces the amount of newborns $\{B(X, \delta X, t)\Delta t + o(\Delta t)\}\delta U(X, \delta X, t)$ in $[t, t + \Delta t)$, where $B(X, \delta X, t)$ means the momental per capita birth rate for the cohort at time t. Besides, as a reasonable modeling assumption, we assume the range of reproductive age as $[a_{min}, a_{max})$, and consider an arbitrary age classification such that

$$X_0 = a_{min} < X_1 < X_2 < \cdots < X_k < \cdots < X_n = a_{max}$$

with $\delta X_i = X_i - X_{i-1}$ ($i = 1, 2, \ldots, n$). Then we can make the following equations about $\Delta U(t)$:

$$\begin{aligned}
\Delta U(t) &= \sum_{i=1}^{n} B(X_i, \delta X_i, t)\delta U(X_i, \delta X_i, t)\Delta t + o(\Delta t) \\
&= \sum_{i=1}^{n} \left[B(X_i, 0, t) + \left.\frac{\partial B(X_i, x, t)}{\partial x}\right|_{x=0} \delta X_i + o(\delta X_i)\right] \\
&\quad \times \left[\left.\frac{\partial U(X, t)}{\partial X}\right|_{X=X_i} \delta X_i + o(\delta X_i)\right]\Delta t + o(\Delta t) \\
&= \sum_{i=1}^{n} \left[B(X_i, 0, t)\left.\frac{\partial U(X, t)}{\partial X}\right|_{X=X_i} \delta X_i + o(\delta X_i)\right]\Delta t + o(\Delta t) \\
&= \sum_{i=1}^{n} \left[B(X_i, 0, t)u(X_i, t)\delta X_i + o(\delta X_i)\right]\Delta t + o(\Delta t). \tag{10.30}
\end{aligned}$$

From (10.29) and (10.30), we find the equation

$$u(0, t) = \sum_{i=1}^{n} \left[b(X_i, t) u(X_i, t) \delta X_i + o(\delta X_i) \right] \tag{10.31}$$

with $b(X, t) := B(X, 0, t)$ that means the momental per capita reproduction rate for the individual of age X. Taking the limit as $n \to \infty$ with $\sup_i \delta X_i \to 0$ for the right side of (10.31), we have

$$u(0, t) = \int_{a_{\min}}^{a_{\max}} b(a, t) u(a, t) \, da \tag{10.32}$$

by the definition of Riemann integral. This equation is mathematically a boundary condition for the partial differential Eq. (10.25). Alternatively we can use the following formula instead of (10.32):

$$u(0, t) = \int_0^{\infty} b(a, t) u(a, t) \, da \tag{10.33}$$

with $b(a, t) = 0$ for any a such that $a < a_{\min}$ or $a > a_{\max}$. Since $b(a, t) \geq 0$ only for $a \in [a_{\min}, a_{\max})$ by this definition, the reproductive age belongs to $[a_{\min}, a_{\max})$. In the modeling in a mathematical sense, we may consider that $a_{\min} = 0$ without taking account of $a_{\max}$, that is, mathematically with $a_{\max} = \infty$.

To completely set a dynamical system about the temporal change of age distribution with von Foerster equation (10.25), we need the other boundary condition in addition to (10.32) or (10.33). From the reasonability of modeling for the age structured population dynamics, it is generally given by $\lim_{a \to \infty} u(a, t) = 0$ or $u(a, t) = 0$ for any $a > a_{\sup}$ at any time t where the supremum age $a_{\sup}$ defines the physiological or ecological life span (Sect. 1.5). The latter boundary condition explicitly introduces a finiteness of the life span. As long as a continuous age density distribution, we can have the condition that $u(a_{\sup}, t) = 0$ for any t. In contrast, the former boundary condition could be regarded as a mathematically approximated one, and must satisfy the other condition for the reasonable modeling such that

$$U(\infty, t) := \int_0^{\infty} u(a, t) \, da < \infty.$$

This is because the integral gives the total population size which must be finite for the reasonable modeling.

Exercise 10.5 When $b(a, t) - \mu(a, t) = r$ with a constant r for any a about the population dynamics by von Foerster equation (10.25), show that the total population size $U(\infty, t)$ follows the Malthus growth with the malthusian coefficient r.

10.2.4 Density Distribution Function on Characteristic Curve

The von Foerster equation (10.25) with the boundary conditions (10.32) or (10.33), and $\lim_{a \to \infty} u(a, t) = 0$ determines a unique solution $u(a, t)$ for each appropriately given initial age density distribution $u(a, 0)$. We can find the mathematical theory related to von Foerster equation (10.25) or the more general Eq. (10.27) about the physiologically structured population dynamics in [4, 13]. Especially, Laplace transformation or the method of *characteristic curve* is typically applied for the analysis on the partial differential Eq. (10.25) or (10.27). For example, in the book of Haberman [7, 8], the partial differential equation appears as a mathematical model for the temporal change of traffic volume, and its mathematical analysis is instructively described in detail.

In this section, we shall describe the application of the method of characteristic curve to obtain the mathematical solution of von Foerster equation (10.25). As seen in the following description, the characteristic curve becomes a line for (10.25), while it is a curve in general for (10.27) with (10.26) according to a general physiologically structured population dynamics.

Mathematial Solution Along Characteristic Curve

Let us focus on a cohort of the same age, and denote the age of the member at time t by $a = a(t) = t - \tau + \xi$ with given (τ, ξ), which indicates that the member is characterized by the age $a = a(\tau) = \xi$ at a specific time $t = \tau$. The temporal change of the cohort must be along the line $a = a(t) = t - \tau + \xi$ on the (t, a)-plane as shown in Fig. 10.2.

Tracking the temporal change of the cohort size along the line $a = t - \tau + \xi$, the corresponding age density distribution $u(a, t)$ is given by $u(a(t), t) = u(t - \tau + \xi, t)$ as a function of time t. The line $a = t - \tau + \xi$ is what is called the *characteristic curve* for the partial differential Eq. (10.25) according to the condition that $a(\tau) = \xi$.

While the characteristic *curve* is given by a line for von Foerster equation (10.25), it is given by a curve in general for the physiologically structured population dynamics with (10.26) and (10.27), determined by the function $g(x, t)$ governing the temporal change of physiological state for the member of cohort.

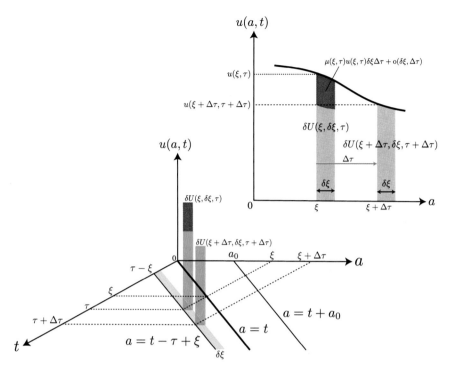

Fig. 10.2 Schematic description of the temporal change of the cohort size along the characteristic curve (line) for von Foerster equation (10.25)

Since

$$\frac{du(a(t), t)}{dt} = \frac{\partial u(a, t)}{\partial a} \cdot \frac{da(t)}{dt} + \frac{\partial u(a, t)}{\partial t} = \frac{\partial u(a, t)}{\partial a} + \frac{\partial u(a, t)}{\partial t}$$

and, from (10.25),

$$\frac{\partial}{\partial t} u(a, t) = -\mu(a, t) u(a, t) - \frac{\partial}{\partial a} u(a, t),$$

we can derive the following equation with respect to the temporal change of the age density $u(t - \tau + \xi, t)$ according to the cohort:

$$\frac{du(t - \tau + \xi, t)}{dt} = -\mu(t - \tau + \xi, t) u(t - \tau + \xi, t). \tag{10.34}$$

This equation is an ordinary differential equation in terms of t. Hence we can obtain the following mathematical solution:

$$u(t - \tau + \xi, t) = u(\xi, \tau) \exp\left[-\int_{\tau}^{t} \mu(s - \tau + \xi, s) \, ds \right]. \tag{10.35}$$

From this result of (10.35), we find that the temporal change of the cohort size $\delta U(X, \delta X, t) = \int_X^{X+\delta X} u(a, t)\,da = u(X, t)\delta X + o(\delta X)$ of individuals in a range of age $[X, X + \delta X)$ with a sufficiently small age span δX along the characteristic curve (that is actually a sufficiently narrow linear stripe on the (t, a)-plane, as schematically shown in Fig. 10.2) follows

$$\delta U(X, \delta X, t) = u(\xi, \tau) \exp\left[- \int_\tau^t \mu(s - \tau + \xi, s)\,ds \right] \delta X + o(\delta X),$$

supposing that $X = \xi$ at a given time $t = \tau$ for the cohort.

Two Kinds of Cohort

As shown in Fig. 10.2, we must distinguish the following two kinds of cohort with respect to the age structured population dynamics:

- Cohort existing from the initial time $t = 0$;
- Cohort emerges at time $t = t_1 > 0$ $(\tau \geq t_1)$.

The latter kind of cohort does not exist for $t < t_1$. For a closed population, the latter kind of cohort is produced by the reproduction of individuals existing until time t_1. We must consider the characteristic curve for these two kinds of cohort respectively.

First, let us consider the cohort existing from the initial time $t = 0$. Supposing that the member of such a cohort has the age ξ at time $\tau \geq 0$, the age of the member a at t is given by $a = a(t) = t - \tau + \xi$ with $a(0) = a_0 = -\tau + \xi \geq 0$. Thus, for such a cohort, it must be satisfied that $\xi \geq \tau$. Inversely, if $\xi \geq \tau$ for the member of a cohort, the cohort must be one that exists at the initial time $t = 0$. Such a cohort existing from $t = 0$, the mathematical solution (10.35) gives

$$u(\xi, \tau) = u(a_0, 0) \exp\left[- \int_0^\tau \mu(s + a_0, s)\,ds \right] \tag{10.36}$$

with $a_0 = -\tau + \xi \geq 0$. This equation holds for any (ξ, τ) such that $\xi \geq \tau$.

Next let us consider a cohort which emerges at time $t = t_1 > 0$. The member's age is zero at $t = t_1$. Since the age a of the member in the cohort at time t is given by $a = t - t_1$, the characteristic curve is given by $a = t - t_1$ for $t \geq t_1$ about the cohort. For such a cohort with age ξ at time τ, it must hold that $\xi < \tau$, as shown by the arguments already described above. Besides, if $\xi < \tau$ for the member of a cohort, the cohort is one emerges at a certain time $t = t_1 > 0$. Therefore, from the mathematical solution (10.35), we can find the following equation for such a cohort:

$$u(\xi, \tau) = u(0, t_1) \exp\left[- \int_{t_1}^\tau \mu(s - t_1, s)\,ds \right] \tag{10.37}$$

with $t_1 = \tau - \xi > 0$. This equation holds for any (ξ, τ) such that $\xi < \tau$.

Consequently, from (10.36) and (10.37), we can have the following mathematical solution of von Foerster equation (10.25):

$$u(a, t) = \begin{cases} u(0, t - a) \exp\left[-\int_0^a \mu(s, s + t - a)\,ds\right] & \text{if } a < t; \\[2em] u(a - t, 0) \exp\left[-\int_0^t \mu(a - t + s, s)\,ds\right] & \text{if } a \geq t. \end{cases} \tag{10.38}$$

10.2.5 Renewal Equation

In the mathematical solution (10.38), $u(0, t - a)$ is determined by the boundary condition (10.33), while $u(a - t, 0)$ is done by the initial condition $u(a, 0)$. Since $u(a, 0)$ is given for the population dynamics (10.25), $u(a - t, 0)$ can be determined a priori by it. In contrast, $u(0, t)$ is determined a posteriori through (10.33) by the population dynamics (10.25) itself, so that $u(0, t - a)$ is not trivial.

Substituting (10.38) for (10.33), we can derive the following equation with respect to $u(0, t)$ of von Foerster equation (10.25) (Exercise 10.6):

$$u(0, t) = F(t) + \int_0^t K(t - \tau, t)u(0, \tau)\,d\tau, \tag{10.39}$$

where

$$F(t) := \int_t^\infty b(a, t) \exp\left[-\int_0^t \mu(a - t + s, s)\,ds\right] u(a - t, 0)\,da;$$

$$K(\zeta, t) := b(\zeta, t) \exp\left[-\int_0^\zeta \mu(s, s + t - \zeta)\,ds\right].$$

Functions $F(t)$ is determined by the initial condition $u(a, 0)$. The Eq. (10.39) is what determines the unknown function $u(0, t)$, called the *renewal equation* for von Foerster equation (10.25). It may be called *Lotka equation*. The mathematical solution (10.38) for von Foerster equation (10.25) gives the unique $u(a, t)$ for any (a, t), with the given initial condition $u(a, 0)$ and $u(0, t)$ determined by the renewal Eq. (10.39).

Exercise 10.6 Derive the renewal Eq. (10.39).

The renewal Eq. (10.39) is one of what is called the *Volterra integral equation of the second kind* in mathematics. The uniqueness of its solution $u(0, t)$ can be proved by the method of Laplace transformation or the method of iteration.

Exercise 10.7 Let us consider the case where μ and b are positive constants independent of age a and time t. Derive the age density distribution function $u(a, t)$ and the age distribution function $U(a, t)$ when the initial condition $u(a, 0)$ is given as $u(a, 0) = \delta(a)$ with Dirac delta function $\delta(x)$ for $x \in \mathbb{R}$, which is a generalized function satisfying now that $\delta(0) = \infty$, and

$$\delta(x) = 0; \quad \int_0^x \delta(\zeta) \, d\zeta = 1; \quad \int_0^x \delta(\zeta) f(\zeta) \, d\zeta = f(0) \quad \text{for any } x > 0.$$

10.2.6   Stationary Age Distribution

Now we shall define the relative age distribution $\varphi(a, t)$ as

$$\varphi(a, t) := \frac{u(a, t)}{\int_0^\infty u(\zeta, t) \, d\zeta} = \frac{u(a, t)}{U(\infty, t)}. \tag{10.40}$$

We have $\int_0^\infty \varphi(a, t) \, da = 1$, which indicates that $\varphi(a, t)$ can be regarded as the distribution given by the normalization of the density distribution function $u(a, t)$. Since

$$\frac{\delta U(X, \delta X, t)}{U(\infty, t)} = \frac{\int_X^{X+\delta X} u(a, t) \, da}{U(\infty, t)} = \frac{u(X, t)\delta X + o(\delta X)}{U(\infty, t)} = \varphi(X, t)\delta X + o(\delta X),$$

$\varphi(a, t)\Delta a + o(\Delta a)$ means the frequency of the subpopulation in the age range $[a, a + \Delta a]$ within the population. Hence the function $\varphi(a, t)$ gives the frequency density distribution with respect to the age in the population.

Let us assume here for von Foerster equation (10.25) that μ and b are independent of time t, given as functions of age a: $\mu = \mu(a)$; $b = b(a)$. Besides, we suppose that the age density distribution function $u(a, t)$ can be expressed as the product of $A(a)$ of age a and $T(t)$ of time t: $u(a, t) = A(a)T(t)$. Then we have

$$U(\infty, t) = \int_0^\infty u(\zeta, t) \, d\zeta = T(t) \int_0^\infty A(\zeta) \, d\zeta,$$

so that we find

$$\varphi(a, t) = \frac{A(a)}{\int_0^\infty A(\zeta)\, d\zeta}$$

from (10.40). Thus, in this case, the relative age distribution is independent of time t. For this reason, the solution of von Foerster equation (10.25) with the renewal Eq. (10.39) which has the form of $u(a, t) = A(a)T(t)$ is called the solution of *stationary age distribution*.

Substituting $u(a, t) = A(a)T(t)$ for von Foerster equation (10.25), we can derive the equation

$$-\mu(a) - \frac{1}{A(a)}\frac{dA(a)}{da} = \frac{1}{T(t)}\frac{dT(t)}{dt}.$$

Since this equation must hold for any a and t, there must exist a constant λ such that

$$\begin{cases} -\mu(a) - \dfrac{1}{A(a)}\dfrac{dA(a)}{da} = \lambda; \\[2mm] \dfrac{1}{T(t)}\dfrac{dT(t)}{dt} = \lambda. \end{cases}$$

Each of these ordinary differential equations can be solved to give the solutions

$$\begin{cases} A(a) = A(\hat{a}) \exp\left[-\lambda(a - \hat{a}) - \displaystyle\int_{\hat{a}}^a \mu(\zeta)\, d\zeta \right]; \\[2mm] T(t) = T(0)\, e^{\lambda t}, \end{cases}$$

where $\hat{a} \geq 0$ is an appropriately chosen age such that $u(\hat{a}, 0) = A(\hat{a})T(0) > 0$. Then we obtain the age density distribution

$$u(a, t) = u(\hat{a}, 0) \exp\left[\lambda(t - a + \hat{a}) - \int_{\hat{a}}^a \mu(\zeta)\, d\zeta \right]. \tag{10.41}$$

If $T(0) = 0$, the initial age density distribution satisfies that $u(a, 0) = A(a)T(0) \equiv 0$ for any a. Then we have $U(\infty, 0) = 0$, which means that the population size is zero at $t = 0$. In such a case, the formula (10.41) becomes nonsense. Especially for a closed population, we have $U(a, t) = 0$ for any a and t in such a case. Clearly this is a nonsense case. Thus we need to consider the case of $U(\infty, 0) > 0$ for a closed population, so that we must assume that $T(0) > 0$. For an open population, it may hold that $U(\infty, t_1) > 0$ for a

(continued)

certain time $t = t_1 > 0$ even if $T(0) = 0$ and $U(\infty, 0) = 0$. In such a case, we can choose the moment $t = t_1$ as the initial time, that is, replace t by $t - t_1$. Without loss of mathematical generality, we can assume that $T(0) > 0$.

Consequently, it is necessary for the reasonable modeling to assume that $U(\infty, 0) > 0$. So there must exist an age $\hat{a} \geq 0$ such that $u(\hat{a}, 0) > 0$. Only when $u(0, 0) > 0$, we can make $\hat{a} = 0$.

Substituting (10.41) for the renewal Eq. (10.39) results in the following equation:

$$Q(\hat{a}) \, e^{\lambda t} = F(t) + Q(\hat{a}) \int_0^t K(t - \tau) \, e^{\lambda \tau} \, d\tau,$$

where

$$Q(\hat{a}) := u(\hat{a}, 0) \exp\left[\lambda \hat{a} - \int_{\hat{a}}^0 \mu(\zeta) \, d\zeta\right];$$

$$F(t) := Q(\hat{a}) \, e^{\lambda t} \int_t^\infty b(a) \exp\left[-\lambda a - \int_0^a \mathfrak{m}(s) \, ds\right] da;$$

$$K(\zeta) := b(\zeta) \exp\left[-\int_0^\zeta \mu(s) \, ds\right].$$

We can derive the following equation to determine λ, independently of $\hat{a}$:

$$1 = \int_t^\infty b(a) \exp\left[-\lambda a - \int_0^a \mu(s) \, ds\right] da$$

$$+ e^{-\lambda t} \int_0^t b(t - \tau) \exp\left[\lambda \tau - \int_0^{t-\tau} \mu(s) \, ds\right] d\tau$$

$$= \int_0^\infty b(a) \exp\left[-\lambda a - \int_0^a \mu(s) \, ds\right] da =: \Psi(\lambda). \tag{10.42}$$

It can be shown that the function $\Psi(\lambda)$ of λ is continuous and monotonically decreasing in terms of λ, satisfying that

$$\lim_{\lambda \to -\infty} \Psi(\lambda) = \infty; \qquad \lim_{\lambda \to \infty} \Psi(\lambda) = 0.$$

Therefore, the Eq. (10.42), $\Psi(\lambda) = 1$, has a unique real root. With the value of λ given by the root, the formula (10.41) gives a solution of von Foerster equation (10.25) with the renewal Eq. (10.39), that is, the solution of stationary age distribution.

Case of Constant Death and Growth Rates

Let us consider here a closed population with constant death rate μ and birth rate b. The renewal Eq. (10.39) for von Foerster equation (10.25) is given by

$$F(t) = b \int_t^\infty e^{-\mu t} u(a-t, 0) \, da = b e^{-\mu t} \int_0^\infty u(\zeta, 0) \, d\zeta = b e^{-\mu t} U(\infty, 0)$$

and $K(\zeta, t) = b e^{-\mu \zeta}$, and becomes

$$u(0, t) = b e^{-\mu t} \left[U(\infty, 0) + \int_0^t e^{\mu \tau} u(0, \tau) \, d\tau \right]. \tag{10.43}$$

Supposing $u(0, t) = \phi(t) e^{-\mu t}$, we find the following equation from (10.43):

$$\phi(t) = b \left[U(\infty, 0) + \int_0^t \phi(\tau) \, d\tau \right], \tag{10.44}$$

and get $\phi(0) = bU(\infty, 0)$. Next, differentiating both sides of (10.44) in terms of t, we can derive

$$\frac{d\phi(t)}{dt} = b\phi(t).$$

Thus, with the initial condition $\phi(0) = bU(\infty, 0)$, we obtain $\phi(t) = bU(\infty, 0) e^{bt}$. The renewal equation results in

$$u(0, t) = bU(\infty, 0) e^{(b-\mu)t}.$$

Then the age density distribution (10.38) becomes

$$u(a, t) = \begin{cases} u(0, t-a) e^{-\mu a} = bU(\infty, 0) e^{(b-\mu)t} e^{-ba} & \text{if } a < t; \\ u(a-t, 0) e^{-\mu t} & \text{if } a \geq t. \end{cases} \tag{10.45}$$

From this result on the age density distribution function $u(a, t)$, we can derive the age distribution function $U(a, t)$ as follows:

$$U(a, t) = \int_0^a u(\zeta, t) \, d\zeta$$

$$= \begin{cases} \displaystyle\int_0^a bU(\infty, 0) e^{(b-\mu)t} e^{-b\zeta} \, d\zeta & \text{if } a < t; \\ \displaystyle\int_0^t bU(\infty, 0) e^{(b-\mu)t} e^{-b\zeta} \, d\zeta + \int_t^a u(\zeta - t, 0) e^{-\mu t} \, d\zeta & \text{if } a \geq t \end{cases}$$

$$= \begin{cases} U(\infty, 0)\,(1 - e^{-ba})\,e^{(b-\mu)t} & \text{if } a < t; \\ e^{(b-\mu)t}\left[U(\infty, 0)\,(1 - e^{-bt}) + U(a - t, 0)\,e^{-bt}\right] & \text{if } a \geq t. \end{cases}$$

Hence we can find that $U(\infty, t) = U(\infty, 0)\,e^{(b-\mu)t}$, which indicates that the total population size follows the Malthus growth with the malthusian coefficient $b - \mu$.

Exercise 10.8 When μ and b are constants independent of age a and time t, derive the ordinary differential equation with respect to $U(\infty, t)$ by integrating both sides of von Foerster equation (10.25) in terms of a over $[0, \infty)$. Then solve it to get the solution of $U(\infty, t)$.

From the obtained $U(\infty, t)$ and the age density distribution (10.45), the relative age distribution (10.40) becomes

$$\varphi(a, t) = \begin{cases} be^{-ba} & \text{if } a < t; \\ \dfrac{u(a - t, 0)}{U(\infty, 0)}\,e^{-bt} = \varphi(a - t, 0)\,e^{-bt} & \text{if } a \geq t. \end{cases} \tag{10.46}$$

This result indicates that the relative age distribution asymptotically approaches a stationary age distribution as $t \to \infty$:

$$\varphi(a, t) \to be^{-ba}. \tag{10.47}$$

Since this convergence is independent of the initial age density distribution $u(a, 0)$, the above stationary age distribution can be called the *stable age (density) distribution*. As a consequence from (10.47), the stable age distribution is an exponential distribution for the closed population with constant death rate μ and birth rate b.

> We remark from (10.46) that the relative density $\varphi(a, t)$ for any age a BECOMES be^{-ba} independent of time t after time $t = a$. It does NOT asymptotically approach to be^{-ba}. This feature indicates that the newborns produced by the population following the age density distribution $u(a, t)$ has the relative age distribution independent of time t.

Density Effect on Death and Growth Rates

In this section, let us consider the general and simplest case of density-dependent birth and death rates when they depends on the total population density given by $U(\infty, t)$: $b = b(U(\infty, t))$; $\mu = \mu(U(\infty, t))$. This is the case where the reproduction and death are independent of age, while they are influenced by the population density within the population, that is, determined under the density effect.

Integrating both sides of von Foerster equation (10.25) in terms of a over $[0, \infty)$, we can get

$$-\mu(U(\infty, t))\, U(\infty, t) = -u(0, t) + \frac{dU(\infty, t)}{dt} \qquad (10.48)$$

with the boundary condition that $\lim\limits_{a \to \infty} u(a, t) = 0$, as done in Exercise 10.5 of Sect. 10.2.3. Besides from (10.33), we have

$$u(0, t) = b(U(\infty, t)) \int_0^\infty u(a, t)\, da = b(U(\infty, t))\, U(\infty, t). \qquad (10.49)$$

Therefore the Eq. (10.48) leads to

$$\frac{dU(\infty, t)}{dt} = \left\{ b(U(\infty, t)) - \mu(U(\infty, t)) \right\} U(\infty, t). \qquad (10.50)$$

From (10.40), (10.49) and the mathematical solution (10.38), we now have

$$\varphi(a, t) = \begin{cases} b(U(\infty, t - a)) \dfrac{U(\infty, t - a)}{U(\infty, t)} \dfrac{M(t)}{M(t - a)} & \text{if } a < t; \\[3mm] \varphi(a - t, 0) M(t) & \text{if } a \geq t, \end{cases} \qquad (10.51)$$

where

$$M(t) := \exp\left[-\int_0^t \mu(U(\infty, s))\, ds \right].$$

Let us consider here a closed population such that the momental per capita reproduction rate b is a constant independent of age and time while the per capita death rate μ is linearly increasing in terms of the population density $U(\infty, t)$: $b(U(\infty, t)) = b_0$; $\mu(U(\infty, t)) = \mu_0 + \beta U(\infty, t)$ with positive constants b_0, μ_0, and β. In this case, the population dynamics about the total population size (10.50) is equivalent to the logistic equation (5.10) with the intrinsic growth rate $r_0 = b_0 - \mu_0$ in Sect. 5.3.

From the Eq. (10.50), we now have

$$\frac{d \ln U(\infty, t)}{dt} = b_0 - \mu_0 - \beta U(\infty, t).$$

Thus, integrating both sides of this equation in terms of t over $[0, t)$, we can derive

$$\beta \int_0^t U(\infty, s)\, ds = (b_0 - \mu_0) t - \ln \frac{U(\infty, t)}{U(\infty, 0)},$$

so that we find that

$$M(t) := \frac{U(\infty, t)}{U(\infty, 0)} e^{-b_0 t}.$$

Therefore, from (10.51), we now obtain

$$\varphi(a, t) = \begin{cases} b_0 e^{-b_0 a} & \text{if } a < t; \\ \varphi(a - t, 0) \dfrac{U(\infty, t)}{U(\infty, 0)} e^{-b_0 t} & \text{if } a \geq t. \end{cases} \tag{10.52}$$

Since $U(\infty, t)$ is the solution of logistic equation, we know that $U(\infty, t) \to (b_0 - \mu_0)/\beta$ if $b_0 > \mu_0$, and $U(\infty, t) \to 0$ if $b_0 < \mu_0$ as $t \to \infty$ (Sect. 5.3). However, independently of which the population persists or goes extinct, this result of the relative age distribution (10.52) shows that it asymptotically approaches the stable age distribution as $t \to \infty$, given by $b_0 e^{-b_0 a}$ that is an exponential distribution.

> This result does not mean that the age distribution of a population growing with a logistic equation necessarily approaches an exponential distribution as the stable age distribution. As seen in the above arguments on (10.51), it depends on the detail of density-dependence for the birth and death rates, b and μ.

10.3 Age Distribution from Death Process

In this section, we shall see the age structured population dynamics derived from the death process described in Sect. 4.3 (for example, refer also to [6]). We consider a closed population again.

As in Sect. 10.2.1, let us consider a cohort of individuals with a range of age $[X, X + \delta X]$ at time t, which size is given by $\delta U(X, \delta X, t) = U(X + \delta X, t) - U(X, t)$. For a given time $t = t_0$ of the birth for an individual of the cohort, the age at time $t > t_0$ can be given by $t - t_0$, so that the time when an individual of the cohort has age X can be given by $X + t_0$. Hence we have $\delta U(X, \delta X, t) = \delta U(X, \delta X, X + t_0)$. Therefore the cohort size can be denoted by a function of age X, $N(X) := \delta U(X, \delta X, X + t_0)$, with a given age span δX and birth time t_0 which characterize the cohort. We hereafter call the age X the representative age for the cohort.

When the representative age changes from a to $a + \Delta a$ with sufficiently small Δa, the decrease of the cohort size $\Delta N(a)$ is given by

$$\Delta N(a) = N(a) - N(a + \Delta a)$$

$$= N(a) - \left\{ N(a) + \frac{dN(a)}{da} \Delta a + \mathrm{o}(\Delta a) \right\} = -\frac{dN(a)}{da} \Delta a - \mathrm{o}(\Delta a),$$

(10.53)

making use of Taylor expansion. Remark that the decrease $\Delta N(a)$ occurs in the time interval $[t_0 + a, t_0 + a + \Delta a)$. Now let us assume that

$$\Delta N(a) = \mu(a)\Delta a \cdot N(a) + \mathrm{o}(\Delta t),$$

(10.54)

where the per capita death rate during $[a, a + \Delta a)$ for an individual which has age a at time $t = a + t_0$ is given by $\mu(a)\Delta a + \mathrm{o}(\Delta a)$. The death rate is now assumed to depend only on the age. The momental per capita death rate $\mu(a)$ may be regarded as corresponding to what is called *hazard function* in the survival analysis of demography.

From (10.53) and (10.54), we can derive the following ordinary differential equation with the limit as $\Delta a \to 0$:

$$\frac{dN(a)}{da} = -\mu(a)N(a).$$

(10.55)

Then we can easily get

$$N(a') = N(a)\, \mathrm{e}^{-\int_a^{a'} \mu(z)\, dz},$$

(10.56)

which gives the cohort size after the representative age changes from a to a' ($> a$).

The Eq. (10.55) can be regarded as equivalent to the death process with a time-dependent death rate for an extinct population described in Sect. 4.3.5. Therefore, from the arguments in Exercise 4.5 of the section, we can find the following frequency density distribution $f_a(T)$ with respect to the rest of life span for an individual of age a:

$$f_a(T) = \mu(a + T)\, \mathrm{e}^{-\int_a^{a+T} \mu(z)\, dz},$$

(10.57)

which subsequently gives the cumulative frequency distribution

$$F_a(T) = 1 - \mathrm{e}^{-\int_a^{a+T} \mu(z)\, dz}.$$

$F_a(T)$ means the probability that an individual of age a has the rest of life span shorter than T. Thus the probability $S_a(T)$ that an individual with age a has the rest of life span longer than T is given by

$$S_a(T) = 1 - F_a(T) = e^{-\int_a^{a+T} \delta(z)\,dz}. \tag{10.58}$$

In the survival analysis of demography, the distribution $F_a(T)$ and the probability $S_a(T)$ may be called *survival distribution* and *survival function* respectively.

Gurney & Nisbet [6] assumed the age-dependent per capita death rate as

$$\mu(a) = \frac{p+1}{a_0}\left(\frac{a}{a_0}\right)^p, \tag{10.59}$$

where a_0 is a positive constant, and parameter p characterizes the age-dependence. As p is larger, the death rate gets large more steeply as the individual becomes older. With the per capita death rate (10.59), $S_a(T)$ given by (10.58) becomes

$$S_a(T) = \exp\left[\left(\frac{a}{a_0}\right)^{p+1} - \left(\frac{a}{a_0} + \frac{T}{a_0}\right)^{p+1}\right]. \tag{10.60}$$

Especially for the newborn, the probability $S_0(T)$ that the life span is longer than T is given by

$$S_0(T) = e^{-(T/a_0)^{p+1}}. \tag{10.61}$$

The probability distribution given by (10.60) or (10.61) is what is called *Weibull distribution*. It appears in the survival analysis for the life span when the death rate gets larger as the individual becomes older. It is sometimes used also for some arguments in medicine, for example, on the duration from the beginning of treatment for a disease to the end with the complete cure or patient's death. It can be applied for the arguments on the service life or durable period of a manufacturing machine or another material which has an error rate increasing as it is kept used.

Until now we have argued the temporal change of the size about a cohort which contains individuals born in a given time interval $[t_0, t_0 + \delta X)$. In contrast, we shall consider next the temporal change of the subpopulation size in a fixed age range $[a, a+\delta a)$. Let us denote the subpopulation size at time t by $n(a, t) = \delta U(a, \delta a, t)$.

From (10.56), we find that the subpopulation size $n(a, t + \Delta t)$ at time $t + \Delta t$ satisfies

$$n(a, t + \Delta t) = n(a - \Delta t, t)\, e^{-\int_{a-\Delta t}^{a} \mu(z)\,dz}. \tag{10.62}$$

The individual of age a at time $t + \Delta t$ had age $a - \Delta t$ at time t. Applying Taylor expansion around (a, t) for the right side of (10.62), we have

$$n(a, t + \Delta t) = \left\{ n(a, t) - \frac{\partial n(a, t)}{\partial a} \Delta t + \mathrm{o}(\Delta t) \right\} \left\{ 1 - \mu(a) \Delta t + \mathrm{o}(\Delta t) \right\}$$

$$= n(a, t) - \frac{\partial n(t, a)}{\partial a} \Delta t - n(a, t) \mu(a) \Delta t + \mathrm{o}(\Delta t). \tag{10.63}$$

Hence we can obtain

$$\frac{n(a, t + \Delta t) - n(a, t)}{\Delta t} = -\frac{\partial n(a, t)}{\partial a} - \mu(a) n(a, t) + \frac{\mathrm{o}(\Delta t)}{\Delta t}.$$

Taking the limit as $\Delta t \to 0$, we finally find von Foerster equation:

$$-\mu(a) n(a, t) = \frac{\partial n(a, t)}{\partial a} + \frac{\partial n(a, t)}{\partial t}.$$

10.4 Leslie Matrix and von Foerster Equation

In this section, we shall consider the mathematical relation between the discrete time model (10.1) with the Leslie matrix (10.2), that is, the Leslie matrix model in Sect. 10.1.1, and the continuous time model with von Foerster equation (10.25) (for example, refer to [4]).

10.4.1 From von Foerster Equation to Leslie Matrix Model

For the age-structured population with continuous age, we shall now divide the population into a finite and sufficiently large number of age classes as in Sect. 10.2.3:

$$X_0 = 0 < X_1 < X_2 < \cdots < X_k < \cdots < X_m = a_{\mathrm{sup}},$$

where the terminal age a_{sup} satisfies that $u(a, t) = 0$ for any $a > a_{\mathrm{sup}}$ at any time t, as already mentioned about the boundary condition for von Foerster equation (10.25) in the last part of Sect. 10.2.3. We may include the case where $a_{\mathrm{sup}} = \infty$ in the following argument, where the age class of range $[X_{m-1}, X_m)$ contains all individuals which has age greater than or equal to X_{m-1}. For this reason, we now assume that the age range $[X_{i-1}, X_i)$ has the same span $X_i - X_{i-1} = \delta X$ for $i = 1, 2, \ldots, m - 1$, while the age range $[X_{m-1}, X_m)$ has the same span δX when $a_{\mathrm{sup}} < \infty$, and is infinity when $a_{\mathrm{sup}} = \infty$. Especially when $a_{\mathrm{sup}} < \infty$, we have $\delta X = a_{\mathrm{sup}}/m$.

The population size of the age class about age range $[X_{i-1}, X_i)$ $(i = 1, 2, \ldots, m)$ is now given by

$$n_i(t) := \int_{X_{i-1}}^{X_i} u(a, t)\, da = \int_{X_i - \delta X}^{X_i} u(a, t)\, da = u(X_i, t)\delta X + o(\delta X). \qquad (10.64)$$

Integrating both sides of von Foerster equation (10.25) over $[X_{i-1}, X_i)$, we can get

$$-\int_{X_{i-1}}^{X_i} \mu(a, t)u(a, t)\, da = \int_{X_{i-1}}^{X_i} \frac{\partial u(a, t)}{\partial a}\, da + \int_{X_{i-1}}^{X_i} \frac{\partial u(a, t)}{\partial t}\, da$$

$$= \int_{X_{i-1}}^{X_i} \frac{\partial u(a, t)}{\partial a}\, da + \frac{d}{dt} \int_{X_{i-1}}^{X_i} u(a, t)\, da$$

$$= u(X_i, t) - u(X_{i-1}, t) + \frac{dn_i(t)}{dt},$$

where the left side can be led to

$$-\int_{X_{i-1}}^{X_i} \mu(a, t)u(a, t)\, da = -\int_{X_{i-1}}^{X_{i-1} + \delta X} \mu(a, t)u(a, t)\, da$$

$$= -\mu(X_{i-1}, t)u(X_{i-1}, t)\delta X + o(\delta X)$$

$$= -\mu(X_{i-1}, t)n_{i-1}(t) + o(\delta X),$$

for which we applied (10.64). Hence we have

$$\frac{dn_i(t)}{dt} = u(X_{i-1}, t) - u(X_i, t) - \mu(X_{i-1}, t)n_{i-1}(t) + o(\delta X). \qquad (10.65)$$

Since

$$n_i(t + \delta X) = n_i(t) + \frac{dn_i(t)}{dt}\delta X + o(\delta X),$$

we can derive the following equation from (10.64) and (10.65):

$$n_i(t + \delta X) = \left\{1 - \mu(X_{i-1}, t)\delta X\right\}n_{i-1}(t) + o(\delta X). \qquad (10.66)$$

We now need to give the equation for the temporal change of $n_0(t) = u(0, t)\delta X + o(\delta X)$ defined by (10.64). From the Eq. (10.32) and (10.64), we now have

$$u(0, t) = \int_0^{a_{\text{sup}}} b(a, t)u(a, t)\, da = \sum_{i=1}^m \int_{X_{i-1}}^{X_i} b(a, t)u(a, t)\, da$$

$$= \sum_{i=1}^m b(X_i, t)u(X_i, t)\delta X + o(\delta X) = \sum_{i=1}^m b(X_i, t)n_i(t) + o(\delta X).$$

Multiplying both sides by δX, we get

$$n_0(t) = \sum_{i=1}^{m} b(X_i, t)\delta X \, n_i(t) + o(\delta X). \tag{10.67}$$

Consequently from (10.66) and (10.67), we can obtain the following equation:

$$n(t + \delta X) = A(t)n(t) + o(\delta X), \tag{10.68}$$

where

$$n(t) \equiv \begin{pmatrix} n_0(t) \\ n_1(t) \\ \vdots \\ n_m(t) \end{pmatrix} ; \quad A(t) \equiv \begin{pmatrix} 0 & b_1(t) & b_2(t) & b_3(t) & b_4(t) & \cdots & b_m(t) \\ a_0(t) & 0 & 0 & 0 & 0 & \cdots & 0 \\ 0 & a_1(t) & 0 & 0 & 0 & \cdots & 0 \\ \vdots & \ddots & \ddots & \ddots & \vdots & & \vdots \\ 0 & \cdots & 0 & a_j(t) & 0 & \cdots & 0 \\ \vdots & & \vdots & \ddots & \ddots & \ddots & \vdots \\ 0 & 0 & 0 & \cdots & 0 & a_{m-1}(t) & 0 \end{pmatrix}$$

with $b_i(t) = b(X_i, t)\delta X$ $(i = 1, 2, \ldots, m)$ and $a_j(t) = 1 - \mu(X_j, t)\delta X$ $(j = 0, 1, \ldots, m - 1)$. Therefore, denoting $n_i(t) = n_{i,k}$ with $t = k\delta X$ $(k = 0, 1, 2, \ldots)$, the discrete time population dynamics (10.68) with time step size δX is clearly approximated by a Leslie matrix model described in Sect. 10.1.1.

10.4.2 From Leslie Matrix Model to von Foerster Equation

As we did in Chap. 3, we shall now introduce the age span for the age class to the Leslie matrix model as $h := x_i - x_{i-1}$ $(i = 1, 2, \ldots, m)$, where $x_0 = 0$ is the moment of birth, x_{i-1} the minimal age in the age class i, x_i the maximal age in the age class i, and x_m the supreme age in the oldest age class for the Leslie matrix model. The subpopulation size $n_{i,k}$ in the Leslie matrix model corresponds to the cohort size of the age class which consists of the individuals with the age in range $[x_{i-1}, x_i) = [(i - 1)h, ih)$ at time $t = kh$. In a sense, this idea could reintroduce the age and the time step into the modeling as continuous values.

Supposing a closed population for the Leslie matrix model with (10.2), the above modeling leads to the following equation derived from (10.1):

$$v(t + h, \alpha + h) = a_h(t, \alpha)v(t, \alpha), \tag{10.69}$$

where $v(t, \alpha) = n_{j,k}$, $v(t+h, \alpha+h) = n_{j+1,k+1}$, and $a_h(t, \alpha) = a_j$ with $\alpha = x_j = jh$ and $t = kh$ $(j = 1, 2, \ldots, m - 1)$ according to the correspondence to the Leslie matrix model (10.1) with (10.2). Thus $v(t, \alpha)$ denotes the subpopulation size with age in range $[\alpha - h, \alpha) = [(j - 1)h, jh)$ at time $t = kh$. Especially $a_h(t, \alpha)$ means the survival rate in the time interval $[t, t + h) = [kh, (k + 1)h)$, which generally depends on time t and age α, further satisfying that $0 \leq a_h \leq 1$. It must depend on the age span h too. As a reasonable assumption, a_h is monotonically decreasing in terms of h, since the longer time step or the larger age span induces the higher likelihood to die before the transition to the next age class. Moreover, as introduced about the time-step-zero limit in Sect. 3.3, it must be satisfied that

$$\lim_{h \to 0} a_h(t, \alpha) = 1 \tag{10.70}$$

for any t and α, since the death cannot occur as the time step is zero.

Now let us consider the age distribution function $U(\alpha, t)$ introduced in Sect. 10.2.1. From its definition, we have the equation

$$v(t, \alpha) = U(\alpha, t) - U(\alpha - h, t). \tag{10.71}$$

Hence we can derive the following equation:

$$\frac{v(t + h, \alpha + h) - v(t, \alpha)}{h^2}$$
$$= \frac{\{U(\alpha + h, t + h) - U(\alpha, t + h)\} - \{U(\alpha, t) - U(\alpha - h, t)\}}{h^2}$$
$$= \frac{1}{h} \left[\frac{U(\alpha + h, t + h) - U(\alpha, t + h)}{h} - \frac{U(\alpha + h, t) - U(\alpha, t)}{h} \right]$$
$$+ \frac{1}{h} \left[\frac{U(\alpha + h, t) - U(\alpha, t)}{h} - \frac{U((\alpha - h) + h, t) - U(\alpha - h, t)}{h} \right]. \tag{10.72}$$

On the other hand, from (10.69) and (10.71), we have

$$\frac{v(t + h, \alpha + h) - v(t, \alpha)}{h^2} = \frac{a_h(t, \alpha) - 1}{h} \frac{U(\alpha, t) - U(\alpha - h, t)}{h}. \tag{10.73}$$

From the condition (10.70), we assume that there exists a function $\mu(\alpha, t)$ such that

$$\lim_{h \to 0} \frac{1 - a_h(t, \alpha)}{h} = \mu(\alpha, t). \tag{10.74}$$

At this limit, we note that $1 - a_h(t, \alpha)$ means the death rate in the time interval $[t, t + h)$, so that the limit in (10.74) actually does the momental death rate at time t for age α.

Lastly by the time-step-zero limit as $h \to 0$ for (10.72) with (10.73) and (10.74), we can derive the following partial differential equation:

$$-\mu(\alpha, t) \frac{\partial U(\alpha, t)}{\partial \alpha} = \frac{\partial}{\partial t} \left\{ \frac{\partial U(\alpha, t)}{\partial \alpha} \right\} + \frac{\partial^2 U(\alpha, t)}{\partial \alpha^2}. \tag{10.75}$$

This is the same as (10.22) derived in Sect. 10.2.1, and indicates that the limiting dynamics follows von Foerster equation (10.25).

We must remark that the Leslie matrix model is defined for a finite number of age classes, which implies a certain supreme age x_m mentioned at the beginning of this part. Hence, in the above arguments, the whole population is classified into x_m/h age classes with the age span h in a conceptual sense. When x_m/h becomes fractional, we may tune x_m as the smallest number not below the actual terminal age after which no individual can survive. With this trick, the limit as $h \to 0$ makes the number of age classes infinity, while the subpopulation size of each age class becomes zero at the same time.

Answer to Exercise

Exercise 10.1 (p. 330)

Since the stable age distribution n^* satisfies (10.5), we have $n_{k+1} = \lambda n^*$ if $n_k = n^*$. Hence we find that

$$\begin{aligned} f_{k+1} &= \frac{1}{\sum_{i=1}^{m} n_{i,k+1}} n_{k+1} = \frac{1}{\sum_{i=1}^{m} \lambda n_{i,k}} \lambda n_k \\ &= \frac{1}{\sum_{i=1}^{m} n_{i,k}} n_k = \frac{1}{\sum_{i=1}^{m} n_{i,k}^*} n^* = f^*. \end{aligned}$$

Therefore, if $f_k = f^*$, then $f_{k+1} = f^*$. That is, the age frequency distribution for the stable age distribution n^* is always given by f^* independently of the generation.

Exercise 10.2 (p. 332)

First we must consider m parallel equations of $n_1^*, n_2^*, \ldots, n_m^*$ which can be derived from (10.5) with a constant matrix A of (10.2) and $\lambda = \lambda_+ > 0$ that is the principal eigenvalue for A. From the parallel equations, we can derive the expression with

n_m^* about each of $n_1^*, n_2^*, \ldots, n_{m-1}^*$, though we here leave the calculation itself to the readers. In the calculation, we need to make use of the characteristic Eq. (10.7) in order to get the formula (10.8) in which b_j ($j = 1, 2, \ldots, m$) does not appear explicitly. Dependence on the parameter b_j is implicitly included in λ_+ and n_m^* of (10.8).

Exercise 10.3 (p. 336)

When a constant transition matrix A has m distinct eigenvalues, the right eigenvectors corresponding to them are linearly independent of each other. So are the left eigenvectors. Hence the m right eigenvectors make the base for the m dimensional linear space, and so do the m left eigenvectors.

Let us introduce the complex conjugate and transposed vector v_j^* of the right eigenvector for eigenvalue λ_j. Multiplying both sides of $Au_i = \lambda_i u_i$ about the right eigenvalue u_i for eigenvalue λ_i by v_j^* from the left makes the equation $v_j^* A u_i = \lambda_i v_j^* u_i$. In the same way, we have $v_j^* A u_i = \lambda_j v_j^* u_i$ which is derived by multiplying $v_j^* A = \lambda_j v_j^*$ about the left eigenvector v_j for λ_j by the right eigenvalue u_i for λ_i. Remark that vectors u_i and v_j are defined here as column vectors.

Since the left sides of them are the same as each other, we find the equation $\lambda_i v_j^* u_i = \lambda_j v_j^* u_i$, so that $v_j^* u_i = 0$ for $i \neq j$ because of $\lambda_i \neq \lambda_j$ for $i \neq j$. This result means that the inner product of any pair of the left and right eigenvalues for different eigenvalues must be zero, that is, they are orthogonal.

On the other hand, since the m right eigenvectors $\{v_j\}$ make the base for the m dimensional linear space, the m dimensional vector u_i must be expressed by a linear combination of m vectors $\{v_j\}$ like $u_i = \sum_{j=1}^m c_j v_j$ with real constants c_j ($j = 1, 2, \ldots, m$). From the orthogonality of v_i and v_j for $i \neq j$, we have

$$v_i^* u_i = \sum_{j=1}^m c_j v_i^* v_j = c_i v_i^* v_i = c_i \|v_i\|^2. \tag{10.76}$$

Since the eigenvector cannot be the zero vector, that is, $\|v_i\| > 0$, $v_i^* u_i$ can become zero only when $c_i = 0$.

Now let us suppose that $c_i = 0$. Then we have $u_i = \sum_{j=1, j\neq i}^m c_j v_j$. Since u_i and v_ℓ for $i \neq \ell$ are orthogonal as shown in the above, and so are u_i and v_ℓ for $i \neq \ell$, we have

$$v_\ell^* u_i = \sum_{j=1, j\neq i}^m c_j v_\ell^* v_j = c_\ell v_\ell^* v_\ell = c_\ell \|v_\ell\|^2 = 0.$$

Thus we have $c_\ell = 0$. This result holds for any $\ell \neq i$, so that $c_j = 0$ for any $j \neq i$. This is contradictory because it means that u_i is the zero vector. Therefore, it must hold that $c_i \neq 0$ for the above linear combination of $\{v_j\}$ for u_i. Consequently from (10.76), we have found that $v_i^* u_i \neq 0$ for any i.

Exercise 10.4 (p. 338)

Making use of the definition of the sensitivity by (10.15), we find that

$$\sum_{k,\ell=1}^{m} e_{i,k\ell} = \sum_{k,\ell=1}^{m} \frac{a_{k\ell}}{\lambda_i} \frac{\partial \lambda_i}{\partial a_{k\ell}} = \sum_{k,\ell=1}^{m} \frac{a_{k\ell}}{\lambda_i} \frac{\bar{v}_{i,k} u_{i,\ell}}{v_i^* u_i} = \frac{\sum_{k,\ell=1}^{m} a_{k\ell} \bar{v}_{i,k} u_{i,\ell}}{\lambda_i v_i^* u_i} = \frac{v_i^* A u_i}{\lambda_i v_i^* u_i}.$$

Since $v_i^* A u_i = (v_i^* A) u_i = (\lambda_i v_i^*) u_i = \lambda_i v_i^* u_i$, the right side of the above equality becomes unity.

Exercise 10.5 (p. 346)

By integrating the Eq. (10.25) in terms of age a over $[0, \infty)$, we have

$$-\int_0^\infty \mu(a, t) u(a, t) \, da = \int_0^\infty \frac{\partial}{\partial a} u(a, t) \, da + \int_0^\infty \frac{\partial}{\partial t} u(a, t) \, da$$

$$= \left[u(a, t) \right]_0^\infty + \frac{d}{dt} \int_0^\infty u(a, t) \, da = -u(0, t) + \frac{d}{dt} U(\infty, t),$$

where we used the boundary condition that $\lim_{a \to \infty} u(a, t) = 0$. Making use of (10.33), we can derive

$$\frac{d}{dt} U(\infty, t) = u(0, t) - \int_0^\infty \mu(a, t) u(a, t) \, da$$

$$= \int_0^\infty b(a, t) u(a, t) \, da - \int_0^\infty \mu(a, t) u(a, t) \, da$$

$$= \int_0^\infty \{b(a, t) - \mu(a, t)\} u(a, t) \, da = r \int_0^\infty u(a, t) \, da = r U(\infty, t).$$

As seen from this result, if $b(a, t) - \mu(a, t) = r - \beta U(\infty, t)$ with positive constants r and β for any a and t, the total population size $U(\infty, t)$ follows the logistic equation. For example, this could be regarded as the case for a closed population such that the momental per capita reproduction rate b is

(continued)

a constant independent of age and time while the per capita death rate μ is linearly increasing in terms of the population density $U(\infty, t)$. We shall revisit such a modeling in the last part of Sect. 10.2.6.

Exercise 10.6 (p. 349)

Substituting (10.38) for (10.33), we can derive

$$
\begin{aligned}
u(0, t) &= \int_0^t b(a, t) \exp\left[-\int_0^a \mu(s, s + t - a)\, ds \right] u(0, t - a)\, da \\
&\quad + \int_t^\infty b(a, t) \exp\left[-\int_0^t \mu(a - t + s, s)\, ds \right] u(a - t, 0)\, da \\
&= \int_0^t b(t - \tau, t) \exp\left[-\int_0^{t - \tau} \mu(s, s + t - (t - \tau))\, ds \right] u(0, \tau)\, d\tau \\
&\quad + \int_t^\infty b(a, t) \exp\left[-\int_0^t \mu(a - t + s, s)\, ds \right] u(a - t, 0)\, da,
\end{aligned}
$$

where we used the variable transformation $\tau = t - a$ for the former integral. This equation becomes (10.39).

Exercise 10.7 (p. 350)

Since μ and b are positive constants independent of age a and time t, we have

$$
\begin{aligned}
F(t) &= b \int_t^\infty e^{-\mu t} u(a - t, 0)\, da \\
&= be^{-\mu t} \int_t^\infty \delta(a - t)\, da = be^{-\mu t} \int_0^\infty \delta(\zeta)\, d\zeta = be^{-\mu t}
\end{aligned}
$$

and $K(\zeta, t) = be^{-\mu \zeta}$ about the renewal Eq. (10.39) for von Foerster equation (10.25), making use of the initial condition given by $u(a, 0) = \delta(x)$ with Dirac delta function $\delta(x)$. Then the renewal Eq. (10.39) gives

$$
u(0, t) = be^{-\mu t} + b \int_0^t e^{-\mu(t - \tau)} u(0, \tau)\, d\tau = be^{-\mu t}\left[1 + \int_0^t e^{\mu \tau} u(0, \tau)\, d\tau \right].
$$

Putting $u(0, t) = \phi(t) e^{-\mu t}$ in this equation, we can get the following equation with respect to $\phi(t)$:

$$\phi(t) = b\left[1 + \int_0^t \phi(\tau) \, d\tau\right].$$

We find that $\phi(0) = b$. Differentiating both sides of this equation in terms of t, we can derive the following ordinary differential equation with respect to $\phi(t)$:

$$\frac{d\phi(t)}{dt} = b\phi(t).$$

We can easily solve this equation with the initial condition $\phi(0) = b$, and find $\phi(t) = be^{bt}$. Therefore we finally obtain

$$u(0, t) = be^{(b-\mu)t}.$$

Making use of this result about $u(0, t)$, the mathematical solution (10.38) becomes

$$u(a, t) = \begin{cases} u(0, t - a) e^{-\mu a} = be^{(b-\mu)t} e^{-ba} & \text{if } a < t; \\ u(a - t, 0) e^{-\mu t} = \delta(a - t) e^{-\mu t} & \text{if } a \geq t. \end{cases} \tag{10.77}$$

Subsequently we can derive

$$U(a, t) = \int_0^a u(\zeta, t) \, d\zeta = \begin{cases} \int_0^a be^{(b-\mu)t} e^{-b\zeta} \, d\zeta & \text{if } a < t; \\ \int_0^t be^{(b-\mu)t} e^{-b\zeta} \, d\zeta + \int_t^a \delta(\zeta - t) e^{-\mu t} \, d\zeta & \text{if } a \geq t, \end{cases}$$

that is,

$$U(a, t) = \begin{cases} (1 - e^{-ba}) e^{(b-\mu)t} & \text{if } a < t; \\ e^{(b-\mu)t} & \text{if } a \geq t. \end{cases}$$

For this model, the total population size at time t is mathematically given by $U(\infty, t) := \lim_{a \to \infty} U(a, t) = e^{(b-\mu)t}$. We find that it exponentially grows with the malthusian coefficient $b - \mu$, that is, it shows a Malthus growth.

From the initial condition $u(a, 0) = \delta(a)$, we have $U(\infty, 0) = 1$ which could be translated as the situation such that there is only one newborn in the population at time $t = 0$. Hence, as long as the initial individual is alive, the possibly oldest individual must be the initial individual which has age t. There is no individual beyond age t for this reason, so that $u(a, t) = 0$ for any $a > t$. Actually from the mathematical solution of $u(a, t)$ given by (10.77), we find that $u(a, t) = 0$ for any $a > t$ because of the nature of Dirac delta function. In the above arguments, we mathematically defined the total population size by $U(\infty, t)$, though it can be given instead by $U(t, t)$ from the meaning of modeling.

We must remark that the above results contain further mathematically special features due to the assumption for the initial condition with Dirac delta function. From the mathematical solution of $u(a, t)$ given by (10.77), we find also that $u(a, a) = \delta(0) = \infty$ for any time $t = a$. From the above arguments, this could be regarded as corresponding to the oldest individual in the population. Whereas it may seem that there could be always an alive oldest individual in the population, this is not valid as the appropriate translation about the mathematical feature, as already mentioned at the end of Sect. 10.2.2. From the above result of $U(a, t)$, we can find that the population goes extinct if $b < \mu$ even though mathematically $u(a, a) = \delta(0) = \infty$ for any time $t = a$.

Exercise 10.8 (p. 354)

When μ and b are constants independent of age a and time t, we can integrate both sides of von Foerster equation (10.25) over $[0, \infty)$ as follows:

$$-\mu\, U(\infty, t) = \int_0^\infty \frac{\partial u(a, t)}{\partial a}\, da + \int_0^\infty \frac{\partial u(a, t)}{\partial t}\, da$$

$$= \int_0^\infty \frac{\partial u(a, t)}{\partial a}\, da + \frac{d}{dt} \int_0^\infty u(a, t)\, da$$

$$= \lim_{\zeta \to \infty} \left[u(a, t) \right]_0^\zeta + \frac{dU(\infty, t)}{dt}$$

$$= \lim_{a \to \infty} u(a, t) - u(0, t) + \frac{dU(\infty, t)}{dt}. \qquad (10.78)$$

From the renewal Eq. (10.39)

$$u(0, t) = b \int_0^\infty u(a, t)\, da = bU(\infty, t)$$

and the boundary condition $\lim_{a \to \infty} u(a, t) = 0$ (refer to Sect. 10.2.3), the above Eq. (10.78) becomes

$$-\mu\, U(\infty, t) = -bU(\infty, t) + \frac{dU(\infty, t)}{dt},$$

that is,

$$\frac{dU(\infty, t)}{dt} = (b - \mu)\, U(\infty, t).$$

Finally, we can easily solve this ordinary differential equation and get $U(\infty, t) = U(\infty, 0)\, e^{(b-\mu)t}$.

Actually, this exercise is on a specific case of what was considered in Exercise 10.5 of Sect. 10.2.3.

References

1. H. Caswell, *Matrix Population Models: Construction, Analysis, and Interpretation*, 2nd edn. (Sinauer Associates, Sunderland, 2001)
2. H. Caswell, *Sensitivity Analysis: Matrix Methods in Demography and Ecology*. Demographic Research Monographs (Springer, Berlin, 2019)
3. B. Charlesworth, *Evolution in Age-Structured Populations*. Cambridge Studies in Mathematical Biology, vol. 14, 2nd edn. (Cambridge University Press, Cambridge, 2008)
4. J.M. Cushing, *An Introduction to Structured Population Dynamics*. CBMS-NSF Regional Conference Series in Applied Mathematics, vol. 71 (Society for Industrial and Applied Mathematics (SIAM), Philadelphia, 1998)
5. R.A. Fisher, *The Genetical Theory of Natural Selection* (Oxford University Press, Oxford, 1930)
6. W.S.C. Gurney, R.M. Nisbet. *Ecological Dynamics* (Oxford University Press, Oxford, 1998)
7. R. Haberman, *Mathematical Models: Mechanical Vibrations, Population Dynamics, and Traffic Flow* (Prentice-Hall, New Jersey, 1977)
8. R. Haberman, *Mathematical Models: Mechanical Vibrations, Population Dynamics, and Traffic Flow*. Classics in Applied Mathematics, vol. 21 (Society for Industrial and Applied Mathematics (SIAM), Philadelphia, 1998)
9. L.P. Lefkovitch, The study of population growth in organisms grouped by stages. Biometrics **21**, 1–18 (1965)
10. P.H. Leslie, On the use of matrices in certain population mathematics. Biometrika **33**, 183–212 (1945)
11. P.H. Leslie, Some further notes on the use of matrices in population mathematics. Biometrika **35**, 213–245 (1948)
12. A.G. McKendrick, Applications of mathematics to medical problems. Proc. Edin. Math. Soc. **44**, 98–130 (1926)
13. J.A.J. Metz, O. Diekmann, *The Dynamics of Physiologically Structured Populations*. Lecture Notes in Biomathematics, vol. 68 (Springer Verlag, Berlin, 1986)
14. E.C. Pielou, *An Introduction to Mathematical Ecology* (John Wiley & Sons, London, 1969)
15. E.C. Pielou, *Mathematical Ecology*, 2nd edn. (John Wiley & Sons, London, 1977)
16. J.W. Silvertown, D. Charlesworth, *Introduction to Plant Population Ecology*, 4th edn. (Blackwell Scientific Publications, Oxford, 2001)
17. E. Trucco, Mathematical models for cellular systems: the von Foerster equation. Bull. Math. Biophys. **27**, 285–305, 449–471 (1965)
18. E. Trucco, Collection functions for non-equivalent cell populations. J. Theor Biol. **15**, 180–189 (1967)
19. H. von Foerster, Some remarks on changing populations, in *The Kinetics of Cellular Proliferation*. ed. by F. Stohlman (Grune and Stratton, New York, 1959), pp. 382–407

Part II
Mathematical Equipments

Chapter 11
Homogeneous Linear Difference Equation

Abstract This chapter describes the fundamentals on the homogeneous linear difference equation, which are closely related to the contents of Chaps. 1 and 2 about the discrete time models.

11.1 Second Order Linear Equation

The general second order linear difference equation (recurrence relation) with constant coefficients is expressed as

$$a_{n+2} = p a_{n+1} + q a_n + r, \tag{11.1}$$

where p, q and r are real constants independent of n with $q \neq 0$. For given initial values a_1 and a_2, the recurrence relation (11.1) determines a unique sequence $\{a_n\}$.

In this section, we consider the general term for the sequence $\{a_n\}$ determined by the homogeneous equation

$$a_{n+2} = p a_{n+1} + q a_n, \tag{11.2}$$

which is one with $r = 0$ in (11.1). Now, suppose that $a_n = \lambda^n$ with a non-zero constant λ as the general term to satisfy (11.2). By substituting $a_n = \lambda^n$, $a_{n+1} = \lambda^{n+1}$ and $a_{n+2} = \lambda^{n+2}$ for (11.2), we obtain the following equation in terms of λ:

$$\lambda^{n+2} = p\lambda^{n+1} + q\lambda^n, \quad \text{that is,} \quad \lambda^2 - p\lambda - q = 0. \tag{11.3}$$

Let denote two roots of this quadratic equation by λ_1 and λ_2. Both of $a_n = \lambda_1^n$ and $a_n = \lambda_2^n$ can satisfy the recurrence relation (11.2). The Eq. (11.3) is called the *characteristic equation* for the recurrence relation (11.2). λ_1^n and λ_2^n are the base solutions for (11.2).

On the other hand, if two formulas $f(n)$ and $g(n)$ of n satisfy the recurrence relation (11.2), the linear combination $sf(n) + tg(n)$ with arbitrary constants s and

t can satisfy (11.2). This is easily seen by substituting $a_n = sf(n) + tg(n)$ for (11.2). Therefore, $a_n = s\lambda_1^n + t\lambda_2^n$ satisfies the recurrence relation (11.2).

Let us consider first the case where $\lambda_1 \neq \lambda_2$. For $a_n = s\lambda_1^n + t\lambda_2^n$, we have $a_1 = s\lambda_1 + t\lambda_2$ and $a_2 = s\lambda_1^2 + t\lambda_2^2$, so that

$$s = \frac{a_1\lambda_2 - a_2}{\lambda_1(\lambda_2 - \lambda_1)}; \quad t = \frac{a_1\lambda_1 - a_2}{\lambda_2(\lambda_1 - \lambda_2)}. \tag{11.4}$$

As a result, we obtain the following unique formula of a_n for the initial condition given by a_1 and a_2 as a function of n:

$$a_n = \frac{a_1\lambda_2 - a_2}{\lambda_1(\lambda_2 - \lambda_1)}\lambda_1^n + \frac{a_1\lambda_1 - a_2}{\lambda_2(\lambda_1 - \lambda_2)}\lambda_2^n. \tag{11.5}$$

That is, this is the general term for a_n as proven in the following arguments.

From the relation of roots to the coefficients in the quadratic Eq. (11.3), we have $p = \lambda_1 + \lambda_2$ and $q = -\lambda_1\lambda_2$. Substituting these for the Eq. (11.2), we can transform the equation to

$$a_{n+2} - \lambda_1 a_{n+1} = \lambda_2(a_{n+1} - \lambda_1 a_n) \tag{11.6}$$

or

$$a_{n+2} - \lambda_2 a_{n+1} = \lambda_1(a_{n+1} - \lambda_2 a_n). \tag{11.7}$$

For the Eq. (11.6), letting $b_n = a_{n+1} - \lambda_1 a_n$, we have $b_{n+1} = \lambda_2 b_n$. Then the sequence $\{b_n\}$ is a geometric progression with the common ratio λ_2. Thus we have $b_n = b_1\lambda_2^{n-1}$, that is,

$$a_{n+1} - \lambda_1 a_n = (a_2 - \lambda_1 a_1)\lambda_2^{n-1}. \tag{11.8}$$

In the same way, we can get the following equation from (11.7):

$$a_{n+1} - \lambda_2 a_n = (a_2 - \lambda_2 a_1)\lambda_1^{n-1}. \tag{11.9}$$

Solving (11.8) and (11.9) with respect to a_{n+1} and a_n, we can get the Eq. (11.5) for a_n.

These arguments are applicable even when the roots λ_1 and λ_2 are imaginary numbers. Since we are considering the sequence of real values, this is to be explained a little more. When the roots λ_1 and λ_2 of are imaginary, they must be conjugate to each other: $\overline{\lambda_1} = \lambda_2$. Let us denote the argument of λ_1 by $\theta = \arg \lambda_1$, and the absolute value by $\rho = |\lambda_1|$. With Euler's formulation, we have the expressions $\lambda_1 = \rho(\cos\theta + i\sin\theta)$ and $\lambda_2 = \rho(\cos\theta - i\sin\theta)$, where i is imaginary unit. From de Moivre's theorem, we can find that $\lambda_1^n = \rho^n(\cos n\theta + i \sin n\theta)$ and $\lambda_2^n = \rho^n(\cos n\theta - i \sin n\theta)$. Hence, in this case, we have

$$a_n = s\lambda_1^n + t\lambda_2^n = \rho^n\{(s+t)\cos n\theta + (s-t)i \sin n\theta\}. \tag{11.10}$$

From (11.4), we can find that

$$s + t = \frac{a_1(\lambda_1 + \lambda_2) - a_2}{\lambda_1\lambda_2};$$

$$(s - t)i = \frac{(\lambda_1^2 + \lambda_2^2)a_1 - (\lambda_1 + \lambda_2)a_2}{\lambda_1\lambda_2} \frac{i(\lambda_2 - \lambda_1)}{(\lambda_2 - \lambda_1)^2}.$$

It can be easily proven that both of $s + t$ and $(s - t)i$ are real, since $\lambda_1 + \lambda_2$, $\lambda_1^2 + \lambda_2^2$, and $\lambda_1\lambda_2$ are real, while $\lambda_2 - \lambda_1$ is purely imaginary from $\overline{\lambda_1} = \lambda_2$. Finally we can see that the right side of the general term (11.10) is real for the case where the characteristic Eq. (11.3) has imaginary roots. These arguments show that the general term (11.5) is applicable even when the characteristic Eq. (11.3) has imaginary roots.

Next let us consider the case where the characteristic Eq. (11.3) has a multiple root, that is, when $\lambda_1 = \lambda_2$. In this case, the general term (11.5) is not applicable. Although λ_1^n is the base solution for the recurrence relation (11.2), we cannot find the general term only with it.

Actually the other base solution is given by $n\lambda_1^n$. Indeed, substituting $a_n = n\lambda_1^n$, $a_{n+1} = (n+1)\lambda_1^{n+1}$ and $a_{n+2} = (n+2)\lambda_1^{n+2}$ for (11.2), we have

$$(n+2)\lambda_1^2 = p(n+1)\lambda_1 + qn. \tag{11.11}$$

Since the characteristic Eq. (11.3) must now become $(\lambda - \lambda_1)^2 = 0$, so that $p = 2\lambda_1$ and $q = -\lambda_1^2$ hold, we can find that the Eq. (11.11) is the identical equation.

Therefore, when the characteristic Eq. (11.3) has a multiple root, the general term for the recurrence relation (11.2) becomes the linear combination of two base solutions λ_1^n and $n\lambda_1^n$. Substituting $p = 2\lambda_1$ and $q = -\lambda_1^2$ for (11.2), we can get

$$a_{n+2} - \lambda_1 a_{n+1} = \lambda_1(a_{n+1} - \lambda_1 a_n),$$

and find that the sequence $\{a_{n+1} - \lambda_1 a_n\}$ is a geometric progression with the common ratio λ_1:

$$a_{n+1} - \lambda_1 a_n = \lambda_1^{n-1}(a_2 - \lambda_1 a_1).$$

Moreover from this equation, we can get the following equation

$$\frac{a_{n+1}}{\lambda_1^{n+1}} - \frac{a_n}{\lambda_1^n} = \frac{a_2 - \lambda_1 a_1}{\lambda_1^2},$$

which indicates that the sequence $\{a_n/\lambda_1^n\}$ is an arithmetic progression with the common difference $(a_2 - \lambda_1 a_1)/\lambda_1^2$. Consequently we can obtain the following general term of a_n:

$$a_n = \left\{ \frac{a_1}{\lambda_1} + \frac{a_2 - \lambda_1 a_1}{\lambda_1^2} (n-1) \right\} \lambda_1^n \qquad (11.12)$$

Those general terms (11.5), (11.10), and (11.12), obtained in the above, we can get the following conclusion:

Theorem 11.1 *The sequence $\{a_n\}$ generated by the second order homogeneous linear difference Eq. (11.2) converges to zero as $n \to \infty$ independently of the initial value if $|\lambda_1| < 1$ and $|\lambda_2| < 1$ hold for the roots λ_1 and λ_2 of the characteristic Eq. (11.3). If $|\lambda_1| > 1$ or $|\lambda_2| > 1$ holds, the sequence $\{a_n\}$ diverges for almost every initial value.*

If $|\lambda_1| = 1$ and $|\lambda_2| < 1$ hold, the sequence $\{a_n\}$ converges to a certain finite value. If $|\lambda_1| = |\lambda_2| = 1$ hold, the sequence $\{a_n\}$ diverges for almost every initial value when λ_1 and λ_2 are real and converges to a permanent oscillatory variation with a finite supremum of the amplitude for almost every initial value when they are imaginary.

Whereas some details in this theorem were not described in this section, readers can easily prove them, making use of those results obtained there.

11.2 Two Dimensional System of First Order Linear Equations

11.2.1 Simultaneous First Order Equations

The following type of two dimensional autonomous system of first order difference equations in terms of (x_n, y_n)

$$\begin{cases} x_{n+1} = f(x_n, y_n) \\ y_{n+1} = g(x_n, y_n) \end{cases} \qquad (11.13)$$

appears as a mathematical model for a variety of phenomena. When both of functions f and g are linear in terms of x_n and y_n with constant coefficients, the general term for (x_n, y_n) can be mathematically obtained as shown in the subsequent sections. In contrast, when they are nonlinear, it is generally difficult to get the explicit solution (i.e., general term), and instead, some mathematical techniques are applied to investigate the qualitative feature of the solution (for example, see [1–6]).

Like (11.13), the system of difference equations which consists of only the dependent variable(s) and its derivative(s) is called *autonomous*. If the system contains some functions of independent variable t in any part of it, it is called *non-autonomous*.

For the system of nonlinear difference Eq. (11.13), the qualitative analysis frequently uses the following type of a system of homogeneous linear difference equations (refer to Sect. 12.2.1):

$$\begin{cases} u_{n+1} = a_{11}\, u_n + a_{12}\, v_n \\ v_{n+1} = a_{21}\, u_n + a_{22}\, v_n, \end{cases} \tag{11.14}$$

where coefficients a_{11}, a_{12}, a_{21} and a_{22} are mathematically determined for each analysis. The system (11.14) can be expressed by the two dimensional column vector $\mathbf{v}_n := \begin{pmatrix} u_n \\ v_n \end{pmatrix}$ and 2×2 matrix $A := \begin{pmatrix} a_{11} & a_{12} \\ a_{21} & a_{22} \end{pmatrix}$ as follows:

$$\mathbf{v}_{n+1} = A\,\mathbf{v}_n. \tag{11.15}$$

In the following part,[1] we discuss the general solution for the system of homogeneous linear difference Eq. (11.14) with constant coefficients, that is, (11.15) with matrix A which elements a_{11}, a_{12}, a_{21} and a_{22} are all constant.

11.2.2 Case of Distinct Real Eigenvalues

The characteristic equation $\det(A - \lambda E) = 0$ for the eigenvalue λ of matrix A with the unit matrix $E := \begin{pmatrix} 1 & 0 \\ 0 & 1 \end{pmatrix}$ becomes the following quadratic equation:

$$\lambda^2 - (\mathrm{tr}\,A)\lambda + \det A = 0, \tag{11.16}$$

where $\det A := a_{11}a_{22} - a_{12}a_{21}$ and $\mathrm{tr}\,A := a_{11} + a_{22}$.

In this section, we consider the case where eigenvalues λ_+ and λ_- given by the roots for the characteristic Eq. (11.16) are real and different from each other. With

---

[1] The arguments in this part are mathematically analogous to those in Sect. 13.2 for the two dimensional system of first order linear *differential* equations. It would be very helpful for readers' clearer understanding to compare one with the other.

column (right) eigenvectors $\mathbf{p}_1 := {}^{\mathsf{T}}(p_{11}\ p_{12})$ and $\mathbf{p}_2 := {}^{\mathsf{T}}(p_{21}\ p_{22})$ respectively for λ_+ and λ_-, let us define the 2×2 matrix

$$P := (\mathbf{p}_1\ \mathbf{p}_2) = \begin{pmatrix} p_{11}\ p_{21} \\ p_{12}\ p_{22} \end{pmatrix}.$$

It can be mathematically proved that matrix P is regular. Now we can diagonalize matrix A as

$$P^{-1}AP = \begin{pmatrix} \lambda_+ & 0 \\ 0 & \lambda_- \end{pmatrix},$$

where P^{-1} is the inverse matrix of P.

Next let us define $\mathbf{w}_n = {}^{\mathsf{T}}(w_{n,1}\ w_{n,2}) := P^{-1}\mathbf{v}_n$. Then we have $\mathbf{v}_n = P\mathbf{w}_n$. From $A\,\mathbf{v}_n = AP\,\mathbf{w}_n$, we can find the following recurrence relation:

$$\mathbf{w}_{n+1} = P^{-1}\mathbf{v}_{n+1} = P^{-1}A\,\mathbf{v}_n = P^{-1}AP\,\mathbf{w}_n = \begin{pmatrix} \lambda_+ & 0 \\ 0 & \lambda_- \end{pmatrix}\mathbf{w}_n.$$

Hence we have

$$\begin{cases} w_{n+1,1} = \lambda_+\, w_{n,1} \\ w_{n+1,2} = \lambda_-\, w_{n,2}, \end{cases}$$

and subsequently find that $w_{n,1} = w_{1,1}\,\lambda_+^{n-1}$ and $w_{n,2} = w_{1,2}\,\lambda_-^{n-1}$. From this result, we can get the general solution for the system of difference Eq. (11.15):

$$\mathbf{v}_n = P\mathbf{w}_n = (\mathbf{p}_1\ \mathbf{p}_2)\begin{pmatrix} w_{n,1} \\ w_{n,2} \end{pmatrix} = w_{1,1}\,\lambda_+^{n-1}\mathbf{p}_1 + w_{1,2}\,\lambda_-^{n-1}\mathbf{p}_2, \qquad (11.17)$$

where $\mathbf{w}_1 = {}^{\mathsf{T}}(w_{1,1}\ w_{1,2})$ is uniquely determined for given $\mathbf{v}_1 = {}^{\mathsf{T}}(u_1\ v_1)$ with $\mathbf{w}_1 = P^{-1}\mathbf{v}_1$.

From these arguments, we find that the system of homogeneous linear difference Eq. (11.15) has the following general form of solution when matrix A has distinct real eigenvalues:

$$\begin{cases} u_n = c_{\mathrm{u}}\,\lambda_+^n + c_{\mathrm{u}}'\,\lambda_-^n \\ v_n = c_{\mathrm{v}}\,\lambda_+^n + c_{\mathrm{v}}'\,\lambda_-^n, \end{cases} \qquad (11.18)$$

where c_{u}, c_{u}', c_{v} and c_{v}' are constants uniquely determined for an initial condition given as (u_1, v_1).

11.2.3 Case of Multiple Eigenvalues

In this section, we consider the case where the roots for the characteristic Eq. (11.16) is degenerated as λ, that is, the case of multiple eigenvalues, $\lambda_1 = \lambda_2 = \lambda$. The case where $A = \lambda E$ is included. In such a case, that is, when $a_{11} = a_{22} = \lambda$ and $a_{12} = a_{21} = 0$, we have

$$\begin{cases} u_{n+1} = \lambda u_n \\ v_{n+1} = \lambda v_n. \end{cases}$$

Then the general solution for the system of difference Eq. (11.15) is given by $u_n = u_1 \lambda^{n-1}$ and $v_n = v_1 \lambda^{n-1}$.

For $A \neq \lambda E$, let us define a column vector $\mathbf{p}'$ such that $(A - \lambda E)\mathbf{p}' = \mathbf{p}$ with eigenvector $\mathbf{p}$ for eigenvalue λ. The vector $\mathbf{p}'$ is called *generalized eigenvector*, and is linearly independent of $\mathbf{p}$. As before, let us define 2×2 matrix $P = (\mathbf{p}\ \mathbf{p}')$. Since $A\mathbf{p} = \lambda\mathbf{p}$ and $A\mathbf{p}' = \mathbf{p} + \lambda\mathbf{p}'$, we have

$$AP = (A\mathbf{p}\ A\mathbf{p}') = (\lambda\mathbf{p}\ \mathbf{p} + \lambda\mathbf{p}') = \lambda\,(\mathbf{p}\ \mathbf{p}') + (\mathbf{0}\ \mathbf{p}) = \lambda P + PN,$$

where $N := \begin{pmatrix} 0 & 1 \\ 0 & 0 \end{pmatrix}$. With $\mathbf{w}_n = P^{-1}\mathbf{v}_n$, we can derive the following recurrence relation as before:

$$\mathbf{w}_{n+1} = P^{-1}AP\,\mathbf{w}_n = P^{-1}(\lambda P + PN)\mathbf{w}_n = (\lambda E + N)\mathbf{w}_n = \begin{pmatrix} \lambda & 1 \\ 0 & \lambda \end{pmatrix}\mathbf{w}_n.$$

Hence we obtain

$$\begin{cases} w_{n+1,1} = \lambda\,w_{n,1} + w_{n,2} \\ w_{n+1,2} = \lambda\,w_{n,2}. \end{cases}$$

From the second equation, we can find that $w_{n,2} = w_{1,2}\,\lambda^{n-1}$. By substituting this for the first equation, we get the following recurrence relation:

$$\frac{w_{n+1,1}}{\lambda^{n+1}} - \frac{w_{n,1}}{\lambda^n} = \frac{w_{1,2}}{\lambda^2}.$$

This relation indicates that the sequence $\{w_{n,1}/\lambda^n\}$ is an arithmetic progression with the common difference $w_{1,2}/\lambda^2$, and thus we can derive

$$w_{n,1} = \left\{ \frac{w_{1,1}}{\lambda} + \frac{w_{1,2}}{\lambda^2}\,(n-1) \right\} \lambda^n.$$

From these arguments, we can obtain the following general solution for the system of difference Eq. (11.15) when matrix A has multiple eigenvalues:

$$\mathbf{v}_n = P\mathbf{w}_n = \left\{ \frac{w_{1,1}}{\lambda} + \frac{w_{1,2}}{\lambda^2}(n-1) \right\} \lambda^n \mathbf{p} + w_{1,2}\lambda^{n-1}\mathbf{p}'. \tag{11.19}$$

As a result, it is shown that the system of homogeneous linear difference Eq. (11.15) has the following general form of solution when matrix A has multiple eigenvalues:

$$\begin{cases} u_n = (c_u + c_u' n)\lambda^n \\ v_n = (c_v + c_v' n)\lambda^n, \end{cases} \tag{11.20}$$

where c_u, c_u', c_v and c_v' are constants uniquely determined for a given initial condition as (u_1, v_1).

11.2.4 Case of Imaginary Eigenvalues

In this section, we consider the case where matrix A has conjugate imaginary eigenvalues λ and $\bar{\lambda}$, for which corresponding eigenvectors are given by $\mathbf{p}$ and $\bar{\mathbf{p}}$. As before, let us define 2×2 matrix $P := (\mathbf{p}\,\bar{\mathbf{p}})$, which is now a complex matrix.

Although vectors $\mathbf{p}$, $\bar{\mathbf{p}}$, and matrix P contain imaginary elements, we can apply the same arguments as for the case where matrix A has distinct real eigenvalues. We can obtain the following general solution for the system of difference Eq. (11.15):

$$\mathbf{v}_n = w_{1,1}\lambda^{n-1}\mathbf{p} + w_{1,2}\bar{\lambda}^{n-1}\bar{\mathbf{p}}, \tag{11.21}$$

where constants $w_{1,1}$ and $w_{1,2}$ are both real or conjugate to each other, because $\mathbf{v}_n$ must be a real vector for any n now. First we shall prove this in the subsequent part.

Since $\mathbf{v}_{n_1}$ must be a real vector for any natural number $n = n_1$ about the system of difference Eq. (11.15), it must hold that $\mathbf{v}_{n_1} = \bar{\mathbf{v}}_{n_1}$. Hence, the following equation must hold:

$$w_{1,1}\lambda^{n_1-1}\mathbf{p} + w_{1,2}\bar{\lambda}^{n_1-1}\bar{\mathbf{p}} = \overline{w}_{1,1}\bar{\lambda}^{n_1-1}\bar{\mathbf{p}} + \overline{w}_{1,2}\lambda^{n_1-1}\mathbf{p},$$

that is,

$$\lambda^{n_1-1}(w_{1,1} - \overline{w}_{1,2})\mathbf{p} = \bar{\lambda}^{n_1-1}(\overline{w}_{1,1} - w_{1,2})\bar{\mathbf{p}}.$$

Since the right side is conjugate to the left side, this equation indicates that $\lambda^{n_1-1}(w_{1,1} - \overline{w}_{1,2})\mathbf{p}$ is a real vector for $n = n_1$. Here λ^{n_1-1} can be regarded as imaginary for almost every natural number n_1. Therefore, it must be satisfied that

$w_{1,1} - \overline{w}_{1,2} = 0$. Thus we find that $w_{1,1} = \overline{w}_{1,2}$, which means that $w_{1,1}$ and $w_{1,2}$ are both real or conjugate to each other. Applying this result for (11.21), we can rewrite the general solution as follows:

$$\mathbf{v}_n = w_{1,1} \lambda^{n-1} \mathbf{p} + \overline{w}_{1,1} \overline{\lambda}^{n-1} \overline{\mathbf{p}}. \tag{11.22}$$

Since the right side is the sum of mutually conjugate vectors, it is real.

Let us put $\lambda = |\lambda| e^{i\theta} = |\lambda|(\cos\theta + i\sin\theta)$ with imaginary unit i, where $|\lambda|$ is the absolute value of imaginary number λ, and $\theta = \arg\lambda$ is the argument of λ. Besides, denote the complex vector $\mathbf{p}$ as $\mathbf{p} = \boldsymbol{\alpha} + i\boldsymbol{\beta}$ with appropriate real vectors $\boldsymbol{\alpha}$ and $\boldsymbol{\beta}$. From Euler's formulation, we have $\lambda^n = e^{in\theta} = \cos n\theta + i\sin n\theta$. Then the general solution (11.22) becomes

$$\begin{aligned}
\mathbf{v}_n &= w_{1,1}|\lambda|^{n-1}\{\cos(n-1)\theta + i\sin(n-1)\theta\}(\boldsymbol{\alpha} + i\boldsymbol{\beta}) \\
&\quad + \overline{w}_{1,1}|\lambda|^{n-1}\{\cos(n-1)\theta - i\sin(n-1)\theta\}(\boldsymbol{\alpha} - i\boldsymbol{\beta}) \\
&= |\lambda|^{n-1}\{c_1\cos(n-1)\theta + c_2\sin(n-1)\theta\}\boldsymbol{\alpha} \\
&\quad + |\lambda|^{n-1}\{c_2\cos(n-1)\theta - c_1\sin(n-1)\theta\}\boldsymbol{\beta}, \tag{11.23}
\end{aligned}$$

where $c_1 = w_{1,1} + \overline{w}_{1,1} = 2\operatorname{Re} w_{1,1}$ and $c_2 = i(w_{1,1} - \overline{w}_{1,1}) = -2\operatorname{Im} w_{1,1}$.

Consequently we find that the system of homogeneous linear difference Eq. (11.15) has the following general form of solution when matrix A has imaginary eigenvalues:

$$\begin{cases}
u_n = |\lambda|^n(c_u\cos n\theta + c'_u\sin n\theta) \\
v_n = |\lambda|^n(c_v\cos n\theta + c'_v\sin n\theta), \tag{11.24}
\end{cases}$$

where c_u, c'_u, c_v and c'_v are real constants uniquely determined for the initial condition given as (u_1, v_1).

11.2.5 Asymptotic Behavior of the Sequence

Consequently from those results in Sects. 11.2.2–11.2.4, we can find the following theorem about the asymptotic behavior of the sequence $\{(u_n, v_n)\}$ generated by (11.14) as $n \to \infty$:

Theorem 11.2 *The sequence $\{(u_n, v_n)\}$ generated by the system of homogeneous linear difference Eq. (11.14) has the following asymptotic behavior as $n \to \infty$:*

- *If every eigenvalue λ of matrix A defined by (11.15) has the absolute value less than one, $|\lambda| < 1$, it converges to $(0, 0)$ as $n \to \infty$ independently of the initial condition.*

- *If matrix A has an eigenvalue λ which absolute value is greater than one, $|\lambda| > 1$, it diverges, that is, $|u_n| \to \infty$ or $|v_n| \to \infty$ for almost every initial condition.*
- *If one eigenvalue of matrix A is 1 and the other has the absolute value less than one, it converges to a certain finite point for almost every initial condition.*
- *If one eigenvalue of matrix A is -1 and the other has the absolute value less than one, it converges to the sequence of a repetition of positive and negative numbers with the same absolute value for almost every initial condition.*
- *If the absolute value of every eigenvalue of matrix A is unity, it diverges for almost every initial condition when the eigenvalues are real, and converges to an oscillatory variation with a finite supremum of the amplitude for almost every initial condition when they are imaginary.*

Some readers may find the similarity of Theorem 11.2 here with Theorem 11.1 for the homogeneous second order linear difference equation in Sect. 11.1. It is mathematically right, because the second order linear difference Eq. (11.2) can be rewritten as the following mathematically equivalent system of homogeneous two dimensional linear difference equations:

$$\begin{cases} a_{n+1} = pa_n + b_n; \\ b_{n+1} = qa_n. \end{cases}$$

It is easy to find that the matrix $A = \begin{pmatrix} p & 1 \\ q & 0 \end{pmatrix}$ has the same characteristic equation as (11.3).

References

1. S. Elaydi, *An Introduction to Difference Equations*, 3rd edn. (Springer, Berlin, 2005)
2. G. Fulford, P. Forrester, A. Jones, *Modelling with Differential and Difference Equations*. Australian Mathematical Society Lecture Series, vol. 10 (Cambridge University Press, Cambridge, 1997)
3. S. Goldberg, *Introduction to Difference Equations: With Illustrative Examples from Economics, Psychology and Sociology* (Dover, New York, 1986)
4. D. Kaplan, L. Glass, *Understanding Nonlinear Dynamics*. Textbooks in Mathematical Sciences (SpringerVerlag, New York, 1995)
5. G. Ledder, *Mathematics for the Life Sciences: Calculus, Modeling, Probability and Dynamical Systems*. Springer Undergraduate Texts in Mathematics and Technology (Springer, New York, 2013)
6. R.E. Mickens, *Difference Equations: Theory Applications and Advanced Topics*. Monographs and Research Notes in Mathematics, 3rd edn. (CRC Press, Boca Raton, 2015)

Chapter 12
Qualitative Analysis for Discrete Time Model

Abstract In this section, some mathematical equipments are introduced for the qualitative analysis on the one dimensional discrete time model, which are closely related to the contents of Chap. 2.

12.1 One Dimensional Discrete Time Model

In this section, some mathematical equipments are introduced for the qualitative analysis on one dimensional discrete time model given by an autonomous difference equation

$$x_{n+1} = g(x_n), \qquad (12.1)$$

which defines the sequence $\{x_n\}$.

12.1.1 Local Stability of Equilibrium

Equilibrium x^* for the model (12.1) is given by the root of the following equation:

$$x^* = g(x^*). \qquad (12.2)$$

The equilibrium means a state of the dynamics such that the value of x_n is unchanged and kept being x^* if $x_1 = x^*$. The *local stability* about equilibrium x^* depends on the behavior of the sequence $\{x_n\}$ when x_1 is different from equilibrium x^* and sufficiently near it.

The equilibrium may be called *equilibrium state, equilibrium point, steady state, steady point,* or *rest point,* while each of those words could be sometimes used in a definition different from that of the above "equilibrium" in a rigorous sense, depending on the context. For example, from the viewpoint of *map* to give a correspondence between elements of different sets, the equilibrium for such a map may be called *fixed point.* Although we do not usually have to distinguish them from each other, we must know the possibility of such a difference in the use. It is to be remarked that the definition of equilibrium or the above words does not have no relation to its stability in general.

Let $x_1 = x^* + \epsilon_1$ with $0 < |\epsilon_1| \ll 1$. The value ϵ_1 is sometimes called *perturbation* from equilibrium x^*. It must be remarked that ϵ_1 can be negative. We shall consider the sequence $\{\epsilon_n\}$ defined as $\epsilon_n = x_n - x^*$. If $|\epsilon_n|$ becomes larger as n gets larger for some ϵ_1, such an equilibrium x^* is called *unstable equilibrium.* This indicates that a slight change of x_n from equilibrium x^* may cause a state transition far away from x^*. In contrast, if $|\epsilon_n|$ becomes smaller toward zero as n gets larger for *any* ϵ_1, it indicates that any slight change of the state from x^* diminishes as n gets larger, so that the state returns to equilibrium x^*. Such an equilibrium x^* is called *locally asymptotically stable.*

Let us now assume that the function $g(x)$ can have the following expansion:

$$g(x) = g(x^*) + g'(x^*)(x - x^*) + o(x - x^*),$$

where $g'(x^*) = dg/dx\big|_{x=x^*}$. This may be derived by Taylor expansion around $x = x^*$ for the differentiable function g.

The mathematical symbol o is called *Landau o* (Landau small o). In the above expansion, it means all the rest terms which have higher order than $x - x^*$, satisfying

$$\lim_{x \to x^*} \frac{o(x - x^*)}{x - x^*} = 0.$$

More generally, $p(x) = o(q(x))$ as $x \to a$ when $\lim_{x \to a} p(x)/q(x) = 0$. It must be remarked that there is the other symbol called *Landau O* (Landau big o) with the different definition: $p(x) = O(q(x))$ as $x \to a$ when $\lim_{x \to a} |p(x)/q(x)| < \infty$.

From (12.1) with $x_n = x^* + \epsilon_n$, we have

$$x^* + \epsilon_{n+1} = g(x^* + \epsilon_n) = g(x^*) + g'(x^*)\epsilon_n + o(\epsilon_n) \approx x^* + g'(x^*)\epsilon_n,$$

since $g(x^*) = x^*$. Hence, the sequence $\{x_n\}$ sufficiently near x^* can be approximately determined by the sequence $\{\widetilde{\epsilon}_n\}$ defined by the following recurrence relation, because it is uniquely determined by the sequence $\{\epsilon_n\}$:

$$\widetilde{\epsilon}_{n+1} = g'(x^*)\widetilde{\epsilon}_n. \tag{12.3}$$

This approximation is applicable only when $g'(x^*) \neq 0$. In case of $g'(x^*) = 0$, since the sequence $\{\epsilon_n\}$ is governed by the term with a higher order than ϵ, the following way to determine the local stability of x^* is useless. Although such a case would be mathematically interesting, we shall not go into such a case because it is singular generally for a mathematical model of population dynamics.

For the local stability analysis on the model with a non-linear function g, it is essential to derive the linear recurrence relation (12.3) from the original (12.1). The derivation is called *linearization* of the model around equilibrium x^*, and the derived linear equation is called *linearized equation*.

Since the linear recurrence Eq. (12.3) is of a geometric progression, we find that, if $|g'(x^*)| < 1$, $\widetilde{\epsilon}_n \to 0$ as $n \to \infty$. Inversely, if $|g'(x^*)| > 1$, $|\widetilde{\epsilon}_n|$ becomes larger as n gets larger. As a consequence of these results about the sequence $\{\widetilde{\epsilon}_n\}$, we can get the following theorem on the local stability of x^* for (12.1):

Theorem 12.1 *Equilibrium x^* for the model* (12.1) *is locally asymptotically stable if* $|g'(x^*)| < 1$, *while it is unstable if* $|g'(x^*)| > 1$.

12.1.2 Cobwebbing Method

In this section, we introduce a method to discuss the qualitative nature of the sequence $\{x_n\}$, well-known as *cobwebbing method*. It is a technique to understand the qualitative nature of the sequence $\{x_n\}$ generated by (12.1), making use of the graph of $y = g(x)$ in the (x, y)-plane (Fig. 12.1). The reader can easily find its introduction in many textbooks about the discrete dynamical system or population dynamics model (for example, [2, 4, 7, 8, 10–12]).

For $x = x_0$, we can get $y = g(x_0) = x_1$, which means that the point (x_0, x_1) is on the curve of $y = g(x)$. In this method, we shall add a line $y = x$ in the same plane as $x_{n+1} = x_n$ shown in Fig. 12.1. The intersection of the horizontal line passing the point (x_0, x_1) with $y = x$ gives the point (x_1, x_1). The intersection of vertical line passing the point (x_1, x_1) with $y = g(x)$ gives the point $(x_1, g(x_1)) = (x_1, x_2)$. With these relations about points, we can draw the orbit $\{x_n\}$ on the graph, repeating the following two steps (see Fig. 12.1):

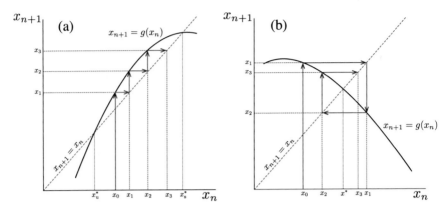

Fig. 12.1 Illustrative examples of the orbit $\{x_n\}$ generated by (12.1), drawn with the cobwebbing method. (**a**) The orbit is monotonically increasing; (**b**) The orbit shows a damped oscillation. See the detail in the main text

1. Draw a horizontal line from $(x_n, g(x_n)) = (x_n, x_{n+1})$ on $y = g(x)$ to the line $y = x$, and find the intersection (x_{n+1}, x_{n+1});
2. Draw a vertical line from (x_{n+1}, x_{n+1}) on $y = x$ to the curve $y = g(x)$, and find the intersection $(x_{n+1}, g(x_{n+1})) = (x_{n+1}, x_{n+2})$.

A resemblance of the process drawing lines to the spider's cobwebbing is the reason of its naming.

It should be remarked that the intersection between the curve $y = g(x)$ and the line $y = x$ indicates the equilibrium for the model (12.1), since it is the point (x^*, x^*) such that $x^* = g(x^*)$. In Fig. 12.1a, the orbit $\{x_n\}$ generated by (12.1) asymptotically approaches equilibrium x_s^* in a monotonic manner. From the above process to draw the orbit, it is clear that the sequence $\{x_n\}$ converges to equilibrium x_s^* as $n \to \infty$. Further, we can easily find that the other equilibrium x_u^* is unstable.

In contrast, Fig. 12.1b gives an example such that the orbit $\{x_n\}$ appears oscillatory around equilibrium x^* in a damping manner. It should be remarked that it is in general difficult to determine only by the cobwebbing method which the orbit asymptotically approaches equilibrium x^* in a damping manner or shows the other behavior, when such an oscillatory orbit appears. As shown in Fig. 12.1a, generally the cobwebbing method is a strong technique to investigate the qualitative nature of the sequence $\{x_n\}$ for the range where $y = g(x)$ is increasing, while it is just a supplementarily useful technique of the qualitative analysis for the range where $y = g(x)$ is decreasing.

> As shown by Fig. 12.1a for the range where $y = g(x)$ is increasing, the cobwebbing method can identify the stability of the equilibrium not

(continued)

necessarily about the neighborhood around it but about a certain range of x. This means a possibility to identify the set of initial values x_0 which will approach the equilibrium as $n \to \infty$. Such a set of initial values is called *basin of attraction* for the equilibrium. Differently from the local stability analysis introduced in Sect. 12.1.1, such a result on the basin of attraction shows the *global stability* about the equilibrium. In Fig. 12.1a, equilibrium x_s^* is *globally asymptotically stable* in a wider sense, with the basin of attraction which contains the interval $(x_u^*, x_s^*]$.

12.1.3 Logistic Map

In this section, we shall consider the model (12.1) with $g(x) = \mathcal{R}_0(1-x)x$ where $\mathcal{R}_0$ is a positive constant:

$$x_{n+1} = \mathcal{R}_0(1-x_n)x_n. \tag{12.4}$$

This is the logistic map introduced by (2.13) in Sect. 2.1.3, to which the recurrence relation (12.4) mathematically corresponds with $x_n = c_n/c_c$.

We can easily apply the cobwebbing method for the qualitative analysis on the sequence $\{x_n\}$ generated by the logistic map (12.4) in case of $0 < \mathcal{R}_0 \le 2$, although we can find only in part the nature of $\{x_n\}$ in case of $\mathcal{R}_0 > 2$ as shown in the following description.

When $0 < \mathcal{R}_0 \le 1$, the graph of $y = g(x)$ is as shown in Fig. 12.2a. We can easily see that it has no intersection with $y = x$ in the first quadrant, and has the

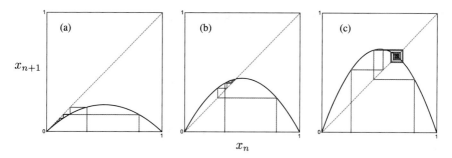

Fig. 12.2 The application of the cobwebbing method for the logistic map (12.4). Numerically drawn for different two initial values. (**a**) $\mathcal{R}_0 = 0.9$; (**b**) $\mathcal{R}_0 = 1.8$; (**c**) $\mathcal{R}_0 = 2.8$. In (**a**), the orbit asymptotically approaches equilibrium 0 in a monotonic manner. In (**b**), it asymptotically approaches equilibrium $1 - 1/\mathcal{R}_0$ in a monotonic manner. In (**c**), it asymptotically approaches equilibrium $1 - 1/\mathcal{R}_0$ with a damped oscillation

intersection at the origin. From the graph of $y = g(x)$ in this case, the cobwebbing method can show the following result:

Result $(0 < \mathcal{R}_0 \leq 1)$ The orbit $\{x_n\}$ asymptotically approaches zero in a monotonic manner for any initial value x_0 such that $0 \leq x_0 \leq 1$: $x_n \to 0$ as $n \to \infty$.

This is the case of population extinction mentioned in Sect. 2.1.3.

With the local stability analysis described in Sect. 12.1.1, it can be easily found that equilibrium 0 is locally asymptotically stable since $0 < g'(0) < 1$ when $0 < \mathcal{R}_0 < 1$. Furthermore, the cobwebbing method can show that it is globally asymptotically stable when $0 < \mathcal{R}_0 \leq 1$.

This result can be proved also in the following way: First, for $x_0 = 0$ or $x_0 = 1$, it is easily shown from (12.4) that $x_k = 0$ for any $k > 0$. Next, for $0 < x_0 < 1$, it can be easily proved from (12.4) that $0 < x_k < 1$, that is, $0 < 1 - x_k < 1$ for any $k > 0$. Then we have $x_{k+1} = \mathcal{R}_0(1 - x_k)x_k < x_k$ for any $k \geq 0$. These arguments clearly show that the sequence $\{x_n\}$ is monotonically decreasing toward zero.

As shown by Fig. 12.2b, in case of $1 < \mathcal{R}_0 \leq 2$, there are two different equilibria 0 and $1 - 1/\mathcal{R}_0$. We find that $f'(0) > 1$ and $0 \leq f'(1 - 1/\mathcal{R}_0) < 1$, so that equilibrium 0 is unstable while $1 - 1/\mathcal{R}_0$ is locally asymptotically stable. In this case, the cobwebbing method can show the following result:

Result $(1 < \mathcal{R}_0 \leq 2)$ For any initial value x_0 such that $0 < x_0 < 1$, $x_n \to 1 - 1/\mathcal{R}_0$ as $n \to \infty$. The sequence $\{x_n\}$ asymptotically approaches $1 - 1/\mathcal{R}_0$ in a monotonic manner when it becomes sufficiently near equilibrium $1 - 1/\mathcal{R}_0$. (See Fig. 2.9 in p. 43)

We can see that equilibrium $1 - 1/\mathcal{R}_0$ is globally asymptotically stable. The qualitative nature of the sequence $\{x_n\}$ is certainly similar to that of the growth curve of logistic Eq. (5.10) (p. 140) in this case, whereas the sequence $\{x_n\}$ can show a variation which can never occur for the logistic equation as numerically demonstrated for $\mathcal{R}_0 = 1.5$ in Fig. 2.9 (p. 43).

When $2 < \mathcal{R}_0 \leq 3$, we can see with the cobwebbing method that the orbit $\{x_n\}$ goes near equilibrium $1 - 1/\mathcal{R}_0$, as shown in Fig. 12.2c. However, only with the cobwebbing method, it is unclear whether it approaches equilibrium $1 - 1/\mathcal{R}_0$ or not. On the other hand, we can easily show that, when $2 < \mathcal{R}_0 < 3$, there exist two equilibria 0 and $1 - 1/\mathcal{R}_0$, where $g'(0) > 1$ and $-1 < g'(1 - 1/\mathcal{R}_0) < 0$. Therefore, when $2 < \mathcal{R}_0 < 3$, equilibrium 0 is unstable while $1 - 1/\mathcal{R}_0$ is locally asymptotically stable.

The case of $\mathcal{R}_0 = 3$ must be treated in a different way. It is impossible to investigate the local stability by the linearization around equilibrium $1 - 1/\mathcal{R}_0$, as described in Sect. 12.1.1, because $g'(1 - 1/\mathcal{R}_0) = -1$ in this case. However, a little expansion of the mathematical arguments for the linearization, we can show that equilibrium $1 - 1/\mathcal{R}_0 = 2/3$ is locally asymptotically stable (Exercise 12.1 below).

From these arguments, we can get the following result:

Result $(2 < \mathcal{R}_0 \leq 3)$ For any initial value x_0 such that $0 < x_0 < 1$, $x_n \to 1-1/\mathcal{R}_0$ as $n \to \infty$. The sequence $\{x_n\}$ asymptotically approaches $1 - 1/\mathcal{R}_0$ with a damped oscillation. (See Fig. 2.9 in p. 43)

Exercise 12.1 When $\mathcal{R}_0 = 3$, show that, equilibrium $1 - 1/\mathcal{R}_0 = 2/3$ for (12.4) is locally asymptotically stable.

12.1.4 Periodic Orbit

When $\mathcal{R}_0 > 3$, it is easily shown by the local stability analysis that equilibrium $1 - 1/\mathcal{R}_0$ for the logistic map (12.4) is unstable, because $g'(1 - 1/\mathcal{R}_0) < -1$. The other equilibrium 0 is unstable too, as easily seen by the cobwebbing method (Fig. 12.3). For the logistic map (12.4), when every equilibrium is unstable, the orbit $\{x_n\}$ approaches a *periodic solution* that repeats a finite sequence of specific values, or causes a *chaotic variation* (Fig. 2.9 in p. 43). The parameter dependence of the existence and stability of such solutions is in general called *bifurcation* of the solution.

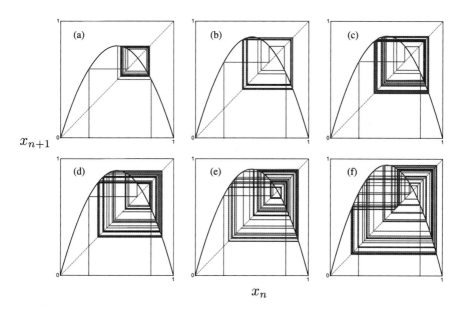

x_{n+1}

x_n

Fig. 12.3 The application of the cobwebbing method for the logistic map (12.4). Numerically drawn for different two initial values. (**a**) $\mathcal{R}_0 = 3.15$; (**b**) $\mathcal{R}_0 = 3.46$; (**c**) $\mathcal{R}_0 = 3.5$; (**d**) $\mathcal{R}_0 = 3.6$; (**e**) $\mathcal{R}_0 = 3.65$; (**f**) $\mathcal{R}_0 = 3.8$. The orbit approaches a period-2 solution in (**a**), a period-4 solution in (**b**), and a certain period-2^k solution in (**c**)

The simplest periodic solution is the period-2 solution, which repeats two different values A and B as $ABABAB\cdots$. The logistic map (12.4) has a period-2 solution of the following nature (Exercise 12.2 below):

Result $(3 < \mathcal{R}_0 \leq 1 + \sqrt{6})$ The orbit $\{x_n\}$ approaches a period-2 solution for any initial value x_0 $(0 < x_0 < 1)$ as $k \to \infty$, which is uniquely determined by the value of $\mathcal{R}_0$.

See Fig. 12.3a in this section and Fig. 2.9 in p. 43.

When the discrete time model (12.1) has a period-2 solution with two different values A and B, it is satisfied that $B = g(A)$ and $A = g(B)$. Thus we have $A = g(g(A))$ and $B = g(g(B))$. This means that, if a period-2 solution exists, it is given by the roots of equation $g(g(x)) = x$ which are different from the root of equation $g(x) = x$. Remark that the former roots necessarily contain the latter. There may not exist such roots, that is, the period-2 solution does not necessarily exist.

Hence when there is a period-2 solution with different values A and B, the values A and B must be equivalent to the equilibria for the discrete dynamical system defined by $y_{n+1} = g(g(y_n)) = g \circ g(y_n)$. The stability of the period-2 solution for (12.1) coincides with the stability of those equilibria A and B for $y_{n+1} = g(g(y_n))$.

When a period-2 solution is locally asymptotically stable, the orbit $\{x_n\}$ for the initial value x_0 sufficiently near it asymptotically approaches it. The sequences $\{x_{2k}\}$ and $\{x_{2k+1}\}$ must converge to those values of the period-2 solution respectively. The stability of a period-2 solution is equivalent to the stability of the equilibria for the discrete dynamical system $y_{n+1} = g(g(y_n))$, which are different from the root of equation $g(x) = x$. Inversely, when the discrete dynamical system $y_{n+1} = g(g(y_n))$ has locally asymptotically stable equilibria different from the root of equation $g(x) = x$, there exists a locally asymptotically stable period-2 solution for (12.1).

Therefore we can investigate the stability of a period-2 solution for (12.1) by considering the stability of the equilibria for the discrete dynamical system $y_{k+1} = g(g(y_k))$. From Theorem 12.1 in Sect. 12.1.1, the local stability can be argued by the absolute value of $dg(g(x))/dx = g'(g(x))g'(x)$ at the equilibria.

If the absolute values of

$$\left|\frac{d}{dx}g(g(x))\right|_{x=A} = \left|g'(g(A))g'(A)\right| = \left|g'(B)g'(A)\right|$$

and

$$\left|\frac{d}{dx}g(g(x))\right|_{x=B} = \left|\frac{d}{dx}g(g(x))\right|_{x=g(A)} = \left|g'(g(g(A)))g'(g(A))\right| = \left|g'(A)g'(B)\right|.$$

are less than one, the period-2 solution is locally asymptotically stable, while it is unstable if they are greater than one. As seen from the above results, the local stability of equilibrium $y = A$ necessarily coincides with that of the other equilibrium $y = B = g(A)$ for the discrete dynamical system $y_{k+1} = g(g(y_k))$, where the values A and $B = f(A)$ define a period-2 solution for (12.1).

Exercise 12.2 Prove the following features of the period-2 solution for the logistic map (12.4):

(a) A period-2 solution exists when and only when $\mathscr{R}_0 > 3$.
(b) The period-2 solution is locally asymptotically stable if $\mathscr{R}_0 < 1 + \sqrt{6}$.
(c) The period-2 solution is unstable if $\mathscr{R}_0 > 1 + \sqrt{6}$.

To mathematically prove the above result on the convergence to a period-2 solution, the local stability analysis on the period-2 solution is clearly unsatisfactory. This is because the local stability analysis cannot prove the convergence to the period-2 solution for any x_0 ($0 < x_0 < 1$). It is impossible to show it even with the cobwebbing method for (12.4) as seen from Fig. 12.3a. Instead, from the above arguments about the relation of a period-2 solution for (12.4) to equilibria for the composed map with respect to the existence and stability, we can investigate the stability of a period-2 solution by the cobwebbing method for the composed map, as shown in Fig. 12.4a.

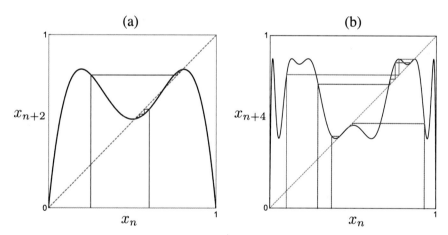

Fig. 12.4 Application of the cobwebbing method for the composed map of (12.4). Numerically drawn some orbits from different initial values. (**a**) double composed map $x_{n+2} = g(g(x_n))$ with $a = 3.2$; (**b**) threefolds composed map $x_{n+4} = g(g(g(g(x_n))))$ with $a = 3.46$. In (**a**), the obits converge to two equilibria which define a period-2 solution for (12.4), while they converge to four equilibria which define a period-4 solution for it

Suppose that a period-k solution exists for (12.1). Its local stability can be determined by the absolute value of the first derivative about the k-fold composed map of g:

$$G_k(x) = \underbrace{g \circ g \circ \cdots \circ g(x)}_{k}. \qquad (12.5)$$

This is because each value of the period-k solution is an equilibrium for the discrete dynamical system given by $y_{i+1} = G_k(y_i)$. By the chain rule of the derivative for the composition function, we have

$$\frac{dG_k(x)}{dx} = \prod_{\ell=1}^{k} g'(G_{\ell-1}(x)), \qquad (12.6)$$

where $G_0(x) = x$. Now for a value $x^\star$ belonging to the period-k solution, all of k values composing it are given by $\{x^\star, G_1(x^\star), G_2(x^\star), \ldots, G_{k-1}(x^\star)\}$. Thus, from (12.6), we can get the following theorem:

Theorem 12.2 *If the product of all the absolute values of the first derivative g' for each value of a period-k solution for (12.1) is less than one, it is locally asymptotically stable. If the product is greater than one, it is unstable.*

12.1.5   Period-Doubling Bifurcation

Logistic map (12.4) manifests its special character when $\mathscr{R}_0 > 1 + \sqrt{6}$. There exists a certain value $\tau_4 > 1 + \sqrt{6}$ such that the orbit $\{x_n\}$ asymptotically approaches a period-4 solution from any initial value x_0 ($0 < x_0 < 1$) when $1 + \sqrt{6} < \mathscr{R}_0 \leq \tau_4$ (Fig. 12.4b). As illustrated by Fig. 2.9 with $\mathscr{R}_0 = 3.5$ (p. 43), the sequence $\{x_n\}$ asymptotically approaches a stationary variation such that specific four different values are repeated as $ABCDABCDABCD\cdots$. Further, when $\tau_4 < \mathscr{R}_0 \leq \tau_8$ with a specific value $\tau_8 > \tau_4$, the orbit $\{x_n\}$ asymptotically approaches a period-8 solution from any initial value x_0 ($0 < x_0 < 1$).

It has been proved about the *bifurcation* of the solution that there exists an infinite sequence $\{\tau_{2^n} \mid 3 \leq \tau_{2^n} < \tau_\infty < 4, \tau_{2^n} < \tau_{2^{n+1}}; \ n = 1, 2, 3, \ldots\}$, which is monotonically increasing toward a finite value $\tau_\infty < 4$, such that the orbit $\{x_n\}$ asymptotically approaches a period-2^{n+1} solution from any initial value x_0 ($0 < x_0 < 1$) when $\tau_{2^n} < \mathscr{R}_0 \leq \tau_{2^{n+1}}$. The limit of τ_{2^n} as $n \to \infty$, that is, τ_∞ is known to be near 3.57, and sometimes called *Feigenbaum point*. Such a parameter dependence of the solution with doubling the period is called *period-*

doubling bifurcation. Moreover, it is one of what is called *pitchfork bifurcation* in the theory of dynamical system [1, 6, 10, 12, 13].

The size of range $(\tau_{2^{n-1}}, \tau_{2^n}]$ shrinks as n gets larger. Mitchell J. Feigenbaum [3] numerically observed

$$\lim_{n \to \infty} \frac{\tau_{2^n} - \tau_{2^{n-1}}}{\tau_{2^{n+1}} - \tau_{2^n}} = 4.669201609 \cdots \quad [\textit{Feigenbaum constant}].$$

Subsequentmathematical researches showed that the Feigenbaum constant appears for a wide class of discrete dynamical system [5, 12], and it is still an attractive mathematical problem.

For $\mathcal{R}_0 > \tau_\infty$, the orbit $\{x_n\}$ may approach a periodic solution with period different from any power of 2. For example, there exist ranges of $\mathcal{R}_0$ for which a period-3 solution or the other with odd number of period appear (Fig. 12.5). The ranges of $\mathcal{R}_0$ for which such solutions with different periods are distributed in a disconnected manner within $\tau_\infty < \mathcal{R}_0 < 4$. For $\mathcal{R}_0$ between nearest ranges for such periodic solutions, the orbit $\{x_n\}$ does not approach any periodic solution or equilibrium for almost every initial value x_0, but continues to take different values, which is today called *chaotic variation* (see Fig. 2.9 with $\mathcal{R}_0 = 3.8$ in p. 43).

Visualization of such a bifurcation of the solution about a dynamical system is given by the *bifurcation diagram* introduced in Sect. 2.1.1 and its subsequent part. Numerically drawn bifurcation diagram for the logistic map (12.4) is shown in Fig. 2.10 (p. 45). It shows the bifurcation structure of the solution with the pitchfork bifurcation toward the chaotic variation in terms of the bifurcation parameter $\mathcal{R}_0$.

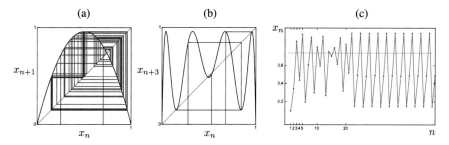

Fig. 12.5   Period-3 solution for (12.4). Numerically drawn with $\mathcal{R}_0 = 3.84$. (**a**) The orbits $\{x_n\}$ by the cobwebbing method for two different initial values; (**b**) The orbits $\{x_{3n}\}$ by the cobwebbing method for three different initial values, generated by the threefolds composed map defined by $x_{n+3} = g(g(g(x_n)))$; (**c**) Plots of the sequence $\{x_n\}$ from $x_1 = 0.1$

As described in Sect. 2.1.1, the numerical drawing of the bifurcation diagram like Fig. 2.10 and the others cannot clearly show any periodic solution with a much long period, since it is drawn only with plots of the value x_n over a limited range of large n. Especially for $\mathscr{R}_0 > \mathfrak{r}_\infty$, there must be a lot of points which are not plotted by the numerical calculation.

For the logistic map (12.4) with $\mathscr{R}_0 = \mathfrak{r} > \mathfrak{r}_\infty$, it is mathematically shown that there exists any periodic solution which appears for the value of $\mathscr{R}_0 < \mathfrak{r}$, as explained below. However, they are necessarily unstable except for the one which is asymptotically stable for $\mathscr{R}_0 = \mathfrak{r}$. This means that, if there exists an asymptotically stable periodic solution for a given value of $\mathscr{R}_0$ about the logistic map (12.4), it is the unique asymptotically stable periodic solution, and there does not exist any other asymptotically stable periodic solution or equilibrium.

For these reasons, such a numerically drawn bifurcation diagram is just an approximation for the mathematically exact one. Hence it may be sometimes called *orbit diagram* instead of bifurcation diagram.

Such a numerically drawn bifurcation diagram can be regarded as a drawing of *ω-limit set* by numerics. The ω-limit set for an initial value x_0 is defined as the set of values which the orbit $\{x_n\}$ asymptotically approaches. When it asymptotically approaches an equilibrium, the ω-limit set is the set containing only a single value of the equilibrium. When it asymptotically approaches a period-k solution, it is the set of k different values which define the period-k solution. Especially, when it shows a chaotic variation, the ω-limit set is the set of infinite number of values which are specifically determined by the dynamical system itself. It is called *strange attractor*. For example, see [5] for more detail, and [1, 6, 10, 13] for the further mathematical contents about it.

As seen in the bifurcation diagram of Fig. 2.10 for the logistic map (12.4), there is a range of $\mathscr{R}_0$ around 3.84 where a period-3 solution appears (see Fig. 12.5, and Fig. 2.9 in p. 43 of Sect. 2.1.3). Such a finite range beyond $\mathfrak{r}_\infty$ for a periodic solution is called *window* in the theory of dynamical system. The following theorem is well-known [9]:

Theorem 12.3 (Li-York Theorem) *When a period-3 solution exists for a one dimensional dynamical system given by (12.1) with a function g continuous over* $\mathbb{R}$, *the system has every period-q solution for any positive integer q.*

From this theorem, if a period-3 solution appears, any other periodic solution exists and they are unstable. More precisely in a mathematical sense, for an initial value which is included in the values of a certain periodic solution, the orbit $\{x_n\}$ becomes the periodic solution even if it is unstable. Since all periodic solutions exist, the reader may think that such a periodic solution would be likely to appear. However, note that there are infinite initial values from which the orbit approaches the period-

3 solution. In an actual numerical calculation with an inevitable rounding error, the orbit on any unstable period solution would eventually become out of it. Thus we can expect that it would at least approximately approach the asymptotically stable period-3 solution even in the numerical calculation.

The above theorem may be regarded as included in the following famous theorem:

Theorem 12.4 (Sharkovskii Theorem) *When a period-p solution exists for a one dimensional dynamical system given by (12.1) with a function g continuous over $\mathbb{R}$, the system has all period-q solutions for q such as $p \rhd q$ in the following order, called Sharkovskii ordering:*

$$3 \rhd 5 \rhd 7 \rhd 9 \rhd \cdots \rhd 2 \cdot 3 \rhd 2 \cdot 5 \rhd 2 \cdot 7 \rhd \cdots \rhd 2^2 \cdot 3 \rhd 2^2 \cdot 5 \rhd 2^2 \cdot 7 \rhd \cdots$$

$$\rhd 2^k \cdot 3 \rhd 2^k \cdot 5 \rhd 2^k \cdot 7 \rhd \cdots \rhd 2^{k+1} \rhd 2^k \rhd 2^{k-1} \rhd \cdots \rhd 2^3 \rhd 2^2 \rhd 2 \rhd 1$$

In this order, if $p_1 \rhd p_2 \rhd p_3$, then $p_1 \rhd p_3$.

As for the mathematical proofs of these theorems, for example, see [1, 12]. As seen in the bifurcation diagram of Fig. 2.10 for the logistic map (12.4), for $\mathscr{R}_0$ less than the window for the period-3 solution, there are windows of periodic solutions with period which has an odd factor.

From Sharkovskii theorem, there are infinite number of different periodic solutions when a periodic solution with an odd period appears. It must be remarked that the theorem does not claim the appearance of stable periodic solution in the Sharkovskii ordering. It is the theorem about the existence of periodic solutions. For example, if a one dimensional discrete dynamical system has a period-12 solution, it cannot have any periodic solution with odd period. Logistic map (12.4) is one of one dimensional discrete dynamical systems for which the asymptotically stable periodic solutions appear in the bifurcation following all the Sharkovskii ordering. In the next section to get the better understanding of the analysis on the nature of one dimensional discrete dynamical system (12.1), we shall see another example for which all periodic solutions exist and they are always unstable.

12.1.6 Tent Map

In this section, we shall consider the one dimensional dynamical system given by (12.1) with

$$g(x) = \begin{cases} 2\alpha x & \left(0 \leq x \leq \dfrac{1}{2}\right) \\[2mm] 2\alpha(1-x) & \left(\dfrac{1}{2} < x \leq 1\right), \end{cases} \tag{12.7}$$

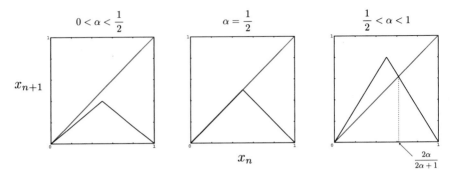

Fig. 12.6 The graph of $x_{n+1} = g(x_n)$ for the tent map (12.1) with (12.7)

where α is a positive constant such that $0 < \alpha \leq 1$. This simple dynamical system is sometimes called *tent map*, and has made an important contribution to the development of the theory of discrete dynamical system [1, 5, 6, 10, 13].

As seen from the above definition of the function g, let us consider the range of the initial value x_0 only in $[0, 1]$. Since $\alpha \in (0, 1]$, the range of g is $[0, a] \subset [0, 1]$. Hence, for any initial value $x_0 \in [0, 1]$, the value x_n remains in $[0, 1]$ for any $n \geq 0$.

Existence of Equilibrium

As shown in Fig. 12.6, the graph of $x_{n+1} = g(x_n)$ in $[0, 1]$ intersects with the line $x_{n+1} = x_n$ only at the origin when $\alpha < 1/2$, while it has two intersections at the origin and at $x = 2\alpha/(2\alpha + 1)$ when $\alpha > 1/2$. That is, the existence of equilibrium x^* for the tent map (12.1) with (12.7) depends on the value of the parameter α:

$$
x^* = \begin{cases} 0 & \text{for } \alpha < \dfrac{1}{2}; \\[2ex] \left[0, \dfrac{1}{2}\right] & \text{for } \alpha = \dfrac{1}{2}; \\[2ex] 0, \dfrac{2\alpha}{2\alpha + 1} & \text{for } \alpha > \dfrac{1}{2}. \end{cases}
$$

The case of $\alpha = 1/2$ is special, because $x_n = x_0$ for any $n \geq 0$ when $x_0 \in [0, 1/2]$, and $x_n = 1 - x_0 \in [0, 1/2]$ for any $n \geq 1$ when $x_0 \in (1/2, 1]$. Hence we may call any point in $[0, 1/2]$ "equilibrium", whereas it depends on the initial value x_0. In general, such "equilibrium" must be distinguished from the equilibrium in the other case. The stability of such "equilibrium" is mathematically categorized as *Lyapunov stable (L-stable), weakly stable,*

(continued)

neutrally stable, or simply stable, as introduced in Sect. 14.3 (see p. 427). Such a stability does not indicate the asymptotical stability since any small perturbation from an "equilibrium" results in a small change of "equilibrium" value. However, such a change of "equilibrium" value is sufficiently small for any sufficiently small perturbation of the previous "equilibrium". In this sense, such an "equilibrium" cannot be classified as "unstable", following the mathematical definition of the unstable equilibrium.

In the following arguments, we shall consider only the case of $\alpha \neq 1/2$. The non-trivial equilibrium $x^* = 2\alpha/(2\alpha + 1)$ exists when and only when $\alpha > 1/2$.

Stability of Equilibrium

The absolute value of the slope of $g(x)$ is 2α for any $x \neq 1/2$. When $\alpha > 1/2$, $|g'(x^*)| = 2\alpha > 1$ for both of equilibria 0 and $2\alpha/(2\alpha + 1)$. Therefore, both of equilibria 0 and $2\alpha/(2\alpha + 1)$ are unstable when $\alpha > 1/2$.

When $\alpha < 1/2$, the absolute value of the slope of $g(x)$, 2α, is less than 1 for any $x \neq 1/2$. Thus, the unique equilibrium $x^* = 0$ is locally asymptotically stable. Further, by the cobwebbing method, it is easy to show that equilibrium $x^* = 0$ is globally asymptotically stable (see Fig. 12.7).

Period-2 Solution

Let us consider the existence and stability of a period-2 solution for the tent map (12.1) with (12.7). If it exists, it must be given by the roots of the equation $g(g(x)) =$

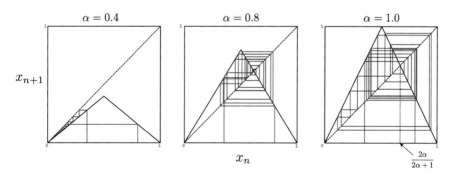

Fig. 12.7 Orbits $\{x_n\}$ by the cobwebbing method for the tent map (12.1) with (12.7). Numerically drawn up to x_{50} with $x_0 = 0.351$ and $x_0 = 0.801$ respectively

x except for the roots of $g(x) = x$ (refer to Sect. 12.1.4). Remark that the former roots necessarily contain the latter.

When $\alpha \leq 1/2$, no period-2 solution exists. The orbit $\{x_n\}$ for any initial value $x_0 \in (0, 1]$ satisfies that $g(x_n) < x_n$ for any $n \geq 0$. Hence it is satisfies that $g(g(x)) < g(x) < x$ for any $x \in (0, 1]$. This means that any period-2 solution cannot exist.

When $\alpha > 1/2$, if there exists a period-2 solution with its values $x^\star$ and $g(x^\star)$ such that $x^\star < g(x^\star)$, it must be satisfied that $x^\star < 2\alpha/(2\alpha + 1) < g(x^\star)$. This is because of the following reason. From Fig. 12.6, we can easily see that, when $\alpha > 1/2$, $x < g(x)$ for $x \in [0, 2\alpha/(2\alpha + 1))$ while $x > g(x)$ for $x \in (2\alpha/(2\alpha + 1), 1]$. Hence, in order to satisfy that $x^\star < g(x^\star)$, it must be satisfied that $x^\star < 2\alpha/(2\alpha + 1)$. Further, since it is satisfied that $g(g(x^\star)) = x^\star$ for the period-2 solution, we can find that $2\alpha/(2\alpha + 1) < g(x^\star)$.

Now suppose that $x^\star > 1/2$. Then we have $g(x^\star) = 2\alpha(1 - x^\star)$. Since $g(x^\star) > 1/2$ from the above arguments, it is satisfied that $g(g(x^\star)) = 2\alpha\{1 - g(x^\star)\}$. Thus, from the equation $g(g(x^\star)) = x^\star$, we get the equation $2\alpha\{1 - 2\alpha(1 - x^\star)\} = x^\star$, and find the root

$$x^\star = \frac{2\alpha(1 - 2\alpha)}{1 - 4\alpha^2} = \frac{2\alpha}{2\alpha + 1}.$$

This is the root of the equation $g(x) = x$, and thus is not the root for the period-2 solution.

Next, suppose that $x^\star < 1/2$. Then we have $g(x^\star) = 2\alpha x^\star$, and $g(g(x^\star)) = 2\alpha x^\star\{1 - g(x^\star)\}$. Thus, from the equation $g(g(x^\star)) = x^\star$, we get the equation $2\alpha(1 - 2\alpha x^\star) = x^\star$. The root of this equation is

$$x^\star = \frac{2\alpha}{4\alpha^2 + 1}. \tag{12.8}$$

The assumption $x^\star < 1/2$ is necessarily satisfied for (12.8) when $\alpha > 1/2$. Further from (12.8), we get

$$g(x^\star) = \frac{4\alpha^2}{4\alpha^2 + 1}.$$

We can easily show that it is always satisfied when $\alpha > 1/2$ that $2\alpha/(2\alpha + 1) < g(x^\star)$, as is required from the above arguments.

For $\alpha > 1/2$, there cannot exist any period-2 solution such that $x^\star = 1/2$. Since $g(1/2) = \alpha$, we have $g(\alpha) = 1/2$ if there exists such a period-2 solution. However, for $\alpha > 1/2$, it is only when $\alpha = 1/2$. This is contradictory.

From these arguments, we can conclude that there exists a period-2 solution with the values $x^\star$ and $g(x^\star)$ defined by (12.8) when and only when $\alpha > 1/2$. Moreover, since $|g'(x)| = 2\alpha > 1$ for any $x \neq 1/2$ when $\alpha > 1/2$, we find that

$|g'(g(x^\star))g'(x^\star)| = 4\alpha^2 > 1$. This means that the period-2 solution is necessarily unstable (refer to Sect. 12.1.4 about the local stability of period-2 solution).

Period-k Solution

Suppose that there exists a period-k solution which does not contain $1/2$. From the above arguments, it is necessary that $\alpha > 1/2$. Same as argued about the period-2 solution in the above, for the value ζ^* ($\neq 1/2$) of the period-k solution, the other $k - 1$ values are given by

$$G_\ell(\zeta^*) = \underbrace{g \circ g \circ \cdots \circ g(\zeta^*)}_{\ell}$$

over $\ell = 1, 2, \ldots, k - 1$. From (12.6), we find that

$$\left|\frac{dG_k(x)}{dx}\right|_{x=\zeta^*} = (2\alpha)^k > 1$$

for any k when $\alpha > 1/2$, because $|g'(x)| = 2\alpha > 1$ for $x \neq 1/2$. This means that any period-k solution which does not contain $1/2$ is necessarily unstable (refer to Sect. 12.1.4).

If a period-k solution contains $1/2$, the above arguments cannot be applied because g is not differentiable at $x = 1/2$. It is the case to be separately considered. Now, let $x_0 = 1/2 + \epsilon$ with $0 < |\epsilon| \ll 1$ when $\alpha > 1/2$. Then we have

$$g(\frac{1}{2} + \epsilon) = \left\{ \begin{array}{ll} 2\alpha(1 - \frac{1}{2} - \epsilon) & (\epsilon > 0) \\ 2\alpha(\frac{1}{2} + \epsilon) & (\epsilon < 0) \end{array} \right\} = \alpha - 2\alpha|\epsilon| > \frac{1}{2}.$$

Since $x = 1/2$ is now an equilibrium for the one dimensional dynamical system

$$x_{i+1} = G_k(x_i) = G_{k-1} \circ g(x_i),$$

we find that

$$\frac{1}{2} = G_k(\frac{1}{2}) = G_{k-1} \circ g(\frac{1}{2}) = G_{k-1}(\alpha).$$

On the other hand, we have

$$G_k(\frac{1}{2} + \epsilon) = G_{k-1} \circ g(\frac{1}{2} + \epsilon) = G_{k-1}(\alpha - 2\alpha|\epsilon|).$$

Since $|2\alpha\epsilon| \ll 1$,

$$G_{k-1}(\alpha - 2\alpha|\epsilon|) = G_{k-1}(\alpha) - \left.\frac{dG_{k-1}(x)}{dx}\right|_{x=\alpha} 2\alpha|\epsilon| + o(|\epsilon|)$$

$$\approx \frac{1}{2} - 2\alpha\left\{\prod_{i=1}^{k-1} g'(G_{i-1}(\alpha))\right\}|\epsilon|,$$

with $\left|g'(G_{i-1}(\alpha))\right| = 2\alpha > 1$ because $G_{i-1}(\alpha)$ $(i = 1, 2, \ldots, k-1)$ is a value of the period-k solution, different from $1/2$. Therefore, we can get

$$\left|G_k(\frac{1}{2} + \epsilon) - \frac{1}{2}\right| \approx \left|2\alpha\left\{\prod_{i=1}^{k-1} g'(G_{i-1}(\alpha))\right\}|\epsilon|\right| = (2\alpha)^k|\epsilon| > |\epsilon|.$$

This means that, if $|x_0 - 1/2| = |\epsilon|$ for $0 < |\epsilon| \ll 1$ when $\alpha > 1/2$, then necessarily $|G_k(x_0) - 1/2| > |\epsilon|$. As a result, we find that equilibrium $x = 1/2$ for the one dimensional dynamical system $x_{i+1} = G_k(x_i)$ is unstable. Finally we can result that any period-k solution which contains $1/2$ for the tent map (12.1) with (12.7) is necessarily unstable if it exists.

Consequently we have shown that any period-k is necessarily unstable if it exists when $\alpha > 1/2$. It is trivial that the existence of a period-k solution depends on the value of α. It can be seen from Fig. 12.8 that, when a period-2 solution exists, the

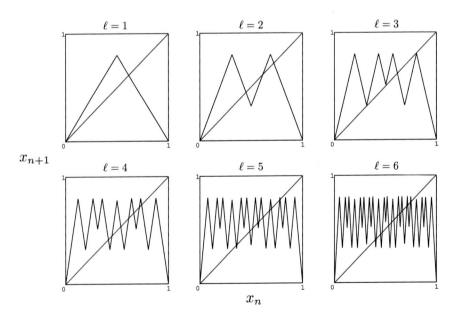

Fig. 12.8 Graphs of the ℓ-fold composed map $G_\ell(x) = g \circ g \circ \cdots \circ g(x)$ $(\ell = 1, 2, 3, 4, 5, 6)$ for the tent map (12.1) with (12.7). Numerically drawn for $\alpha = 0.8$

existence of the other periodic solution depends on the value of α. In the special case of $\alpha = 1$, any periodic solution exists. Since we can get the explicit simple formula for the k-fold composed function G_k for any k, it can be easily proved by using its graph for the general k. We skip the proof here.

Chaotic Variation

From the above arguments about the tent map (12.1) with (12.7), we have seen that $x_n \to 0$ as $n \to \infty$ for $\alpha < 1/2$, while the orbit $\{x_n\}$ for $\alpha > 1/2$ does not converge any equilibrium or periodic solution for almost every initial value x_0 except for the case where x_0 is the value of an equilibrium or periodic solution. In such a case, the orbit $\{x_n\}$ shows a variation of values different from each other. As shown in Fig. 12.9, it becomes a *chaotic* variation. Figure 12.10 shows the bifurcation diagram numerically drawn for $\alpha > 1/2$ about the tent map (12.1) with (12.7). Especially for $\alpha = 1$, The ω-limit set becomes dense in the interval $[0, 1]$ with a non-uniform specific distribution.

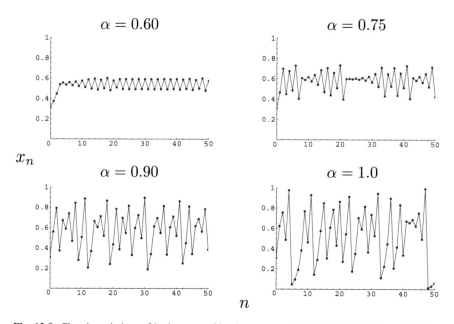

Fig. 12.9 Chaotic variations of $\{x_n\}$ generated by the tent map (12.1) with (12.7). Numerical plots from $x_0 = 0.311$ to x_{50} for each value of α

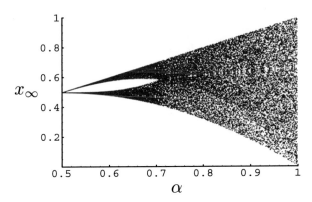

Fig. 12.10 The bifurcation diagram numerically drawn for $\alpha > 1/2$ about the tent map (12.1) with (12.7), with $x_0 = 0.4$. The plots are from x_{300} to x_{400}. This may be regarded as a figure to show the ω-limit set numerically approximated for each value of α (refer to Sect. 12.1.5)

12.2 Two Dimensional Discrete Time Model

12.2.1 Linearization Around Equilibrium

In this section, we introduce the linearization for the two dimensional system of first order difference Eq. (11.13) (Sect. 11.2.1) around an equilibrium (x^*, y^*):

$$\begin{cases} x_{n+1} = f(x_n, y_n) \\ y_{n+1} = g(x_n, y_n) \end{cases} \qquad (11.13)$$

It is similar with the linearization for the one dimensional case described in Sect. 12.1.1. The application of Taylor expansion for two variable functions $f(x, y)$ and $g(x, y)$ around an equilibrium (x^*, y^*) gives

$$\begin{aligned} f(x_n, y_n) &= f(x^*, y^*) + f_x(x^*, y^*)(x_n - x^*) + f_y(x^*, y^*)(y_n - y^*) \\ &\quad + \mathrm{o}(x_n - x^*, y_n - y^*); \\ g(x_n, y_n) &= g(x^*, y^*) + g_x(x^*, y^*)(x_n - x^*) + g_y(x^*, y^*)(y_n - y^*) \\ &\quad + \mathrm{o}(x_n - x^*, y_n - y^*), \end{aligned} \qquad (12.9)$$

where $f_x(x, y) := \partial f(x, y)/\partial x$, $f_y(x, y) := \partial f(x, y)/\partial y$, $g_x(x, y) := \partial g(x, y)/\partial x$, and $g_y(x, y) := \partial g(x, y)/\partial y$. The residual terms of higher order expressed by $\mathrm{o}(x_n - x^*, y_n - y^*)$ satisfy that

$$\lim_{(x_n, y_n) \to (x^*, y^*)} \frac{\mathrm{o}(x_n - x^*, y_n - y^*)}{x_n - x^*} = 0; \qquad \lim_{(x_n, y_n) \to (x^*, y^*)} \frac{\mathrm{o}(x_n - x^*, y_n - y^*)}{y_n - y^*} = 0.$$

Since $f(x^*, y^*) = x^*$ and $g(x^*, y^*) = y^*$, we can derive the following system of first order linear difference equations with respect to the perturbation of (x, y) from equilibrium (x^*, y^*), $u_n = x_n - x^*$ and $v_n = y_n - y^*$, from (11.13):

$$\begin{cases} u_{n+1} = f_x(x^*, y^*)u_n + f_y(x^*, y^*)v_n + o(u_n, v_n); \\ \\ v_{n+1} = g_x(x^*, y^*)u_n + g_y(x^*, y^*)v_n + o(u_n, v_n). \end{cases} \tag{12.10}$$

If and only if $(u_n, v_n) \to (0, 0)$ as $n \to \infty$ for any (u_0, v_0) such that $0 < |u_0| \ll 1$ and $0 < |v_0| \ll 1$, equilibrium (x^*, y^*) is called *locally asymptotically stable*.

Similarly with the arguments for the one dimensional case in Sect. 12.1.1, we focus on the behavior of the solution (x_n, y_n) for (11.13) in a neighborhood of equilibrium (x^*, y^*). Thus, supposing that $|x_n - x^*| = |u_n| \ll 1$ and $|y_n - y^*| = |v_n| \ll 1$, we consider the following system of $(\widetilde{u}_n, \widetilde{v}_n)$ which temporal variation approximates that of (u_n, v_n) by (12.10) in a neighborhood of (x^*, y^*):

$$\begin{cases} \widetilde{u}_{n+1} = f_x(x^*, y^*)\widetilde{u}_n + f_y(x^*, y^*)\widetilde{v}_n; \\ \\ \widetilde{v}_{n+1} = g_x(x^*, y^*)\widetilde{u}_n + g_y(x^*, y^*)\widetilde{v}_n. \end{cases} \tag{12.11}$$

This is called *linearized system* around equilibrium (x^*, y^*) for (11.13).

The linearized system (12.11) for (11.13) corresponds to (11.14) for (11.13) in Sect. 11.2, and matrix A defined by (11.15) now becomes

$$A = \begin{pmatrix} f_x(x^*, y^*) & f_y(x^*, y^*) \\ g_x(x^*, y^*) & g_y(x^*, y^*) \end{pmatrix}. \tag{12.12}$$

This matrix defined by partial derivatives of f and g is called *Jacobian matrix* at the point (x^*, y^*). The determinant of A, det A, is called *Jacobian determinant*. The behavior of $(\widetilde{u}_n, \widetilde{v}_n)$ by (12.11) is determined by the eigenvalues of A, as described in Sect. 11.2. This means that the behavior of (u_n, v_n) by (12.10) in a neighborhood of $(0, 0)$ corresponding to equilibrium (x^*, y^*) is determined by the eigenvalues of Jacobian matrix A at equilibrium (x^*, y^*). In conclusion, from Theorem 11.2 in Sect. 11.2, the local stability of equilibrium (x^*, y^*) for the system (11.13) is determined as follows:

Theorem 12.5 *If every eigenvalue λ of Jacobian matrix A for an equilibrium (x^*, y^*) of the system (11.13) has the absolute value less than one, $|\lambda| < 1$, equilibrium (x^*, y^*) is locally asymptotically stable. If the absolute value of an eigenvalue λ is greater than one, $|\lambda| > 1$, it is unstable.*

Table 12.1 Classification of the equilibrium and its stability for the two dimensional system of first order difference Eq. (11.13). For the case of purely imaginary eigenvalue, the local stability must be investigated with some additional analysis as mentioned in the main text

Eigenvalues	Classification of equilibrium		Local stability
Real and $\lvert\lambda_1\rvert \geq \lvert\lambda_2\rvert > 1$	Source	Unstable node	Unstable
Imaginary and $\lvert\lambda_1\rvert = \lvert\lambda_2\rvert > 1$		Unstable spiral/focus	
Real and $\lvert\lambda_1\rvert < 1 < \lvert\lambda_2\rvert$	Saddle		
Real and $\lvert\lambda_1\rvert \leq \lvert\lambda_2\rvert < 1$	Sink	Stable node	Locally asymptotically stable
Imaginary and $\lvert\lambda_1\rvert = \lvert\lambda_2\rvert < 1$		Stable spiral/focus	
Purely imaginary	Center		$\left(\begin{array}{c}\text{Lyapunov stable}\\\text{Neutrally stable}\end{array}\right)$

In the next section,[1] we will see the classification of equilibrium with respect to the local stability.

12.2.2   Classification of Equiliblium

As already described in the previous section, the local stability of an equilibrium about the model with a two dimensional system of first order difference Eq. (11.13) can be investigated by analyzing the behavior of the system of first order linear difference Eq. (12.11) derived by the linearization of (11.13) around the equilibrium. Technically the eigenvalues of Jacobian matrix A defined by (12.12) determine the asymptotic behavior of $(\widetilde{u}_n, \widetilde{v}_n)$ by (12.11), and subsequently the asymptotic behavior of (u_n, v_n) by (12.10). Therefore, as described in Sect. 11.2, the eigenvalues of Jacobian matrix A defined by (12.12) determine the local stability of equilibrium (x^*, y^*) for the system (11.13) as described above in Theorem 12.5, and classify it as shown in Table 12.1.

If the following condition is satisfied for an equilibrium $\mathbf{x}^* = (x^*, y^*)$ of the system (11.13), equilibrium $\mathbf{x}^*$ is called *Lyapunov stable, L-stable, neutrally stable, weakly stable*, or sometimes simply *stable*:

- For any $\varepsilon > 0$, there exists a sufficiently small positive δ such that $\lvert\lvert\mathbf{x}_n - \mathbf{x}^*\rvert\rvert < \varepsilon$ for all $n > 0$ about $\mathbf{x}_n = (x_n, y_n)$ generated by (11.13) with any initial condition $\mathbf{x}_0 = (x_0, y_0)$ satisfying that $\lvert\lvert\mathbf{x}_0 - \mathbf{x}^*\rvert\rvert < \delta$.

---

[1] The arguments in the section are mathematically analogous to those in Sect. 14.3 for the two dimensional system of first order linear *differential* equations. It would be very helpful for readers' clearer understanding to compare one with the other.

The mathematical definition of this stability means that the sequence $\{(x_n, y_n)\}$ generated by (11.13) stays in a neighborhood of equilibrium (x^*, y^*) if the perturbation is sufficiently small. In other words, the distance of point $\mathbf{x}_n$ from equilibrium $\mathbf{x}^*$ keeps sufficiently small for any n if the perturbation is sufficiently small. We must remark that the definition does *not require* an asymptotic approach of the sequence toward the equilibrium.

If the following condition about such an asymptotic approach of the sequence toward the equilibrium is satisfied in addition, equilibrium $\mathbf{x}^*$ is called *locally asymptotically stable*:

- With a sufficiently small positive δ, $||\mathbf{x}_n - \mathbf{x}^*|| \to 0$ as $n \to \infty$ for all $\mathbf{x}_0$ such that $||\mathbf{x}_0 - \mathbf{x}^*|| < \delta$.

We must remark here the treatment of Lyapunov stability, as mentioned also at p. 221 about the stability of an equilibrium for a continuous time model, Lotka-Volterra prey-predator model (8.15) in Sect. 8.4. If an equilibrium $\mathbf{x}^*$ for (11.13) is Lyapunov stable, both of two eigenvalues for the equilibrium have the absolute value 1. However, the inverse is not true. That is, the equilibrium cannot be identified as Lyapunov stable only from the fact that both of two eigenvalues have the absolute value 1. *Even when they have the absolute value 1, the equilibrium may not be Lyapunov stable.*

This is because the local stability analysis is based on the linearized system (12.11) only. It is a sort of linear approximation of the system (11.13) in a neighborhood of equilibrium $\mathbf{x}^*$. Hence, when both of two eigenvalues have the absolute value 1, the terms of higher order in (12.10) determine the actual behavior of the sequence near the equilibrium, which may cause its asymptotic approach toward it or its leaving from it. In a mathematical sense, two eigenvalues with the absolute value 1 only imply the possibility of Lyapunov stability about the equilibrium. A further investigation in addition to the local stability analysis is necessary to get the result on the stability.

12.2.3 Jury Stability Test

Every root of the following equation has the absolute value less than one as $|\lambda| < 1$,

$$f(\lambda) = \lambda^n + a_1\lambda^{n-1} + \cdots + a_{n-1}\lambda + a_n = 0,$$

if and only if every value of $J_{i,0}$ ($i = 0, 1, 2, \ldots, n$) defined below is positive:

$$J_{0,j} := a_j; \quad J_{i,j} := \frac{1}{J_{i-1,0}} \begin{vmatrix} J_{i-1,0} & J_{i-1,n+1-i-j} \\ J_{i-1,n+1-i} & J_{i-1,j} \end{vmatrix},$$

where $i = 1, 2, \ldots, n$, $j = 0, 1, \ldots, n-i$, and $a_0 = 1$. This necessary and sufficient condition is called *Jury stability test*, and practically used to investigate the local stability (may be called the linear stability as well) about an equilibrium state for a dynamical system, though it is rarely applied for the analytical proof for the higher or general dimensional system because of a complicated mathematical expression of inequalities about the condition. We note that there are some different formulas to show Jury stability test and mathematically equivalent condition, which has been studied and derived until now in applied mathematics. In most cases of its use, the condition is checked by an iterative calculation following the definition of $J_{i,0}$ given in the above.

From the above condition, we can derive the following conditions respectively about the coefficients of the above equation $f(\lambda) = 0$ for low dimensional cases of $n = 2$ and $n = 3$:

$$\text{For } n = 2: \quad a_2^2 < 1, \; a_1^2 - 2a_2 - a_2^2 < 1.$$

$$\text{For } n = 3: \quad \begin{cases} a_3^2 < 1; \\ a_2^2 + (2 + a_1^2)a_3^2 - a_3^4 - 2a_1a_2a_3 < 1; \\ a_1^2 - a_2^2 - 2a_2 + 2a_1a_3 + a_3^2 < 1. \end{cases}$$

It is easy to see that the condition must appear complicated even for the case of $n = 3$, though the condition must be useful to numerically estimate the local stability of an equilibrium with a specific set of parameter values in the dynamical system under the analysis.

Answer to Exercise

Exercise 12.1 (p. 389)

Let us consider the perturbation of x_n from equilibrium $1 - 1/\mathscr{R}_0 = 2/3$: $\epsilon_k = x_n - 2/3$. Substituting this into (12.4) with $\mathscr{R}_0 = 3$, we can get

$$\frac{2}{3} + \epsilon_{n+1} = 3\left(1 - \frac{2}{3} - \epsilon_n\right)\left(\frac{2}{3} + \epsilon_n\right) = \frac{2}{3} - \epsilon_k - 3\epsilon_k^2,$$

that is,

$$\epsilon_{n+1} = -(1 + 3\epsilon_n)\epsilon_n.$$

From this recurrence relation, we can further get

$$\epsilon_{n+2} = -(1 + 3\epsilon_{n+1})\epsilon_{n+1}$$

$$= \{1 - 3(1 + 3\epsilon_n)\epsilon_n\}(1 + 3\epsilon_n)\epsilon_n = (1 - 18\epsilon_n^2 - 27\epsilon_n^3)\epsilon_n.$$

Since $|1 - 18\epsilon_n^2 - 27\epsilon_n^3| < 1$ for $|\epsilon_n| \ll 1$, we find that $|\epsilon_{n+2}| < |\epsilon_n|$. Therefore, $\epsilon_{2n} \to 0$ and $\epsilon_{2n+1} \to 0$ as $n \to \infty$. This means that, when $\mathscr{R}_0 = 3$, equilibrium $1 - 1/\mathscr{R}_0 = 2/3$ for (12.4) is locally asymptotically stable.

As for the local stability of an equilibrium x^* with $g'(x^*) = -1$ about the general discrete time model (12.1), we can apply the above arguments with Taylor expansion. Let us consider the perturbation $\epsilon_n = x_n - x^*$ as before. Taylor expansion around x^* leads to

$$x^* + \epsilon_{n+2} = g(x_{n+1}) = g(g(x_n)) = g(g(x^* + \epsilon_n))$$

$$= x^* + \epsilon_n - \frac{1}{6}\left[2g'''(x^*) + 3\{g''(x^*)\}^2\right]\epsilon_n^3 + o(\epsilon_n^3),$$

that is,

$$\epsilon_{n+2} = \left(1 - \frac{1}{6}\left[2g'''(x^*) + 3\{g''(x^*)\}^2\right]\epsilon_n^2 + o(\epsilon_n^2)\right)\epsilon_n.$$

Therefore, when $|\epsilon_n| \ll 1$, if

$$2g'''(x^*) + 3\{g''(x^*)\}^2 > 0, \tag{12.13}$$

we have $|\epsilon_{k+2}| < |\epsilon_k|$. Consequently we can get the following result:

Theorem 12.6 *For an equilibrium x^* with $g'(x^*) = -1$ about the discrete time model (12.1), if the condition (12.13) is satisfied, equilibrium x^* is locally asymptotically stable. If the condition with the inverse inequality in (12.13) is satisfied, it is unstable.*

For the logistic map (12.4), the condition (12.13) is necessarily satisfied since $g''(x^*) \neq 0$ and $g'''(x) = 0$. Similarly to the logistic map, we can find the following more general result about the discrete time model (12.1) with a quadratic polynomial function g:

Corollary 12.1 *For the discrete time model (12.1) with a quadratic polynomial function g, any equilibrium x^* with $g'(x^*) = -1$ is necessarily locally asymptotically stable.*

This corollary can be proved simply for the general quadratic polynomial function $g(x) = ax^2 + bx + c \ (a \neq 0)$. Indeed, the condition (12.13) is necessarily satisfied since $g''(x^*) = 2a \neq 0$ and $g'''(x) = 0$.

Exercise 12.2 (p. 391)

(a) As described in the main text, when the logistic map (12.4) has a period-2 solution defined by different two values A and B, these values are equivalent to the equilibria for the discrete dynamical system $y_{n+1} = g \circ g(y_n)$. That is, A and B are given by the roots for equation

$$y = \mathcal{R}_0\{1 - g(y)\}g(y) = \mathcal{R}_0\{1 - \mathcal{R}_0(1 - y)y\}\mathcal{R}_0(1 - y)y.$$

The equilibria 0 and $1 - 1/\mathcal{R}_0$ for the logistic map (12.4) must be the roots for the above equation, since it holds that $g \circ g(x^*) = g(g(x^*)) = g(x^*) = x^*$ for equilibrium x^* with $g(x^*) = x^*$ for (12.4). This fact helps us to factorize the above equation as follows:

$$y\left\{y - \left(1 - \frac{1}{\mathcal{R}_0}\right)\right\}\left\{\mathcal{R}_0 y^2 - (\mathcal{R}_0 + 1)y + 1 + \frac{1}{\mathcal{R}_0}\right\} = 0.$$

Thus the values A and B of the period-2 solution are given by the roots of equation

$$\mathcal{R}_0 y^2 - (\mathcal{R}_0 + 1)y + 1 + \frac{1}{\mathcal{R}_0} = 0 \qquad (12.14)$$

such that they are different from each other and in $(0, 1)$. It is easy to show that the necessary and sufficient condition for the existence of such roots about the above equation is given by $\mathcal{R}_0 > 3$. Further we can explicitly get the values A and B by solving the above quadratic equation:

$$\frac{\mathcal{R}_0 + 1 \pm \sqrt{(\mathcal{R}_0 + 1)(\mathcal{R}_0 - 3)}}{2\mathcal{R}_0}. \qquad (12.15)$$

(b) From the arguments in p. 390 of Sect. 12.1.4, a period-2 solution with the values A and B is locally asymptotically stable if $|g'(A)g'(B)| < 1$. For the logistic map (12.4), the condition becomes $\mathcal{R}_0^2|(1 - 2A)(1 - 2B)| < 1$. Since these values A and B are the roots of the quadratic Eq. (12.14), given by (12.15), they satisfy that $A + B = 1 + 1/\mathcal{R}_0$ and $AB = (1 + 1/\mathcal{R}_0)/\mathcal{R}_0$. With these relations, the sufficient condition for the local stability of the period-2 solution, $\mathcal{R}_0^2|(1 - 2A)(1 - 2B)| < 1$, becomes

$$|\mathcal{R}_0^2 - 2\mathcal{R}_0 - 4| < 1,$$

which can be solved under the condition for the existence of a period-2 solution that $\mathscr{R}_0 > 3$. From this inequality, we can finally get the sufficient condition for the local stability that $\mathscr{R}_0 < 1 + \sqrt{6}$.

(c) From the arguments in (b), when $\mathscr{R}_0 > 1 + \sqrt{6}$, we have $\left|\mathscr{R}_0^2 - 2\mathscr{R}_0 - 4\right| > 1$, so that $|g'(A)g'(B)| > 1$. This indicates that the equilibria A and B for the discrete dynamical system $y_{n+1} = g \circ g(y_n)$ are unstable. Therefore the period-2 solution with the values A and B is unstable when $\mathscr{R}_0 > 1 + \sqrt{6}$.

References

1. R.L. Devaney, *An Introduction to Chaotic Dynamical Systems*, 3rd edn. (Chapman and Hall/CRC, Boca Raton, 2021)
2. S. Elaydi, *An Introduction to Difference Equations*, 3rd edn. (Springer, Berlin, 2005)
3. M. Feigenbaum, Quantitative universality of a class of non-linear transformations. J. Stat. Phys. **19**, 25–52 (1978)
4. G. Fulford, P. Forrester, A. Jones, *Modelling with Differential and Difference Equations*. Australian Mathematical Society Lecture Series, vol. 10 (Cambridge University Press, Cambridge, 1997)
5. J. Gleick, *Chaos—Making A New Science* (Penguin Books, New York, 1987)
6. M.W. Hirsch, S. Smale, R.L. Devaney, *Differential Equations, Dynamical Systems, and an Introduction to Chaos*, 3rd edn. (Elsevier/Academic Press, Waltham, 2012)
7. D. Kaplan, L. Glass, *Understanding Nonlinear Dynamics*. Textbooks in Mathematical Sciences (SpringerVerlag, New York, 1995)
8. G. Ledder, *Mathematics for the Life Sciences: Calculus, Modeling, Probability and Dynamical Systems*. Springer Undergraduate Texts in Mathematics and Technology (Springer, New York, 2013)
9. T.Y. Li, J.A. Yorke, Period three implies chaos. Am. Math. Mon. **82**, 985–992 (1975)
10. M. Martelli, *Discrete Dynamical Systems and Chaos*, 1st edn. (CRC Press LLC, Boca Raton, 1992)
11. R.E. Mickens, *Difference Equations: Theory Applications and Advanced Topics*. Monographs and Research Notes in Mathematics, 3rd edn. (CRC Press, Boca Raton, 2015)
12. R.C. Robinson, *An Introduction to Dynamical Systems: Continuous and Discrete*. Pure and Applied Undergraduate Texts, vol. 19, 2nd edn. (American Mathematical Society, Providence, 2012)
13. S.H. Strogatz, *Nonlinear Dynamics and Chaos: With Applications to Physics, Biology Chemistry and Engineering*, 2nd edn. (Westview Press, Boulder, 2015)

Chapter 13
First Order Linear Ordinary Differential Equation

Abstract This chapter describes the fundamentals on the first order linear ordinary differential equation, which are closely related to many parts in the main text especially about the continuous time model.

13.1 One Dimensional First Order Linear Equation

13.1.1 First Order Ordinary Differential Equation

One dimensional differential equation is a mathematical expression of the relation about an unknown function and its derivative(s). A function identically satisfying a differential equation is called *solution* of it. Analysis to find a solution is called "solving the differential equation". In this section, we shall describe the least fundamentals about it.

The most general expression of first order ordinary differential equation with respect to an unknown function $y(t)$ of an independent variable t is given by a function F of t, y and $y' := dy(t)/dt$:

$$F(t, y, y') = 0. \tag{13.1}$$

The simplest form of first order ordinary differential equation has the following normal form:

$$\frac{dy(t)}{dt} = f(t), \tag{13.2}$$

where a function $f(t)$ is given a priori for the unknown function $y(t)$. The general solution can be expressed by the improper integral:

$$y(t) = \int f(t)\, dt,$$

since the equation means that the first derivative of $y(t)$ is given by a known function $f(t)$. For example, about the first order ordinary differential Eq. (13.2) has the general solution $y(t) = \sin t + C$ with an arbitrary constant C when $f(t) = \cos t$.

To find the solution of (13.1) which satisfying that $y(t_0) = y_0$ is called *initial value problem* or *Cauchy problem*, and the condition $y(t_0) = y_0$ is called *initial condition* for (13.1). For example, the first order ordinary differential Eq. (13.2) with $f(t) = \cos t$ and the initial condition $y(0) = 1$ leads to the solution $y(t) = \sin t + 1$, which is one of what is called *special solutions* for (13.2).

The initial value problem for a first order ordinary differential equation may have no solution or multiple solutions. The following theorem is well-known and most fundamental about the existence and uniqueness of solution for the initial value problem:

Theorem 13.1 (Cauchy Theorem) *Consider the following initial value problem:*

$$\frac{dy(t)}{dt} = f(t, y), \quad y(t_0) = y_0. \tag{13.3}$$

If f is continuous in a domain $D := \{(t, y) \mid |t - t_0| \le a, \ |y - y_0| \le b\}$, and there exists a positive constant L such that $|f(t, z) - f(t, y)| \le L|z - y|$ for any (t, z) and (t, y) in D (what is called Lipschitz condition), then the solution of the initial value problem (13.3) uniquely exists for $|t - t_0| \le \min\{a, \frac{b}{M}\}$ with the maximal value M of $|f(t, y)|$ in D.

This theorem assures the existence and uniqueness of solution for t sufficiently near t_0 for the initial value problem satisfying the condition given there.

13.1.2   Separation of Variables

For the first order ordinary differential equation

$$\frac{dy(t)}{dt} = f(t, y), \tag{13.4}$$

if the function $f(t, y)$ can be expressed by the product of single variable functions $T = T(t)$ of t and $Y = Y(y)$ of y, that is, if it can be expressed as

$$\frac{dy(t)}{dt} = T(t)Y(y), \tag{13.5}$$

it is called *separation of variables* for (13.4).

Firstly let us consider the solution satisfying $Y(y) \ne 0$. From (13.5), we have

$$\frac{1}{Y(y)} \frac{dy}{dt} = T(t).$$

Taking the improper integral for both sides,

$$\int \frac{1}{Y(y)} \frac{dy}{dt} \, dt = \int T(t) \, dt,$$

and changing the variable for the integrand of left side, we have

$$\int \frac{1}{Y(y)} \, dy = \int T(t) \, dt. \tag{13.6}$$

We may obtain the general solution by calculating the integral of (13.6).

Next, if there exists a value of $y = y_s$ such that $Y(y) = 0$, it is easily proved that the constant function $y(t) \equiv y_s$ satisfies (13.5), so that $y(t) \equiv y_s$ is a solution of (13.5). This solution may be included as a special case in the solution obtained for $Y(y) \neq 0$. When it is not included, it may be called *singular solution* of (13.5).

Some of first order ordinary differential equations which is not the separation of variables may be transformed to the separation of variables by appropriate variable changes. It could be solved by the method of integration as above [2].

13.1.3 Linear Ordinary Differential Equation

The nth order linear ordinary differential equation is expressed as

$$y^{(n)} + p_{n-1}(t) y^{(n-1)} + \cdots + p_1(t) y' + p_0(t) y = r(t), \tag{13.7}$$

where $p_i(t)$ $(i = 0, 1, \ldots, n-1)$ and $r(t)$ are defined in a region Ω, and $y^{(k)} = d^k y(t)/dt^k$. When $r(t) = 0$, it is called *homogeneous* linear ordinary differential equation.

The following theorem is on the existence and uniqueness of the initial value problem about (13.7) with the initial condition

$$y(t_0) = c_0, \ y'(t_0) = c_1, \ \ldots, \ y^{(n-2)}(t_0) = c_{n-2}, \ y^{(n-1)}(t_0) = c_{n-1}. \tag{13.8}$$

Theorem 13.2 *The initial value problem of (13.7) with (13.8) has the unique solution of C^1 class (i.e., differentiable, and the derivative is continuous) for arbitrarily given t_0 and c_i $(i = 0, 1, \ldots, n-1)$ if $p_i(t)$ $(i = 0, 1, \ldots, n-1)$ and $r(t)$ are continuous in Ω.*

Now let us consider the following first order equation with $p(t)$ and $r(t)$ defined in a certain region:

$$\frac{dy(t)}{dt} + p(t)y(t) = r(t). \tag{13.9}$$

We shall derive the general solution of (13.9). Firstly let us consider the corresponding homogeneous equation:

$$\frac{dy(t)}{dt} = -p(t)y(t), \tag{13.10}$$

which is clearly the separation of variables. Thus, from (13.6), the general solution of (13.10) mathematically becomes

$$y(t) = C\,e^{-\int p(t)\,dt} \tag{13.11}$$

with an arbitrary constant C.

To get the solution of the corresponding non-homogeneous Eq. (13.9), we suppose the following form about it:

$$y(t) = u(t)\,e^{-\int p(t)\,dt}, \tag{13.12}$$

which can be derived by the substitution of the arbitrary constant C in (13.11) by an unknown function $u(t)$ of t. This is called *method of variation of constants*. Substituting (13.12) for (13.9), we can get the following first order ordinary differential equation about $u(t)$:

$$\frac{du(t)}{dt} = r(t)\,e^{\int p(t)\,dt}.$$

This can be easily solved by the integration, and we can get the general solution as

$$u(t) = \int r(t)\,e^{\int p(t)\,dt}dt + C$$

with an arbitrary constant C. Hence, from (13.12), we can get the following general solution of (13.9):

$$y(t) = \frac{1}{\Xi(t)}\left[\int r(t)\,\Xi(t)\,dt + C\right], \tag{13.13}$$

where $\Xi(t) := \exp\left[\int p(t)\,dt\right]$, and C is an arbitrary constant.

13.1.4 Bernoulli Equation

The following first order ordinary differential equation is called *Bernoulli equation*:

$$\frac{dx(t)}{dt} = a(t)x(t) + b(t)[x(t)]^q,$$

where $a(t)$ and $b(t)$ are appropriate functions of t, and the exponent q is an arbitrarily given real number. When $q = 0$ or $q = 1$, it becomes a linear ordinary differential equation, and can be solved as described in Sect. 13.1.3.

When $q \neq 0$ and $q \neq 1$, we can apply the variable transformation such that $y(t) = [x(t)]^{1-q}$, and subsequently obtain the following linear equation with respect to $y(t)$:

$$\frac{dy(t)}{dt} = (1 - q)a(t)y(t) + (1 - q)b(t).$$

Since this is a first order linear ordinary differential equation, we can get the solution $y(t)$ as before. Then we can derive the solution $x(t)$ with $x(t) = [y(t)]^{1/(1-q)}$. As mentioned in Sect. 5.3, the logistic equation is a Bernoulli equation.

13.2 Two Dimensional System of First Order Linear Equations

13.2.1 Simultaneous First Order Equations

A two dimensional system of first order ordinary differential equations in terms of (u, v)

$$\begin{cases} \dfrac{dx(t)}{dt} = f(x, y) \\ \dfrac{dy(t)}{dt} = g(x, y) \end{cases} \tag{13.14}$$

appears as a mathematical model for a variety of phenomena. When both of functions f and g are linear in terms of x and y, the general solution for (x, y) can be found in most cases. In contrast, when f or g is nonlinear, it is generally difficult to get the explicit solution, and instead some mathematical techniques are applied to investigate the qualitative feature of the solution (for example, see [1, 3–5]).

For the system of nonlinear ordinary differential Eq. (13.14), the qualitative analysis frequently uses the following system of homogeneous linear ordinary differential equations:

$$\begin{cases} \dfrac{du(t)}{dt} = a_{11} u(t) + a_{12} v(t) \\ \dfrac{dv(t)}{dt} = a_{21} u(t) + a_{22} v(t), \end{cases} \tag{13.15}$$

where coefficients a_{11}, a_{12}, a_{21} and a_{22} are mathematically determined for each analysis on the original system (13.14). This system can be expressed by two dimensional column vector $\mathbf{v}(t) := \begin{pmatrix} u(t) \\ v(t) \end{pmatrix}$ and 2×2 matrix $A := \begin{pmatrix} a_{11} & a_{12} \\ a_{21} & a_{22} \end{pmatrix}$ as follows:

$$\frac{d}{dt} \mathbf{v}(t) = A \mathbf{v}(t). \tag{13.16}$$

In the following part,[1] we discuss the general solution for the system of homogeneous linear ordinary differential Eq. (13.15) with constant coefficients, that is, (13.16) with matrix A which elements a_{11}, a_{12}, a_{21} and a_{22} are all constant.

13.2.2   Case of Distinct Real Eigenvalues

The characteristic equation $\det(A - \lambda E) = 0$ for the eigenvalue λ of matrix A with the unit matrix $E := \begin{pmatrix} 1 & 0 \\ 0 & 1 \end{pmatrix}$ becomes the quadratic equation (11.16) in Sect. 11.2.2

Here we consider the case where the eigenvalues λ_+ and λ_- given by the roots for the characteristic equation (11.16) are real and different from each other. With column eigenvectors $\mathbf{p}_1 := {}^T(p_{11} \; p_{12})$ and $\mathbf{p}_2 := {}^T(p_{21} \; p_{22})$ respectively for λ_+ and λ_-, let us define the 2×2 matrix

$$P = (\mathbf{p}_1 \; \mathbf{p}_2) = \begin{pmatrix} p_{11} & p_{21} \\ p_{12} & p_{22} \end{pmatrix}.$$

---

[1] The arguments in this part are mathematically analogous to those in Sect. 11.2 for the two dimensional system of first order linear *difference* equations. It would be very helpful for readers' clearer understanding to compare one with the other.

It can be mathematically proved that matrix P is regular. We can diagonalize matrix A as

$$P^{-1}AP = \begin{pmatrix} \lambda_+ & 0 \\ 0 & \lambda_- \end{pmatrix},$$

where P^{-1} is the inverse matrix of P.

Now let $\mathbf{w}(t) = P^{-1}\mathbf{v}(t) = {}^{\mathrm{T}}(w_1(t)\ w_2(t))$. Then we have $\mathbf{v}(t) = P\mathbf{w}(t)$. From

$$\frac{d}{dt}\mathbf{v}(t) = \frac{d}{dt}P\mathbf{w}(t) = P\frac{d}{dt}\mathbf{w}(t)$$

and $A\mathbf{v}(t) = AP\mathbf{w}(t)$, we can find the following ordinary differential equation:

$$\frac{d}{dt}\mathbf{w}(t) = P^{-1}AP\mathbf{w}(t) = \begin{pmatrix} \lambda_+ & 0 \\ 0 & \lambda_- \end{pmatrix}\mathbf{w}(t).$$

Then we have

$$\begin{cases} \dfrac{dw_1(t)}{dt} = \lambda_+ w_1(t) \\ \dfrac{dw_2(t)}{dt} = \lambda_- w_2(t), \end{cases}$$

and find that $w_1(t) = C_1 e^{\lambda_+ t}$ and $w_2(t) = C_2 e^{\lambda_- t}$ with arbitrary constants C_1 and C_2. From this result, we can get the general solution for the system of ordinary differential Eq. (13.16):

$$\mathbf{v}(t) = P\mathbf{w}(t) = (\mathbf{p}_1\ \mathbf{p}_2)\begin{pmatrix} w_1(t) \\ w_2(t) \end{pmatrix} = C_1 e^{\lambda_+ t}\mathbf{p}_1 + C_2 e^{\lambda_- t}\mathbf{p}_2. \tag{13.17}$$

From these arguments, we find that the system of homogeneous linear ordinary differential Eq. (13.16) has the following general form of solution when matrix A has distinct real eigenvalues:

$$\begin{cases} u(t) = c_{\mathrm{u}} e^{\lambda_+ t} + c'_{\mathrm{u}} e^{\lambda_- t} \\ v(t) = c_{\mathrm{v}} e^{\lambda_+ t} + c'_{\mathrm{v}} e^{\lambda_- t}, \end{cases} \tag{13.18}$$

where c_{u}, c'_{u}, c_{v} and c'_{v} are constants uniquely determined for a given initial condition $(u(0), v(0))$.

13.2.3 Case of Multiple Eigenvalues

In this section, we consider the case where the roots for the characteristic equation (11.16) is degenerated as λ, that is, the case of multiple eigenvalues, $\lambda_1 = \lambda_2 = \lambda$. The case where $A = \lambda E$ is included. In such a case, that is, when $a_{11} = a_{22} = \lambda$ and $a_{12} = a_{21} = 0$, we have

$$\begin{cases} \dfrac{du(t)}{dt} = \lambda u(t) \\[2mm] \dfrac{dv(t)}{dt} = \lambda v(t). \end{cases}$$

Then the general solution for the system of ordinary differential Eq. (13.16) is given by $u(t) = C_1 e^{\lambda t}$ and $v(t) = C_2 e^{\lambda t}$ with arbitrary constants C_1 and C_2.

For $A \neq \lambda E$, let us define a column vector $\mathbf{p}'$ such that $(A - \lambda E)\mathbf{p}' = \mathbf{p}$ with eigenvector $\mathbf{p}$ for eigenvalue λ. The vector $\mathbf{p}'$ is called *generalized eigenvector*, and is linearly independent of $\mathbf{p}$. Similarly as before, let us define 2×2 matrix $P = (\mathbf{p}\ \mathbf{p}')$. Since $A\mathbf{p} = \lambda\mathbf{p}$ and $A\mathbf{p}' = \mathbf{p} + \lambda\mathbf{p}'$, we have

$$AP = (A\mathbf{p}\ A\mathbf{p}') = (\lambda\mathbf{p}\ \mathbf{p} + \lambda\mathbf{p}') = \lambda\,(\mathbf{p}\ \mathbf{p}') + (\mathbf{0}\ \mathbf{p}) = \lambda P + PN,$$

where $N := \begin{pmatrix} 0 & 1 \\ 0 & 0 \end{pmatrix}$. With $\mathbf{w}(t) = P^{-1}\mathbf{v}(t)$, we can derive the following ordinary differential equation as before:

$$\frac{d}{dt}\mathbf{w}(t) = P^{-1}AP\,\mathbf{w}(t) = P^{-1}(\lambda P + PN)\mathbf{w}(t) = (\lambda E + N)\mathbf{w}(t) = \begin{pmatrix} \lambda & 1 \\ 0 & \lambda \end{pmatrix}\mathbf{w}(t).$$

Hence we obtain

$$\begin{cases} \dfrac{dw_1(t)}{dt} = \lambda w_1(t) + w_2(t) \\[2mm] \dfrac{dw_2(t)}{dt} = \lambda w_2(t). \end{cases}$$

From the second equation, we can find that $w_2(t) = C_2 e^{\lambda t}$ with an arbitrary constant C_2. By substituting this into the first equation, we get the following ordinary differential equation:

$$\frac{dw_1(t)}{dt} - \lambda w_1(t) = C_2 e^{\lambda t}.$$

We can solve this equation with (13.13), and obtain the solution $w_1(t) = (C_1 + C_2\,t)\,e^{\lambda t}$ with an additional arbitrary constant C_1.

From these arguments, we can obtain the following general solution for the system of ordinary differential Eq. (13.16) when matrix A has multiple eigenvalues:

$$\mathbf{v}(t) = P\mathbf{w}(t) = (C_1 + C_2 t)\,e^{\lambda t}\mathbf{p} + C_2 e^{\lambda t}\mathbf{p}'. \qquad (13.19)$$

As a result, it is shown that the system of homogeneous linear ordinary differential Eq. (13.16) has the following general form of solution when matrix A has multiple eigenvalues:

$$\begin{cases} u(t) = (c_{\mathrm{u}} + c_{\mathrm{u}}' t)e^{\lambda t} \\ v(t) = (c_{\mathrm{v}} + c_{\mathrm{v}}' t)e^{\lambda t}, \end{cases} \qquad (13.20)$$

where c_{u}, c_{u}', c_{v} and c_{v}' are constants uniquely determined for a given initial condition $(u(0), v(0))$.

13.2.4 Case of Imaginary Eigenvalues

In this section, we consider the case where matrix A has conjugate imaginary eigenvalues λ and $\bar{\lambda}$, for which corresponding eigenvectors are given by $\mathbf{p}$ and $\bar{\mathbf{p}}$. As before, let us define 2×2 matrix $P = (\mathbf{p}\ \bar{\mathbf{p}})$, which is now a complex matrix.

Although vectors $\mathbf{p}$, $\bar{\mathbf{p}}$, and matrix P contain imaginary elements, we can apply the same arguments as for the case where matrix A has distinct real eigenvalues. We can obtain the following general solution for the system of differential Eq. (13.16):

$$\mathbf{v}(t) = C_1 e^{\lambda t}\mathbf{p} + C_2 e^{\bar{\lambda} t}\bar{\mathbf{p}}, \qquad (13.21)$$

where constants C_1 and C_2 are imaginary numbers to make the right side a real vector for any time t because $\mathbf{v}(t)$ must be a real vector for any time t now. First we shall prove this in the subsequent part.

Since $\mathbf{v}(t_1)$ must be a real vector for any real number $t = t_1$ about the system of ordinary differential Eq. (13.16), it must hold that $\mathbf{v}(t_1) = \bar{\mathbf{v}}(t_1)$. Hence, the following equation must hold:

$$C_1 e^{\lambda t_1}\mathbf{p} + C_2 e^{\bar{\lambda} t_1}\bar{\mathbf{p}} = \overline{C}_1 e^{\bar{\lambda} t_1}\bar{\mathbf{p}} + \overline{C}_2 e^{\lambda t_1}\mathbf{p},$$

that is,

$$e^{\lambda t_1}(C_1 - \overline{C}_2)\mathbf{p} = e^{\bar{\lambda} t_1}(\overline{C}_1 - C_2)\bar{\mathbf{p}}.$$

Since the right side is conjugate to the left side, this equation indicates that $e^{\lambda t_1}(C_1 - \overline{C}_2)\mathbf{p}$ is a real vector for any real number $t = t_1$. For $t_1 \neq 0$, $e^{\lambda t_1}$ becomes imaginary. Hence, as $e^{\lambda t_1}(C_1 - \overline{C}_2)\mathbf{p}$ is a real vector for any real number $t = t_1$, it must be satisfied that $C_1 - \overline{C}_2 = 0$. Therefore we find that $C_1 = \overline{C}_2$, which means that C_1 and C_2 are conjugate to each other. Applying this result for (13.21), we can rewrite the general solution as follows:

$$\mathbf{v}(t) = C_1 e^{\lambda t}\mathbf{p} + \overline{C}_1 e^{\overline{\lambda} t}\overline{\mathbf{p}}. \tag{13.22}$$

Since the right side is the sum of mutually conjugate vectors for any imaginary number C_1, it is real.

Let us put $\lambda = |\lambda|e^{i\theta} = |\lambda|(\cos\theta + i\sin\theta)$ with imaginary unit i, where $|\lambda|$ is the absolute value of imaginary number λ, and $\theta = \arg\lambda$ is the argument of λ. Besides, denote the complex vector $\mathbf{p}$ as $\mathbf{p} = \boldsymbol{\alpha} + i\boldsymbol{\beta}$ with appropriate real vectors $\boldsymbol{\alpha}$ and $\boldsymbol{\beta}$. From Euler's formulation, we have $\lambda^n = e^{in\theta} = \cos n\theta + i\sin n\theta$. Then the general solution (13.22) becomes

$$\mathbf{v}(t) = C_1 e^{\rho t}(\cos\omega t + i\sin\omega t)(\boldsymbol{\alpha} + i\boldsymbol{\beta}) + \overline{C}_1 e^{\rho t}(\cos\omega t - i\sin\omega t)(\boldsymbol{\alpha} - i\boldsymbol{\beta})$$
$$= e^{\rho t}(c_1 \cos\omega t + c_2 \sin\omega t)\boldsymbol{\alpha} + e^{\rho t}(c_2 \cos\omega t - c_1 \sin\omega t)\boldsymbol{\beta}, \tag{13.23}$$

where $c_1 = C_1 + \overline{C}_1 = 2\,\mathrm{Re}\,C_1$ and $c_2 = i(C_1 - \overline{C}_1) = -2\,\mathrm{Im}\,C_1$. Since C_1 is an arbitrary imaginary number, c_1 and c_2 are arbitrary real numbers independent of each other.

Consequently we find that the system of homogeneous linear ordinary differential Eq. (13.16) has the following general form of solution when matrix A has imaginary eigenvalues:

$$\begin{cases} u(t) = e^{\rho t}(c_u \cos\omega t + c'_u \sin\omega t) \\ v(t) = e^{\rho t}(c_v \cos\omega t + c'_v \sin\omega t), \end{cases} \tag{13.24}$$

where c_u, c'_u, c_v and c'_v are real constants uniquely determined for a given initial condition $(u(0), v(0))$.

13.2.5 Asymptotic Behavior of the Solution

Consequently from those results in Sects. 13.2.2–13.2.4, we can find the following theorem about the asymptotic behavior of the solution $(u(t), v(t))$ for (13.15) as $t \to \infty$:

Theorem 13.3 *The solution $(u(t), v(t))$ for the system of homogeneous linear differential Eq. (13.15) has the following asymptotic behavior as $t \to \infty$:*

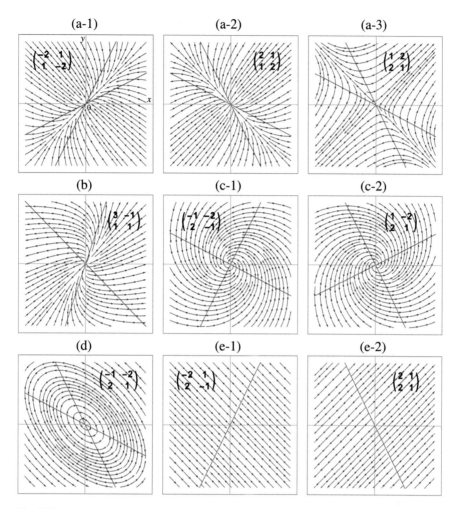

Fig. 13.1 Numerically drawn vector flows in the (u, v)-phase plane for the system of homogeneous linear differential Eq. (13.15) with matrix A defined by (13.16). Matrix A has (**a**) non-zero and distinct real eigenvalues; (**b**) multiple positive eigenvalues; (**c**) imaginary eigenvalues with a non-zero real part; (**d**) purely imaginary eigenvalues; (**e**) zero and non-zero real eigenvalues

- *If every eigenvalue λ of matrix A defined by (13.16) has a negative real part, $\operatorname{Re} \lambda < 0$, it converges to $(0, 0)$ as $t \to \infty$ independently of the initial condition (Fig. 13.1a-1 and c-1).*
- *If matrix A has an eigenvalue λ with a positive real part, $\operatorname{Re} \lambda > 0$, it diverges, that is, $|u(t)| \to \infty$ or $|v(t)| \to \infty$ for almost every initial condition (Fig. 13.1a-2, a-3, b, c-2, and e-2).*
- *If one eigenvalue of matrix A is 0 and the other is negative, it converges to a certain finite point for almost every initial condition (Fig. 13.1e-1).*

- *If every eigenvalue of matrix A is purely imaginary, it converges to an oscillatory variation with a finite supremum of the amplitude for almost every initial condition (Fig. 13.1d).*

References

1. R.L. Devaney, *An Introduction to Chaotic Dynamical Systems*, 3rd edn. (Chapman and Hall/CRC, Boca Raton, 2021)
2. E. Kreyszig, *Advanced Engineering Mathematics*, 10th edn. (John Wiley & Sons, New York, 2011)
3. G. Ledder, *Mathematics for the Life Sciences: Calculus, Modeling, Probability and Dynamical Systems*. Springer Undergraduate Texts in Mathematics and Technology (Springer, New York, 2013)
4. R.C. Robinson, *An Introduction to Dynamical Systems: Continuous and Discrete*. Pure and Applied Undergraduate Texts, vol. 19, 2nd edn. (American Mathematical Society, Providence, 2012)
5. H.R. Thieme, *Mathematics in Population Biology* (Princeton University Press, Princeton, 2003)

Chapter 14
Qualitative Analysis for Continuous Time Model

Abstract This chapter provides the important least knowledge about the fundamental mathematical theories to understand or find the dynamical nature of continuous time model.

14.1 Local Stability Analysis for One Dimensional Model

The qualitative analysis on the behavior of solution in the neighborhood of an equilibrium is generally called *local stability analysis* about the equilibrium. The stability determined by the analysis is called *local stability* of the equilibrium. In this section, we shall see how the local stability analysis is applied for one dimensional continuous time model given by the following ordinary autonomous differential equation:

$$\frac{dN(t)}{dt} = g(N(t)). \tag{14.1}$$

Let N^* be an equilibrium for (14.1), satisfying that $g(N^*) = 0$. We consider now the temporal change of the difference $n(t) = N(t) - N^*$, which uniquely determines the temporal change of $N(t)$ itself. Especially the sufficiently small difference n is sometimes called *perturbation* of N from equilibrium N^*. Since $dN/dt = d(n + N^*)/dt = dn/dt$, the temporal change of $n(t)$ is governed by

$$\frac{dn(t)}{dt} = g(n(t) + N^*). \tag{14.2}$$

We must remark that the value of n may be negative. When and only when $n(t) \to 0$ as $t \to \infty$, we have $N(t) \to N^*$ as $t \to \infty$. If and only if $n(t) \to 0$ as $t \to \infty$ for any $n(0)$ such that $0 < |n(0)| \ll 1$, equilibrium N^* is called *locally asymptotically stable* (refer also to Sect. 12.1.1). Hence it is necessary and sufficient for the locally asymptotic stability of equilibrium N^* that $n(t) \to 0$ as $t \to \infty$ for any $n(0)$ such that $0 < |n(0)| \ll 1$.

© The Author(s), under exclusive license to Springer Nature Singapore Pte Ltd. 2022
H. Seno, *A Primer on Population Dynamics Modeling*, Theoretical Biology,
https://doi.org/10.1007/978-981-19-6016-1_14

By Taylor expansion of the function $g(x)$ around $x = N^*$,

$$g(x) = g(N^*) + g'(N^*)(x - N^*) + \mathrm{o}(x - N^*)$$

with $g'(N^*) = dg/dx|_{x=N^*}$, we have

$$g(n(t) + N^*) = g(N^*) + g'(N^*)n(t) + \mathrm{o}(n(t)),$$

since $g(N^*) = 0$. As long as considering the temporal change of $N(t)$ in a neighborhood of equilibrium N^*, we can use the assumption that $|N(t) - N^*| = |n(t)| \ll 1$ and $|\mathrm{o}(n(t))| \ll |n(t)|$. Therefore, when $g'(N^*) \neq 0$, we have $g(n(t) + N^*) \approx g'(N^*)n(t)$. Then we find that the dynamics of (14.2) in a neighborhood of equilibrium N^* for (14.1) can be approximated well by the temporal change of $\widetilde{n}(t)$ governed by

$$\frac{d\widetilde{n}(t)}{dt} = g'(N^*)\widetilde{n}(t). \tag{14.3}$$

Since $g'(N^*)$ is a constant, this is mathematically equivalent to Malthus model (5.3).

This approximation is valid only when $g'(N^*) \neq 0$. When $g'(N^*) = 0$, the term of the higher order in the above Taylor expansion determines the temporal change of $n(t)$. The arguments in this section to investigate the local stability are not applicable. Such a case could be regarded as singular usually for a model with (14.1), and we shall not go into such a case any more in this book.

From (14.3), we can get the following result (Exercise 14.1), taking into account that the value of $\widetilde{n}$ may be negative:

Lemma 14.1 *If $g'(N^*) < 0$, then $\widetilde{n}(t)$ monotonically approaches 0 for $t > 0$, while, if $g'(N^*) > 0$, then $|\widetilde{n}(t)|$ monotonically increases for $t > 0$.*

Exercise 14.1 Prove Lemma 14.1.

From Lemma 14.1, we can immediately get the following theorem about the local stability of equilibrium N^* for (14.1):

Theorem 14.1 *If $g'(N^*) < 0$, equilibrium N^* for (14.1) is locally asymptotically stable. If $g'(N^*) > 0$, it is unstable.*

In the local stability analysis described here, it is essential to derive the ordinary differential Eq. (14.3) from the original Eq. (14.1). Especially it is called *linearization* of (14.1) around equilibrium N^* to derive the approximate linear differential Eq. (14.3) in order to investigate the local stability of N^*. The Eq. (14.3) is called *linearized equation* for (14.1) around equilibrium N^*.

14.2   Linearization of Two Dimensional System around Equilibrium

In this section, let us consider the linearization of the following two dimensional system of autonomous ordinary differential equations around an equilibrium (x^*, y^*):

$$\begin{cases} \dfrac{dx(t)}{dt} = f(x, y); \\[2mm] \dfrac{dy(t)}{dt} = g(x, y). \end{cases} \tag{14.4}$$

The way is similar with that for the one dimensional case in the previous section. The application of Taylor expansion for two variable functions $f(x, y)$ and $g(x, y)$ around equilibrium (x^*, y^*) gives

$$f(x, y) = f_x(x^*, y^*)(x - x^*) + f_y(x^*, y^*)(y - y^*) + o(x - x^*, y - y^*);$$

$$g(x, y) = g_x(x^*, y^*)(x - x^*) + g_y(x^*, y^*)(y - y^*) + o(x - x^*, y - y^*), \tag{14.5}$$

where $f_x(x, y) := \partial f(x, y)/\partial x$, $f_y(x, y) := \partial f(x, y)/\partial y$, $g_x(x, y) := \partial g(x, y)/\partial x$, and $g_y(x, y) := \partial g(x, y)/\partial y$. We used the relation that $f(x^*, y^*) = 0$ and $g(x^*, y^*) = 0$. The residual terms of higher order expressed by $o(x - x^*, y - y^*)$ satisfy that

$$\lim_{(x,y)\to(x^*,y^*)} \frac{o(x - x^*, y - y^*)}{x - x^*} = 0; \qquad \lim_{(x,y)\to(x^*,y^*)} \frac{o(x - x^*, y - y^*)}{y - y^*} = 0.$$

Then from (14.4), we can derive the following system of ordinary differential equations with respect to the perturbation of (x, y) from equilibrium (x^*, y^*), $u(t) = x(t) - x^*$ and $v(t) = y(t) - y^*$:

$$\begin{cases} \dfrac{du(t)}{dt} = f_x(x^*, y^*)u(t) + f_y(x^*, y^*)v(t) + o(u, v); \\[2mm] \dfrac{dv(t)}{dt} = g_x(x^*, y^*)u(t) + g_y(x^*, y^*)v(t) + o(u, v). \end{cases} \tag{14.6}$$

If and only if $(u(t), v(t)) \to (0, 0)$ as $t \to \infty$ for any $(u(0), v(0))$ such that $0 < |u(0)| \ll 1$ and $0 < |v(0)| \ll 1$, equilibrium (x^*, y^*) is called *locally asymptotically stable*.

Similarly with the arguments for the one dimensional case in the previous section, we focus on the behavior of the solution $(x(t), y(t))$ for (14.4) in a neighborhood of equilibrium (x^*, y^*). Thus, supposing that $|x(t) - x^*| = |u(t)| \ll 1$ and $|y(t) - y^*| = |v(t)| \ll 1$, we consider the following system of $(\tilde{u}(t), \tilde{v}(t))$

which temporal change approximates that of (u, v) by (14.6) in a neighborhood of (x^*, y^*):

$$
\begin{cases}
\dfrac{d\widetilde{u}(t)}{dt} = f_x(x^*, y^*)\widetilde{u}(t) + f_y(x^*, y^*)\widetilde{v}(t) \\[2mm]
\dfrac{d\widetilde{v}(t)}{dt} = g_x(x^*, y^*)\widetilde{u}(t) + g_y(x^*, y^*)\widetilde{v}(t)
\end{cases}
\tag{14.7}
$$

This is called *linearized system* for (14.4) around equilibrium (x^*, y^*).

The linearized system (14.7) for (14.4) corresponds to (13.15) for (13.14) in Sect. 13.2.1, and matrix A defined by (13.16) now becomes *Jacobian matrix* at the point (x^*, y^*) (refer to Sect. 12.2.1):

$$
A = \begin{pmatrix} f_x(x^*, y^*) & f_y(x^*, y^*) \\ g_x(x^*, y^*) & g_y(x^*, y^*) \end{pmatrix}.
\tag{14.8}
$$

The behavior of $(\widetilde{u}(t), \widetilde{v}(t))$ by (14.7) is determined by the eigenvalues of A, as described in Sect. 13.2. This means that the behavior of $(u(t), v(t))$ by (14.6) in a neighborhood of $(0, 0)$, that is, the behavior of $(x(t), y(t))$ by (14.4) in a neighborhood of equilibrium (x^*, y^*) is determined by the eigenvalues of Jacobian matrix A for the equilibrium. Finally, from Theorem 13.3 in Sect. 13.2, the local stability of equilibrium (x^*, y^*) for the system (14.4) is determined as follows:

Theorem 14.2 *If every eigenvalue λ of Jacobian matrix A for equilibrium (x^*, y^*) has negative real part, $\mathrm{Re}\,\lambda < 0$, equilibrium (x^*, y^*) for the system (14.4) is locally asymptotically stable. If the real part of an eigenvalue λ is positive, $\mathrm{Re}\,\lambda > 0$, it is unstable.*

In the next section,[1] we will see the classification of equilibrium with respect to the local stability.

14.3   Classification of Equiliblium

As already mentioned in the previous section, the local stability of an equilibrium about the model with a two dimensional system of autonomous ordinary differential Eq. (14.4) can be investigated by analyzing the behavior of the system of homogeneous linear ordinary differential Eq. (14.7) derived by the linearization of (14.4) around the equilibrium. Technically the eigenvalues of Jacobian matrix A defined by (14.8) determine the asymptotic behavior of $(\widetilde{u}(t), \widetilde{v}(t))$ by (14.7),

---

[1] The arguments in the section are mathematically analogous to those in Sect. 12.2.2 for the two dimensional system of first order linear *difference* equations. It would be very helpful for readers' clearer understanding to compare one with the other.

and subsequently the asymptotic behavior of $(u(t), v(t))$ by (14.6). Therefore, as described in Sect. 13.2, the eigenvalues of Jacobian matrix A defined by (14.8) determine the local stability of the equilibrium, and classify it as shown in Table 14.1 (refer also to Fig. 13.1 in Sect. 13.2).

If the following two conditions are satisfied for equilibrium $\mathbf{x}^* = (x^*, y^*)$ about the system (14.4), equilibrium $\mathbf{x}^*$ is called *Lyapunov stable, L-stable, neutrally stable, weakly stable,* or sometimes simply *stable:*

(i) With a sufficiently small $\rho > 0$, the solution of (14.4), $\mathbf{x}(t) = (x(t), y(t))$, exists for all $t > 0$ and for any initial condition $\mathbf{x}_0 = (x(0), y(0))$ such that $||\mathbf{x}_0 - \mathbf{x}^*|| < \rho$.

(ii) For any $\varepsilon > 0$, there exists a sufficiently small positive $\delta < \rho$ such that $||\mathbf{x}(t) - \mathbf{x}^*|| < \varepsilon$ for all $t > 0$ about $\mathbf{x}_0$ satisfying that $||\mathbf{x}_0 - \mathbf{x}^*|| < \delta$.

The mathematical definition of this stability means that stays for all $t > 0$ in a neighborhood of equilibrium (x^*, y^*) if the perturbation is sufficiently small. In other words, the distance of point $\mathbf{x}(t)$ from equilibrium $\mathbf{x}^*$ keeps sufficiently small for all $t > 0$ if the perturbation is sufficiently small. We must remark that the definition does *not require* an asymptotic approach of the solution toward the equilibrium.

If the following condition about such an asymptotic approach of the solution toward the equilibrium is satisfied in addition, equilibrium $\mathbf{x}^*$ is called *asymptotically stable:*

(iii) With a sufficiently small δ such that $0 < \delta < \rho$, $||\mathbf{x}(t) - \mathbf{x}^*|| \to 0$ as $t \to \infty$ for all $\mathbf{x}_0$ such that $||\mathbf{x}_0 - \mathbf{x}^*|| < \delta$.

We must remark the treatment of Lyapunov stability, as mentioned also at p. 221 about the stability of an equilibrium for Lotka-Volterra prey-predator model (8.15) in Sect. 8.4. If an equilibrium (x^*, y^*) for (14.4) is Lyapunov stable, two eigenvalues for the equilibrium are purely imaginary. However, the inverse is not true. That is, only from the fact that two eigenvalues are purely imaginary, the equilibrium cannot be identified as Lyapunov stable only from the fact that two eigenvalues are purely imaginary. *Even when two eigenvalues are purely imaginary, the equilibrium may not be Lyapunov stable.* This is because the local stability analysis is based on the linearized system (14.7) only. It is a sort of linear approximation of the system (14.4) in the neighborhood of equilibrium (x^*, y^*). Hence, even when two eigenvalues are purely imaginary, the terms of higher order in (14.6) determines the actual behavior of the solution near the equilibrium, which may cause its asymptotic approach toward it or its leaving from it. In a mathematical sense, two purely imaginary eigenvalues only imply the possibility of Lyapunov stability about the equilibrium. A further investigation in addition to the local stability analysis is necessary to get the result on the stability.

Table 14.1 Classification of the equilibrium and its stability for the two dimensional system of autonomous ordinary differential Eq. (14.4). For the case of purely imaginary eigenvalue, the local stability must be investigated with some additional analysis as mentioned in the main text

Eigenvalues	Classification of equilibrium		Asymptotic behavior of solution in the neighborhood of equilibrium	Local stability
Only positive real	Source	Unstable node		Unstable
Imaginary with positive real part		Unstable spiral/focus		
Positive and negative real	Saddle			

	Sink		Locally asymptotically stable
Only negative real	Stable node		
Imaginary with negative real part	Stable spiral/focus		
Purely imaginary	Center		$\left(\begin{array}{c}\text{Lyapunov stable}\\ \text{Neutrally stable}\end{array}\right)$

14.4   Lotka-Volterra Two Species Competition Model

In this part, let us see an actual local stability analysis on the equilibrium of Lotka-Volterra two species competition model (7.4) in Sect. 7.1.2 as an example for it. The linearized system (14.7) now becomes

$$
\begin{cases}
\dfrac{d\tilde{n}_1(t)}{dt} = (r_1 - \beta_1 N_1^* - \gamma_{12} N_2^*)\tilde{n}_1(t) - \big\{\beta_1 \tilde{n}_1(t) + \gamma_{12}\tilde{n}_2(t)\big\} N_1^*; \\[2mm]
\dfrac{d\tilde{n}_2(t)}{dt} = (r_2 - \beta_2 N_2^* - \gamma_{21} N_1^*)\tilde{n}_2(t) - \big\{\beta_2 \tilde{n}_2(t) + \gamma_{21}\tilde{n}_1(t)\big\} N_2^*
\end{cases}
\tag{14.9}
$$

with respect to equilibrium (N_1^*, N_2^*), where $\tilde{n}_i$ $(i = 1, 2)$ is the approximation for the perturbation $n_i(t) = N_i(t) - N_i^*$ from the equilibrium value N_i^* for (7.4).

Now let us consider equilibrium E_3 defined by (7.5) as (N_1^*, N_2^*), supposing the condition that it exists as given in Exercise 7.1 of p. 195. The characteristic equation $\det(A - \lambda E) = 0$ with unit matrix E for the eigenvalue λ of Jacobian matrix A defined by (14.8) can be easily derived as

$$
\lambda^2 + (\beta_1 N_1^* + \beta_2 N_2^*)\lambda + (\beta_1 \beta_2 - \gamma_{12}\gamma_{21}) N_1^* N_2^* = 0.
$$

Since the discriminant is positive from

$$
(\beta_1 N_1^* + \beta_2 N_2^*)^2 - 4(\beta_1\beta_2 - \gamma_{12}\gamma_{21})N_1^* N_2^* = (\beta_1 N_1^* - \beta_2 N_2^*)^2 + 4\gamma_{12}\gamma_{21} N_1^* N_2^* > 0,
$$

we find that every eigenvalue must be real. From the relation of the coefficients of quadratic equation to the roots λ_1 and λ_2, we can find that there is at least one negative eigenvalue since $\beta_1 N_1^* + \beta_2 N_2^* = -(\lambda_1 + \lambda_2) > 0$. Hence, if the constant term satisfies that $(\beta_1\beta_2 - \gamma_{12}\gamma_{21})N_1^* N_2^* = \lambda_1\lambda_2 > 0$, every eigenvalue is negative, so that equilibrium E_3 is locally asymptotically stable (stable node). In contrast, if $(\beta_1\beta_2 - \gamma_{12}\gamma_{21})N_1^* N_2^* = \lambda_1\lambda_2 < 0$, one eigenvalue must be positive and the other negative. In this case, equilibrium E_3 is unstable (saddle). From these results and the condition for the existence of E_3, we can get the conclusion about the local stability of E_3 given by (7.6) in Sect. 7.1.2. Remark that the condition $\beta_1\beta_2 - \gamma_{12}\gamma_{21} > 0$ coincides with $\mathcal{R}_1\mathcal{R}_2 > 1$ of (7.6) with the definitions of $\mathcal{R}_1$ and $\mathcal{R}_2$ in Exercise 7.1 (p. 195 of Sect. 7.1.2).

14.5   Rosenzweig-MacArthur Model

In this part, we describe the local stability analysis on the coexistent equilibrium E_2 given by (8.41) for the Rosenzweig-MacArthur model (8.40) in Sect. 8.6. Jacobian matrix A for equilibrium E_2 becomes

$$A = \begin{pmatrix} \widetilde{a}^* & -\mu/k \\ \widetilde{c}^* & 0 \end{pmatrix}$$

from the linearized system around E_2 for (8.40), where

$$\widetilde{a}^* := 1 - 2\widetilde{H}_+^* - \frac{1 - \widetilde{H}_+^*}{1 + \eta\widetilde{H}_+^*}; \quad \widetilde{c}^* := \frac{k(1 - \widetilde{H}_+^*)}{1 + \eta\widetilde{H}_+^*}.$$

Hence the characteristic equation $\det(A - \lambda E) = 0$ for the eigenvalue λ becomes

$$\lambda^2 - \widetilde{a}^*\lambda + \frac{\mu\,\widetilde{c}^*}{k} = 0.$$

Since $\mu\widetilde{c}^*/k > 0$ when equilibrium E_2 exists, we find that the roots of this characteristic equation, that is, the eigenvalues are alternatively real with the same sign or imaginary, from the relation between the roots and constant term. Therefore, from the relation between the roots and coefficient, if $\widetilde{a}^* > 0$, the eigenvalues are both positive or imaginary with a positive real part. If $\widetilde{a}^* < 0$, they are both negative or imaginary with a negative real part.

Consequently, equilibrium E_2 is unstable as a source if $\widetilde{a}^* > 0$, and locally asymptotically stable as a sink if $\widetilde{a}^* < 0$. For the critical case where $\widetilde{a}^* = 0$ and $\widetilde{c}^* > 0$, that is,

$$\eta > 1 \text{ and } \frac{k}{\mu} - \eta = 2 + \frac{2}{\eta - 1}, \tag{14.10}$$

the eigenvalues for E_2 are purely imaginary, so that it is a center point. These results lead to the stability of E_2 given by (8.44) in Sect. 8.6.

14.6 Routh-Hurwitz Criterion

Although Sects. 14.1–14.3 describe only the local stability analysis on the equilibrium for one or two dimensional model with autonomous ordinary differential equations, the similar local stability analysis is applicable also for the model with the system of ordinary differential equations of more than two dimension. However, it is very likely that the derivation of explicit eigenvalues for Jacobian matrix is difficult. In such a case, since the local stability of equilibrium is essentially determined by the sign of the real part of eigenvalue, another qualitative analysis could be used to estimate it. The *Routh-Hurwitz criterion* is very popular as such a way for the qualitative analysis, introduced in this section, whereas it provides in general a set of more complicated conditions than those obtained by the eigenvalue analysis on the corresponding Jacobian matrix for the local stability of equilibrium.

The Routh-Hurwitz criterion is not specified to the eigenvalue analysis or the stability analysis on the dynamical system. It is a specific condition on the root of polynomial equation, which is really useful for the eigenvalue analysis in a variety of mathematical contexts. In this section, we shall give the criterion for the general nth order polynomial equation.

Theorem 14.3 (Routh-Hurwitz Criterion) *Every root of the nth order equation in terms of λ*

$$\lambda^n + a_1\lambda^{n-1} + \cdots + a_{n-1}\lambda + a_n = 0 \qquad (14.11)$$

has a negative real part, $\operatorname{Re}\lambda < 0$, if and only if the following condition is satisfied for any $k = 1, 2, \ldots, n$:

$$\Delta_k := \begin{vmatrix} a_1 & a_0 & 0 & 0 & 0 & 0 & \cdots & 0 \\ a_3 & a_2 & a_1 & a_0 & 0 & 0 & \cdots & 0 \\ a_5 & a_4 & a_3 & a_2 & a_1 & a_0 & \cdots & 0 \\ \vdots & \vdots & \vdots & \vdots & \vdots & \vdots & \ddots & \vdots \\ a_{2k-1} & a_{2k-2} & a_{2k-3} & a_{2k-4} & a_{2k-5} & a_{2k-6} & \cdots & a_k \end{vmatrix} > 0, \qquad (14.12)$$

where let $a_m = 0$ for $m > n$ and $a_0 = 1$.

The necessary and sufficient condition mathematically consists of n inequalities for the nth order equation. For example, when $n = 3$, that is, about the cubic equation, the above condition is described as

$$\Delta_1 = a_1 > 0; \quad \Delta_2 = \begin{vmatrix} a_1 & a_0 \\ a_3 & a_2 \end{vmatrix} = \begin{vmatrix} a_1 & 1 \\ a_3 & a_2 \end{vmatrix} > 0; \quad \Delta_3 = \begin{vmatrix} a_1 & a_0 & 0 \\ a_3 & a_2 & a_1 \\ a_5 & a_4 & a_3 \end{vmatrix} = \begin{vmatrix} a_1 & 1 & 0 \\ a_3 & a_2 & a_1 \\ 0 & 0 & a_3 \end{vmatrix} > 0,$$

which can be easily proved equivalent to the following simpler condition:

$$a_1 > 0; \quad a_3 > 0; \quad a_1a_2 - a_3 > 0.$$

When and only when these three inequalities are satisfied, the cubic Eq. (14.11) has only roots with negative real part. In the same way, the conditions for $n = 4$ and $n = 5$ are expressed as follows:

When $n = 4$,

$$a_1 > 0; \quad a_3 > 0; \quad a_4 > 0; \quad a_1a_2a_3 > a_3^2 + a_1^2a_4.$$

When $n = 5$,

$$a_1 > 0; \quad a_3 > 0; \quad a_5 > 0; \quad a_1 a_2 a_3 > a_3^2 + a_1^2 a_4 - a_1 a_5;$$

$$(a_1 a_4 - a_5)(a_1 a_2 a_3 - a_3^2 - a_1^2 a_4 + a_1 a_5) > a_5 (a_1 a_2 - a_3)^2.$$

As seen from these examples, as the degree n gets larger, the condition by the Routh-Hurwitz criterion becomes less convenient because of its complexity in the formulas, even though the criterion may be very useful for the mathematical analysis in some specific cases, or for a numerical estimation of the local stability about an equilibrium with a specific set of parameter values in the system under the analysis.

The Routh-Hurwitz criterion has been studied and given by some other expressions mathematically equivalent to it. One of them is the *Liénard-Chipart criterion* which is sometimes more useful than the Routh-Hurwitz criterion shown in the above.

Theorem 14.4 (Liénard-Chipart Criterion) *Every root of the nth order Eq. (14.11) has a negative real part,* Re $\lambda < 0$, *if and only if one of the following four conditions for* $k = 1, 2, \ldots, \left[\frac{n}{2}\right] + 1$ *is satisfied:*

(i) $a_n > 0$, $a_{n-2k} > 0$, *and* $\Delta_{2k-1} > 0$;
(ii) $a_n > 0$, $a_{n-2k} > 0$, *and* $\Delta_{2k} > 0$;
(iii) $a_n > 0$, $a_{n-2k+1} > 0$, *and* $\Delta_{2k-1} > 0$;
(iv) $a_n > 0$, $a_{n-2k+1} > 0$, *and* $\Delta_{2k} > 0$,

where let $a_m = 0$ *for* $m > n$ *and* $a_0 = 1$. *The notation* $[x]$ *is the Gauss symbol to indicate the maximal integer less than or equal to* x.

In brief, the Liénard-Chipart criterion is such that all odd (or even) numbered coefficients and determinants of $\{\Delta_1, \Delta_2, \ldots, \Delta_n\}$ defined by (14.12) are positive with $a_n > 0$. In comparison to the Routh-Hurwitz criterion, the Liénard-Chipart criterion requires almost half number of inequalities.

Anyway it is clear that, for the eigenvalue analysis on an equilibrium about the n dimensional system of autonomous ordinary differential equations, the Routh-Hurwitz/Liénard-Chipart criterion could provide the sufficient condition for its local stability.

14.7 Isocline Method

In this section, we describe a method of qualitative analysis on the existence and stability of equilibrium for the two dimensional system of autonomous ordinary differential equations like (14.4) in p. 425 of Sect. 14.2. It is called *isocline method*, which can be regarded as a *phase plane method*, or more generally *phase space method*. It could provide some important qualitative and general results on the existence and stability of equilibrium for the dynamics of a continuous time model.

Here to show the actual way of applying it for the system, we shall use Lotka-Volterra two species competition model (7.4) of Sect. 7.1.2. Shortly saying, the isocline method uses only the signs of $dN_1(t)/dt$ and $dN_2(t)/dt$ determined by (7.4).

Let us consider the set of states (N_1, N_2) satisfying that $dN_1(t)/dt = 0$ in (7.4). It is expressed as

$$\{(N_1, N_2) \mid (r_1 - \beta_1 N_1 - \gamma_{12} N_2)N_1 = 0\}.$$

This set defines two lines on which the velocity of the temporal change of N_1 becomes zero in the (N_1, N_2)-phase plane. Such lines (curves in the general case) are called *nullcline* for N_1. The nullcline becomes the boundary line/curve to divide the phase plane into the regions of positive and negative $dN_1(t)/dt$. Hence with the sign of $dN_1(t)/dt$, the region in which N_1 increases (resp. decreases) is identified in the phase plane.

For Lotka-Volterra two species competition model (7.4), the nullclines for N_1 correspond to two lines in the (N_1, N_2)-phase plane, $N_1 = 0$ and $r_1 - \beta_1 N_1 - \gamma_{12} N_2 = 0$ as shown in Fig. 7.2 in p. 196 of Sect. 7.1.2. The former nullcline coincides to the axis of N_2 in the (N_1, N_2)-phase plane. In the same way, the nullclines for N_2 correspond to two lines in the (N_1, N_2)-phase plane, $N_2 = 0$ and $r_2 - \beta_2 N_2 - \gamma_{21} N_1 = 0$.

As shown in Fig. 7.2, we can find that there are four different cases with respect to the spatial configuration of those nullclines for N_1 and N_2 in the (N_1, N_2)-phase plane. Nullclines divide the first quadrant of the (N_1, N_2)-phase plane into three or four regions with respect to the combination of signs of dN_1/dt and dN_2/dt. Intersections between nullclines for N_1 and N_2 indicate the equilibria for the system (7.4), since they correspond to the pairs of values N_1 and N_2 to satisfy that $dN_1/dt = 0$ and $dN_2/dt = 0$.

With the regions about the combination of signs of dN_1/dt and dN_2/dt, we can find the qualitative direction of vector $(dN_1/dt, dN_2/dt)$ in each region bounded by the nullclines. In this method, the qualitative direction in a region is one of upper rightward, down rightward, upper leftward, and down leftward. Especially on the nullcline, the vector direction becomes vertical or horizontal, since no temporal change of N_1 or N_2 occurs on it. Even with such rough informations on the qualitative direction of the state change in the phase plane, we could get some mathematical results on the behavior of the system, though they would not be definitive but be supplementarily useful for the arguments on the existence and stability of the equilibria. Fig. 14.1 illustratively shows the procedure of this method.

For example, in the case of Fig. 7.2a and d, we can find that the trajectory eventually enters in the middle region, and subsequently approaches equilibrium E_1 or E_2. In the case of Fig. 7.2c, the trajectory eventually approaches the coexistent equilibrium E_3 in the first quadrant. In the case of Fig. 7.2b, both of equilibria E_1 and E_2 on the axes are approachable, which indicates a bistable situation.

The isocline method cannot mathematically reveal all nature of the behavior of Lotka-Volterra two species competition model (7.4), while it is successful to

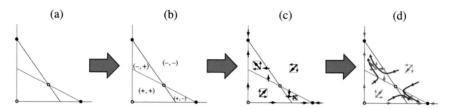

Fig. 14.1 Illustrative procedure of the isocline method. (**a**) to draw the nullclines in the (N_1, N_2)-phase plane; (**b**) to identify the combination of signs of dN_1/dt and dN_2/dt in each region bounded by the nullclines; (**c**) to identify the direction of vector $(dN_1/dt, dN_2/dt)$ in each region; (**d**) to find the possible shape of trajectories of (N_1, N_2) coincident with the direction of vectors

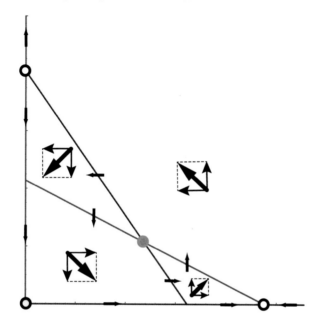

Fig. 14.2 An illustrative example of the phase plane in which the isocline method gives only supplementary informations on the stability of equilibrium, where the stability of the interior equilibrium cannot be estimated by it. Remark that this case does not occur for Lotka-Volterra two species competition model (7.4)

give satisfactory mathematical informations on the behavior. Actually, numerical examples of the trajectories and vector flow in the (N_1, N_2)-phase plane shown in Figs. 7.3 and 7.4 indicate its success.

On the other hand, for example, in the case shown in Fig. 14.2, the isocline method can indicate that every equilibrium on the axes is unstable, though it cannot give any information on the stability of the equilibrium in the first quadrant. The isocline method is certainly useful, and becomes more useful to understand the dynamical nature of two dimensional system when it is used in addition to some other mathematical informations obtained by another analytical method, for example, the local stability analysis (refer to Sects. 14.2 and 14.3).

In the case of Fig. 14.2 no determinative information on the stability of equilibrium in the first quadrant is available only by the isocline method. However, since the isocline method implies that the trajectory tends to spiral around it, the eigenvalue analysis on its local stability would be expected to give the imaginary eigenvalue.

14.8   Lyapunov Function

In this section, as one of mathematical approaches to show the asymptotical stability of an equilibrium, we shall see the method by *Lyapunov function*. This is a strong method, though there is no general way to find or construct a Lyapunov function for a given dynamical system. It is found out or constructed by some mathematical analogies or by a mathematically intuitive step-by-step process. This means that there is no mathematically general proof of its existence for a dynamical system. Once a Lyapunov function is found out for an equilibrium, its stability can be investigated not necessarily in the sense of local stability but also in that of global stability.

First let us see a well-known Lyapunov function for the coexistent equilibrium E_2 about Lotka-Volterra prey-predator system (8.22) in Sect. 8.4. The following arguments are under the condition for the existence of E_2. Making use of the equilibrium values at $E_2(H^*, P^*) = (\delta/(\kappa\gamma), r/\gamma - \beta\delta/(\kappa\gamma^2))$, we shall rewrite the system (8.22) as follows:

$$\begin{cases} \dfrac{dH(t)}{dt} = \gamma\left[P^* - P(t) + \dfrac{\beta}{\gamma}\{H^* - H(t)\}\right]H(t) \\ \dfrac{dP(t)}{dt} = \kappa\gamma\{H(t) - H^*\}P(t) \end{cases} \tag{14.13}$$

This rewriting is convenient for the following arguments, although it is not essential.

Let us introduce the following function $V(H, P)$ of $H = H(t)$ and $P = P(t)$:

$$V(H, P) := \kappa\{H - H^* - H^*(\ln H - \ln H^*)\} + P - P^* - P^*(\ln P - \ln P^*). \tag{14.14}$$

Fig. 14.3 Numerically drawn contour map in terms of the value of Lyapunov function (14.14) in the (H, P)-phase plane about Lotka-Volterra prey-predator system (8.22). Nullclines for the system is shown too. $r = 0.2$; $\beta = 0.05$; $\gamma = 0.1$; $\delta = 0.1$; $\kappa = 0.8$

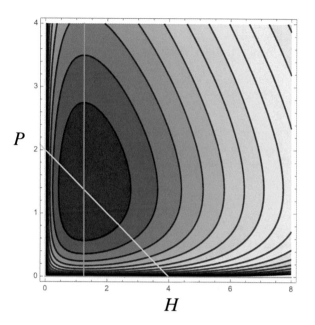

It is clear that $V(H^*, P^*) = 0$. Further we have

$$V_H(H, P) = \frac{\partial V(H, P)}{\partial H} = \kappa\left(1 - \frac{H^*}{H}\right); \quad V_P(H, P) = \frac{\partial V(H, P)}{\partial P} = 1 - \frac{P^*}{P};$$

$$V_{HH}(H, P) = \frac{\partial^2 V(H, P)}{\partial H^2} = \frac{\kappa H^*}{H^2}; \quad V_{PP}(H, P) = \frac{\partial^2 V(H, P)}{\partial P^2} = \frac{P^*}{P^2};$$

$$V_{HP}(H, P) = \frac{\partial^2 V(H, P)}{\partial H \partial P} = \frac{\partial^2 V(H, P)}{\partial P \partial H} = 0.$$

Since $V_H(H^*, P^*) = 0$, $V_P(H^*, P^*) = 0$, $V_{HH}(H^*, P^*) = \kappa/H^* > 0$, and

$$V_{HH}(H^*, P^*)V_{PP}(H^*, P^*) - \left\{V_{HP}(H^*, P^*)\right\}^2 = \frac{\kappa}{H^* P^*} > 0,$$

it is mathematically proved that the function $V(H, P)$ takes its extremal minimum $V(H^*, P^*) = 0$ at $(H, P) = (H^*, P^*)$. The function $V(H, P)$ is continuous for $H > 0$ and $P > 0$. Therefore, it becomes zero only at $(H, P) = (H^*, P^*)$, while it takes positive value, that is, $V(H, P) > 0$ for any $H > 0$ and $P > 0$ except for $(H, P) = (H^*, P^*)$. In other mathematical words, $V(H, P)$ is positive definite in $\mathbb{R}_+^2 \backslash \{(H^*, P^*)\}$. Hence, the function $V(H, P)$ has the unique minimum $V(H^*, P^*) = 0$ at the unique point $E_2(H^*, P^*)$ (see Fig. 14.3).

On the other hand, making use of (14.13), we can derive

$$\frac{dV(H(t), P(t))}{dt} = \frac{\partial V(H, P)}{\partial H} \cdot \frac{dH(t)}{dt} + \frac{\partial V(H, P)}{\partial P} \cdot \frac{dP(t)}{dt}$$

$$= -\kappa\beta\{H(t) - H^*\}^2 \leq 0$$

This means that the value of $V(H(t), P(t))$ monotonically decreases with the passage of time as long as $H(t) \neq H^*$. From (14.13), even if $H(t_1) = H^*$ at a certain time t_1, we have

$$\left.\frac{dH(t)}{dt}\right|_{t=t_1} = \gamma H^*\{P^* - P(t_1)\} \neq 0$$

unless $P(t_1) = P^*$. Thus $H(t)$ must change from H^* at $t = t_1$. As a result, we can find that $V(H(t), P(t))$ is temporally decreasing unless $P(t_1) = P^*$ even when $H(t_1) = H^*$ at any time t_1. Therefore, $V(H(t), P(t))$ is temporally decreasing as long as $(H(t), P(t)) \neq (H^*, P^*)$.

From these arguments, we can result that $V(H(t), P(t))$ approaches the minimum as $t \rightarrow \infty$. This means that $(H(t), V(t)) \rightarrow E_2(H^*, P^*)$ as $t \rightarrow \infty$. Thus, when equilibrium E_2 exists, the solution of the system (14.13) from any positive initial point approaches E_2 with the passage of time. Consequently, the coexistent equilibrium E_2 is *globally asymptotically stable* when it exists.

The function $V(H(t), P(t))$ defined by (14.14) can be called the *Lyapunov function* with respect to equilibrium E_2 of the system (14.13).

Definition 14.1 For the system of first order nonlinear ordinary differential Eq. (14.4) for (x, y), a real-valued function $V(x, y)$ defined in a region $D \subseteq \mathbb{R}^2$, which contains an equilibrium (x^*, y^*) of (14.4) is called *weak Lyapunov function* for equilibrium (x^*, y^*) if the following conditions (i) and (ii) are satisfied:

(i) $V(x, y) > V(x^*, y^*)$ for any point $(x, y) \neq (x^*, y^*)$ in D;
(ii) For any $(x(t), y(t))$ in D,

$$\frac{dV(x(t), y(t))}{dt} \leq 0.$$

If the following condition is further satisfied, $V(x, y)$ is called *strict Lyapunov function*, or simply *Lyapunov function* for equilibrium (x^*, y^*) of (14.4):

(iii) For any $(x(t), y(t)) \neq (x^*, y^*)$ in D,

$$\frac{dV(x(t), y(t))}{dt} < 0 \quad \text{and} \quad \left.\frac{dV(x(t), y(t))}{dt}\right|_{(x(t),y(t))=(x^*,y^*)} = 0.$$

Following these definitions, the function $V(H(t), P(t))$ defined by (14.14) is just a weak Lyapunov function for equilibrium E_2 of the system (14.13). As shown by the above arguments, since $V(H(t), P(t))$ is temporally decreasing for any $(H, P) \neq (H^*, P^*)$ in $\mathbb{R}_+^2$, we may regard it as a Lyapunov function in a wider sense. This corresponds to its generalized definition with the replacement of the above condition (iii) by

(iii') For any $(x(t), y(t)) \neq (x^*, y^*)$ in D and any $\Delta t > 0$,

$$V(x(t + \Delta t), y(t + \Delta t)) < V(x(t), y(t))$$

and

$$\left. \frac{dV(x(t), y(t))}{dt} \right|_{(x(t), y(t)) = (x^*, y^*)} = 0.$$

Exercise 14.2 For Lotka-Volterra competition system (7.4) in Sect. 7.1, when the coexistent equilibrium $E_3(N_1^*, N_2^*)$ given by (7.5) exists and is locally asymptotically stable, show that the following function becomes a Lyapunov function for E_3:

$$V(N_1, N_2) := \frac{\beta_1}{\gamma_{12}} (N_1 - N_1^*)^2 + \frac{\beta_2}{\gamma_{21}} (N_2 - N_2^*)^2 + 2(N_1 - N_1^*)(N_2 - N_2^*).$$

$$(14.15)$$

In the above, we showed the definition of Lyapunov function for the two dimensional system of ordinary differential equations. Lyapunov function can be defined for the higher dimensional system in the same way (for example, see [1, 7]). Especially, when the so-called Lotka-Volterra system of ordinary differential equations

$$\frac{dx_i(t)}{dt} = \left\{ r_i + \sum_{j=1}^{n} \gamma_{ij} x_j(t) \right\} x_i(t), \qquad (14.16)$$

where the signs of coefficients r_i and γ_{ij} are arbitrary, has a unique locally asymptotically stable equilibrium $(x_1^*, x_2^*, \ldots, x_n^*)$ in $\mathbb{R}_+^n$, it is shown by

(continued)

Goh [4] that there exists the following form of a Lyapunov function for the equilibrium:

$$V(x_1, x_2, \ldots, x_n) := \sum_{i=1}^{n} c_i \{(x_i - x_i^*) - x_i^*(\ln x_i - \ln x_i^*)\}, \qquad (14.17)$$

where the positive coefficient c_i can be appropriately determined by the parameters of the system (14.16).

For example, for Lotka-Volterra competition system (7.4) in Sect. 7.1, when the coexistent equilibrium $E_3(N_1^*, N_2^*)$ exists and is locally asymptotically stable with $\beta_1\beta_2 - \gamma_{12}\gamma_{21} > 0$, the function $V(N_1, N_2)$ by (14.17) for $n = 2$ becomes a Lyapunov function for E_3 in a wider sense, taking

$$c_1 = \frac{\sqrt{\beta_1\beta_2} - \sqrt{\beta_1\beta_2 - \gamma_{12}\gamma_{21}}}{\gamma_{12}}; \quad c_2 = \frac{\sqrt{\beta_1\beta_2} + \sqrt{\beta_1\beta_2 - \gamma_{12}\gamma_{21}}}{\gamma_{21}},$$
$$(14.18)$$

which are determined from $c_1\gamma_{12} + c_2\gamma_{21} = 2\sqrt{\beta_1\beta_2}$ and $c_1c_2 = 1$. This Lyapunov function is different from that shown in Exercise 14.2 for equilibrium E_3 of the same system (7.4) (see Fig. 14.4). This is an example to indicate that Lyapunov function for an equilibrium is not necessarily unique even though one is found out.

As known in the above, finding a Lyapunov function for an equilibrium leads to getting an important information about its stability:

Theorem 14.5 *For a region $D \subseteq \mathbb{R}^2$ including equilibrium (x^*, y^*) for the two dimensional system of autonomous ordinary differential Eq. (14.4),*

(i) *the existence of a weak Lyapunov function $V(x, y)$ indicates that equilibrium (x^*, y^*) is Lyapunov stable (refer also to Sect. 14.3);*

(ii) *the existence of a strict Lyapunov function $V(x, y)$ indicates that equilibrium (x^*, y^*) is asymptotically stable in D.*

As already seen in the above description about equilibrium E_2 of Lotka-Volterra prey-predator system (14.13), even when we have only a weak Lyapunov function for an equilibrium, we may be able to find its global stability with an additional mathematical arguments on the dynamical behavior of system or with the generalized definition of Lyapunov function.

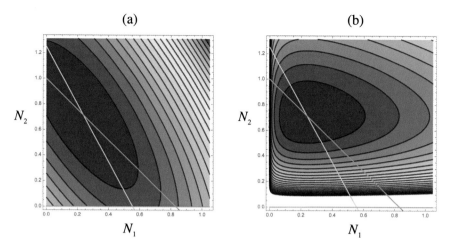

Fig. 14.4 Numerically drawn contour map of the value of Lyapunov function in the (N_1, N_2)-phase plane about Lotka-Volterra competition system (7.4) in Sect. 7.1. Nullclines for the system is shown too. $r_1 = r_2 = 1.0$; $\beta_1 = 1.8$; $\beta_2 = 1.0$; $\gamma_{12} = 0.8$; $\gamma_{21} = 1.2$. (**a**) Lyapunov function (14.15); (**b**) Lyapunov function (14.17) for $n = 2$ with (14.18). This is the case where the coexistent equilibrium is asymptotically stable, corresponding to Figs. 7.1c, 7.2c, and 7.3c

14.9 Poincaré-Bendixson Theorem

In this section, let us see a theorem about the set of points to which approaches the trajectory $\Gamma(\mathbf{x}_0, t)$ of the autonomous two dimensional system (14.4) for the initial point $\mathbf{x}_0 := (x_0, y_0)$, called *$\omega$-limit set* for $\mathbf{x}_0$ (refer also to Sect. 12.1.5). It was presented by Jules H. Poincaré (1854–1912) and Ivar O. Bendixson (1861–1935), and is known well in the theory of dynamical system today:

Theorem 14.6 (Poincaré-Bendixson Theorem) *When the trajectory $\Gamma(\mathbf{x}_0, t)$ for the autonomous two dimensional system (14.4) is a finite closed set, and its ω-limit set does not contain any equilibrium, one of the followings holds:*

(i) The trajectory $\Gamma(\mathbf{x}_0, t)$ forms a periodic orbit;
(ii) The ω-limit set for $\mathbf{x}_0$ is a periodic orbit.

The former (i) is the case where the initial point $\mathbf{x}_0$ belongs to the ω-limit set. The latter (ii) is the case where $\mathbf{x}_0$ is not included in the ω-limit set. The finite closeness of the trajectory $\Gamma(\mathbf{x}_0, t)$ is essential for the theorem. This theorem shows that, when the trajectory $\Gamma(\mathbf{x}_0, t)$ for $\mathbf{x}_0$ always remains in a finite domain Ω of (x, y)-phase plane, if there is no asymptotically stable equilibrium in Ω, then the trajectory necessarily becomes or approaches a periodic orbit. Especially in the latter case (ii) when the trajectory asymptotically approaches a periodic orbit, the periodic orbit is called *limit cycle* too.

As a consequence from the above Poincaré-Bendixson Theorem, the following theorem holds, which is sometimes called *Poincaré-Bendixson Trichotomy*:

Theorem 14.7 (Poincaré-Bendixson Trichotomy) *When the trajectory $\Gamma(\mathbf{x}_0, t)$ of the autonomous two dimensional system* (14.4) *always remains in a finite closed domain Ω which contains a finite number of equilibria, the ω-limit set for $\mathbf{x}_0$ becomes one of the following three:*

(i) an equilibrium;
(ii) a periodic orbit;
(iii) a cycle graph consisting of a finite number of equilibria and some component trajectories each of which has one of those equilibria as the ω-limit set or the α-limit set.

The α-limit set is defined as the set of points which the trajectory $\Gamma(\mathbf{x}_0, t)$ for $\mathbf{x}_0$ approaches as $t \to -\infty$. It can be regarded as the ω-limit set for the system (14.4) with the inverse sign in the right hand. In comparison to Poincaré-Bendixson Trichotomy, the previous theorem may be called *Poincaré-Bendixson Dichotomy*.

In the third case of the above Poincaré-Bendixson Trichotomy, the cycle graph means a set which contains *homoclinic orbit(s)* and *heteroclinic orbit(s)* in the domain Ω. The homoclinic orbit is a trajectory of closed curve containing only one of equilibria in Ω. The heteroclinic orbit is a trajectory of curve connecting two different equilibria in Ω.

Since Poincaré-Bendixson Theorem states that the trajectory of the autonomous two dimensional system (14.4) approaches alternatively an equilibrium or a periodic orbit if it does not diverge, that is, if it remains in a certain bounded domain. Therefore, it is mathematically assured that the autonomous two dimensional system (14.4) cannot show any chaotic variation. This mathematical nature is applicable only for the autonomous two dimensional system (14.4). For more than two dimensional system, there are a lot of examples to show a chaotic variation [2, 3, 5, 6, 8] (see Fig. 14.5).

It is possible to mathematically show the existence of a finite closed domain Ω of positive region in the phase plane about Rosenzweig-MacArthur model (8.38), equivalently the system (8.40), in Sect. 8.6, such that the trajectory for any initial point in Ω always remains in it (Exercise 14.3). Then Poincaré-Bendixson Theorem and Poincaré-Bendixson Trichotomy can be applied for the system. It results that the existence of an asymptotically stable periodic solution can be mathematically assured for Rosenzweig-MacArthur model (8.38), as mentioned in Sect. 8.6.

Exercise 14.3 Consider the trajectory of Rosenzweig-MacArthur model (8.40) in the $(\tilde{H}, \tilde{P})$-phase plane for the initial condition $\mathbf{x}_0 = (\tilde{H}_0, \tilde{P}_0)$ in the following triangle closed domain of the first quadrant

$$\Omega(Q) := \left\{ (\tilde{H}, \tilde{P}) \mid \tilde{H} \geq 0, \ \tilde{P} \geq 0, \ k\tilde{H} + \tilde{P} \leq Q \right\}$$

with an arbitrarily chosen Q such that

$$Q > Q_c := \frac{k(1+\mu)^2}{4\mu}.$$

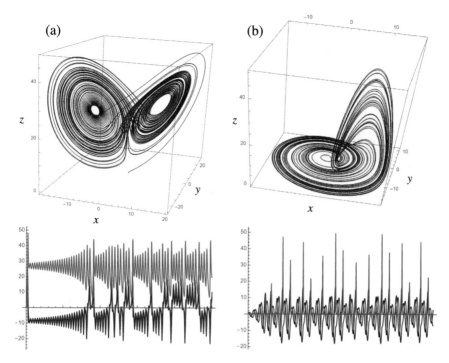

Fig. 14.5 Examples of a chaotic variation by the autonomous three dimensional system. The trajectory in the three dimensional phase space and the temporal change are numerically shown. The trajectory approaches an ω-limit set called *strange attractor*. (**a**) Lorenz equations: $dx/dt = -10x + 10y$, $dy/dt = 28x - y - xz$, $dz/dt = -8z/3 + xy$; (**b**) Rössler equations: $dx/dt = -y - z$, $dy/dt = x + 0.3$, $dz/dt = 2 + xz - 10z$. Numerical calculation commonly for the initial condition $(x(0), y(0), z(0)) = (1.0, 0.0, 0.0)$

Show that the trajectory always remains in the domain $\Omega(Q)$, that is, any point $(\widetilde{H}(\tau), \widetilde{P}(\tau))$ on the trajectory is included in the domain $\Omega(Q)$ for any time $\tau > 0$.

> The meaning of Poincaré-Bendixson Theorem must not be regarded as an implication that a chaotic variation could occur only with more than two dimensional dynamical system. As seen in Sects. 2.1, 12.1.5, and 12.1.6, there are many discrete time one dimensional dynamical system to cause a chaotic variation. For a phenomenon which could have a chaotic variation, Poincaré-Bendixson Theorem implies that a more than two dimensional autonomous dynamical system of ordinary differential equations would be necessary to model it. Such a mathematical viewpoint may be valuable to theoretically consider the kinetics of such a variation, since the implication

(continued)

could be regarded as the necessity of more than two variable factors to drive the dynamics. However, this is not always an appropriate direction of theoretical/mathematical consideration on such a chaotic dynamics, because the discrete time dynamical system of lower dimension could cause it as mentioned before. For these reasons, it is important that we must design a reasonable modeling to clarify a scientific standing point about which the modeling is applied for the dynamics in the theoretical/mathematical consideration with a mathematical model. It must include which the model becomes of discrete or continuous time.

Answer to Exercise

Exercise 14.1 (p. 424)

The solution of ordinary differential Eq. (14.3) is given by

$$\widetilde{n}(t) = \widetilde{n}(0)\, e^{g'(N^*)t}.$$

Thus, if $\widetilde{n}(0) > 0$, we have $\widetilde{n}(t) > 0$ for any $t > 0$, and if $\widetilde{n}(0) < 0$, $\widetilde{n}(t) < 0$ for any $t > 0$. Hence from the above solution, we can find that, if $g'(N^*) < 0$, then $\widetilde{n}(t)$ monotonically approaches 0, independently of the sign of $\widetilde{n}(0)$. On the other hand, since

$$|\widetilde{n}(t)| = |\widetilde{n}(0)|\, e^{g'(N^*)t},$$

$|\widetilde{n}(t)|$ monotonically increases for $t > 0$ if $g'(N^*) > 0$.

Exercise 14.2 (p. 439)

We can derive the following partial derivatives about $V(N_1, N_2)$ defined by (14.15):

$$\frac{\partial V(N_1, N_2)}{\partial N_1} = \frac{2\beta_1}{\gamma_{12}}\,(N_1 - N_1^*) + 2(N_2 - N_2^*);$$

$$\frac{\partial V(N_1, N_2)}{\partial N_2} = \frac{2\beta_2}{\gamma_{21}}\,(N_2 - N_2^*) + 2(N_1 - N_1^*);$$

$$\frac{\partial^2 V(N_1, N_2)}{\partial N_1^2} = \frac{2\beta_1}{\gamma_{12}}; \quad \frac{\partial^2 V(N_1, N_2)}{\partial N_2^2} = \frac{2\beta_2}{\gamma_{21}}; \quad \frac{\partial^2 V(N_1, N_2)}{\partial N_2 \partial N_1} = \frac{\partial^2 V(N_1, N_2)}{\partial N_1 \partial N_2} = 2.$$

Making use of the condition that $\beta_1\beta_2 - \gamma_{12}\gamma_{21} \neq 0$, satisfied when the coexistent equilibrium E_3 exists, we can find that the rest point satisfying that $V_{N_1}(N_1, N_2) = 0$ and $V_{N_2}(N_1, N_2) = 0$ uniquely exists and is given by $(N_1, N_2) = (N_1^*, N_2^*)$. Then we have $V_{N_1 N_1}(N_1^*, N_2^*) = 2\beta_1/\gamma_{12} > 0$ and

$$V_{N_1 N_1}(N_1^*, N_2^*)V_{N_2 N_2}(N_1^*, N_2^*) - \{V_{N_1 N_2}(N_1^*, N_2^*)\}^2 = \frac{4(\beta_1\beta_2 - \gamma_{12}\gamma_{21})}{\gamma_{12}\gamma_{21}}.$$

Hence we can result that $V(N_1^*, N_2^*) = 0$ is

$$\begin{cases} \text{the extremal minimum if } \beta_1\beta_2 - \gamma_{12}\gamma_{21} > 0; \\ \text{the extremal maximum if } \beta_1\beta_2 - \gamma_{12}\gamma_{21} < 0. \end{cases}$$

Now let us consider the case of $\beta_1\beta_2 - \gamma_{12}\gamma_{21} > 0$ when $V(N_1^*, N_2^*) = 0$ is the extremal minimum of $V(N_1, N_2)$ in $\mathbb{R}_+^2$. Since

$$V(N_1, N_2) = \frac{\beta_1}{\gamma_{12}}\left\{(N_1 - N_1^*) + \frac{\gamma_{12}}{\beta_1}(N_2 - N_2^*)\right\}^2 + \frac{\beta_1\beta_2 - \gamma_{12}\gamma_{21}}{\beta_1\gamma_{21}}(N_2 - N_2^*)^2,$$

the function $V(N_1, N_2)$ is definitely positive for any $(N_1, N_2) \neq (N_1^*, N_2^*)$ in $\mathbb{R}_+^2$, and it takes the minimum ($= 0$) at (N_1^*, N_2^*).

Next we can derive the following:

$$\frac{dV(N_1(t), N_2(t))}{dt} = \frac{\partial V(N_1, N_2)}{\partial N_1} \cdot \frac{dN_1(t)}{dt} + \frac{\partial V(N_1, N_2)}{\partial N_2} \cdot \frac{dN_2(t)}{dt}$$

$$= \left\{\frac{2\beta_1}{\gamma_{12}}(N_1 - N_1^*) + 2(N_2 - N_2^*)\right\}\frac{dN_1(t)}{dt}$$

$$+ \left\{\frac{2\beta_2}{\gamma_{21}}(N_2 - N_2^*) + 2(N_1 - N_1^*)\right\}\frac{dN_2(t)}{dt}$$

$$= -\frac{2}{\gamma_{12}}\left\{\beta_1(N_1 - N_1^*) + \gamma_{12}(N_2 - N_2^*)\right\}^2 N_1$$

$$- \frac{2}{\gamma_{21}}\left\{\beta_2(N_2 - N_2^*) + \gamma_{21}(N_1 - N_1^*)\right\}^2 N_2$$

$$\leq 0.$$

Therefore we find that $dV(N_1(t), N_2(t))/dt$ is negative at any $(N_1, N_2) \neq (N_1^*, N_2^*)$ in $\mathbb{R}_+^2$, and it becomes zero at (N_1^*, N_2^*).

With these arguments, from the definition of Lyapunov function in p. 438, the function $V(N_1, N_2)$ given by (14.15) is a strict Lyapunov function for equilibrium E_3 of Lotka-Volterra competition system (7.4). Consequently, when the coexistent equilibrium E_3 exists with $\beta_1\beta_2 - \gamma_{12}\gamma_{21} > 0$, it is globally asymptotically stable in $\mathbb{R}_+^2$. This is the other description of the case (c) for Lotka-Volterra competition system (7.4) in p. 193.

Besides the condition that $\beta_1\beta_2 - \gamma_{12}\gamma_{21} > 0$ coincides with the necessary condition for the locally asymptotic stability of E_3 that $\mathcal{R}_1\mathcal{R}_2 > 1$ given by (7.6) in Sect. 7.1.2 (refer to Sect. 14.4). Therefore, the above result on the Lyapunov function (14.15) mathematically indicates that the coexistent equilibrium E_3 is globally asymptotically stable whenever it is locally asymptotically stable, as already found by the isocline method in Sect. 7.1.2.

Exercise 14.3 (p. 442)

First the system (8.40) can be expressed in the following form:

$$
\begin{cases}
\dfrac{1}{\widetilde{H}(\tau)}\dfrac{d\widetilde{H}(\tau)}{d\tau} = \mathscr{F}(\widetilde{H}(\tau), \widetilde{P}(\tau)) := 1 - \widetilde{H}(\tau) - \dfrac{\widetilde{P}(\tau)}{1 + \eta\widetilde{H}(\tau)}; \\[3mm]
\dfrac{1}{\widetilde{P}(\tau)}\dfrac{d\widetilde{P}(\tau)}{d\tau} = \mathscr{G}(\widetilde{H}(\tau), \widetilde{P}(\tau)) := -\mu + k\dfrac{\widetilde{H}(\tau)}{1 + \eta\widetilde{H}(\tau)}.
\end{cases} \tag{14.19}
$$

Hence we mathematically have

$$
\begin{cases}
\widetilde{H}(\tau) = \widetilde{H}_0 \exp\left[\displaystyle\int_0^\tau \mathscr{F}(\widetilde{H}(s), \widetilde{P}(s))\, ds\right]; \\[3mm]
\widetilde{P}(\tau) = \widetilde{P}_0 \exp\left[\displaystyle\int_0^\tau \mathscr{G}(\widetilde{H}(s), \widetilde{P}(s))\, ds\right],
\end{cases} \tag{14.20}
$$

and we can say that the temporal change of point $(\widetilde{H}(\tau), \widetilde{P}(\tau))$ on the trajectory for the initial condition $\mathbf{v}_0 = (\widetilde{H}_0, \widetilde{P}_0)$ in the $(\widetilde{H}, \widetilde{P})$-phase plane follows the equations of (14.20). The equations of (14.20) indicate that $\widetilde{H}(\tau) \geq 0$ for any $\tau > 0$ if $\widetilde{H}_0 \geq 0$, and $\widetilde{P}(\tau) \geq 0$ for any $\tau > 0$ if $\widetilde{P}_0 \geq 0$. The same mathematical logic is used also in Exercise 8.5 (p. 225) of Sect. 8.4.3.

Next, from (8.40), we find that

$$
\frac{d}{d\tau}\{k\widetilde{H}(\tau) + \widetilde{P}(\tau)\} = k\frac{d\widetilde{H}(\tau)}{d\tau} + \frac{d\widetilde{P}(\tau)}{d\tau} = k\{1 - \widetilde{H}(\tau)\}\widetilde{H}(\tau) - \mu\widetilde{P}(\tau).
$$

Hence, if $k\widetilde{H}(\tau_0) + \widetilde{P}(\tau_0) > Q_c$ at $\tau = \tau_0 \geq 0$, then we have

$$\frac{d}{d\tau}\{k\widetilde{H}(\tau) + \widetilde{P}(\tau)\}\Big|_{\tau=\tau_0} < k\{1 - \widetilde{H}(\tau_0)\}\widetilde{H}(\tau_0) - \mu\{Q_c - k\widetilde{H}(\tau_0)\}$$

$$= k\left[\frac{(1+\mu)^2}{4} - \left\{\widetilde{H}(\tau_0) - \frac{1+\mu}{2}\right\}^2\right] - \mu Q_c$$

$$\leq \frac{k(1+\mu)^2}{4} - \mu Q_c = 0,$$

since $\widetilde{P}(\tau_0) > Q_c - k\widetilde{H}(\tau_0)$. Therefore we have shown that

$$\frac{d}{d\tau}\{k\widetilde{H}(\tau) + \widetilde{P}(\tau)\} < 0$$

if $k\widetilde{H}(\tau) + \widetilde{P}(\tau) > Q_c$. This means that, if $k\widetilde{H}(\tau) + \widetilde{P}(\tau) > Q_c$, $k\widetilde{H}(\tau) + \widetilde{P}(\tau)$ is monotonically decreasing in terms of τ.

As a result, for the initial condition $\mathbf{v}_0 = (\widetilde{H}_0, \widetilde{P}_0)$ such that $k\widetilde{H}_0 + \widetilde{P}_0 > Q_c$, we have $k\widetilde{H}(\tau) + \widetilde{P}(\tau) < k\widetilde{H}_0 + \widetilde{P}_0$ for any time $\tau > 0$. Subsequently, if $\mathbf{v}_0$ satisfies that $k\widetilde{H}_0 + \widetilde{P}_0 \leq Q$ for a $Q > Q_c$, then it cannot be satisfied for any $\tau > 0$ that $k\widetilde{H}(\tau) + \widetilde{P}(\tau) \geq Q$, and in other words, it is satisfied for any $\tau > 0$ that $k\widetilde{H}(\tau) + \widetilde{P}(\tau) < Q$. These arguments show that, for the initial condition $\mathbf{v}_0$ belonging to $\Omega(Q)$ with $Q > Q_c$, the trajectory of (8.40) must always remain in $\Omega(Q)$.

References

1. L.J.S. Allen, *An Introduction to Mathematical Biology* (Pearson Prentice Hall, Upper Saddle River, 2007)
2. R.L. Devaney, *An Introduction to Chaotic Dynamical Systems*, 3rd edn. (Chapman and Hall/CRC, Boca Raton, 2021)
3. J. Gleick, *Chaos—Making A New Science* (Penguin Books, New York, 1987)
4. B.S. Goh, Global stability in many-species systems. Am. Nat. **111**, 135–143 (1977)
5. M.W. Hirsch, S. Smale, R.L. Devaney, *Differential Equations, Dynamical Systems, and an Introduction to Chaos*, 3rd edn. (Elsevier/Academic Press, Waltham, 2012)
6. M. Martelli, *Discrete Dynamical Systems and Chaos*, 1st edn. (CRC Press LLC, Boca Raton, 1992)
7. R.C. Robinson, *An Introduction to Dynamical Systems: Continuous and Discrete*. Pure and Applied Undergraduate Texts, vol. 19, 2nd edn. (American Mathematical Society, Providence, 2012)
8. S.H. Strogatz, *Nonlinear Dynamics and Chaos: With Applications to Physics, Biology Chemistry and Engineering*, 2nd edn. (Westview Press, Boulder, 2015)

Chapter 15
Essentials of Poisson Process/Distribution

Abstract As the simplest and most important stochastic process for the mathematical modeling about population dynamics, the Poisson process is introduced and used in some parts of this book. This chapter serves to provide the mathematical fundamentals about it.

15.1 Poisson Process

Let us assume the probability that an event occurs just one time in a sufficiently short interval $[t, t + \Delta t]$ from time t by $\lambda(t)\Delta t + o(\Delta t)$ with a non-negative function $\lambda(t)$ of t. The stochastic process for the event with this probability is called *Poisson process*, and λ is called *intensity* (intensity parameter) of Poisson process [1–4].

> For Poisson process defined as above, the probability that the event occurs more than one times in a sufficiently short interval $[t, t + \Delta t]$ is given by $o(\Delta t)$. For example, when the event occurs just two times in the interval, suppose a time t_1 between the first and second occurrences of the event, and each event occurs respectively in the interval $[t, t_1]$ and $[t_1, t + \Delta t]$. Thus the probability of each occurrence is given by $\lambda(t)(t_1 - t) + o(t_1 - t)$ and $\lambda(t_1)(t + \Delta t - t_1) + o(t + \Delta t - t_1)$ from the above definition, because we have $t_1 - t \le \Delta t$ and $t + \Delta t - t_1 \le \Delta t$ with $t \le t_1 \le t + \Delta t$. The probability that the event occurs just two times in $[t, t + \Delta t]$ is given by the product of them, and becomes $o(\Delta t)$. The same arguments are applicable for the probability that the event occurs more than two times in $[t, t + \Delta t]$.

Poisson process with a time-dependent parameter $\lambda = \lambda(t)$ is called *non-homogeneous* Poisson process. For example, the probability of a biological event could seasonally change under a seasonally variable environment.

When the parameter λ is constant independently of time t, Poisson process is called *homogeneous*. The event occurrence with the homogeneous Poisson

H. Seno, *A Primer on Population Dynamics Modeling*, Theoretical Biology,
https://doi.org/10.1007/978-981-19-6016-1_15

process has a specific randomness, called *independent increments* and *stationary independent increments* defined as follows:

Definition 15.1 A stochastic process is said to have *independent increments* if the stochastic variable $X(t)$ is continuous in terms of $t > 0$ and satisfies that differences $X(t_1) - X(t_0)$ and $X(t_3) - X(t_2)$ are independent of each other for any time t_0, t_1, t_2 and t_3 such that $t_0 < t_1 \leq t_2 < t_3$.

Definition 15.2 A stochastic process is said to have *stationary independent increments* if the stochastic variable $\{X(t), \ 0 \leq t < \infty\}$ is continuous in terms of t and satisfies that the difference $X(t + h) - X(t' + h)$ follows the same probability distribution as $X(t) - X(t')$ does for any time t, t', and any interval $h > 0$.

The number of event occurrences $Q(t)$ until time t and $Q(s + t) - Q(s)$ in an interval $[s, s+t]$ are discrete stochastic variables. The stochastic process about $Q(t)$ is called *counting process*. Poisson process can define a counting process. When Poisson process has independent increments, the stochastic variable $Q(t+s) - Q(s)$ follows the same probability distribution for any $t, s \geq 0$.

15.2   Poisson Distribution

Let us denote by $P(n, t)$ the probability that an event occurs just n times until time t. With the probability $\lambda(t)\Delta t + o(\Delta t)$ that an event occurs just one time in a sufficiently short interval $[t, t + \Delta t]$, we can derive the following equations to govern the transition of the probabilities $\{P(n, t) \mid n = 0, 1, 2, \ldots\}$ in $[t, t + \Delta t]$:

$$P(0, t + \Delta t) = \left[1 - \{\lambda(t)\Delta t + o(\Delta t)\}\right] P(0, t);$$

$$P(1, t + \Delta t) = \left[1 - \{\lambda(t)\Delta t + o(\Delta t)\}\right] P(1, t) + \{\lambda(t)\Delta t + o(\Delta t)\} P(0, t);$$

$$P(n, t + \Delta t) = \left[1 - \{\lambda(t)\Delta t + o(\Delta t)\}\right] P(n, t) + \{\lambda(t)\Delta t + o(\Delta t)\} P(n - 1, t)$$
$$+ \sum_{k=0}^{n-2} o(\Delta t) P(k, t) \qquad (n = 2, 3, \ldots).$$

The factor $1 - \{\lambda(t)\Delta t + o(\Delta t)\}$ in the first term of the right side means the probability that the event does not occur in $[t, t + \Delta t]$. The factor $\lambda(t)\Delta t + o(\Delta t)$ in the second term of the right side about the second and third equations means the probability that the event occur only once in $[t, t + \Delta t]$. We used the mathematically conventional expression $o(\Delta t)$ as the probability that the events occur more than once in $[t, t + \Delta t]$, as before.

From the above equations, we have

$$\frac{P(0, t + \Delta t) - P(0, t)}{\Delta t} = -\left\{\lambda(t) + \frac{o(\Delta t)}{\Delta t}\right\} P(0, t);$$

$$\frac{P(1, t + \Delta t) - P(1, t)}{\Delta t} = -\left\{\lambda(t) + \frac{o(\Delta t)}{\Delta t}\right\} P(1, t) + \left\{\lambda(t) + \frac{o(\Delta t)}{\Delta t}\right\} P(0, t);$$

$$\frac{P(n, t + \Delta t) - P(n, t)}{\Delta t} = -\left\{\lambda(t) + \frac{o(\Delta t)}{\Delta t}\right\} P(n, t) + \left\{\lambda(t) + \frac{o(\Delta t)}{\Delta t}\right\} P(n - 1, t)$$

$$+ \sum_{k=0}^{n-2} \frac{o(\Delta t)}{\Delta t} P(k, t) \qquad (n = 2, 3, \dots),$$

and, taking the limit as $\Delta t \to 0$, subsequently

$$\frac{dP(0, t)}{dt} = -\lambda(t) P(0, t);$$

$$\frac{dP(n, t)}{dt} = -\lambda(t) P(n, t) + \lambda(t) P(n - 1, t) \qquad (n = 1, 2, \dots). \tag{15.1}$$

Let us consider this system of non-homogeneous linear differential equations with the initial condition that $P(n, 0) = \delta_{n0}$, where δ_{nm} denotes what is called Kronecker delta which satisfies that $\delta_{nn} = 1$ and $\delta_{nm} = 0$ for $n \neq m$. That is, $P(0, 0) = 1$ and $P(n, 0) = 0$ for any $n > 0$. Thus, the initial condition means that the event does not occur at the initial time $t = 0$ from which we start to observe the event occurrence. We may consider that the observation of the event occurrence starts just after a moment of its occurrence as $t = 0$.

From the first equation of (15.1), we can easily get the solution

$$P(0, t) = \exp\left[-\int_0^t \lambda(\tau)\, d\tau \right], \tag{15.2}$$

making use of the initial value $P(0, 0) = 1$ mentioned in the above. Next we suppose

$$P(n, t) = u_n(t) \exp\left[-\int_0^t \lambda(\tau)\, d\tau \right] \qquad (n = 1, 2, \dots), \tag{15.3}$$

and substitute this for the second equation of (15.1). Since it becomes as follows:

$$-u_n(t)\lambda(t) \exp\left[-\int_0^t \lambda(\tau)\, d\tau \right] + \frac{du_n(t)}{dt} \exp\left[-\int_0^t \lambda(\tau)\, d\tau \right]$$

$$= -\lambda(t) u_n(t) \exp\left[-\int_0^t \lambda(\tau)\, d\tau \right] + \lambda(t) u_{n-1}(t) \exp\left[-\int_0^t \lambda(\tau)\, d\tau \right],$$

we can obtain the following system of ordinary differential equations with respect to $\{u_n(t)\}$:

$$\frac{du_n(t)}{dt} = \lambda(t)u_{n-1}(t) \qquad (n = 1, 2, \ldots).$$

From the initial value $P(n, 0) = 0$ for any $n > 0$, it must be satisfied that $u_n(0) = 0$ $(n = 1, 2, \ldots)$. The above differential equations can be mathematically transformed into the following integral recurrence relation:

$$u_n(t) = \int_0^t \lambda(\tau)u_{n-1}(\tau)\,d\tau \qquad (n = 1, 2, \ldots).$$

Now, since $u_0(t) = P(0, t) = 1$ from (15.2), we can immediately find

$$u_1(t) = \int_0^t \lambda(\tau)\,d\tau.$$

Subsequently we have

$$u_2(t) = \int_0^t \lambda(\tau_2)u_1(\tau_2)\,d\tau_2 = \int_0^t \lambda(\tau_2)\left\{\int_0^{\tau_2} \lambda(\tau_1)\,d\tau_1\right\}d\tau_2$$

$$= \int_0^t \frac{1}{2}\frac{d}{d\tau_2}\left\{\int_0^{\tau_2} \lambda(\tau_1)\,d\tau_1\right\}^2 d\tau_2 = \frac{1}{2}\left\{\int_0^t \lambda(\tau)\,d\tau\right\}^2.$$

We can prove by the mathematical induction that

$$u_n(t) = \frac{1}{n!}\left\{\int_0^t \lambda(\tau)\,d\tau\right\}^n \qquad (n = 1, 2, \ldots).$$

Consequently, from (15.3), we get the solution of $P(n, t)$:

$$P(n, t) = \frac{(\langle\lambda\rangle_t t)^n}{n!} e^{-\langle\lambda\rangle_t t} \qquad (n = 1, 2, \ldots), \tag{15.4}$$

where

$$\langle\lambda\rangle_t := \frac{1}{t}\int_0^t \lambda(\tau)\,d\tau.$$

The value of $\langle\lambda\rangle_t$ means the mean of λ over the interval $[0, t]$. When λ is constant independently of time t, $\langle\lambda\rangle_t$ becomes equivalent to the constant. Making use of the mathematically conventional definition as $0! = 1$, the solution (15.4) can include (15.2), so that (15.4) can be regarded as the solution for $n = 0, 1, 2, \ldots$ It is easy to prove that $\sum_{n=0}^{\infty} P(n, t) = 1$, making use of the equation $\sum_{n=0}^{\infty} x^n/n! = e^x$. The

Fig. 15.1 Poisson distribution with intensity ν,

$$\Pi(j; \nu) = \frac{\nu^j}{j!} e^{-\nu}$$

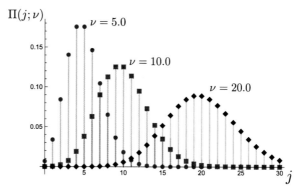

probability distribution $\{P(n, t) \mid n = 0, 1, 2, \ldots\}$ defined by (15.2) and (15.4) is called *Poisson distribution* (Fig. 15.1).

The expected number $\langle n \rangle_t$ of events which occur until time t with the probability distribution $\{P(n, t)\}$ is defined as

$$\langle n \rangle_t := \sum_{n=0}^{\infty} n P(n, t). \tag{15.5}$$

For Poisson distribution defined by (15.2) and (15.4), it becomes

$$\langle n \rangle_t = \sum_{n=0}^{\infty} n \cdot \frac{(\langle \lambda \rangle_t t)^n}{n!} e^{-\langle \lambda \rangle_t t} = e^{-\langle \lambda \rangle_t t} \sum_{n=1}^{\infty} \frac{(\langle \lambda \rangle_t t)^n}{(n-1)!}$$

$$= e^{-\langle \lambda \rangle_t t} \langle \lambda \rangle_t t \sum_{n=1}^{\infty} \frac{(\langle \lambda \rangle_t t)^{n-1}}{(n-1)!} = e^{-\langle \lambda \rangle_t t} \langle \lambda \rangle_t t \sum_{k=0}^{\infty} \frac{(\langle \lambda \rangle_t t)^k}{k!}$$

$$= e^{-\langle \lambda \rangle_t t} \langle \lambda \rangle_t t \cdot e^{\langle \lambda \rangle_t t} = \langle \lambda \rangle_t t = \int_0^t \lambda(\tau) \, d\tau. \tag{15.6}$$

As is intuitively expected, the expected number $\langle n \rangle_t$ is monotonically increasing in terms of t, since $\lambda(t)$ takes non-negative value. Especially when λ is a positive constant, the expected number $\langle n \rangle_t$ is proportional to time t.

15.3 Interarrival Time

In Poisson process, the length Y_n of the time interval from the nth event to the $n + 1$th one is a continuous stochastic variable. The sequence of time lengths $\{Y_n \mid n = 1, 2, \ldots\}$ is called *interarrival time*.

Let us denote by Prob$\{Y_1 \leq t\}$ the probability that the first event occurs until time t. Here Prob$\{Y_1 > t\}$ means the probability that the first event does not occur until time t, which is equivalent to $P(0, t)$ in the previous section. Hence, for Poisson process, we have

$$\text{Prob}\{Y_1 \leq t\} = 1 - \text{Prob}\{Y_1 > t\} = 1 - P(0, t) = 1 - e^{-(\lambda)_t t}$$

$$= \int_0^t \lambda(\tau) \exp\left[-\int_0^\tau \lambda(s)\, ds\right] d\tau$$

with Poisson distribution $\{P(n, t) \mid n = 0, 1, 2, \ldots\}$ defined by (15.2) and (15.4). This result indicates that the probability Prob$\{Y_1 \leq t\}$ is given by an exponential distribution determined by parameter λ.

Especially when λ is a positive constant, we have Prob$\{Y_1 \leq t\} = 1 - e^{-\lambda t}$. As described in Sect. 15.1, in this case, Poisson process is homogeneous and has stationary independent increments defined by Definition 15.2. We can obtain the conditional probability Prob$\{Y_2 \leq t \mid Y_1 = s\}$ that, when the first event occurs at time s, the second event occurs until time $t > s$:

$$\begin{aligned}
\text{Prob}\{Y_2 \leq t \mid Y_1 = s\} &= 1 - \text{Prob}\{Y_2 > t \mid Y_1 = s\} \\
&= 1 - \text{Prob}\{Q(s + t) - Q(s) = 0 \mid Y_1 = s\} \\
&= 1 - \text{Prob}\{Q(s + t) - Q(s) = 0\} \\
&= 1 - \text{Prob}\{Q(t) - Q(0) = 0\} \\
&= 1 - P(0, t) = 1 - e^{-\lambda t} \left(= \int_0^t \lambda e^{-\lambda \tau}\, d\tau\right),
\end{aligned}$$

where $Q(t)$ is the number of events which occurs until time t as introduced in Sect. 15.1. The equation $Q(s + t) - Q(s) = 0$ means that no event occurs in the interval $[s, s + t]$, in other words, the second event occurs after time $s + t$. Since the conditional probability Prob$\{Y_2 \leq t \mid Y_1 = s\}$ is independent of s, it is clearly shown that the probability distribution for Prob$\{Y_2 \leq t \mid Y_1 = s\}$ is independent of that for Prob$\{Y_1 \leq s\}$, while they are the same as the cumulative exponential distribution $F(t) = 1 - e^{-\lambda t}$. Consequently, we can find the following theorem:

Theorem 15.1 *For the homogeneous Poisson process with intensity λ, the interarrival time $\{Y_n \mid n = 1, 2, \ldots\}$ is independent of each other, and follows the same exponential distribution $\lambda e^{-\lambda t}$ with the mean $1/\lambda$.*

When a counting process has the interarrival times independent of each other, and follows the same distribution $F(t)$, it is called *renewal process*. The homogeneous Poisson process with a constant λ is a renewal process which has the interarrival time following the exponential distribution $\lambda e^{-\lambda t}$ and the cumulative exponential distribution $F(t) = 1 - e^{-\lambda t}$.

Next let us consider the time at which the nth event occurs, $S_n = Y_1 + Y_2 + \cdots + Y_n$. The probability $\text{Prob}\{S_n \leq t\}$ that at least n events occur until time t is the cumulative probability for S_n, which is now given by

$$\text{Prob}\{S_n \leq t\} = \sum_{k=n}^{\infty} P(k, t) = \sum_{k=n}^{\infty} \frac{(\langle\lambda\rangle_t t)^k}{k!} e^{-\langle\lambda\rangle_t t}.$$

Since $\text{Prob}\{S_1 \leq t\} = \text{Prob}\{Y_1 \leq t\}$, we can prove by mathematical induction that

$$\text{Prob}\{S_n \leq t\} = \int_0^t \lambda(t) \frac{(\langle\lambda\rangle_\tau \tau)^{n-1}}{(n-1)!} e^{-\langle\lambda\rangle_\tau \tau} d\tau, \tag{15.7}$$

making use of the relation $\text{Prob}\{S_{n+1} \leq t\} = \text{Prob}\{S_n \leq t\} - P(n, t)$. Especially when λ is a constant independent of time, we find from (15.7) that the probability density function for S_n is given by

$$\lambda \frac{(\lambda t)^{n-1}}{(n-1)!} e^{-\lambda t}.$$

This is a kind of gamma distribution, called *Erlang distribution* (of phase n).

On the other hand, we can derive the following conditional probability $\text{Prob}\{Y_1 > t + \tau \mid Y_1 > t\}$ that, when no event occurs until time t, no event occurs until $t + \tau$ either:

$$\text{Prob}\{Y_1 > t + \tau \mid Y_1 > t\} = \frac{\text{Prob}\{Y_1 > t + \tau \text{ and } Y_1 > t\}}{\text{Prob}\{Y_1 > t\}} = \frac{\text{Prob}\{Y_1 > t + \tau\}}{\text{Prob}\{Y_1 > t\}}$$

$$= \frac{e^{-\langle\lambda\rangle_{t+\tau}(t+\tau)}}{e^{-\langle\lambda\rangle_t t}} = \exp\left[-\int_t^{t+\tau} \lambda(s) \, ds\right].$$

Therefore we find that, when λ is a constant independent of time t, the conditional probability $\text{Prob}\{Y_1 > t + \tau \mid Y_1 > t\}$ is independent of time t and determined only by the time interval τ. This feature of stochastic process is called *memoryless property*. The memoryless property essentially indicates that $\text{Prob}\{Y > t + \tau\}/\text{Prob}\{Y > t\}$ is independent of time t and determined only by the time interval τ. It has been mathematically proven that the memoryless property holds only when the probability $\text{Prob}\{Y > t\}$ follows the exponential distribution.

References

1. L.J.S. Allen, *An Introduction to Stochastic Processes with Applications to Biology* 2nd edn. (Chapman & Hall/CRC, Boca Raton, 2010)
2. W. Feller, *An Introduction to Probability Theory and Its Applications*, vol I, 3rd edn. (John Wiley & Sons, New York, 1968)
3. J. Medhi, *Stochastic Models in Queueing Theory*, 2nd edn. (Academic Press, Amsterdam, 2002)
4. S.M. Ross, *Introduction to Probability Models*, 12th edn. (Academic Press, Amsterdam, 2019)

Correction to: Influence From Surrounding

Correction to:
Chapter 2 in: H. Seno, *A Primer on Population Dynamics Modeling*, Theoretical Biology,
https://doi.org/10.1007/978-981-19-6016-1_2

The original version of the book was updated after publication.

In chapter 2, page 43, footnote 1 has been corrected to read as "Rigorously saying in a mathematical sense, the extinction occurs except when the initial value belongs to a measure–zero set. However, in numerical calculations to get the sequence $\{c_n\}$, it eventually occurs (refer to the arguments in p. 37)."

The updated online version of this chapter can be found at
https://doi.org/10.1007/978-981-19-6016-1_2

© The Author(s), under exclusive license to Springer Nature Singapore Pte Ltd. 2022
H. Seno, *A Primer on Population Dynamics Modeling*, Theoretical Biology,
https://doi.org/10.1007/978-981-19-6016-1_16

Index

© The Author(s), under exclusive license to Springer Nature Singapore Pte Ltd. 2022   457
H. Seno, *A Primer on Population Dynamics Modeling*, Theoretical Biology,
https://doi.org/10.1007/978-981-19-6016-1

Printed in the United States
by Baker & Taylor Publisher Services